NELSON

OUTDOOR & ENVIRONMENTAL STUDIES

VCE UNITS 1–4

Belinda Dalziel / Leigh Park / Jarrod Paine

Nelson Outdoor & Environmental Studies VCE Units 1-4
5th Edition
Leigh Park
Belinda Dalziel
Jarrod Paine
ISBN 9780170477123

Publisher: Caroline Williams
Editors: Pete Cruttenden, Alex Chambers
Proofreader: Julie Wicks
First Nations' content reviewer: Dr Mark Lock
Reviewer: Andrew Mannion
Series cover design: Danielle Maccarone
Series text design: Danielle Maccarone
Series designer: Danielle Maccarone
Cover images: Adobe Stock/Juda (sunset); Adobe Stock/Sweta (moth icon); Adobe Stock/Gillian Vann/Stocksy (back cover); Unit 1 opener: Adobe Stock/Jed Forster; Unit 2 opener: Adobe Stock/electra kay-smith; Unit 3 opener: Adobe Stock/Denham; Unit 4 opener: Adobe Stock/Danielle
Permissions researcher: Corrina Gilbert
Production controller: Jess Lovell
Typeset by: MPS Limited

Any URLs contained in this publication were checked for currency during the production process. Note, however, that the publisher cannot vouch for the ongoing currency of URLs.

Acknowledgements
Aboriginal and Torres Strait Islander readers are advised that this textbook (and the resources that support it) may contain images, names and voices of deceased persons. Showing images and voices of people who have passed away can cause distress for Aboriginal and Torres Strait Islander peoples as they believe that it distributes the spirits of those who have died. This is a part of the process of developing a greater understanding, respect and stronger relationships with the Indigenous communities across Australia.

Extracts from the VCE Outdoor and Environmental Studies Study Design (2024 -2028) are used by permission, © VCAA. VCE® is a registered trademark of the VCAA. The VCAA does not endorse or make any warranties regarding this study resource. Current VCE Study Designs, past VCE exams and related content can be accessed directly at www.vcaa.vic.edu.au

For product information and technology assistance,
in Australia call **1300 790 853**;
in New Zealand call **0800 449 725**

For permission to use material from this text or product, please email **aust.permissions@cengage.com**

National Library of Australia Cataloguing-in-Publication Data
A catalogue record for this book is available from the National Library of Australia.

Cengage Learning Australia
Level 5, 80 Dorcas Street
Southbank VIC 3006 Australia

For learning solutions, visit **cengage.com.au**

Printed in China by 1010 Printing International Limited.
3 4 5 6 7 27 26 25 24

CONTENTS

ABOUT THIS BOOK

Nelson Outdoor and Environmental Studies VCE Units 1–4 5ed has been designed to inspire and engage students as they build an understanding of how our relationships with outdoor environments shape us as individuals and as a society. Critically, it also examines how the ways we interact with and relate to outdoor environments over time has shaped, and continues to shape, our impact on these spaces. This new edition aligns explicitly with the revised VCE Outdoor and Environmental Studies Study Design (2024–2028) and guides students through both the practical and theoretical components of this course.

Chapter opener pages

- Each chapter covers one Area of Study in the Outdoor and Environmental Studies Study Design (2024–28). The chapters are divided into modules that address each key knowledge and associated key skill/s.
- The chapter opener pages provide an overview of the chapter in the form of an infographic, and include a list of the key knowledge, key skills and key terms, along with a summary of student resources that appear on Nelson MindTap to support the chapter.

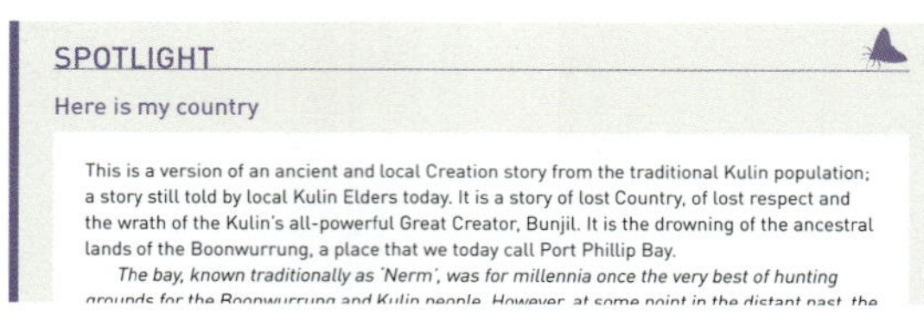

SPOTLIGHT

Here is my country

This is a version of an ancient and local Creation story from the traditional Kulin population; a story still told by local Kulin Elders today. It is a story of lost Country, of lost respect and the wrath of the Kulin's all-powerful Great Creator, Bunjil. It is the drowning of the ancestral lands of the Boonwurrung, a place that we today call Port Phillip Bay.

The bay, known traditionally as 'Nerm', was for millennia once the very best of hunting

Spotlight boxes

Spotlights set the scene for the chapter by introducing a key theme or concept and prompting students to consider the theme or concept before they begin working through the chapter.

CASE STUDY

BETHANY HAMILTON

Bethany Hamilton Authors Children's Book *Surfing Past Fear*

Bethany Hamilton became a household name long ago. The life of this surfer, author and public speaker has been told and retold across a swath of media, from documentaries to dramas to talk shows.

But Hamilton is also a prolific content creator herself. With her latest work, a children's book titled *Surfing Past Fear* (her tenth book), Hamilton's aim is to inspire young girls with a message of hope and overcoming life's inevitable obstacles.

Many children's books aim for that goal, of course, but Hamilton's life has proven her an unofficial world champion in confronting fear — the story of her run-in with a 14-foot tiger shark at 13, and then her drive and perseverance to surf again, has inspired thousands already.

Figure 1.32 Bethany Hamilton based her latest book *Surfing Past Fear* on her experience of surviving a shark attack while surfing.

Case studies

These include real-life examples and are designed to broaden student understanding and provide links to real-world applications of the topics and concepts being covered in the main text. The case studies often include a stimulus – such as a profile, extract or newspaper article – and questions to deepen students' understanding of the topic and concepts.

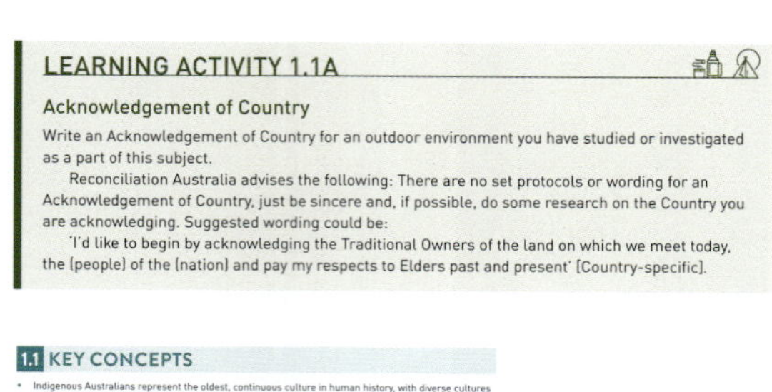

LEARNING ACTIVITY 1.1A

Acknowledgement of Country

Write an Acknowledgement of Country for an outdoor environment you have studied or investigated as a part of this subject.

Reconciliation Australia advises the following: There are no set protocols or wording for an Acknowledgement of Country, just be sincere and, if possible, do some research on the Country you are acknowledging. Suggested wording could be:

'I'd like to begin by acknowledging the Traditional Owners of the land on which we meet today, the (people) of the (nation) and pay my respects to Elders past and present' (Country-specific).

1.1 KEY CONCEPTS

- Indigenous Australians represent the oldest, continuous culture in human history, with diverse cultures and unique, complex knowledge systems.
- Kinship is a person's relationship and responsibilities to other people, to their Country and to natural resources.
- Nature is the living things, the ecosystems and the processes that form them, and the places in which we find all of these.
- Country is the term often used by Indigenous peoples to describe the lands, waterways and seas to

Learning Activities

Engaging learning activities appear throughout the chapter to address and reinforce learning outcomes. Corresponding worksheets and Additional Learning Activities are available on Nelson MindTap.

Key concepts summaries

Summaries of key concepts appear at the end of each module and are ideal for revision.

Concept questions

Questions that check for understanding and provide opportunities for students to apply their knowledge appear at the end of each module. Designed to support differentiation and based on the Bloom's taxonomy, they include an extension challenge for students who wish to investigate a topic in more depth. These questions are also available on Nelson MindTap. Teachers can assign questions to individual students and track their responses through Gradebook on the platform.

1.1 CONCEPT QUESTIONS

REMEMBERING

1 What is meant by Indigenous and non-Indigenous peoples?
2 How does the Indigenous meaning of Country differ from the non-Indigenous meaning?
3 Using an example of each, distinguish between urban environments and built environments.

UNDERSTANDING

4 Explain the role Kinship plays in First Nations cultures.
5 Compare three different ways the word 'nature' may be used.
6 Discuss the value of participating in experiences in outdoor environments.
7 Explain the difference between a national park and a state park.

APPLYING

8 Research and describe the work that has been done by Trust for Nature in supporting a private land owner in the protection of a specific outdoor environment.
9 Create a one-page visual display on what the word 'wilderness' means to you.

EXTENSION CHALLENGE

10 Select an outdoor environment you have visited (or will visit) as part of an outdoor experience. Construct a timeline of events from the deep past to the present, either electronically or on a poster. Include information about the geological history, key historical events, images and major changes in landforms, climate, vegetation, animals and human interactions.

Exam-style questions

Practice exam-style questions for each Area of Study are included at the end of each chapter. Exemplar responses and marking advice for these questions are provided on Nelson MindTap.

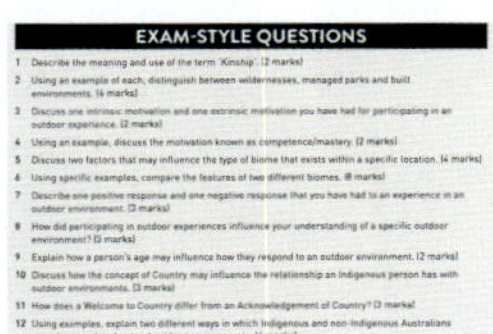

EXAM-STYLE QUESTIONS

1 Describe the meaning and use of the term 'Kinship'. (2 marks)
2 Using an example of each, distinguish between wildernesses, managed parks and built environments. (4 marks)
3 Discuss one intrinsic motivation and one extrinsic motivation you have had for participating in an outdoor experience. (2 marks)
4 Using an example, discuss the motivation known as competence/mastery. (2 marks)
5 Discuss two factors that may influence the type of biome that exists within a specific location. (4 marks)
6 Using specific examples, compare the features of two different biomes. (8 marks)
7 Describe one positive response and one negative response that you have had to an experience in an outdoor environment. (3 marks)
8 How did participating in outdoor experiences influence your understanding of a specific outdoor environment? (3 marks)
9 Explain how a person's age may influence how they respond to an outdoor environment. (2 marks)
10 Discuss how the concept of Country may influence the relationship an Indigenous person has with outdoor environments. (3 marks)
11 How does a Welcome to Country differ from an Acknowledgement of Country? (3 marks)
12 Using examples, explain two different ways in which Indigenous and non-Indigenous Australians experience and understand outdoor environments. (6 marks)

Key Knowledge and Skills Checklists

These checklists include all the key knowledge and key skills covered in the chapter and invite students to rate their understanding against each point. This provides students with the ability to track their own progress and to identify where they need to seek additional support or further revision. These checklists are also provided on Nelson MindTap for teachers to track student confidence and identify any gaps in their comprehension.

CHECK YOUR KNOWLEDGE AND SKILLS

Checklist for students to mark and date, making judgments of their knowledge and understanding (D = developing C = consolidating ER = exam ready)

D	C	ER	Check your knowledge and skills
☐	☐	☐	Indigenous peoples ways of knowing outdoor environments
☐	☐	☐	Non-Indigenous peoples ways of knowing outdoor environments
			Describe the meaning and use of the following terms:
☐	☐	☐	• Kinship
☐	☐	☐	• nature
☐	☐	☐	• Country
☐	☐	☐	• outdoor environments
☐	☐	☐	• private land
☐	☐	☐	• wilderness
☐	☐	☐	• managed parks
☐	☐	☐	• urban environments
☐	☐	☐	• built environments
☐	☐	☐	• outdoor experiences
			Describe the features of the following types of biomes:
☐	☐	☐	• alpine
☐	☐	☐	• coastal
☐	☐	☐	• inland waterways
☐	☐	☐	• grassland
☐	☐	☐	• heathland
☐	☐	☐	• forest
☐	☐	☐	• marine
☐	☐	☐	• arid
☐	☐	☐	The range of motivations for experiencing outdoor environments
☐	☐	☐	Responses to outdoor environments such as fear and appreciation
			Compare the variety of ways in which Indigenous and non-Indigenous peoples experience and understand outdoor environments:
☐	☐	☐	• through custodianship
☐	☐	☐	• as recreation
☐	☐	☐	• as a resource
☐	☐	☐	• as spiritual connection
☐	☐	☐	• as a study site

Extended learning activities

Extended learning activities appear at the end of the chapter and provide an additional opportunity for students to apply key knowledge and key skills.

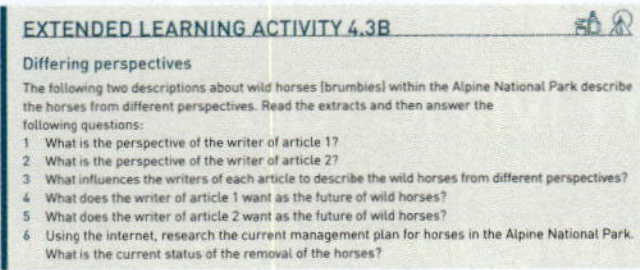

EXTENDED LEARNING ACTIVITY 4.3B

Differing perspectives

The following two descriptions about wild horses (brumbies) within the Alpine National Park describe the horses from different perspectives. Read the extracts and then answer the following questions:

1 What is the perspective of the writer of article 1?
2 What is the perspective of the writer of article 2?
3 What influences the writers of each article to describe the wild horses from different perspectives?
4 What does the writer of article 1 want as the future of wild horses?
5 What does the writer of article 2 want as the future of wild horses?
6 Using the internet, research the current management plan for horses in the Alpine National Park. What is the current status of the removal of the horses?

Indigenous content

The text includes detailed explorations of Aboriginal and Torres Strait Islander peoples' perspectives on the ways humans can relate to and sustainably manage outdoor environments in both historical and contemporary contexts. The concepts of Kinship, Country and customary law are explored, then discussed in the context of traditional environmental management practices including cultural burning, agriculture, aquaculture, and hunting and gathering approaches.

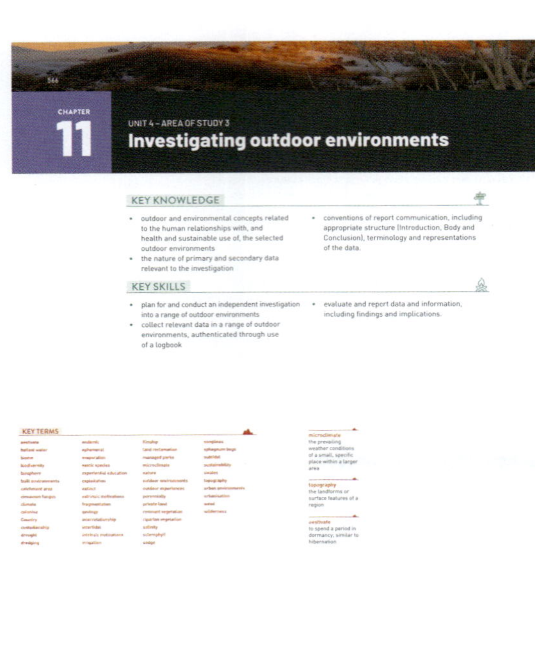

Templates for planning and recording outdoor experiences

Chapters 6 and 11 focus on Area of Study 3 in Unit 1 and Units 3 and 4 respectively. These chapters include advice and templates for students to record their observations and experiences, and to assist teachers with their planning and designing these outdoor experiences. Downloadable and editable copies of these templates are also available on Nelson MindTap.

Key terms

Key terms are highlighted in the text with definitions in the page margins to assist students with their understanding. A glossary with the key terms and definitions for each chapter can be found at the end of the student book with downloadable chapter glossaries available on Nelson MindTap.

FAST FACT

NOTES FOR THE EXAM

For the exam, you should:

- plan and conduct an ongoing investigation, reflecting on four key knowledge points across Units 3 and 4
- visit two outdoor environments where you have documented primary data into your logbook
- supplement your logbook with secondary data
- evaluate your report data, discuss findings and state conclusions (and possibly implications or recommendations).

Fast Fact margin box and Notes for the exam box

Features from the previous edition, including Fast Facts and Notes for the Exam, have been retained.

Nelson MindTap icon

Nelson MindTap icons direct students to additional resources on the online platform, including worksheets, templates and weblinks.

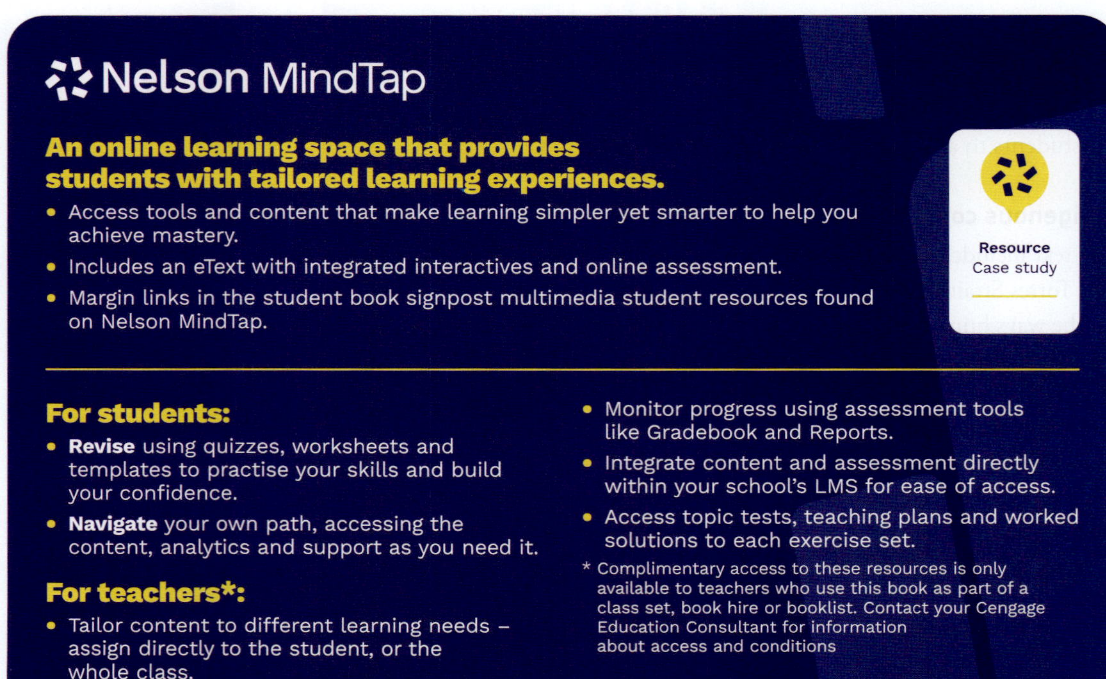

ABOUT THE AUTHORS

Belinda Dalziel has been teaching VCE Outdoor and Environmental Studies for 15 years and is passionate about teaching and learning in this subject. She has authored a variety of materials, including examinations, school-assessed coursework, study guides and course outlines. With over 13 years of experience as a VCAA examination assessor, as well as leading examination assessment roles, Belinda has extensive knowledge of Outdoor and Environmental Studies. She has played a key role in the review panel for the new VCE Outdoor and Environmental Study Design for 2024, along with teacher support materials. Belinda has shared her expertise by conducting teacher and student development sessions for over a decade for different organisations including VCAA, ACHPER and other teacher associations. Belinda enjoys facilitating learning in the outdoors with her students, particularly sea kayaking in Western Port Bay and cross-country skiing and snow camping at Mt Stirling in the Victorian Alps.

Jarrod Paine has taught and delivered the practical component of VCE Outdoor and Environmental Studies for over 20 years in a residential outdoor school, and more recently in a completely online school. He has worked in the design, innovation and implementation of outdoor education programs from pre-school to VCE, and increasingly spends more time working as a writer (including children's picture story books and VCE study designs) as opposed to being backcountry on a ski or mountain bike trip with students. Jarrod lives and breathes education in the outdoors, and strives to connect people of all ages to nature and the essential need to care for it – both for itself and for ourselves. That's of course when he's not working as a professional snake catcher!

Leigh Park has taught Outdoor and Environmental Studies for the past 25 years, and has designed and implemented outdoor education programs across both junior and senior schools for the past 30 years. He has presented at numerous state, national and international conferences and has run examination preparation sessions for teachers and students. Leigh has been an Outdoor and Environmental Studies state reviewer and a VCAA assessor since 2004. He is passionate about involving students in their learning to discover more about themselves, to explore the world around them, and to develop empathy for others.

Unit 1 Connections with Outdoor Environments

Unit 1 – Introduction

Unit 1 of the VCE Outdoor and Environmental Studies study design titled 'Connections with outdoor environments' examines some of the ways in which Indigenous peoples and non-Indigenous peoples understand and relate to nature through experiencing outdoor environments. We explore the range of motivations for interacting with outdoor environments, the factors that affect an individual's access to experiencing *outdoor environments*, and how they connect with outdoor environments.

Chapter 1 (Our place in outdoor environments) examines how humans connect with outdoor environments and why these connections are important.

Chapter 2 (Exploring outdoor environments) considers how our personal responses to outdoor environments are influenced by media portrayals of outdoor environments and perceptions of risk involved in outdoor experiences.

Chapter 3 (Safe and sustainable participation in outdoor experiences) focuses on planning and participating in outdoor experiences.

Through their studies of Unit 1 students will develop practical skills and knowledge to participate safely and sustainably in outdoor experiences and use their experiences and observations as the basis for reflection and analysis of the key skills and knowledge.

CHAPTER 1

UNIT 1 – AREA OF STUDY 1

Our place in outdoor environments

KEY KNOWLEDGE

- Indigenous peoples' and non-Indigenous peoples' ways of knowing outdoor environments, including the meaning and use of terms such as Kinship, nature, Country, outdoor environments, private land, wilderness, managed parks, urban environments, built environments and outdoor experiences
- features of biomes, including alpine, coastal, inland waterways, grassland, heathland, forest, marine and arid
- the range of motivations for experiencing outdoor environments and responses to outdoor environments, such as fear and appreciation
- the variety of ways in which Indigenous and non-Indigenous peoples experience and understand outdoor environments:
  - through custodianship
  - as recreation
  - as a resource
  - as spiritual connection
  - as a study site.

KEY SKILLS

- describe the meaning and use of terms including Kinship, nature, Country, outdoor environments, private land, wilderness, managed parks, urban environments, built environments and outdoor experiences
- describe the features of a range of biomes
- analyse motivations for experiencing outdoor environments and responses to experiencing outdoor environments
- compare the variety of ways in which Indigenous and non-Indigenous peoples experience and understand outdoor environments
- safely participate in specific outdoor experiences.

Indigenous peoples' and non-Indigenous peoples' ways of knowing outdoor environments:

- Kinship
- Nature
- Country
- Outdoor environments
- Private land
- Wilderness
- Managed parks
- Urban environments
- Built environments
- Outdoor experiences

Features of biomes, including:

- Alpine
- Coastal
- Inland waterway
- Grassland
- Heathland
- Forest
- Marine
- Arid

The range of motivations for experiencing outdoor environments and responses to outdoor environments, such as:

- Fear
- Appreciation

The variety of ways in which Indigenous and non-Indigenous peoples experience and understand outdoor environments:

- Through custodianship
- As recreation
- As a resource
- As spiritual connection
- As a study site

KEY TERMS

aestivate
ballast water
biome
biodiversity
biosphere
built environments
catchment area
cinnamon fungus
climate
colonise
Country
custodianship
drought
dredging
endemic
ephemeral
evaporation
exotic species
experiential education
exploitation
extinct
extrinsic motivations
fragmentation
geology
interrelationship
intertidal
intrinsic motivations
irrigation
Kinship
land reclamation
managed parks
microclimate
nature
outdoor environments
outdoor experiences
perennially
private land
remnant vegetation
riparian vegetation
salinity
sclerophyll
sedge
songlines
sphagnum bogs
subtidal
sustainability
swales
topography
urban environments
urbanisation
weed
wilderness

Worksheets

- **1.1a** Acknowledgement of Country **p. 10**
- **1.1b** Analysing and synthesising GMO debates **p. 12**
- **1.1c** Representation of an outdoor environment **p. 19**
- **1.1d** Logbook **p. 20**
- **1.1e** Ways you know outdoor environments **p. 21**
- **1.1f** Perspective challenge **p. 22**
- **1.1** Key concept **p. 23**
- **1.2a** Wetland ecosystems **p. 29**
- **1.2b** Features of biomes **p. 35**
- **1.2** Key concepts **p. 36**
- **1.3a** What is a motivation? **p. 38**
- **1.3b** Poetry of motivations **p. 40**
- **1.3c** Personal motivaitons **p. 41**
- **1.3d** Factors affecting personal responses **p. 44**
- **1.3** Key concept **p. 45**
- **1.4a** Indigenous weather calendar **p. 47**
- **1.4b** Passive and active outdoor recreational experiences **p. 49**
- **1.4c** Environment as a resource **p. 50**
- **1.4d** Fictional farm **p. 51**
- **1.4e** Creative response **p. 52**
- **1.4f** Experience in song **p. 52**
- **1.4g** Collage of images **p. 54**
- **1.4** Key concepts **p. 55**

Weblink – Videos

- The Kinship system **p. 8**
- The land owns us **p. 10**
- Wilderness **p. 16**
- River Country Spirit Ceremony (Murrundi Ruwe Pangari Ringbalin) **p. 48**

Resources and Templates

- Bethany Hamilton **p. 43**
- Jane Goodall - primatologist **p. 53**
- Glossary – Chapter 1 **p. 56**
- Key knowledge and skills checklist **p. 57**

Assessments

- End of chapter exam **p. 56**
- Glossary test **p. 56**

To access resources above, visit **cengage.com.au/nelsonmindtap**

SPOTLIGHT

Here is my country

This is a version of an ancient and local Creation story from the traditional Kulin population; a story still told by local Kulin Elders today. It is a story of lost Country, of lost respect and the wrath of the Kulin's all-powerful Great Creator, Bunjil. It is the drowning of the ancestral lands of the Boonwurrung, a place that we today call Port Phillip Bay.

The bay, known traditionally as 'Nerm', was for millennia once the very best of hunting grounds for the Boonwurrung and Kulin people. However, at some point in the distant past, the population had become complacent and arrogant to the seemingly boundless gifts of the land.

They fought among themselves and, more and more, they began to over-exploit, to waste and to deeply disrespect their precious country, their home – they believed that the land and Bunjil would always provide.

Bunjil's anger swelled and surged, until in retribution to such disrespect, he called the sea and allowed the waters to engulf the land and threatened to drown all the people in a great flood.

The Kulin cried out, realising too late their profound mistakes. Bunjil, however, heeding their desperation, strode out into the churning inundation and told the waters to stop. With two rocks, Bunjil made the Port Phillip Bay heads and told the water to run out and reunite with the sea.

Bunjil's rage and pain was quenched but for the survivors, Bunjil's retribution was remembered, and grave lessons learnt. The Kulin people returned to their old values, to respect the law and lore of Bunjil, and to always honour and never harm the land or its children; for both are our future, as they are our past.

Dean Stewart (Wemba Wemba-Wergaia NW Victoria)
Director of Aboriginal Tours and Education Melbourne, 2016

1.1 INDIGENOUS PEOPLES' AND NON-INDIGENOUS PEOPLES' WAYS OF KNOWING OUTDOOR ENVIRONMENTS

KEY KNOWLEDGE

- Indigenous peoples' and non-Indigenous peoples' ways of knowing outdoor environments, including the meaning and use of terms such as Kinship, nature, Country, outdoor environments, private land, wilderness, managed parks, urban environments, built environments and outdoor experiences

KEY SKILLS

- describe the meaning and use of terms including Kinship, nature, Country, outdoor environments, private land, wilderness, managed parks, urban environments, built environments and outdoor experiences

Bunjil is a significant figure in Aboriginal Australian mythology, particularly in the culture of the Kulin nation, which includes Aboriginal groups in the south-eastern part of Australia, including the Wurundjeri, Boonwurrung, Taungurung, Dja Dja Wurrung and Wathaurong peoples. Bunjil (depicted as a wedge-tailed eagle) is often seen as a creator of landscapes, animals and people. He is also considered a giver of laws and cultural knowledge, and he plays a crucial role in shaping the moral and ethical frameworks of the Kulin nation. Bunjil is often depicted in rock art, stories and ceremonies as a powerful figure who guides and shapes the lives of Indigenous peoples.

Dreaming or Creation stories are used by Aboriginal and Torres Strait Islander peoples to pass on knowledge, cultural values and belief systems to the next generations. They are an important way in which the Indigenous peoples share their connection with their outdoor environments. Indigenous Australians represent the oldest continuous living culture in human history, with diverse cultures and unique, complex knowledge systems. During your studies in VCE Outdoor and Environmental Studies, you will have opportunities to develop understandings of the significant contributions of Indigenous Australians' connection to Country, place and culture through their custodianship of the land.

When we refer to Indigenous Australians, we do so to encompass both Aboriginal peoples and Torres Strait Islander peoples. Aboriginal (meaning first or earliest known) peoples are the custodians of mainland Australia, while Torres Strait Islander peoples are custodians of 274 small islands between Australia's northern tip and Papua New Guinea. We use the plural 'peoples' to recognise that there are many different Aboriginal and Torres Strait Islander cultures and language groups. Alternatively, referring to 'First Nations' respectfully encompasses the diversity of Aboriginal and Torres Strait Islander cultures and identities.

This first chapter of this textbook examines how Indigenous and non-Indigenous peoples connect with outdoor environments and why these connections are important.

Ways of knowing outdoor environments

To 'know something' is to be aware through observation or investigation; to have developed a relationship through meeting and spending time with it; and to have learnt about and to possess information and an understanding of it. In getting to know outdoor environments, we need to observe and enquire about them, spend time in them, and gather information and learn about them in order to understand them to some degree.

iStockphoto/SDI Productions

Figure 1.1 People can experience the outdoor environment in many ways.

We can experience outdoor environments in many ways – through recreational activities; in media such as television, the internet and social media; in literature such as research studies, books and poems; and in education.

It is through their everyday experiences living with the land and listening to the ancient Dreaming stories that Indigenous peoples developed their knowledge of their environment. It was important to know their outdoor environment as it sustained their lives in every aspect: spiritually, physically, socially and culturally.

This first section of the chapter examines Indigenous peoples' perspectives on the ways humans connect with outdoor environments and why these connections are important through the concepts of Kinship and Country. We then investigate concepts associated with non-Indigenous peoples' ways of knowing outdoor environments, including private land, wilderness, managed parks, urban environments and built environments.

Indigenous ways of knowing

KINSHIP

Dreamtime stories demonstrate Indigenous Australians' deep spiritual attachment to the land and Kinship connection to Country. **Kinship** is at the heart of Aboriginal and Torres Strait Islander societies. A person's position in the kinship system establishes their relationship to others, to their Country and to

Kinship
an Indigenous person's relationship and responsibilities to other people, to their Country and to natural resources

the universe, prescribing their responsibilities towards other people, the land and natural resources. The following quote explains Kinship and the meaning of family from an Indigenous Australian perspective.

> Aboriginal and Torres Strait Islander peoples have strong family values. The family system has an extended family structure, as opposed to the nuclear or immediate family structure which is common in Western society. This means the child-rearing responsibilities extend beyond the immediate family group and may include aunts, uncles, cousins and grandparents. ... Over time, Aboriginal and Torres Strait Islander children find out about their family connections and where they belong in relation to others, including who they are related to, who they can and cannot marry, and socially acceptable ways of interacting with others. Aboriginal and Torres Strait Islander cultures thrive through knowledge of family and kin, connection to country and community.
>
> Queensland Government, Child Practice Safety Manual, 'The meaning of family in Aboriginal and Torres Strait Islander cultures', Published 18/11/2019. Licensed under https://creativecommons.org/licenses/by/4.0/

Weblink
Video: The Kinship system

SPOTLIGHT

The Kinship system

Watch the 'Family and Kinship' video by Reconciliation Australia to see a visual explanation of the Kinship system.

Reconciliation Australia

Figure 1.2 Aboriginal and Torres Strait Islander kinship structures are based on their close connection to the land and each other.

COUNTRY

> Coming to Mungo I get a different sense of feeling, that I'm home. You seem to know when you're back in your own Country. It's not taught to you, it's built in you. It's in your soul, that that's your Country.
>
> Roy Kennedy, Ngiyampaa Elder

iStockphoto/donaldyip

Figure 1.3 The full moon rises while the sun sets at the ancient sandstone rock formations known as the Walls of China in Mungo National Park, NSW.

In non-Indigenous Australian culture, people often see land as something that is owned, which they can buy and sell, or as a resource for making money. For Aboriginal and Torres Strait Islander peoples, the land is an integral part of their identity, both as individuals and as communities. The following quote and extract explain this connection to **Country** from an Indigenous Australian perspective.

Penny Tweedie / Alamy Stock Photo

Figure 1.4 Country is a place of learning.

Country
the term often used by Indigenous peoples to describe the lands, waterways and seas to which they are connected. The term contains complex ideas about law, place, custom, language, spiritual belief, cultural practice, material sustenance, family and identity (Source: AIATSIS)

> For Aboriginal peoples, country is much more than a place. Rock, tree, river, hill, animal, human – all were formed of the same substance by the Ancestors who continue to live in land, water, sky. Country is filled with relations speaking language and following Law, no matter whether the shape of that relation is human, rock, crow, wattle. Country is loved, needed and cared for, and country loves, needs and cares for her peoples in turn. Country is family, culture, identity. Country is self.

«Ambelin Kwaymullina, Palyku woman. From 'Seeing the Light: Aboriginal Law, Learning and Sustainable Living in Country', Ambelin Kwaymullina, Indigenous Law Bulletin May/June 2005, Volume 6, Issue 11 «

COUNTRY

When First Nations people use the English word 'Country' it is meant in a special way. For Aboriginal and Torres Strait Islander people, culture, nature and land are all linked. Aboriginal communities have a cultural connection to the land, which is based on each community's distinct culture, traditions and laws.

Country takes in everything within the landscape – landforms, waters, air, trees, rocks, plants, animals, foods, medicines, minerals, stories and special places. Community connections include cultural practices, knowledge, songs, stories and art, as well as all people: past, present and future. People have custodial responsibilities to care for their Country, to ensure that it continues in proper order and provides physical sustenance and spiritual nourishment. These custodial relationships may determine who can speak for particular Country.

These concepts are central to Aboriginal spirituality and continue to contribute to Aboriginal identity. Aboriginal communities associate natural resources with the use and benefit of traditional foods and medicines, caring for the land, passing on cultural knowledge and strengthening social bonds.

Share Mungo Culture, Aboriginal Country' from Mungo National Park, www.visitmungo.com.au.

SONGLINES

songlines ancient paths that crisscross the land, often spanning vast distances, and are believed to have been created by ancestral beings during the Dreamtime or creation period

Songlines are ancient paths that crisscross the land, often spanning vast distances, and are believed to have been created by ancestral beings during the Dreamtime or creation period. Songlines contain essential cultural knowledge, including creation stories, historical events, sacred rituals and the relationships between different elements of the landscape. They are called songlines because the knowledge associated with these paths is often passed down through songs and stories that are sung or recited while travelling along the paths. The landmarks, natural features and other elements mentioned in the songs help guide travellers on their journeys. Songlines are a means for Indigenous people to maintain a deep spiritual connection to their ancestral lands.

One example of songlines is that of the Seven Sisters. As the Seven Sisters leave Roeburn in the west of Australia, they are pursued by an evil shape-shifting spirit called Wati Nyiru or Yurlu, who drives the sisters east across the land and into the night sky – where they become the Pleiades star cluster.

Weblink
Video: The land owns us

SPOTLIGHT

The land owns us

Watch the video 'The Land Owns Us' to see Bob Randall, a Yankunytjatjara elder and traditional owner of Uluru, explain how the connectedness of every living thing to every other living thing is not just an idea but a way of living. This connectedness includes all beings as part of a vast family and calls us to be responsible for this family and care for the land with unconditional love and responsibility. You can watch the video on the Global Oneness Project's channel on YouTube.

Worksheet
1.1a Acknowledgement of Country

LEARNING ACTIVITY 1.1A

Acknowledgement of Country

Write an Acknowledgement of Country for an outdoor environment you have studied or investigated as a part of this subject.

Reconciliation Australia advises the following: There are no set protocols or wording for an Acknowledgement of Country, just be sincere and, if possible, do some research on the Country you are acknowledging. Suggested wording could be:

'I'd like to begin by acknowledging the Traditional Owners of the land on which we meet today, the (people) of the (nation) and pay my respects to Elders past and present' [Country-specific].

Nature

Shutterstock.com/THP Creative

Figure 1.5 The most amazing thing about nature is its infinite variety.

Shutterstock.com/Greg Brave

Figure 1.6 A common use of the term 'nature' is in relation to the natural world.

The word 'nature' can also be used in a wide variety of other contexts and situations, such as:

- human nature – a term used to talk about the innate qualities of humans; that is, the characteristics of our behaviour that we're born with. Some people say that aggression is part of human nature, or that sadness is part of human nature. This relates to the notion of something being 'natural'; that is, human nature is the way humans naturally are.
- nature versus nurture – the concept that refers to a common debate over a key aspect of human development: your genetic make-up (nature) versus the environment you are raised in (nurture). Nature here is taken to mean the genetic code that is found in your cells and governs how your cells and body develop and operate.
- the laws of nature – often discussed when talking about science and physics, and refers to descriptions (often mathematical) of the way the physical world works, and the way parts of the physical world interact with, and relate to, each other. You may have heard of Newton's law of universal gravitation, which is a mathematical description of how gravity works between two objects. This is a law of nature.
- the nature of … – we often use the expression 'the nature of [something]' when describing how something works, or some important feature or characteristic of something. If someone asked you about the nature of gravity, for example, they are really asking: 'What's this thing called gravity all about, how does it work and what does it do?'
- Mother Nature – a term that refers to the Earth's **biosphere**; that is, all of the living things on Earth, and the processes and systems that are part of, or related to, these living things.

Nature is the living things, the ecosystems and the processes that form them, and the places in which we find all of these.

biosphere
the parts of Earth where life dwells

nature
the living things, the ecosystems and the processes that form them, and the places in which we find all of these

NATURAL, UNNATURAL AND ARTIFICIAL

Nature often comes in a kind of continuum – such as when we talk about something that is natural or something that is unnatural, or even something that is artificial.

Something that is 'natural' is said to occur 'out there' somewhere or comes from nature itself. 'Unnatural' usually refers to an object or process that humans have influenced in some way. 'Artificial', at the end of the continuum, refers to things that are created by humans. Examples include:

- natural – a eucalyptus tree, an echidna or a tropical rainforest are all examples of natural things.
- unnatural – the introduction of rabbits into Australia, the rose garden in someone's front yard or a downhill ski slope are all examples of unnatural things, since each of these occurs only because of human impact in (or on) a particular place.
- artificial – a car, a computer or the chemical sweetener aspartame in a diet soft drink are all examples of artificial things.

Shutterstock.com/Troy Wegman

Figure 1.7 Beauty can be found in nature in unlikely objects, such as the decay of human-made objects.

An interesting debate that has occurred in recent years is around the development of food that is genetically modified – called genetically modified organisms (GMOs). One example of a GMO is golden rice (called this because of its yellow colour). Golden rice was engineered to include beta-carotene, a form of vitamin A, because of deficiencies in many developing world diets. It is estimated that around 670000 children under the age of five die each year because of this deficiency. Other GMOs include genes for herbicide resistance, lowering cholesterol, virus protection, growth hormones and fruit decay.

Are GMOs natural? The genes that are spliced into GMO foods are from other organisms, so they could be called natural. But when you put genes from one organism into another, is that natural? Many people don't think so, and the debate has become very emotional. GMOs aren't recent, though. Scientists from Peru have found a genetically modified sweet potato from approximately 8000 years ago. Would that be considered natural today or not?

Worksheet
1.1b Analysing and synthesising GMO debates

LEARNING ACTIVITY 1.1B

Analysing and synthesising GMO debates

Find out about some of the debates concerning the development and use of GMOs. What people or groups are involved in the debates? What are some of the arguments used by people and groups who are for or against the ongoing development of GMOs?

By combining and synthesising some of the arguments, develop your own arguments:

- in support of using GMOs
- against the use of GMOs.

Show your arguments to a variety of different people – friends, family and classmates. Which argument do they find more convincing? Why?

1.1.1 OUTDOOR ENVIRONMENTS

outdoor environments areas of the natural world, as a whole or in a particular geographical area

Outdoor environments can be simply defined as locations that are outside of a building. However, as part of this subject we associate outdoor environments as areas of the natural world, either as a whole or in a particular geographical area. Non-permanent structures such as tents may also be considered as part of an outdoor environment when environmental factors are not controlled. This lack of control of environmental factors is an interesting key distinction. When we are in indoor environments, we are able to modify and change the environmental conditions by closing doors or turning on heaters or air conditioners to make ourselves comfortable. However, at the same time this disconnects us from nature. The influence of technologies on how we experience outdoor environments is explored further in Chapter 2.

In the context of this subject we consider outdoor environments to be learning environments, enabling immersive learning experiences where students develop key knowledge authentically and tangibly. There is a wide range of outdoor environments, ranging from those that have experienced minimal human influence through to those which have undergone significant human intervention.

We are fortunate in Victoria to have access to a very wide range of types of outdoor environments, and there are many ways to classify these environments. The most common way uses the concept of **biomes** to distinguish between different environments. A biome can be defined as a large, naturally occurring community of flora and fauna occupying a major habitat. Biomes are differentiated by their climate, geological features, the plant and animal communities found within them, and the spacing of plants, among other things. In this course, we investigate the biome types that are present in Victoria. Each of these biomes is explored in detail later in this chapter.

biomes
a large, naturally occurring community of flora and fauna occupying a major habitat

Like the natural–unnatural–artificial continuum, there is also a sort of progression or hierarchy of places to which we give the label 'outdoor environments'. Outdoor environments such as local parks might have native Australian plants mixed with gardens of non-native flowers, shrubs and trees. They might also have some sportsgrounds or ovals, as well as nearby car parks, a bike path and perhaps a children's playground. Depending on where you live, local parks might even have a dam, a lake or some other similar water feature, or they might be built around or bordering a creek or river. For many Australians, particularly those who live in big cities, using these sorts of parks might be their most common experience of outdoor environments. Beyond these local parks are the larger and 'wilder' protected places: the state and national parks. It is often these places that we think of when we talk about outdoor environments.

Fairfax Photo/Chris Hopkins

Figure 1.8 For many people, playgrounds such as this one at Royal Park are a common experience of outdoor environments.

Historically, Indigenous peoples modified outdoor environments on a small scale, but since colonisation, Australian outdoor environments have been altered to meet commercial, conservation and recreation needs, as well as to feed a growing population. Today, outdoor environments remain an important aspect of Australian identity and continue to be used by industry while also being places of adventure, recreation, scientific study, social action and enterprise. Outdoor environments also provide space for connectedness with nature and opportunities to reflect upon the past, live in the present and take action for sustainable futures.

So far in this chapter we have examined a range of perspectives on the ways Indigenous peoples connect with outdoor environments and, through the concepts of Kinship and Country, why these connections are important. We now move on to investigate concepts associated with non-Indigenous peoples' ways of knowing outdoor environments, including private land, wilderness, managed parks, urban environments and built environments.

Private land

Private land land that is not owned by a government

The total land area of Victoria is 22 744 400 hectares, plus a further 1 021 300 hectares of coastal area. The majority of this is privately owned by non-Indigenous individuals, families, companies and other entities. **Private land** is any land that is not owned by a government. The remainder is referred to as public land, or Crown land, and is managed by various government authorities. When Australia was colonised on behalf of King George III in 1788, all land was claimed for him and managed for the British Government by its colonial representatives. Hence, the term 'Crown land' was used. Since then, as arable land has been granted to settlers or sold to private individuals, Crown land has decreased in size. Much that remains is found alongside roads and railways, in national and state parks, and in hospitals, schools and other government institutions. In total, public land covers around 39% of Victoria. The following map shows areas of private and public land within Victoria.

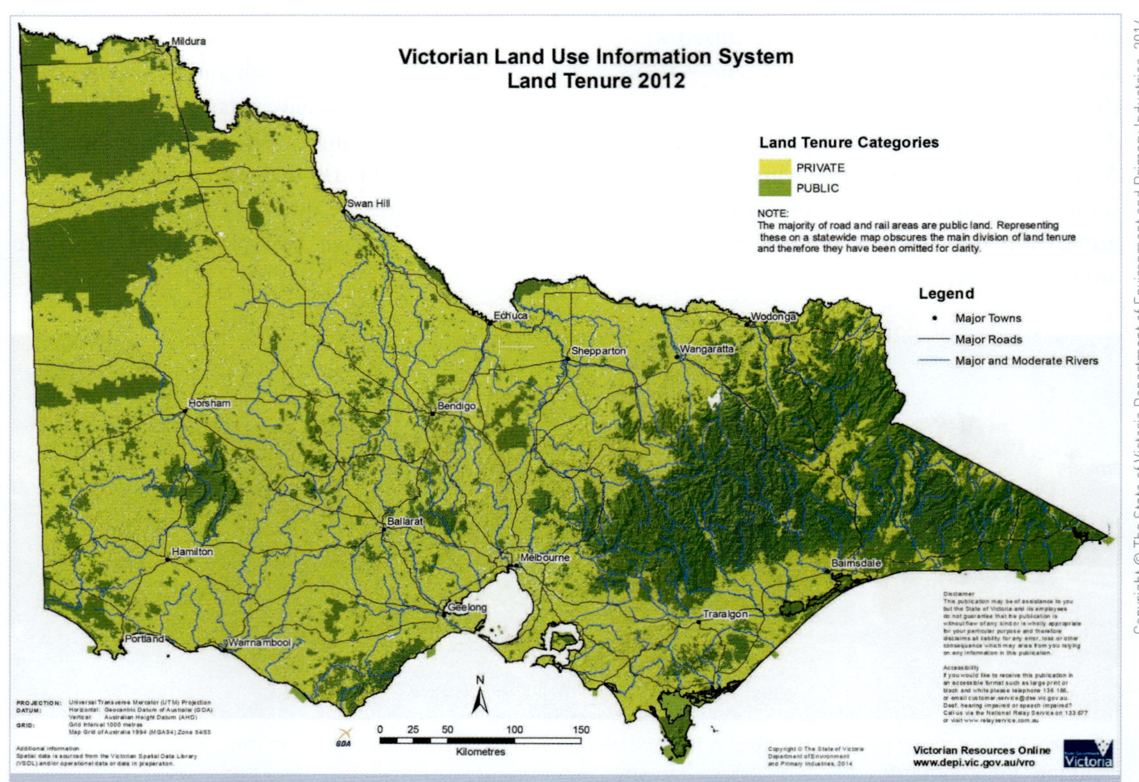

Figure 1.9 Private and public land in Victoria mapped using the Victorian Land Use Information System (VLUIS)

While ultimately controlled by the land owner, the management of private land is also influenced by government policies and regulation. For example, there are rules about land clearing, and some community-led organisations aim to improve the health of outdoor environments through education and awareness campaigns, as well as lobbying governments to increase protections and employ effective environmental policies.

Since the arrival of the non-Indigenous peoples, approximately 66% of Victoria's native habitat has been cleared mainly for food production, with most of this occurring on private land. This has led to a significant reduction in native biodiversity, although some private land includes large areas with significant conservation value.

Trust for Nature is a not-for-profit organisation that works to protect native plants and wildlife in cooperation with private landowners. It was established under the *Victorian Conservation Trust Act 1972* (Vic) to enable people to contribute to nature conservation by donating land or money. Trust for Nature is now one of Victoria's primary land conservation organisations, helping private land owners protect biodiversity. Across the state, more than 1450 private landholders have protected threatened woodlands, wetlands and grasslands. These places are home to some of Victoria's most threatened species, including the helmeted honeyeater and the plains-wanderer.

Over a 20-year period, Trust for Nature has transformed Ned's Corner Station, a 30 000-hectare property near Mildura, from an area degraded by a century of heavy grazing and cropping to one where native vegetation is flourishing.

Wilderness

In everyday usage, the term **wilderness** often refers to natural, untouched places – wild places. A wilderness environment is:

- big
- remote
- untouched, or relatively untouched, by humans.

wilderness
an environment that is big, remote and untouched (or relatively untouched) by humans

Unsplash/Erin Baker

Figure 1.10 The Big Desert Wilderness Park in the state's west is big, remote and untouched.

CHARACTERISTICS OF WILDERNESS

Big

Like many protected places, wilderness areas can begin as smaller environments that are then gradually expanded over time. That said, remote places that become wilderness areas often start out large, simply because of their remoteness. The size required to be considered a wilderness can vary, but a lower limit is often set at about 2000 hectares. More generally, a wilderness area needs to be big enough to be self-sufficient in maintaining its ecosystem processes and biological diversity. In Victoria, two of the biggest wilderness areas are the Avon Wilderness Park in the Victorian Alps (at just under 40000 hectares) and the Big Desert Wilderness Park in the state's west (at over 142000 hectares).

Remote

Remoteness refers to the ease with which humans can access a place. We would typically expect a remote place to be some distance from large population centres such as Melbourne, Geelong, Ballarat, Bendigo and other large cities or towns. We would also expect that a remote place doesn't necessarily have easy transportation access – public transport will be limited, probably with no air, bus or rail connections. To get to a wilderness, you probably have to drive – likely on unmade roads – and you may also need to walk or cycle to get the last part of the way.

Untouched

There are some places on Earth that humans have yet to travel to, live in or alter in some way – but there aren't many. In Australia, there aren't any of these places – Indigenous peoples, European settlers and more recent generations of non-Indigenous Australians have walked, ridden or driven pretty much everywhere across the continent. So, strictly speaking, there aren't likely to be any 'untouched' places. The Wilderness Society defines wilderness as follows:

> A wilderness area is defined as an area that is, or is capable of, being restored to be:
> - of sufficient size to enable long-term preservation of its natural systems and biological diversity;
>
> The Wilderness Society (Australia)

Critiques of wilderness

Wilderness is a modern concept – a big and remote place, untouched by humans. Many Indigenous communities that have long-standing and traditional connections with outdoor places see these places as a part of their communities – as a part of their homes – and reject the notion that we can set these places aside from human interactions. They would argue that these places have been interacted with for as long as humans have been in them, and to set them apart from other places ignores the rich connections people have built with these areas over tens of thousands of years.

In some situations, designating an area as a wilderness environment means that people are restricted from certain activities in these places. Hunting, for example, would not be an appropriate activity in a wilderness environment, at least based on the definitions we looked at earlier, and yet many Indigenous communities have hunted in these places in the past, and some still wish to do so, in order to keep their culture alive.

> Before agriculture ... humans were in the wilderness. We had no concept of 'wilderness' because everything was wilderness and we were a part of it. But with irrigation ditches, crop surpluses and permanent villages, we became apart from the natural world ... Between the wilderness that created us and the civilization created by us grew an ever-widening rift.
>
> Dave Foreman, 'The Great New Wilderness Debate', (eds. Callicott, J.B. and Nelson, M.P), 1998, University of Georgia Press, Georgia

Weblink
Video:
Wilderness

SPOTLIGHT

Wilderness

Watch the video 'Wilderness' by Alastair Humphreys, which explores the appeal and importance of wilderness.

Managed parks

In the previous section we looked briefly at wilderness areas, but there are a number of other types of protected areas across Australia and other parts of the world. **Managed parks** are areas of public land that are controlled by and are the responsibility of governments. In Victoria, managed parks fall within the responsibility of the government agency called Parks Victoria. It is responsible for managing a diverse estate of more than 4 million hectares of public land, including 3000 land and marine parks and reserves making up 18% of Victoria's landmass, 75% of Victoria's wetlands and 70% of Victoria's coastline. Victoria's parks are home to more than 4300 native plants and around 1000 native animal species.

Shutterstock.com/FiledIMAGE

Figure 1.11 Gariwerd (the Grampians National Park) receives approximately a million visitors each year. This requires careful management to ensure its health and sustainability.

Managed parks areas of public land that are controlled by and are the responsibility of governments

> Parks Victoria's estate attracts more than 100 million visits every year and we are committed to providing accessible, enjoyable, diverse programs and destinations while protecting and enhancing environmental and cultural values. It is our primary responsibility to ensure parks are healthy and resilient for current and future generations.
>
> Parks Victoria 

The protection of outdoor environments in Victoria and the rest of Australia (and in many other parts of the world) has led to a type of hierarchy of levels of protection. The following identifies many of these and roughly runs from the highest levels of protection down to lower levels of protection.

WORLD HERITAGE PROTECTED AREAS

These are places deemed so significant for humans and for environmental protection that they are given a status beyond the borders of the state or country. Australia has 20 sites on the World Heritage List, including Uluru-Kata Tjuta, Kakadu and the Great Barrier Reef. Australia has more natural World Heritage sites than most other countries.

NATIONAL PARKS

There are currently 45 national parks in Victoria; together they cover about 11% of the state's total area.

WILDERNESS PARKS

There are three wilderness parks in Victoria:

- Avon Wilderness Park (40,000 hectares) in the Victorian Alps near Licola, Dargo and Lake Tali Karng
- Big Desert Wilderness Park (142,000 hectares) on the border with South Australia
- Wabba Wilderness Park (20,000 hectares) in the Victorian Alps near Corryong.

MARINE PARKS

Victoria has protected just over 5% of its coastal areas in 13 marine national parks and 11 marine sanctuaries – a world first in coastal area protection.

STATE PARKS

There are 26 state parks in Victoria covering over 3 million hectares. They are managed to supplement the national park system, provide access to a range of natural resources (such as timber) and offer a more extensive array of recreational activities than are permissible in the national park network.

LOCAL AND METROPOLITAN PARKS

Victoria currently has over 3000 managed parks and reserves that are part of the protection system. Many of these are local, metropolitan and regional parks.

Urban and built environments

urban environments areas of permanent infrastructure designed to support higher population densities, such as cities and towns

built environments areas that have been created or modified by people, including buildings, parks and transport systems

Just as the term 'artificial' is the opposite of 'natural', the terms '**urban environments**' and '**built environments**' can be considered to be the opposite of natural environments. Built environments are those places that have been created by, or modified by, people and include buildings, parks and transport systems such as roads and railways. Urban environments are built environments that have a high density of human population – in other words, cities.

Indoor rock-climbing centres, artificial surfing reefs and indoor snow-skiing centres are all examples of built environments that attempt to re-create conditions found in natural environments.

Is a backyard (or front yard) garden or a school garden heavily planted with natives an urban environment or a natural environment? This is where trying to divide places into these two groups can break down. A school is clearly an urban, developed place, although you could argue that it also could be a natural environment if it's big enough and has a variety of plant species in it.

When considered in relation to outdoor environments, an interesting aspect of built environments is the cost of maintaining them. Asphalt, concrete and many of the other materials that we use in our built environments degrade over time and need to either be replaced or repaired.

FAST FACT

It's estimated that around 46% of the world's landmass is wilderness, but only about 20% of this is protected as designated wilderness areas.

AAP Image/Julian Smith

Figure 1.12 Urbansurf, an expensive built environment in an urban area of Melbourne, replicates a natural environment. It's loads of fun too!

Shutterstock.com/FiledIMAGE

Figure 1.13 Nature is also the parks and the gardens within our built environments.

LEARNING ACTIVITY 1.1C

Representation of an outdoor environment

1 Create a poster or multimedia representation of an outdoor environment you have visited. Include in your poster or multimedia representation at least three of the following:
- images and/or photos
- factual descriptions
- audio sounds (bird songs, frog croaks, water flows, etc.)
- hand-drawn sketches
- fictional descriptions (such as a poem, short story or song lyrics).

Include a statement about what the key or important aspects of this place are for you.

2 Examples of nature – choose one of the following places:
- zoo
- aquarium
- residential garden.

Argue whether or not this place should be considered an outdoor environment. Include at least ONE argument supporting this place as an outdoor environment. Include at least ONE argument that opposes the idea of this place as an outdoor environment. Give your own judgement based on the arguments you include.

3 Artificial nature – produce a report on an artificial or simulated environment such as an artificial reef, indoor climbing wall or indoor skiing centre. Include in your report answers to the following questions:
- What are some reasons why people would construct this type of artificial environment? (Give at least TWO reasons.)
- Briefly describe an example of this type of artificial environment.
- What might be some of the difficulties involved in constructing this type of artificial environment?
- What might be some of the impacts (good and/or bad) of constructing and using one of these artificial environments?

Worksheet 1.1c Representation of an outdoor environment

Outdoor experiences

One of the highlights of your journey through the VCE Outdoor and Environmental Studies course is the opportunity to participate in a range of outdoor experiences. These may range from short investigations within your nearby local environment to multiday, journey-based activities. Through participation in these outdoor experiences, you will learn to respect and value these landscapes and their living cultural history. These experiences are examples of experiential learning, the process of learning by doing and then reflecting on the experience. **Experiential education** is the foundation of VCE Outdoor and Environmental Studies.

iStockphoto/Paul Feikema

Figure 1.14 The best way to know a place is to experience it for yourself.

Outdoor experiences refer to activities completed outside, most commonly in natural settings. These activities will vary depending on the outdoor environment in which they are being carried out. The VCE Outdoor and Environmental Studies Study Design states that outdoor experiences may include guided activities in areas such as farms, mining or logging sites, interpretation centres, coastal areas, rivers, mountains, bushland, forests, urban parks, cultural and historical sites, and state or national parks.

experiential education an engaged learning process whereby students 'learn by doing' and by reflecting on the experience

outdoor experiences activities completed outside, most commonly in natural settings

There is a vast array of activities that can be considered to be outdoor activities, including bushwalking, surfing, cross-country skiing, caving, fishing, canoe touring, rock climbing, orienteering, conservation and restoration activities, marine exploration, running, photography and participation in community projects.

Outdoor environments also provide space for connectedness with nature and opportunities to reflect upon the past, live in the present and take action for sustainable futures. By spending extended periods of time in outdoor environments to support experiential development of theoretical understandings, you will learn to assess the health of, and evaluate the importance of, healthy outdoor environments.

Outdoor experiences allow the development of understandings of outdoor environments from various perspectives, including through:

- experiential knowledge
- environmental and natural history
- ecological, social and economic perspectives.

EXPERIENTIAL KNOWLEDGE

Experiential knowledge is essentially obtaining knowledge and understanding through actively engaging in an environment. It is a personal experience, such as visiting a location and/or getting involved in an activity. These hands-on experiences are extremely important in allowing us to form a relationship with the environment, and feature heavily throughout the Outdoor and Environmental Studies course. People who come to know the environment through their own individual experiences often have a deep and clear understanding of it.

Worksheet
1.1d Logbook

LEARNING ACTIVITY 1.1D

Logbook

While on an outdoor experience, keep a logbook of your encounters with the outdoor environment. Include details such as:

- location
- environment type
- flora and fauna
- outdoor activity(s) undertaken and how they made you feel
- sustainability measures
- observation of key knowledge relevant to the experience chosen by the teacher
- observation of key skills
- what you learnt about yourself
- what you learnt about the outdoor environment
- impacts you had on the outdoor environment
- the relationship you developed as a result of the experiences.

1.1.2 ENVIRONMENTAL AND NATURAL HISTORY

Environmental and natural history is based on land formations, climate and weather events, changes to the landscape and the animals that inhabit it, and a basic knowledge of what has occurred in a specific environment throughout a period of time. Those who encounter outdoor environments through the discovery of history have the ability to reflect on what has changed and why it might have changed, and then try to make predictions about the future.

Alamy/Danita Delimont

Figure 1.15 Bunjil's Shelter is home to the only known rock art depiction of the Aboriginal creator spirit Bunjil in Gariwerd (Grampians National Park), Victoria.

> Bunjil Shelter sits within the Gariwerd, a cultural landscape that supports people both physically and spiritually. Bunjil created our land, our people, the plants and animals, our religion and the laws by which we live. He is the leading figure in our spiritual life, essential in teaching young people the importance of laws and beliefs.
>
> Levi Lovett, local custodian, Parks Victoria

LEARNING ACTIVITY 1.1E

Ways you know outdoor environments

1. Compile a list of all the ways you have come to know and understand outdoor environments. Think about trips you might have been on, something you read about, a television show you watched, a conversation you had with a neighbour or what you like to do outdoors.
2. Once your list is complete, compare it with a friend and discuss what you know about outdoor environments.

Worksheet
1.1e Ways you know outdoor environments

1.1.3 ECOLOGICAL, SOCIAL AND ECONOMIC PERSPECTIVES

Ecological

The ecological perspective is grounded in biology and is concerned with the interrelationships between living organisms and their physical surroundings. It plays an important role in ensuring the adequate functioning of ecosystems within an environment. Ecological perspectives can lead to a level of knowledge and understanding about the environment that is not always immediately obvious, and the development of an appreciation that goes beyond what we can simply observe with the naked eye.

Social

Human interactions with outdoor environments are the foundation of the social perspective. They refer to the actions of society as a collective or community, rather than just focusing on individuals. By encountering the environment through a social perspective, we can gather knowledge about which activities and behaviours have been successful and which ones have not, thereby leading to a greater understanding about specific environments.

Economic

The economic perspective is directly linked to profit, and is therefore associated with what the outdoor environment offers in terms of its resources and income opportunities, such as those in tourism, farming, timber harvesting, water harvesting, mining and commercial fishing. People who come to know the outdoor environment through this perspective may be largely profit driven, but many also understand the value of protecting the very asset they require – the outdoor environment itself.

Shutterstock.com/Morgan DDL

Figure 1.16 Commercial fishers may have an economic perspective of the environment.

Worksheet
1.1f Perspective challenge

LEARNING ACTIVITY 1.1F

Perspective challenge

Your teacher will divide the class into small groups and ask each group to investigate a particular perspective (ecological, social or economic) of an outdoor environment. When researching the outdoor environment, you should be doing so from your given perspective only. Your group should challenge itself with the following questions:

1 What does knowing an environment mean from this perspective?
2 What is valuable or important about this environment in relation to your given perspective?
3 How would people come to know and understand the environment from this perspective?

Present your group's perspective in a way that reflects its meaning. For example, the group researching an economic perspective may choose to present their information graphically; those researching a social perspective may undertake a role-play; and those researching an ecological perspective may put together a short animation. Once all groups have presented, discuss how different people may come to know and understand the same environment.

1.1 KEY CONCEPTS

- Indigenous Australians represent the oldest, continuous culture in human history, with diverse cultures and unique, complex knowledge systems.
- Kinship is a person's relationship and responsibilities to other people, to their Country and to natural resources.
- Nature is the living things, the ecosystems and the processes that form them, and the places in which we find all of these.
- Country is the term often used by Indigenous peoples to describe the lands, waterways and seas to which they are connected.
- Outdoor environments are areas of the natural world, as a whole or in a particular geographical area.
- Private land refers to land that is not owned by a government
- Wilderness is an environment that is big, remote and untouched (or relatively untouched) by humans.
- Managed parks are areas of public land that are controlled by and are the responsibility of governments.
- Urban environments are areas of permanent infrastructure designed to support higher population densities, such as cities and towns.
- Built environments are areas that have been created or modified by people and include buildings, parks and transport systems.
- Outdoor experiences are activities completed outside, most commonly in natural settings.

1.1 CONCEPT QUESTIONS

Worksheet
1.1 Key concepts

REMEMBERING

1 What is meant by Indigenous and non-Indigenous peoples?
2 How does the Indigenous meaning of Country differ from the non-Indigenous meaning?
3 Using an example of each, distinguish between urban environments and built environments.

UNDERSTANDING

4 Explain the role Kinship plays in First Nations cultures.
5 Compare three different ways the word 'nature' may be used.
6 Discuss the value of participating in experiences in outdoor environments.
7 Explain the difference between a national park and a state park.

APPLYING

8 Research and describe the work that has been done by Trust for Nature in supporting a private land owner in the protection of a specific outdoor environment.
9 Create a one-page visual display on what the word 'wilderness' means to you.

EXTENSION CHALLENGE

10 Select an outdoor environment you have visited (or will visit) as part of an outdoor experience. Construct a timeline of events from the deep past to the present, either electronically or on a poster. Include information about the geological history, key historical events, images and major changes in landforms, climate, vegetation, animals and human interactions.

1.2 FEATURES OF BIOMES

KEY KNOWLEDGE

- features of biomes, including alpine, coastal, inland waterways, grassland, heathland, forest, marine and arid

KEY SKILLS

- describe the features of a range of biomes

> No matter how far I have traversed around this Earth, I have yet to find another location that rivals Australia. Nowhere else on Earth can you find such spectacular landscapes, such unique and fascinating animals, and such warm friendly people. This is why I will always call Australia home.
>
> Adam Cropp, conservationist and filmmaker

Australia is the planet's sixth-largest country after Russia, Canada, China, the United States and Brazil. It comprises a land area of about 7.692 million square kilometres, which accounts for 5% of the world's land area. It is the flattest and (with the exception of Antarctica) the driest continent in the world. It has varying biomes, ranging from deserts and tropical rainforests to cool-temperature forests and snow-covered mountains. Long periods of erosion, inundation, deposition and glacial action have resulted in the range of landforms we now experience. It is one of the most biologically diverse countries on the planet.

biodiversity the number and variety of organisms found within a specified area

Victoria's various outdoor environments support a high level of **biodiversity**, including at least 3140 native species of vascular plants, 900 lichens, 750 mosses and liverworts, 111 mammals, 447 birds, 46 freshwater and 600 marine fish, 133 reptiles, 33 amphibians and an unknown number of invertebrates, fungi and algae.

Leigh Park

Figure 1.17 Mount Stapylton in Gariwerd (the Grampians) at sunset

One of the greatest benefits of the VCE Outdoor and Environmental Studies course is learning about and experiencing the amazing array of environments we have outside our classrooms. Outdoor environments within the Earth's biosphere can be classified into sub-groups called biomes. Biomes are large naturally occurring communities of flora and fauna occupying a major habitat; for example, forest or grasslands. Developing knowledge of biomes is crucial so we can understand their importance and how our actions impact the health and **sustainability** of our unique environment.

sustainability
the ongoing capacity of Earth to maintain all life

1.2.1 FACTORS AFFECTING BIOMES

The type of biome that exists within a specific location is the result of the interaction of three factors: geology, **climate**, and position and aspect.

climate
the prevailing weather conditions of a region

Geology

The influence of **geology** in an area can be seen in the type of rock material found in a particular location, soil characteristics and drainage. Soil develops over time through interaction between the parent rock, the elements (such as wind, water and sun), living and decomposing plant and animal matter, and groundwater. As the Australian landscape is very flat, with an average elevation of only about 325 metres, many of Australia's major rivers are slow-flowing. Geology is also relevant and important when we consider topography, or the shape of the land. This can be seen in places such as Gariwerd (the Grampians), where the upswept rocks provide clear evidence of previous geological history.

geology
the scientific study of the origin, history and structure of the Earth

Climate

Annual rainfall, extremes in temperature and average daylight hours are examples of climatic factors that can affect a landscape. Other aspects of climate may include wind patterns, **evaporation**, ground temperature, frost frequency and snow cover.

evaporation
the change of a liquid into a vapour

Two-thirds of Australia receives less than 500 millimetres of rain each year, and very few areas receive more than 1000 millimetres. The distribution of rainfall is usually seasonal and erratic, resulting in extended periods of **drought** to the extent that drought is considered part of the normal climatic pattern in many areas. In addition to 'normal' conditions, Australian flora and fauna have adapted to survive during sustained dry or wet periods.

drought
a long period of abnormally low rainfall, especially one that adversely affects growing or living conditions

The Australian continent spans almost 30 degrees of latitude and includes deserts, mountain ranges, tropical rainforests and extensive coastal areas. Temperatures can vary widely, and extremes are intensified by dry and cloudless skies. According to the Bureau of Meteorology, the highest temperature ever recorded in Australia was 50.7°C at Oodnadatta Airport, South Australia, on 2 January 1960. The lowest temperature ever recorded in Australia was –23.0°C at Charlotte Pass, NSW, on 29 June 1994. Many of the widest ranges in temperature occur in the dry regions. The frequency and intensity of frost is critical for many plants that cannot survive when exposed to freezing temperatures. Temperature variations also affect evaporation and the level of moisture in soil.

Position and aspect

Geographical location is an important determinant in the development of biome type. In Victoria, the vegetation on a hillside facing south (a southerly aspect) will typically have more shade, greater levels of soil moisture and lower temperatures overall when compared with a slope on the same mountain that faces north. A northerly aspect will typically result in drier soils and warmer temperatures, and potentially quite different vegetation. For example, depressions in the high plains or on lowland plains can result in a localised effect on plant communities, leading to a lack of vegetation in these **perennially** cooler areas.

perennially
lasting or active through the year or through many years

These and other aspects of outdoor environments interact constantly with one another to determine the types and characteristics of the flora and fauna that are present in a location. Biome types are generally recognised by either the landform (such as coastal and alpine biomes) or the characteristic vegetation they support (such as rainforest or grassland). In this section of the course, we investigate the features of a range of biomes, including alpine, coastal, inland waterways, grassland, heathland, forest, marine and arid biomes. Biomes are distinguished by their geological features, climate, plant and animal communities and the influence of humans.

1.2.2 ALPINE BIOMES

The word 'alpine' is often used to describe any high mountain area. Technically, the term refers to areas above a certain altitude that are treeless because of prolonged low temperatures. In Australia, the environments that regularly experience snow for an extended period occur in the elevated regions of the Australian Alps on the mainland, and the Central Plateau and other areas of Tasmania. This snow-covered region occupies approximately 5200 square kilometres on the mainland and 6500 square kilometres in Tasmania. The combined area is 0.15% of Australia.

FAST FACT

The Australian Alps are part of the Great Dividing Range, the series of mountains and highlands that extend about 3000 km from northern Queensland into north-eastern Victoria.

Victorian distribution

The Victorian Alps encompass approximately 500000 hectares of the north-east and east of the state. The range extends along the Great Dividing Range, including some isolated plateaus such as Lake Mountain, Mount Baw Baw and Mount Buffalo.

Elevation and climate

Alpine biomes in Victoria are mostly above 1300 metres, with rainfall usually exceeding 1400 millimetres per year. The highest mountain in Australia is Mount Kosciuszko at 2228 metres and the highest peak in Victoria is Mount Bogong at 1986 metres. Climatic conditions in the alpine biomes are harsh, with a covering of snow for more than a third of the year.

Alamy/Chris Mellor

Figure 1.18 *Eucalyptus pauciflora* (snow gum) is the hardiest of all eucalyptus species and is able to survive severe winter temperatures.

Biodiversity and common flora and fauna

sphagnum bogs species of mosses; alpine sphagnum bogs are found in permanently wet sites in high rainfall alpine, sub-alpine and montane areas of NSW, ACT, Victoria and Tasmania

Unlike other alpine regions around the world, Australia's alpine areas have been eroded over 500 million years to form rounded mountains and plateaus. They are typically vegetated by a range of heath, herb and grass species, along with unique **sphagnum bogs** that have adapted to specific soil types and climatic conditions. The predominant tree species at higher elevations is the snow gum Eucalyptus pauciflora, which is well known for its distinctively gnarled form, shaped by wind and snow.

The highest areas consist of communities of heathland, grassland and alpine bog that have adapted to the particular characteristics of their environment, including soils, water availability, **microclimate** and **topography**. The region can usually expect to be covered in snow for three to four months of the year, and most species take advantage of the short growing season during spring and summer. Plants and animals found in these areas have evolved over many years to survive in the unique alpine living conditions, and several species are only found in this environment, including the Baw Baw frog and the mountain pygmy possum. Migratory species such as the flame robin and spine-tailed swift make the most of the abundant food resources available during the short growing season, while millions of Bogong moths **aestivate** in rock crevices in the cooler summer climate.

microclimate the prevailing weather conditions of a small, specific place within a larger area

topography the landforms or surface features of a region

aestivate to spend a period in dormancy; similar to hibernation

Human influence and threats

The Victorian Alps are located mostly within the Alpine National Park, which was established in 1989 after many years of lobbying and campaigning, but they are still under threat from several directions. One of the biggest effects on this environment is from the continued development of ski resorts, particularly in areas such as Falls Creek, Mount Hotham and Mount Buller. These resorts are situated on leasehold land and operated by boards of management, and, in some cases, are surrounded by national park or other reserved land. Other issues affecting the Victorian Alps today include the continued summer grazing of cattle, predation by feral cats and dogs, water catchment pollution and environmental **weeds**. An issue of increasing concern is the effect of global warming on the viability of this unique habitat, particularly for those species of flora and fauna that rely on annual snowfall and cooler temperatures.

weed a type of plant that can invade and establish itself in an ecosystem where it was not originally present, regardless of whether it is a native or non-native species, and deprive other plants of space and food

1.2.3 COASTAL BIOMES

Coastal biomes relate to areas linking the land and the sea; however, the characteristics of these landscapes vary greatly – from beaches and dune systems to woodland, dry forests and rocky coastal cliffs. The coast is constantly changing in response to its interaction with the sculpting effects of wind, rain and waves. Much of the Victorian coast has been reserved as public land. The west coast is sometimes exposed to gale-force winds that have contributed to the spectacular scenery visible along the Great Ocean Road. Some regions, such as Wilsons Promontory, Croajingolong and the Twelve Apostles, have been national parks for some years. In 2002, the Victorian Government declared a system of marine national parks designed to further protect these areas, particularly the adjacent marine waters and the ecosystems within them. The waters and foreshores are also important for the economic opportunities they provide in supporting fishing, tourism, shipping, ports, and social and recreational pursuits.

Victorian distribution

Victoria has 2000 kilometres of coastline, ranging from sheltered bays and inlets to rugged eroded cliffs. There are about 123 bays, inlets and estuaries – varying in size from approximately 1 square kilometre to 2000 square kilometres.

Elevation and climate

Coastal biomes are located in flat landscapes at low altitudes from 0 to 200 metres above sea level, with a rainfall from 700 to 1200 millimetres per year.

Biodiversity and common flora and fauna

colonise to populate an area

swales shallow troughs between sand dunes

Some of the factors that influence coastal vegetation are related to wind, salt and natural land instability. Primary dunes are **colonised** by grasses and herbs such as the hairy spinifex, while salt marsh and mangroves inhabit the mudflats. These plants play a crucial role in holding together the subsoils in these environments. In less exposed areas, such as the lee side of dunes and in **swales**, coastal scrub (consisting of species such as melaleuca, casuarina heaths and banksia) has developed.

Birds are the foremost faunal species living in the immediate environs of the coast. For example, the endangered orange-bellied parrot migrates from south-western Tasmania to spend winter in the coastal salt marshes in Victoria. Little penguins (also known as fairy penguins) and short-tailed shearwaters nest in dunes, along with a large number of migratory birds from Siberia, Japan and the north Pacific Ocean, including the hooded plover and the little tern. Fur seals can also be seen in some areas, such as Phillip Island, raising their young.

Human influence and threats

More than 85% of Australians live less than 50 kilometres from the coast; therefore, the largest threat to coastal areas comes from urban development and the associated issues of introduced weeds and animals, which readily adapt to this ecosystem. Ground-nesting birds and small mammals are threatened by these problems, as well as the significant recreational use of the coast. The escalating use of 4WD vehicles has increased coastal degradation and the spread of weeds in more isolated areas. Additional impacts include rising sea levels and increases in the frequency of storms due to climate change.

1.2.4 INLAND WATERWAY BIOMES

Water is the foundation of all life and therefore is one of our most valuable resources. Water is vital to the health of our environment and our people. Our inland waterway biomes, which include rivers, lakes and wetlands, are the lifeblood of the Australian landscape.

ephemeral temporary or intermittent

Victoria has a rich variety of inland waterway biomes. These waters can be perennial (lasting throughout the year) or **ephemeral** (drying up periodically), such as the intermittently flooded wetlands and red gum floodplains of Barmah National Park in the state's north. Due to Australia's climate being highly variable, droughts and floods are common occurrences and have a major impact on inland waterways. In addition, the increased demand for water from population increases and economic growth has created significant challenges for how our inland waterways are managed. Inland waterways support natural processes that purify water while cycling nutrients and sediments.

Alamy/Hans Wismeijer

Figure 1.19 Barmah National Park is a part of the largest river red gum forest in the world and is internationally protected under the Ramsar Convention on wetlands.

Major functions of inland waterway biomes include water purification, flood prevention, bushfire prevention and carbon storage. Because inland waterways are ideal venues for recreational pursuits such as fishing, swimming, boating, hunting and water-skiing, they can be placed under enormous pressure from human impacts. Naturally, this affects the native flora and fauna that inhabit an area, and careful management is essential. Land use systems must encompass the entire **catchment area** of inland waterways if they are to be successfully conserved.

catchment area the area of land where water from precipitation drains into a body of water

Victorian distribution

Victoria is blessed with a vast array of inland waterways from 29 major river basins, including the mighty Murray River bordering the state's north, the substantial network of lakes, marshes and lagoons of the Gippsland Lakes, the more than 17000 inland wetlands linked to rivers such as those at Kerang, and small freshwater meadows and lakes such as Cherry Lake in Altona. Several of Victoria's inland waterways have been recognised as being of international significance because of the bird species that migrate to them from the northern hemisphere. These sites are covered by fresh or salt water and support diverse vegetation, birdlife and wildlife, and include Port Phillip Bay and the Bellarine Peninsula, Western District Lakes, Hattah Lakes, Barmah National Park and the Gippsland Lakes.

Elevation and climate

Inland waterways are located in all landscapes across the state, from low altitudes near the coast such as the mouth of the Aire River to the majestic Rocky Valley Lake near Falls Creek, which at 1600 metres is the highest significant body of water in Australia. Inland waterway biomes have a variable climate depending on seasonal changes.

Biodiversity and common flora and fauna

Many of the species inhabiting inland waterways (such as birds, frogs, fish and plants) cannot exist anywhere else and rely on these ecosystems as their breeding grounds. Often the **riparian vegetation** that occurs beside rivers and wetlands – such as tall eucalypts, large wattles and broad-leafed shrubs – provide vital breeding sites for native birds such as kingfishers and swallows, and other native animals that inhabit wetlands such as platypuses, frogs, fish, water rats, skinks and snakes.

Human influence and threats

Inland waterways are among the most threatened ecosystems in Australia. Many wetland environments have been significantly degraded or destroyed by human activities such as **irrigation** and drainage for agriculture, **urbanisation** and introduced species. It is vital to maintain adequate ecological flow regimes for wetland ecosystems to be maintained. Many rivers in Victoria have been dammed, dredged or channelled, and severely polluted. Water from many of the rivers is redirected via channels to be used for pasture and crop irrigation. In addition, approximately 35% of wetlands have been drained for **land reclamation**. These processes result in the removal of streamside vegetation and habitat, and the interruption of natural flows, all of which leave the ecosystems in poor health. Effects such as these – alongside land clearing and degradation, water extraction, salt disposal, the construction of dams, weirs and locks, increased nutrient levels from agricultural runoff, de-snagging, predation and competition from **exotic species** (such as European carp) – will continue to degrade the quality of these ecosystems.

riparian vegetation plant habitats and communities along a waterway's margins and banks

irrigation the artificial application of water to arable land for agricultural use

urbanisation the physical growth of urban areas as a result of rural migration

land reclamation the process of creating new land from the ocean, riverbeds or lakes

exotic species a species living outside its native distributional range

LEARNING ACTIVITY 1.2A

Wetland ecosystems

To learn more about the critical role wetland ecosystems play in Australia, visit the Wetlands page on the Australian Department of Climate Change, Energy, the Environment and Water website.

Worksheet 1.2a Wetland ecosystems

1.2.5 GRASSLAND BIOMES

Grasslands are defined as ecological communities where grass species dominate the area and there is less than 10% natural tree or shrub cover.

Victorian distribution

Prior to European settlers arriving in Victoria, extensive grasslands covered the plains between the Murray Valley and the Great Dividing Range. Today, grasslands are found in patches in the northern and western parts of the state and in some parts of Gippsland. Some of the most valuable areas of native grassland are on the western and northern outskirts of metropolitan Melbourne.

Elevation and climate

Grasslands are located in flat to gently undulating landscapes at altitudes below 700 metres, with low to medium rainfall areas of 400 to 1000 millimetres per year.

Biodiversity and common flora and fauna

FAST FACT

Grassland biomes are the most critically endangered ecosystem in Australia. Since European colonisation, 95% of these grasslands have been destroyed.

Grasslands are among the most species-rich plant communities in Australia, containing many grass species and an even greater diversity of other herbs. Grasslands contain a variety of floral species, but are best known for kangaroo grasses, wallaby grasses and spear grasses. In areas where the soils are deeper, grassy woodlands are populated by yellow gums, river red gums, casuarinas and acacias.

Grasslands play host to many faunal species that have adapted to life in this environment, such as kangaroos, bandicoots, possums, bats, skinks, lizards, snakes and a large number of birds, including magpies, swallows, wagtails, ibis, rosellas, parrots and galahs. The bush stone-curlew and the plains-wanderer are examples of ground-dwelling birds whose numbers have declined as a result of habitat loss and predation by feral species.

Human influence and threats

Prior to European colonisation, Indigenous peoples used fire to maintain the open nature of landscape and to promote new growth, thus attracting feeding animals. The open nature of grasslands made this type of environment particularly attractive to European explorers and settlers seeking suitable grazing, cropping and pasturelands. Because much of the grassland environment had little timber and was seen to be resource-poor (in terms of mining and water), it was readily available to free settlers. The introduction of sheep and cattle increased the degradation and, hence, decreased the biodiversity of grassland biomes.

Leigh Park

Figure 1.20 Mulla Mulla Grasslands at Mount Cottrell, a threatened native grassy ecosystem of the Victorian Volcanic Plains

As a result of this extensive settlement, less than 1% of Victoria's native grasslands remain intact today – these mostly occur in small areas, which are at risk from weed invasion, urban development, salinity and agricultural practices.

The greatest natural threat to the health of grasslands is climate change. With an increase in temperatures and drought conditions, shallow-rooted grasses are at great risk.

1.2.6 HEATHLAND BIOMES

Heathland is one of the oldest recognised ecosystems in the world. The heathland biomes is characteristically low and shrubby, with trees twisted and gnarled by the typically dry winds in nutrient-poor sandy soils.

Victorian distribution

Heathlands are found throughout Victoria, from the coast and hinterland to the mountains; however, they are particularly prevalent near the coast and in the south-west of the state.

Shutterstock.com/EQRoy

Figure 1.21 The heathland biome of the Great Otway National Park, with its characteristic dense, low shrubs and scattered, twisted trees

Elevation and climate

The annual rainfall of heathland environments varies from 600 to 1100 millimetres per year and the altitude they can be found at ranges from 50 to 300 metres above sea level.

Biodiversity and common flora and fauna

Nutrient levels in the soils in heathland environments are generally low and the soils are acidic. Many species in heathlands have a close interrelationship with fire. While some may re-sprout following burning, others will die, but their hard, woody fruit pods will open under intense heat and release seeds. The distinctive grass tree, Xanthorrhoea australis, will respond to fire with a display of new growth.

Heathlands tend to be dominated by hard-leaved plants such as banksias, hakeas, peas, bottlebrushes, melaleucas, tea trees and eucalypts. As the name suggests, they are also populated by a range of heaths, including the common heath, cranberry heath, daphne heath and flame heath. Over 80 species of birds inhabit this ecosystem, along with small marsupials such as dunnarts, potoroos and bandicoots, and placental mammals such as native rats and native mice.

Human influence and threats

Approximately half of all Victorian heathland environments have been permanently cleared for agriculture, urban development or mining. Several areas of heathland have been conserved in Victoria's park system, but coastal development, plant diseases and environmental weeds threaten others. Heathlands are also particularly prone to **cinnamon fungus** and invasion by woody native species. If fire occurs too infrequently or is too large in scale, it is a threat to a heathland environment.

cinnamon fungus a soil-borne water mould that produces an infection that causes root rot or dieback

1.2.7 FOREST BIOMES

A forest is an area of land dominated by trees that have a height of at least 2 metres and a tree canopy cover of at least 20%. Australia has 125 million hectares of forest, which is 16% of Australia's land area. More than 80% of Australia's native forest area is dominated by eucalypt forest and acacia forest. These trees are separated, allowing light to reach the ground and enabling a rich variety of hard-leaved shrubs, ferns, herbs and grasses to flourish. The Great Dividing Range, which was formed approximately 3 million years ago, forms a barrier across Victoria. It serves to protect many of the north-facing slopes from the cool, moisture-laden winds sweeping in from Bass Strait and the Southern Ocean. This results in the northern slopes and foothills being relatively dry, which supports a variety of dry forests and woodlands. The vast majority of forests in Australia are known as dry **sclerophyll**. Approximately 80% of the plants in these forests are eucalypt trees with hard, short and often spiky leaves, which is a condition closely associated with low soil fertility.

sclerophyll an Australian vegetation type having plants (typically eucalypts, wattles and banksias) with hard, short and often spiky leaves

Museums Victoria/Photographer: Rodney Start

Figure 1.22 The box-ironbark forest in Greytown, central Victoria, is a dry forest environment.

Victorian distribution

There are many types of forests in Victoria. Stringybark forests dominate the landscape east of Western Port; grassy woodlands are scattered throughout the Western District; box-ironbark forests cover the area west of Stawell to the east of Wangaratta; and temperate rainforests stretch from East Gippsland and the Strzelecki Ranges to Wilsons Promontory, the Central Highlands and the Otways.

Elevation and climate

Forests are found mostly between 200 and 1000 metres above sea level, receiving limited amounts of rain, varying between 550 and 1000 millimetres per year.Forests are found mostly between 200 and 1000 metres above sea level, receiving limited amounts of rain, varying between 550 and 1000 millimetres per year.

Biodiversity and common flora and fauna

Forests are biologically diverse and support a variety of plants and animals, including some of Victoria's rarer flowers and birds.

Open forests comprise a variety of eucalypts, casuarinas and cypress pines, and they have grassy, heathy **sedge** and herb-rich understoreys. Red gum forests dominate major waterways, such as the Goulburn and Murray rivers, in the north of Victoria. Stringybark forests populate the landscape east of Western Port; remnant grassy woodlands are scattered throughout the Western District; and very little of the once extensive box-ironbark forests lie across the middle of Victoria. Fire is an important component in this type of forest as some native trees, such as the banksia, require the heat of a fire to crack open seedpods to free the seed inside.

sedge a grass-like plant with triangular stems and inconspicuous flowers, typically growing in wet ground

Dry forests are home to a vast variety of fauna including wallabies, kangaroos, koalas, wombats, possums, echidnas, parrots and gliders.

Human influence and threats

Much of the forest biomes that would have been found in Victoria around the beginning of the 19th century have diminished, mainly due to extensive clearing during the gold rushes and settlement when the demand for timber for housing, mining and railways was large. Further decline has occurred as a result of agricultural practices and urban development.

Major threats to the health of forest biomes include habitat modification and **fragmentation**, vegetation clearance, overgrazing, weed invasion, cinnamon fungus, feral predators and the loss of hollow-bearing trees.

fragmentation the reduction or breaking up of one area of habitat into several smaller separate areas

1.2.8 MARINE BIOMES

Marine biomes include a wide variety of ecosystems such as **subtidal** and **intertidal** rocky reefs, sponge gardens, kelp forests, mangroves, mudflats, open water, sandy plains and seagrass beds. Victoria's marine biomes are among the most biologically diverse in the world, with over 90% of Victoria's marine organisms found nowhere else on Earth. This high biodiversity is a result of the relatively low level of nutrients in the water compared with other areas of the world. This limits the production by plants, algae and bacteria of food, leading to many species evolving over time to adapt to the specialised conditions, with no one species able to dominate.

subtidal
an area that is permanently covered with water

intertidal
an area that is above water at low tide and under water at high tide

Victorian distribution

Victorian marine biomes extend 3 nautical miles offshore. Victoria's marine waters cover approximately 10000 square kilometres. This includes rocky reefs, sandy seafloors, spectacular underwater canyons, intertidal mudflats, sheltered bays, estuaries and waters adjacent to the open ocean.

Elevation and climate

Most Victorian marine waters are shallow; however, some extend to depths of 90 metres or more. The rainfall is variable, from 700 to 1200 millimetres per year.

Biodiversity and common flora and fauna

Marine waters contain a diversity of habitats such as seagrass meadows, mudflats, intertidal and subtidal rocky reefs, mangrove kelp forests and pelagic systems. Over 12000 marine animals and plants live in these waters, contributing to their high levels of biodiversity as well as being important representations of species **endemic** to the Southern Ocean. Towering kelp forests, seagrass meadows and a wide variety of fish, sponges and other animals from tiny organisms to large sea mammals are contained within these environments.

endemic
a feature or species that is unique to a defined geographic location

Shutterstock.com/Willyam Bradberry

Figure 1.23 Victorian marine ecosystems support more than 12000 species of marine animals and plants.

Alamy Stock Photo/imageBROKER

Figure 1.24 The spectacular Twelve Apostles Marine National Park on the south-western coast of Victoria

Human influence and threats

Threats to marine biomes include:

- overfishing
- increases in sediment from **dredging**, smothering marine growth
- increases in sediment entering the seas as a result of land clearing
- pollution from factories and stormwater drains
- fertilisers, pesticides and herbicides used in farming practices causing poisonous algal blooms, which impede the growth of marine plants
- introduced species such as the Northern Pacific seastar (Japanese starfish) from **ballast water**.

dredging
an excavation activity or operation usually carried out to gather up bottom sediments and dispose of them at a different location

ballast water
water taken on board to provide stability for ships

1.2.9 ARID BIOMES

Arid biomes are characterised by a severe lack of available water, which hinders the growth and development of plant and animal life. Approximately 4 million years ago, a vast inland sea covered what we now call the Mallee and the north-western part of Victoria. After a succession of arid and wet phases, the area has been left with a legacy of sand and shallow soils that cover the area today.

In Australia, arid and semi-arid biomes are located in the interior (or middle) of the continent, with infertile soils, a highly erratic rainfall, extremes of long dry periods and occasional flooding.

Victorian distribution

In Victoria, the name 'Mallee' is used to describe the arid and semi-arid north-western parts of the state. The key environmental features of the Mallee are high summer temperatures, relatively infertile soils and low, unreliable rainfall.

Elevation and climate

Victorian arid biomes have an annual rainfall of about 250 to 400 millimetres and an altitude range of between 50 and 200 metres above sea level.

Shutterstock.com/Photodigitaal.nl

Figure 1.25 An arid biome

Biodiversity and common flora and fauna

The landscape tends to be dominated by low Mallee scrub made up of small mallee eucalypts that can withstand prolonged dry periods and harsh conditions. Deeper soils are home to heathland with desert banksia, scrub pine and casuarina. Saltbush grows in more saline areas, and spear grass and cypress pine intermingle with salt pans. The flora and fauna in this seemingly lifeless landscape are remarkably diverse. Numerous reptile and bird species inhabit the area, along with a range of native grasses and shrubs.

Victorian arid biomes are home to the malleefowl (which constructs huge mounds of sand and litter to incubate its eggs), parrots, kangaroos, geckos, skinks, snakes and a range of small ground-dwelling mammals.

Human influence and threats

About one-third of all Victorian arid biomes have been cleared for agriculture, mining or urban development. Agricultural practices and subsequent issues such as salinity have affected significant areas of arid and semi-arid environments, particularly in the northern parts of Victoria. At least 12 species of mammals have become **extinct** in recent times, including the pig-footed bandicoot, eastern hare-wallaby, lesser stick nest rat and rabbit-eared tree rat. **Remnant vegetation** exists in small, fragmented areas, threatened by environmental weeds, predation by feral animals (such as rabbits and goats) and intense fire events.

extinct
no longer existing or living

Remnant vegetation
small patches of native plants that remain after conversion of landscapes to agricultural or other use

LEARNING ACTIVITY 1.2B

Features of biomes

Working individually or in small groups, produce posters for each of the different types of biomes identified in the text. Include the following information on your poster:

1 the type of biome
2 distribution of your biome on a map of Victoria
3 the biome's elevation, annual rainfall and examples of unique flora and fauna
4 the influence humans and other threats have had on the health of this biome
5 pictures displaying a range of unique features of your biome.

Using information from all of the posters, complete the table below in your workbook.

Features of biomes

Type of Biome	Distribution within Victoria	Elevation	Climate	Biodiversity and common flora and fauna	Human influence and threats

Worksheet
1.2b Features of biomes

1.2 KEY CONCEPTS

- Victoria's various outdoor environments support a high level of biodiversity.
- Outdoor environments within the Earth's biosphere can be classified into sub-groups called biomes.
- Biomes are large naturally occurring communities of flora and fauna occupying a major habitat.
- Developing knowledge of biomes is crucial so we can understand their importance and how our actions impact the health and sustainability of our unique environment.
- The type of biome that exists within a specific location is the result of the interaction of three factors: geology, climate, and position and aspect.
- Biome types are generally recognised by either the landform (such as coastal and alpine biomes) or the characteristic vegetation they support.
- Victoria has a rich and diverse range of biomes, including alpine, coastal, inland waterways, grassland, heathland, forest, marine and arid biomes.
- Climate change, native vegetation clearing, urban development and introduced species have in the past, and continue to have, a significant impact on a range of Victorian biomes.

Worksheet
1.2 Key concepts

1.2 CONCEPT QUESTIONS

1 Using examples of each, distinguish between the terms 'biomes' and 'biodiversity'.
2 Identify and describe three factors that influence the type of biome that is present in an outdoor environment.

UNDERSTANDING

3 On a map of Australia, identify the locations of the highest and lowest ever recorded temperatures and rainfall. Explain why extremes in temperatures and rainfall occur in these areas.
4 Discuss how two different biomes may be interrelated.

APPLYING

5 Produce an infographic that compares the features of two different biomes.
6 Research Australia's changing climate. Create graphs of the annual temperature and rainfall for the past 150 years. Identify and discuss any patterns that you notice and the reasons for these changes.

EXTENSION CHALLENGE

7 Many urban areas in Victoria have been built on land that was originally forested. Which forest type was cleared to make way for the area in which you live?
8 What strategies have been implemented to preserve remnant forest in your area?
9 Apart from the dominant forest type, what other flora and fauna may have lived in your area before it was urbanised?
10 Find a place where some remnant forest exists in your area. Visit and describe its current condition, detailing the plants and animals it contains.

1.3 MOTIVATIONS FOR EXPERIENCING AND RESPONSES TO OUTDOOR ENVIRONMENTS

KEY KNOWLEDGE

- the range of motivations for experiencing outdoor environments and responses to outdoor environments, such as fear and appreciation

KEY SKILLS

- analyse motivations for experiencing outdoor environments and responses to experiencing outdoor environments

> The mountains are calling and I must go ...
>
> John Muir

The quote above was written by John Muir (1838–1914), a Scottish naturalist and mountaineer. It has become an iconic quote for outdoor enthusiasts around the world. The full quote – 'The mountains are calling and I must go and I will work on while I can, studying incessantly' – was in reference to his inner drive to experience, observe, understand and preserve the Yosemite Valley in California.

Alamy/Matthew Baugh Photography

Figure 1.26 Yosemite National Park, California, USA

'Do you want to go to the beach?' 'Would you like to hang out at the park after school?' 'Want to go shopping?' 'How about we go camping?' Your responses to these and other similar questions are based on your motivations – your reasons for saying yes. Just like John Muir, you have a range of motivations to participate in activities you enjoy. In this section, we investigate why people venture out into outdoor environments and how may they respond to their outdoor experiences.

1.3.1 MOTIVATIONS FOR SEEKING OUTDOOR EXPERIENCES

What is motivation?

Motivation is described as the driving force or reasoning behind a person's desire to do something. It is why a person acts or behaves in a particular way. It drives people's willingness and desire to achieve a particular feat, complete a task or achieve a dream. Sometimes we experience high levels of motivation when participating in an exciting and enjoyable activity, and at other times we procrastinate for what seems like ages and may find it almost impossible to motivate ourselves. Motivations can be broken into two main types: intrinsic and extrinsic.

Alamy Stock Photo/LOOK Die Bildagentur der Fotografen GmbH

Figure 1.27 What motivates a person to go rafting? Many adventure sports, including rafting, feature intrinsic motivations such as fun, adrenaline rushes and personal satisfaction.

Intrinsic motivations

Intrinsic motivations motivations to engage in an activity that we get from within ourselves

Intrinsic motivations refer to the motivations to engage in an activity that we get from within ourselves. It is an internal desire to perform a task, such as completing a difficult hike up a mountain. What motivates us in this way? Think of something challenging that you enjoy doing, and ask yourself: 'What is it inside me that helps me to keep going?' The chances are it will be the good feeling you get when you do this thing: the adrenaline rush from scoring a goal, or from reaching the top of a difficult rock climb. If it's not a feeling, it might be a more general sense of personal satisfaction that you've done something worthwhile. 'Peak baggers' enjoy ticking (or bagging) mountains (peaks) off a list. There's a sense of personal satisfaction at having completed a list and this helps to motivate them to continue. Many bird-watchers tick off bird sightings from a similar sort of list, again generating a sense of satisfaction and therefore motivation to continue.

Extrinsic motivations

Extrinsic motivation motivations to engage in an activity because we want to earn a reward or avoid punishment.

Extrinsic motivation is when we are motivated to engage in an activity because we want to earn a reward or avoid a punishment. Extrinsic motivation is based on factors external to the individual, such as receiving praise from our classmates for making it up a difficult rock climb. An extrinsic motivation to engage in an activity may be for money. Many of the things that people do in the outdoors – from professional snowboarders and surfers to farmers and Victoria Parks rangers – they do because, at some point, they will get paid to do these things. That isn't to say this is the only motivation. Often, motivations start with the intrinsic. Snowboarders love to snowboard as it makes them feel good, and park rangers love the environment and the sense they get of caring for it. Their motivation continues with the extrinsic motivation of getting paid. Some people are also motivated because of external physical rewards (other than money), such as in competitions when we are challenged and/or gain the respect of others.

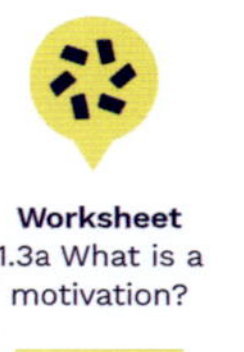

Worksheet 1.3a What is a motivation?

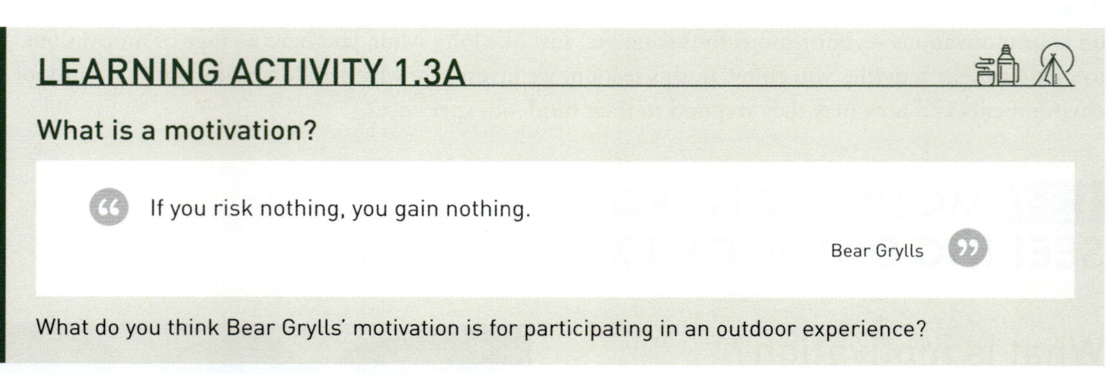

LEARNING ACTIVITY 1.3A

What is a motivation?

> If you risk nothing, you gain nothing.
>
> Bear Grylls

What do you think Bear Grylls' motivation is for participating in an outdoor experience?

1.3.2 MOTIVATIONS FOR OUTDOOR EXPERIENCES

What motivates people to participate in experiences in outdoor environments? Why would you want to head to the hills during winter where there is a possibility of getting cold, wet and uncomfortable?

THE PSYCHOLOGY OF SKIING AND SNOWBOARDING

Joanna Pantazi, in her blog titled 'The Psychology of Skiing and Snowboarding', asks the question, 'Why would someone realistically risk their physical health on quite a high price in order to slide on icy slopes, with one long stick under each foot or a big board under both?' She continues, 'By learning advanced skills that enable you to progress, you can subsequently challenge yourself gradually more every time: go to a steeper slope, climb to a higher peak, go into deeper snow, stay longer, achieve more.' The outcome or motivation Joanna identifies as 'this deep sense of accomplishment and confronting challenges is extremely powerful and it increases your self-esteem and self-efficacy. It fuels your motivation to achieve even more.'

As Joanna points out, participating in outdoor experiences has a range of personal benefits for the individual. Of course, there are many other reasons or motivations for individuals wanting to participate in outdoor experiences, such as for the fun, to learn a new skill, to escape the rigours and stressors of everyday life, to set a record, to feel a sense of achievement, for the adrenaline rush, to hang out with friends, etc. These and other motivations can be grouped into the following four categories: competence/mastery, stimulus avoidance, socialisation and cognitive reward.

Joanna Pantazi, 'The Psychology of Skiing and Snowboarding', youniverse, www.youniversetherapy.com

Table 1.1 Motivations

Competence/Mastery	Stimulus Avoidance
Competence/mastery is the motivation to experience outdoor environments to develop our skills, confidence and performance in an activity. Examples of this type of motivation include: • being the first to achieve something (e.g. climbing Mount Everest) • building strength through participating in the experience • being confident in participating in an activity • setting a record or other achievement (e.g. 'bagging' a peak).	Stimulus avoidance is the motivation to experience outdoor environments so something unpleasant or uncomfortable is avoided. Examples of this type of motivation include: • as an escape from the stressful aspects of our everyday lives • to clear our minds to benefit our health and wellbeing • wanting to avoid unwanted social situations.
Socialisation	**Cognitive Reward**
Socialisation is the motivation to experience outdoor environments for the social benefits. Examples of this type of motivation include: • personal and social development and support • the feeling of community from doing things with other people • to promote a sense of safety, belonging and security.	Cognitive reward is the motivation to experience outdoor environments for the mental benefits. Examples of this type of motivation include: • learning a new skill • achieving something new, different or risky • experiencing an adrenalin rush • proving something to yourself or others • developing resilience • achieving a state of 'flow'.

'Flow' – an example of motivation

Developed in the mid-1970s by Hungarian psychologist Mihaly Csikszentmihalyi, the 'flow model' of concentration and engagement basically suggests that people are at their happiest when they are fully and completely engaged in an activity; that is, when they are in a state of flow.

> [Flow is] being completely involved in an activity for its own sake. The ego falls away. Time flies. Every action, movement, and thought follows inevitably from the previous one, like playing jazz. Your whole being is involved, and you're using your skills to the utmost.
>
> Mihaly Csikszentmihalyi 

When someone is doing something with such concentration that nothing else matters, time seems to move incredibly quickly. When you're in a 'flow' experience:

- you feel at one with the world – your sense of being an individual disappears
- you let go of worries and problems
- you're completely focused
- you feel satisfied with what you're doing
- you're happy, although you probably don't notice it as you are completely engrossed in what you're doing.

This is flow and it is possibly one of the key motivations for participants who are experienced in outdoor activities.

The Buddhist practice known as 'mindfulness' creates a similar effect and is something many adventurers and recreationists practise.

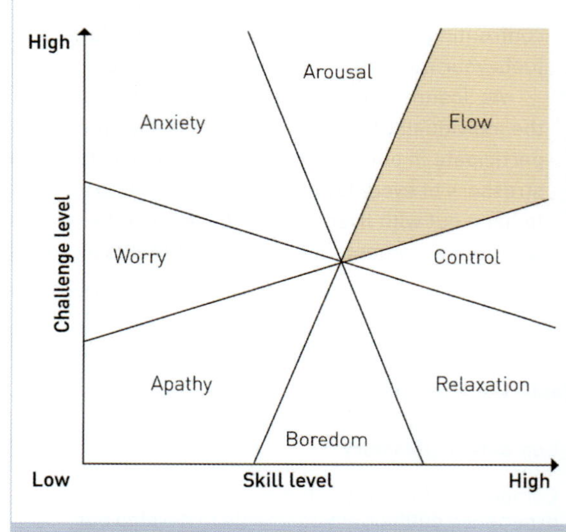

Figure 1.28 Csikszentmihalyi's flow model

> Mindfulness means paying attention in a particular way: on purpose, in the present moment, and non-judgmentally.
>
> Jon Kabat-Zinn, *Wherever You Go, There You Are*, 1994 

Worksheet 1.3b Poetry of motivations

LEARNING ACTIVITY 1.3B

Poetry of motivations

> I went to the woods because I wished to live deliberately, to front only the essential facts of life, and see if I could not learn what it had to teach, and not, when I came to die, discover that I had not lived.
>
> Henry David Thoreau, Walden, 1854 

Find some examples of poetic or literary quotes about outdoor environments and/or outdoor experiences, similar to the quote from Henry David Thoreau. Describe some of the motivations you can find in these quotes.

LEARNING ACTIVITY 1.3C

Personal motivations

Visit Billabong's YouTube channel to watch 'I surf because' videos, which feature individuals describing their motivations to surf. Sort these motivations into the different categories described in this section.

Worksheet
1.3c Personal motivations

1.3.3 PERSONAL RESPONSES TO OUTDOOR ENVIRONMENTS

The previous section looked at why people visit and interact with outdoor environments. In this section, we look at the results of these interactions – how we respond to our experiences in and with outdoor environments.

'Response' refers to the feeling or emotion that an outdoor environment or outdoor experience creates in your mind. Responses will typically lead to certain behaviours. The way you respond to or feel about an environment is likely to influence the way you act in and with that environment.

Types of responses

Personal responses can be divided into two main groups based on the types of behaviours or actions that they may influence: positive responses and negative responses.

POSITIVE RESPONSES

A positive response to an outdoor environment is one that will probably result in positive behaviours towards that place – behaviours that might protect, preserve or enhance that place in some way. Positive responses include the following:

- appreciation – a recognition of value and significance in an outdoor environment
- awe – a feeling of wonder or admiration for an outdoor environment
- contemplation – a feeling that engenders long and thoughtful observation or a deep reflection about an outdoor environment
- inspiration – a feeling about the outdoor environment that leads a person to want to do something or create something
- exhilaration – a feeling of excitement or happiness, particularly resulting from an outdoor experience
- connection – a feeling that we are a part of a place or connected to that place; it can come with spiritual feelings or feelings of the wondrous nature of an environment
- curiosity – a feeling of wanting to know more and wanting to understand an outdoor environment in more detail or in other ways.

Shutterstock.com/michelangeloop

Figure 1.29 It can be incredibly hard to get to, but the beauty of an outdoor place can inspire awe and appreciation.

NEGATIVE RESPONSES

A negative response is one that will probably lead to damaging behaviours or dangerous actions with respect to an outdoor environment, such as the desire to clear it, change it or remove things from it. Negative responses include the following:

- fear – an unpleasant feeling resulting from a belief that something about an outdoor environment (or an activity in an environment) is dangerous
- revulsion – a sense of disgust and loathing; an environment could elicit a revulsion because of something negative that a person sees in an environment, such as one animal hunting and killing another
- curiosity – but in this case leading to a negative response, such as a person's curiosity leading them to a dangerous place or situation.

Alamy Stock Photo/Bjorn Svensson

Figure 1.30 A sense of achievement, wonder ... or exhaustion?

Alamy/ Corey Rich / Aurora Photos

Figure 1.31 Alex Honnold is an American rock climber best known for his free solo ascents of big walls. Why would a person willingly participate in an adventure when there is a significant level of fear and danger involved?

Resource
Case study: Bethany Hamilton

CASE STUDY

BETHANY HAMILTON

Bethany Hamilton Authors Children's Book *Surfing Past Fear*

Bethany Hamilton became a household name long ago. The life of this surfer, author and public speaker has been told and retold across a swath of media, from documentaries to dramas to talk shows.

But Hamilton is also a prolific content creator herself. With her latest work, a children's book titled *Surfing Past Fear* (her tenth book), Hamilton's aim is to inspire young girls with a message of hope and overcoming life's inevitable obstacles.

Fairfax Photo/Andrew De La Rue

Figure 1.32 Bethany Hamilton based her latest book *Surfing Past Fear* on her experience of surviving a shark attack while surfing.

Many children's books aim for that goal, of course, but Hamilton's life has proven her an unofficial world champion in confronting fear — the story of her run-in with a 14-foot tiger shark at 13, and then her drive and perseverance to surf again, has inspired thousands already.

Hamilton's is a heavy story that involved trauma and then extreme resilience. And that's why her new book relies on a 'family of friendly cartoon characters' to 'lighten the heavy load,' a press release said. Aiming the book at children as young as four, Hamilton wanted to ensure they leave the story with a feeling of hope.

'I'm so excited for everyone to be able to experience this story, and I pray that it encourages you to be resilient and adaptable when life isn't perfect, and to live in faith, rather than fear,' Hamilton said.

Her book comes with two games that all kids can participate in: 'Fearing Freaky Food' and 'Cowabunga.' According to Hamilton, the games are meant to teach two lessons: 'the root of fear is a lack of trust' and 'the fear of failure can stop our progress in life.'

'Women Who Survived: Bethany lost her arm in a shark attack at 13 and became a pro surfer' Libby-Jane Charleston', Honey.

QUESTIONS

Complete the following questions using the article and any additional research.

1 Describe the incident that occurred to Bethany while surfing.
2 What would be a typical emotional and behavioural response to someone who experienced this incident?
3 Describe how this incident has influenced Bethany's experiences in and with outdoor environments.
4 How has Bethany used this incident to inspire others?
5 Using examples, compare two different ways fear may influence how people respond to outdoor environments

1.3.4 VARYING RESPONSES TO OUTDOOR ENVIRONMENTS

Like everything else about people, responses to environments can vary significantly. Just as some people will relish eating fast food and others will not, one particular place may elicit a positive response in one person but a negative response in another. Additionally, the ways we respond to these places can change over time.

As we age

As a person ages, and as their character, personality, experiences and values change, the ways they respond to a particular outdoor environment may change. A place that a teenager may have responded to positively may elicit a completely different response in them many years later (or vice versa).

As society changes

Society changes over time, and this can lead to society-wide changes in the way we respond to particular places. As an example, in the early days of European colonisationof Australia, the bush was seen as a harsh, hostile and scary place. As European explorers and settlers opened up the country to settlement and development, for many people this fear disappeared, and today fear of the bush is a less common response.

The ways in which we demonstrate our responses can also vary, and may relate to the sorts of practices we are familiar with from other parts of our lives. For example, a writer may blog about their responses, while a painter may depict them or a musician may sing about them. For many Australians, photographs can be a powerful way of capturing a place or an experience to help them revisit and demonstrate their responses. The ease with which we can take, store and transmit images helps this particular way of exhibiting responses.

Factors influencing personal responses to outdoor environments

When we are in particular places or doing particular things, why do we respond in the ways we do?

Everyone responds in different ways for different reasons, but certainly there are some common factors that influence our responses. These can include our:

- age
- background
- education
- experience
- culture
- religion
- socioeconomic background
- media use.

These factors are discussed in further detail in the next chapter.

Worksheet 1.3d Factors affecting personal responses

LEARNING ACTIVITY 1.3D

Factors affecting personal responses

Choose one or more of the factors from the list above and write about how they affect your response to outdoor environments. Ask other people how they think their responses might be affected by these factors. Come up with some other factors that you think might also affect personal responses to environments.

1.3 KEY CONCEPTS

- Motivation is described as the driving force or reasoning behind a person's desire to do something.
- Intrinsic motivations refer to the motivations to engage in an activity that we get from within ourselves.
- Extrinsic motivations refer to the motivations to engage in an activity because we want to earn a reward or avoid punishment.
- There are four main categories of motivations for individuals to experience outdoor environments.
    - Competence/mastery is the motivation to experience outdoor environments to develop our skills, confidence and performance in an activity.
    - Stimulus avoidance is the motivation to experience outdoor environments so something unpleasant or uncomfortable is avoided.
    - Socialisation is the motivation to experience outdoor environments for the social benefits.
    - Cognitive reward is the motivation to experience outdoor environments for the mental benefits.
- A sense of 'flow' may be a key motivation for participants who are experienced in outdoor activities.
- Response refers to the feeling or emotion that an outdoor environment or outdoor experience creates in your mind.
- The way you respond to or feel about an environment is likely to influence the way you act in and with that environment.
- The positive response to an outdoor environment is a recognition of its value and significance.
- A positive response will probably result in positive behaviours towards that place – behaviours that might protect, preserve or enhance that place in some way.
- The negative response to an outdoor environment is a fear that something about an outdoor environment (or an activity in an environment) is dangerous and may harm you.
- A negative response may lead to damaging behaviours or dangerous actions with respect to an outdoor environment, such as the desire to clear it, change it or remove things from it.
- There is a range of factors that influence the ways we respond to the outdoors, including one's age, culture, education and media use

1.3 CONCEPT QUESTIONS

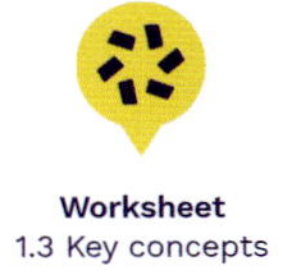

Worksheet
1.3 Key concepts

REMEMBERING

1 Define the term 'motivation'.
2 Outline the difference between intrinsic and extrinsic motivations for experiencing outdoor environments.

UNDERSTANDING

3 Identify, describe and provide an example of each of the four categories of motivation for experiencing outdoor environments.
4 Explain two different ways in which you have responded to experiences within a specific outdoor environment.
5 Outline two factors that may influence the way individuals respond to experiences in outdoor environments.

APPLYING

6 Using examples from your own experiences, analyse how the way you responded to a specific environment influenced how you interacted with that environment.
7 Why would a person willingly participate in an adventure when there is a significant level of fear or danger involved?

EXTENSION CHALLENGE

8 Research an outdoor adventurer/sports person. Create a poster about this person and their achievements within their chosen field. Include information about their motivations and responses to experiencing outdoor environments.

1.4 WAYS TO EXPERIENCE AND UNDERSTAND OUTDOOR ENVIRONMENTS

KEY KNOWLEDGE

- the variety of ways in which Indigenous and non-Indigenous peoples experience and understand outdoor environments

KEY SKILLS

- compare the variety of ways in which Indigenous and non-Indigenous peoples experience and understand outdoor environments

Outdoor environments, and the humans that interact with them, are complex and highly diverse. In this section, we compare the variety of ways in which Indigenous and non-Indigenous peoples experience and understand outdoor environments.

The way people view or value an environment is dependent upon the interactions they are likely to undertake, and what they believe they can gain from the environment, as well as the context of their experience. For example, if someone has grown up on a farm from a young age, they may be more likely to view the environment as a resource – somewhere to raise animals until such time they can sell them for a profit. On the other hand, an Indigenous Australian may be more likely to have a deep spiritual connection with the land – they may feel 'at one' with the environment based on shared stories of the Dreaming and their cultural understandings of the land. Of course, someone who grew up on a farm could develop a spiritual connection with their land, and an Indigenous Australian could also see their land as a resource. Experiences of places and how we understand the outdoor environment can be very complex.

There are many ways in which Indigenous and non-Indigenous Australians experience and understand outdoor environments, including:

- through custodianship
- as recreation
- as a resource
- through a spiritual connection
- as a study site.

1.4.1 OUTDOOR ENVIRONMENTS THROUGH CUSTODIANSHIP

The term '**custodianship**' is commonly associated with Indigenous peoples' deep spiritual relationship with the outdoor environment, reflected in their belief that their ancestors make up every atom of the land and resulting in a deep care and responsibility for the land. 'Country' is the term often used by Indigenous peoples to describe the lands, waterways and seas to which they are connected. The term contains complex ideas about law, place, custom, language, spiritual belief, cultural practice, material sustenance, family and identity.

iStockphoto.com/Jonathan Steinbeck

Figure 1.33 The coldest months of the year are identified as Cockatoo season in the Gariwerd seasonal calendar.

custodianship the responsibility for taking care of or protecting something

Indigenous peoples have nurtured and managed the natural resources of Australia for tens of thousands of years, and have strong cultural connections to Country that include a deep respect for all creatures. Through their longstanding experiences of Country, Indigenous peoples have developed an extensive knowledge and understanding of the natural systems of our continent.

One example of this in Victoria is the Djab Wurrung and Jardwadjali peoples, whose traditional lands include the area surrounding Gariwerd (the Grampians). They recognise six distinct weather periods that relate to climatic features as well as referencing environmental events such as plant flowering, fruiting and animal behaviour patterns. Refer to the Indigenous Weather Calendar in Chapter 4 (Figure 4.15). This understanding has guided the Djab Wurrung and Jardwadjali in their custodianship of the land and their use of sustainable practices to care for and protect the health of outdoor environments into the future.

LEARNING ACTIVITY 1.4A

Indigenous weather calendar

Research an Indigenous weather calendar for the region of an outdoor environment you have investigated and/or visited as a part of this subject. The Bureau of Meteorology's Indigenous Weather Knowledge website is an excellent resource for this.

Create a display that outlines the names, times and characteristics of each season.

How has your understanding of the specific outdoor environment changed due to exploring the Indigenous weather calendar?

Worksheet 1.4a Indigenous weather calendar

Weblink
Video: River Country Spirit Ceremony (Murrundi Ruwe Pangari Ringbalin)

SPOTLIGHT

River Country Spirit Ceremony

Watch the video 'River Country Spirit Ceremony (Murrundi Ruwe Pangari Ringbalin)' to learn about the Ngarrindjeri people's connection to their Country and the Murray River.

Fairfax Photos/Justin Mcmanus

Figure 1.34 Aboriginal people along the Murray River in south-eastern Australia participate in a River Country Spirit Ceremony to heal the river after a severe drought.

1.4.2 OUTDOOR ENVIRONMENT AS RECREATION

Recreation is associated with non-Indigenous peoples' ways to experience and understand outdoor environments. Recreation is an activity that is done for enjoyment, amusement or pleasure, and is considered to be fun by the participant. Many experiences within outdoor environments are recreationally based and these experiences usually provide time to reflect, appreciate and respect the environment you are in. Sitting by the campfire with your peers sharing stories of the day's thrills and spills is often a joyous and memorable experience.

© Richard Carey | Dreamstime.com

Figure 1.35 A great way for individuals to gain an appreciation and understanding of outdoor environments is to experience it themselves.

Recreation activities can be challenging and can encourage people to push themselves to achieve a particular goal. They can include passive activities such as wildlife observation, painting, photography and walking, or active pursuits such as hiking, rock climbing, mountain biking and surfing.

Through participating in recreational activities, participants gain a greater understanding of themselves and the characteristics of the outdoor environment. Reading or viewing footage of people snorkelling is very different from experiencing the sensation of floating in the water and discovering the beauty and uniqueness of the ocean floor for yourself. Both passive and active recreational activities are often undertaken by school groups because of the personal and social benefits gained through these experiences.

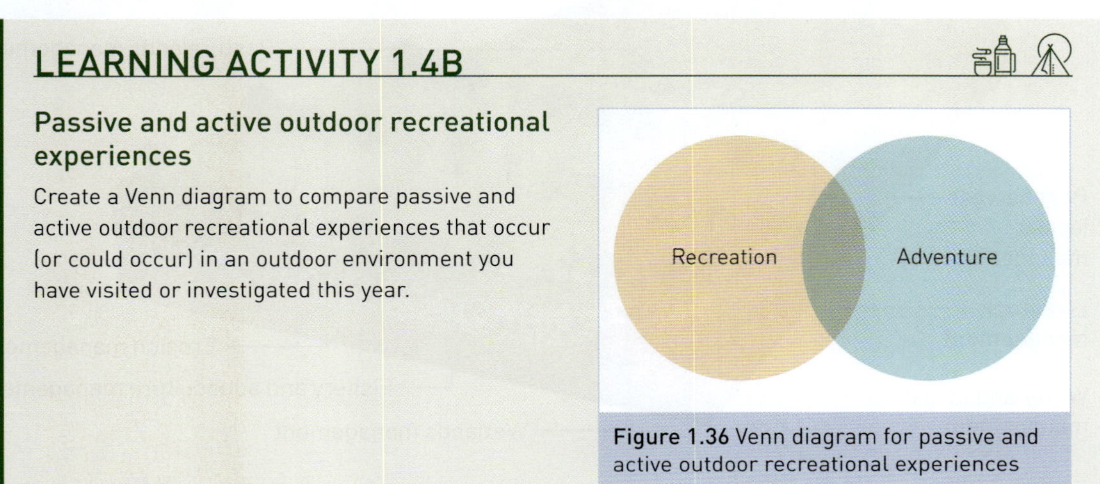

LEARNING ACTIVITY 1.4B

Passive and active outdoor recreational experiences

Create a Venn diagram to compare passive and active outdoor recreational experiences that occur (or could occur) in an outdoor environment you have visited or investigated this year.

Figure 1.36 Venn diagram for passive and active outdoor recreational experiences

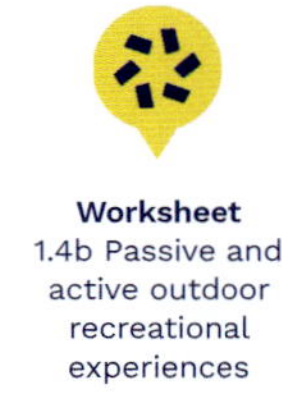

Worksheet
1.4b Passive and active outdoor recreational experiences

1.4.3 OUTDOOR ENVIRONMENT AS A RESOURCE

In Outdoor and Environmental Studies, a resource can be described as something from the environment that supplies, supports or aids humans in some way, and is often a source of income. The environment has always played a part in providing resources in order for humans to survive, whether it is a source of food and water, or for materials such as rock and timber for shelter.

Using the term 'resource' has connotations of taking from the environment to make money. This is very much a non-Indigenous way to experience and understand outdoor environments. Indigenous Australians have carefully and sustainably managed the land through fire-stick farming, fish traps and sanctuary zones to provide the resources they needed to survive for over 50,000 years with very little impact. However, when European settlers arrived just over 230 years ago, they perceived the land as 'terra nullius', meaning a land belonging to no one. There has been a race to secure the natural resources of the land ever since.

The plentiful resources provided by the Australian environment have led non-Indigenous Australians to take advantage of them, developing many ways of exploiting these resources through practices such as timber and water harvesting, agriculture and horticulture, mining, grazing and other farming methods. Because of their experiences, non-Indigenous landowners and land managers often experience and understand environments as a resource. Farmers manage just over 50% of Australia's landscape, giving them a critical role in managing our outdoor environments. As farmers live and at times battle against the effects of climate change, they have developed an extensive understanding of how outdoor environments are impacted by the extremes of drought and flood, and how they need to be carefully managed to remain as a productive resource.

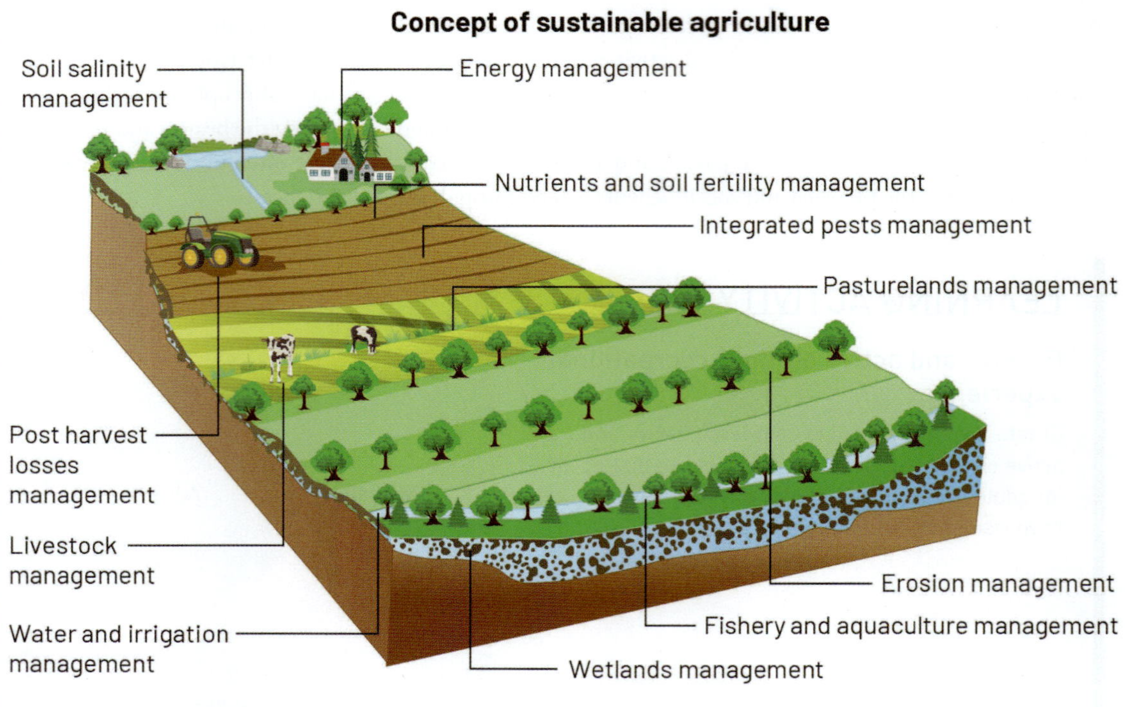

Concept of sustainable agriculture

Soil salinity management
Energy management
Nutrients and soil fertility management
Integrated pests management
Pasturelands management
Post harvest losses management
Livestock management
Water and irrigation management
Erosion management
Fishery and aquaculture management
Wetlands management

Figure 1.37 Farmers are implementing a range of sustainable agricultural practices to enhance productivity and to protect outdoor environment for future generations.

More recently, tourism and education have also become aspects of the outdoor environment that people have been able to harness to make money. Seeing the land as a resource could be a negative response if it leads to increased **exploitation** and destruction, and certainly this has been the case in the past. Over time, though, non-Indigenous Australians have become more aware of their role in the destruction of the natural environment. The exploitation of natural resources has caused a great deal of damage, including soil erosion, high **salinity** levels and deforestation, which places pressure on native flora and fauna. Indigenous peoples' concepts of sustainability are ways of thinking and acting with the aim to conserve and mindfully distribute resources in the hope of prolonging the availability of them for us and for future generations alike.

exploitation making use of and benefitting from resources, often in an unsustainable way and accompanied by environmental degradation

salinity the concentration of dissolved salts in water or soil

Figure 1.38 Farmers are increasingly implementing sustainable agriculture practices through a better understanding of the impacts of climate change.

Worksheet 1.4c Environment as a resource

LEARNING ACTIVITY 1.4C

Environment as a resource

While on an outdoor experience, keep a journal of all the specific ways the environment can be used as a resource. What are some of the ways in which you and your class are using the environment as a resource? Think about people other than yourself and include some details about the kinds of people who would use it as a resource now, and how they may have used it as resource in the past.

LEARNING ACTIVITY 1.4D

Worksheet 1.4d Fictional farm

Fictional farm

Create a fictional farm and build up a case study about it. While this activity requires some creative thinking, you should also include realistic information where possible. Your case study should include the following sections:

1 An introduction that includes general details such as the name of the farm, where it is located, how long it has been operating for and what product(s) are farmed and sold.
2 A description of the destructive nature of the farming practices that may have been used in the past and how farmers may have viewed the environment at that time.
3 A description of the sustainable nature of farming practices that may be used today and how farmers may now view the environment.

1.4.4 OUTDOOR ENVIRONMENTS THROUGH A SPIRITUAL CONNECTION

For over 50,000 years, Indigenous peoples have experienced and understood outdoor environments through a spiritual connection. Describing or defining the term 'spiritual connection' is difficult. Some people say it's connecting to something on a deeper level – feeling close to or having faith in something. Others say it's being in sync or aligned with something, like puzzle pieces fitting together. Spirituality affects a person's feelings and beliefs, and a connection can be defined as the relationship between two things. Therefore, to say a person has a spiritual connection could mean they have a relationship with something that affects their feelings and beliefs.

It is, of course, possible for non-Indigenous people to develop spiritual connections with the outdoor environment. It could manifest as a sense of peace, where a person can put their life into perspective, or a deeper connection that cannot be achieved through participation in a recreational activity – you may visit an environment just to 'be'. Taking time out from the city and the bustle of everyday life to be among nature and immersing yourself in the environment are examples of spiritual experiences and connections.

Alamy Stock Photo/Kurt Lackovic

Figure 1.39 Admiring one of the many views at Gariwerd (the Grampians National Park)

Worksheet 1.4e Creative response

LEARNING ACTIVITY 1.4E

Creative response

In small groups, brainstorm a list of ways that people can form spiritual connections with outdoor environments. Present your ideas in a creative way, such as creating a song, a poem, a cartoon strip, or a short radio/television segment or advertisement.

Worksheet 1.4f Experiences in song

LEARNING ACTIVITY 1.4F

Experiences in song

Xavier Rudd's music video for his song *Follow the Sun* was filmed at Stradbroke Island and along the coast of Queensland and New South Wales. It shows the local Minjerribah people and some of their interactions with the outdoor environment, as well as the interactions of some non-Indigenous Australians. Watch his video and think about what the images are suggesting about people's experiences with the environment. Discuss your thoughts with your fellow class members.

Non-Indigenous people are often seen as simply using environments for their own personal gain, or for relaxation, recreation or study. However, this is a stereotype or a generalisation, and doesn't account for individual differences in a society that is becoming more aware of the importance of the environment. Having a spiritual connection with the environment could include sleeping on the ground, tasting the air after rain, watching the wind rush through the trees, or closing your eyes and listening to the sound of birds.

In some historical cultures, including those of many Indigenous communities, particular places such as certain mountains, rivers, beaches, forests or individual rocks or trees had very important meanings. We might call these meanings religious, or some people use the term 'numinous', which means something that has a mysterious, holy or spiritual quality. Today, many people who travel in the outdoors describe numinous experiences when standing on the top of a mountain, or watching the cloud lift up from a valley early in the morning, or even when listening to the sound of waves crashing on rocks. Developing a spiritual connection adds to our experiences and understandings of the many benefits and opportunities that exist within outdoor environments.

1.4.5 OUTDOOR ENVIRONMENTS AS A STUDY SITE

A study site is a location or place where investigation, analysis and other activities occur in the pursuit of knowledge. There are many people, both Indigenous non-Indigenous, who utilise the environment as a study site, such as scientists, students, land managers, volunteers and outdoor education companies. They may undertake observation, exploration, testing, monitoring, recording and reporting in order to better understand the environment, how and why it has changed over time, and the nature of human **interrelationships** with it.

interrelationships the way in which two or more things affect each other because they are related in some way

Scientific studies of the environment have led to the development of over 50000 plant species worldwide for medicinal purposes. Unfortunately, studies have also identified that, of these, about 15000 species are threatened with extinction from human activities such as timber harvesting.

Through experiencing outdoor environments as study sites, we have come to understand the world and our place within it, the significant impacts of our actions on the environment and the need to act in a more sustainable manner.

Resource
Case study:
Jane Goodall - Primatologist

CASE STUDY

JANE GOODALL – PRIMATOLOGIST

Jane Goodall: Change Starts Within

Everett Collection Inc / Alamy Stock Photo

Figure 1.40 'Even in the middle of the city, every breath we breathe is from the natural world.' Jane Goodall – primatologist

When Jane Goodall entered the forest of Gombe for the first time, the world knew very little about chimpanzees. Equipped with little more than a notebook, binoculars and her fascination for wildlife, she braved a realm of unknowns to give the world a remarkable window into humankind's closest living relatives. She took an unorthodox approach in her field research, immersing herself in their habitat and their lives to experience their complex society as a neighbour rather than a distant observer, and coming to understand them not only as a species, but also as individuals with emotions and long-term bonds. Her discovery in 1960 that chimpanzees make and use tools is considered one of the greatest achievements of 20th-century scholarship. Through more than 50 years of ground-breaking work, Dr Jane Goodall has shown us the urgent need to protect chimpanzees from extinction and redefined species conservation to include the needs of local people and the environment.

'Jane Goodall: Change starts within', Anne-Marie Hoeve, © Imagine5 2023

Worksheet
1.4g Collage of images

LEARNING ACTIVITY 1.4G

Collage of images

Create a collage of images that represent each of the ways in which Indigenous and non-Indigenous peoples experience and understand outdoor environments:

- through custodianship
- as recreation
- as a resource
- through a spiritual connection
- as a study site.

1.4 KEY CONCEPTS

- Outdoor environments, and the humans that interact with them, are complex and highly diverse.
- There are a variety of ways in which Indigenous and non-Indigenous Australians experience and understand outdoor environments.
- Custodianship can be used to describe this sense of responsibility for taking care of and protecting Country.
- Indigenous peoples have nurtured and managed the natural resources of Australia for tens of thousands of years, and through their experiences have developed an extensive knowledge and understanding of the natural systems of our world.
- Recreation is an activity that is done for enjoyment, amusement or pleasure, and is considered to be fun by the participant.
- Through participating in recreational activities, people gain a greater understanding of themselves and the characteristics of the outdoor environment.
- A resource can be described as something from the environment that supplies, supports or aids humans in some way, and is often a source of income.
- Through their daily experiences on the land, farmers have developed an extensive understanding of how outdoor environments are impacted by the extremes of drought and flood, and how they need to be carefully managed to remain a productive resource.
- Farmers are implementing a range of sustainable agricultural practices to enhance productivity and to protect the outdoor environment for future generations.
- It is possible to experience and understand the outdoor environment through a spiritual connection, which means to have a relationship with something that affects a person's feelings and beliefs.
- In terms of the outdoor environment, a spiritual connection could be where you find a sense of peace and can put your life into perspective, or a deeper connection.
- A study site is a location or place where investigation, analysis and other activities occur in the pursuit of knowledge.
- Through people experiencing outdoor environments as study sites, we have come to understand the world and our place within it.

1.4 CONCEPT QUESTIONS

Worksheet
1.4 Key concepts

REMEMBERING

1 Define and provide an example of the term 'custodianship'.
2 Describe what is means to have a spiritual connection with an outdoor environment.

UNDERSTANDING

3 Discuss how your understanding of a specific outdoor environment changed due to your participation in a recreational activity in that environment.
4 Outline one sustainable agricultural practice farmers are implementing to enhance the productivity and protection of outdoor environments.

APPLYING

5 Annotate an image of a study site from an environment you have visited and/or investigated this year. Your annotations should include new knowledge and understandings about the environment you gained from your personal experiences.
6 Sketch one example of a traditional Indigenous practice that has been utilised to nurture and/or manage outdoor environments.

EXTENSION CHALLENGE

7 Investigate the work a specific type of scientist – such as a geologist, ecologist, archaeologist, anthropologist or botanist – does with outdoor environments. Produce a PowerPoint presentation (or similar) on their use of, experience with and understandings of the environment.

Resource
Glossary – Chapter 1

Assessments
End of chapter exam

Glossary test

EXAM-STYLE QUESTIONS

1 Describe the meaning and use of the term 'Kinship'. (2 marks)

2 Using an example of each, distinguish between wildernesses, managed parks and built environments. (6 marks)

3 Discuss one intrinsic motivation and one extrinsic motivation you have had for participating in an outdoor experience. (2 marks)

4 Using an example, discuss the motivation known as competence/mastery. (2 marks)

5 Discuss two factors that may influence the type of biome that exists within a specific location. (4 marks)

6 Using specific examples, compare the features of two different biomes. (8 marks)

7 Describe one positive response and one negative response that you have had to an experience in an outdoor environment. (3 marks)

8 How did participating in outdoor experiences influence your understanding of a specific outdoor environment? (3 marks)

9 Explain how a person's age may influence how they respond to an outdoor environment. (2 marks)

10 Discuss how the concept of Country may influence the relationship an Indigenous person has with outdoor environments. (3 marks)

11 How does a Welcome to Country differ from an Acknowledgement of Country? (3 marks)

12 Using examples, explain two different ways in which Indigenous and non-Indigenous Australians experience and understand outdoor environments. (6 marks)

CHECK YOUR KNOWLEDGE AND SKILLS

Checklist for students to mark and date, making judgments of their knowledge and understanding.
(**D** = developing **C** = consolidating **ER** = exam ready)

Resources
Key knowledge and skills checklist

D	C	ER	Check your knowledge and skills
☐	☐	☐	Indigenous peoples ways of knowing outdoor environments
☐	☐	☐	Non-Indigenous peoples' ways of knowing outdoor environments
			Describe the meaning and use of the following terms:
☐	☐	☐	• Kinship
☐	☐	☐	• nature
☐	☐	☐	• Country
☐	☐	☐	• outdoor environments
☐	☐	☐	• private land
☐	☐	☐	• wilderness
☐	☐	☐	• managed parks
☐	☐	☐	• urban environments
☐	☐	☐	• built environments
☐	☐	☐	• outdoor experiences
			Describe the features of the following types of biomes:
☐	☐	☐	• alpine
☐	☐	☐	• coastal
☐	☐	☐	• inland waterways
☐	☐	☐	• grassland
☐	☐	☐	• heathland
☐	☐	☐	• forest
☐	☐	☐	• marine
☐	☐	☐	• arid
☐	☐	☐	The range of motivations for experiencing outdoor environments
☐	☐	☐	Responses to outdoor environments such as fear and appreciation
			Compare the variety of ways in which Indigenous and non-Indigenous peoples experience and understand outdoor environments:
☐	☐	☐	• through custodianship
☐	☐	☐	• as recreation
☐	☐	☐	• as a resource
☐	☐	☐	• as spiritual connection
☐	☐	☐	• as a study site

CHAPTER

2

UNIT 1 – AREA OF STUDY 2

Exploring outdoor environments

KEY KNOWLEDGE

- the influence of depictions of experiencing outdoor environments on personal responses, such as in the mainstream media, social media, music, art, writing and advertising
- factors that affect access to experiencing outdoor environments, including socioeconomic status, cultural background, age, gender and physical ability
- relevant technologies and their influences on outdoor experiences
- the variety of personal responses to risk when experiencing outdoor environments, including the interplay between competence, perceived risk and real risk.

KEY SKILLS

- analyse the depictions of experiencing outdoor environments on personal responses
- analyse factors that affect access to experiencing outdoor environments
- explain the influence of relevant technologies on experiencing outdoor environments
- compare a range of personal responses to risk when experiencing outdoor environments.

VCE Outdoor and Environmental Studies Study Design (2024–2028) p. 15.

The influence of depictions of experiencing outdoor environments on personal responses

This section examines how the depiction of experiences in outdoor environments can influence personal responses.

Factors that affect access to experiencing outdoor environments

A number of factors may affect how easy or difficult it is to participate in an outdoor experience, including socioeconomic status, cultural background, age, gender and physical ability.

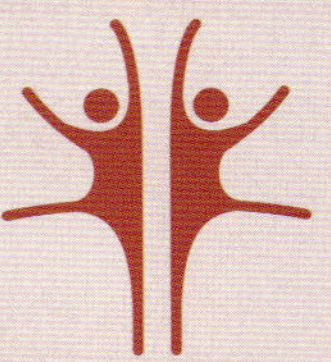

Relevant technologies and their influences on outdoor experiences

Technological developments such as advancements and modifications made to clothing and equipment have enhanced our participation in the ways we experience outdoor environments. It could also be argued that technologies have caused a disconnection between people and their environment, making challenging activities too easy, de-skilling individuals and causing an over-reliance on devices that can fail in the outdoors.

Personal responses to risk

Risk is an integral part of many experiences in outdoor environments, how we respond to this risk as individuals and as a society, is dependent on factors including level of competence, confidence and experience.

KEY TERMS

absolute risk
advertising
competence
contemporary
cultural background
deep ecology
depiction
gender
mainstream media
perceived risk
physical ability
portrayal
real risk
relationship
response
risk
social media
socioeconomic status
sustainability
technology

Worksheets

Resources and Templates

Assessments

Nelson MindTap

To access resources above, visit **cengage.com.au/nelsonmindtap**

SPOTLIGHT

The Pillars in Mt Martha to close after residents complain of visitors trashing the cliff-diving site

People have been urged to stay away from a popular cliff-jumping site over the long weekend – despite the fact that no signs or fences indicating the recent controversial ban have yet been installed. The mayor of Mornington Peninsula Shire – which this week voted to ban public access to The Pillars in Mt Martha – pleaded with people to keep away from the site. But residents are bracing themselves for an influx of cliff jumpers this holiday weekend after discussion on the controversial decision exploded across social media on Wednesday.

Fairfax Photo/Penny Stephens

Figure 2.1 Divers jump into the bay at The Pillars, Mt Martha.

Earlier, fans of the popular cliff-jumping spot reacted with mixed emotions after a shock decision to close the scenic spot, which was 'trashed' after a social media phenomenon. Many people are tagging friends on Facebook and Instagram, suggesting they head to the Mornington Peninsula and Mt Martha's world-renowned The Pillars over the Australia Day weekend, while locals are expressing delight that the scenic spot will soon be off-limits to 'the feral masses'.

Previously a close-guarded secret among locals and famed for its Mediterranean-style good looks, the site swiftly rose to fame via social media and attracted 'hordes' of visitors parking illegally near the Esplanade site, along with serious concerns about traffic congestion, extreme littering, public drinking and offensive behaviour. A packed council meeting last night heard the site – once a local secret – was now a 'must-do' for many visitors to the peninsula, thanks to posts on global social media.

The urgent meeting, called to consider options for the future of the site, heard there had been five serious accidents there in the past year, leading to three air ambulance extractions and two rope rescues. More rubbish rests out of reach down a hole in the cliffs. One swimmer had struck a cliff while backflipping into the water, and there were also injuries from a jet ski collision in the water below the cliff. There had also been many near-misses on the Esplanade, along which visitors walk to reach the cliffs.

The Pillars is not a formal location for visitors and has no parking, footpaths, road crossing area, toilets or rubbish bins.

Fairfax Photo/Penny Stephens

Figure 2.2 Tourists at The Pillars, Mt Martha

The Pillars area includes Aboriginal middens and artefacts and falls within the *Aboriginal Heritage Act 2006*. The council will also ask Aboriginal Affairs Victoria how to reinforce and protect the heritage values, and seek advice from the State Government on the best way to completely ban the public from the site.

'The Pillars in Mt Martha to close after residents complain of visitors trashing the cliff-diving site', Allison Harding, *Herald Sun*, 26 January 2017

You may have heard of the saying, 'loving something to death'. The news article above discussing the environmental impacts at The Pillars in Mount Martha caused by the increase in popularity of an outdoor environment is a prime example of this saying. Unfortunately, encouraging people to visit, enjoy and appreciate our unique outdoor environments also produces negative outcomes such as overcrowding and adverse environmental impacts.

Sharing picturesque images of unique landscapes and death-defying leaps off jagged cliffs into crystal-clear waters on social media will influence how people respond to outdoor environments. This chapter considers how these responses are influenced by media depictions of outdoor environments and perceptions of risk involved in outdoor experiences.

We first explore how outdoor environments are depicted in the media, music, art, writing and advertising, and the influence of these depictions on our responses. We then investigate the factors that affect access and the influence of technologies on outdoor experiences. The chapter finishes by looking at the variety of personal responses to risk when experiencing outdoor environments.

2.1 THE INFLUENCE OF DEPICTIONS OF EXPERIENCING OUTDOOR ENVIRONMENTS ON PERSONAL RESPONSE

KEY KNOWLEDGE

- the influence of depictions of experiencing outdoor environments on personal responses, such as in the mainstream media, social media, music, art, writing and advertising

KEY SKILLS

- analyse the depictions of experiencing outdoor environments on personal responses

Take a moment to look through the following four images. Discuss each image with the person sitting next to you. Use the following questions to guide your discussion.

- How does each image make you feel?
- How is the outdoor environment portrayed?
- Which of the presented outdoor environments would you like to visit? Why?
- Would you like to participate in the outdoor experiences presented? Why or why not?
- How do these images influence our responses to outdoor experiences and outdoor environments?
- Did the way the outdoor experience was presented in the image have an influence on your response?
- How do you think the publisher of each image would want people to respond to it?

Everett Collection Inc / Alamy Stock Photo Carter NewsFairfax Photo/Joe Armao, Newspix/Nathan Richter

Figure 2.3 (top left) Free solo climber Alex Honnold climbing Yosemite's 2308-metre El Capitan without a rope; (top right) A surfer comes face-to-face with what looks like a large shark within the wave, but it's actually a dolphin. (bottom left) Picking strawberries – fill your container (then your mouth); (bottom right) Sometimes the only thing we see of an outdoor adventure is the rescue efforts when things go wrong.

The images are from a range of sources, including social media, advertising and news reports. In this section, we investigate how different media sources depict experiences in outdoor environments and how this may influence personal responses. **Depictions** are how someone or something is represented in words or images.

depiction
how someone or something is represented in words or images

2.1.1 OUR RESPONSES TO OUTDOOR ENVIRONMENTS

A **response** refers to the feelings or emotions that an outdoor environment or outdoor experience creates in your mind. Responses will typically lead to certain behaviours. The way you respond to or feel about an environment is likely to influence the way you act within that environment. In Chapter 1, we identified that personal responses can be divided into two main groups based on the types of behaviours or actions that they may influence: positive and negative responses. Positive responses such as appreciation, awe and exhilaration will probably result in positive behaviours towards that place – behaviours that might protect, preserve or enhance that place in some way. Negative responses such as fear, aversion or dislike may lead to damaging behaviours or dangerous actions, such as the desire to clear it, change it or remove things from it.

response
the feeling or emotion that an outdoor environment or outdoor experience creates in your mind

We can also consider a response to be how we perceive and/or interact with an outdoor environment. In Outdoor and Environmental Studies, we group perceptions, interactions and impacts together as a way of understanding our relationships with outdoor environments. A **relationship** is the way in which two or more people or things are connected, or the state of being connected. How we perceive an outdoor environment will have an influence on our interactions with the environment, which then may have an impact on the environment. Relationships with outdoor environments are explored further in Units 3 and 4.

relationship
the way in which two or more people or things are connected, or the state of being connected

2.1.2 PORTRAYALS OF OUTDOOR ENVIRONMENTS

We are bombarded by thousands of images each week via social media, television, magazines and websites. It is common for many of these images to contain outdoor environments. How we respond to these outdoor environments may be influenced by how they are presented or portrayed in the image or writing. A news article about beaches with dangerous rips may evoke a negative response and lead us to think all beaches are dangerous. Likewise, a social media post about a scenic beach may evoke a positive response and influence people to develop an appreciation for our unique outdoor environments and motivate them to experience it for themselves.

Outdoor experiences and environments can be depicted in certain ways, depending on the message that a person or group is trying to get across. The message may reflect an opinion or attitude about that outdoor environment, or indicate its perceived purpose.

A **portrayal** refers to the way in which something is represented. Specific environments are portrayed in certain ways depending on the message that a person or group is trying to make. It can reflect an opinion or attitude about that environment or it can indicate the perceived purpose of that outdoor environment.

portrayal
the way in which something is represented

Different people or groups can portray the same outdoor environment in different ways, depending on what they see as the purpose of the land. Although there are countless possibilities for how this can be done, the following points outline some different ways the outdoor environment, and experiences in them, can be portrayed. The outdoor environment might be portrayed as:

- an adversary
- a gymnasium
- a resource
- a museum
- a cathedral.

The outdoor environment as an adversary

When the outdoor environment is portrayed as an adversary, it is being considered as something that is fundamentally working against us. This requires us to overcome the challenge that the environment sets for us. This could include portraying the environment as a threat to our safety, or even to our life.

A characteristic of the Australian environment is its climatic variation, including its susceptibility to long periods of low rainfall. The existence of drought has traditionally provided Australian farmers and agriculturalists with significant challenges to their productivity. For this reason, the Australian environment is often portrayed as harsh and unforgiving, and a threat to the livelihood of those trying to make a living from it.

The challenge that elements of the outdoor environment can present during recreation can also be portrayed as the thing to be overcome. It might be a feature such as a rock or cliff face, a high-altitude mountain, severe conditions or anything that poses a threat to the safety of the participant.

Resource
Case study: Gunnamatta Beach

CASE STUDY

GUNNAMATTA BEACH

BEACH

General beach hazard rating: 8/10 (highly hazardous)

Alamy/Phil Holden

Figure 2.4 Gunnamatta Beach

Gunnamatta Beach is an exposed, high-energy beach with a wide, rip-dominated surf zone. It is located in the Mornington Peninsula National Park and is part of the 30 km long sandy and rocky coast that extends from Cape Schanck to Point Nepean. The Gunnamatta section is 3 km long, with extensive intertidal calcarenite reefs and rocks forming the boundaries, with some smaller reefs on the beach and in the surf. Truemans Road runs out through the dunes to the beach, where there is a large car park and the surf lifesaving club. The beach faces south-west, exposing it to high westerly winds and waves. The waves average 1.9 m and combine with the medium sand to produce a 150 m wide single bar surf zone. The bar is cut by strong rips every 300 m, together with additional permanent rips next to major reefs and rocks. The rips intensify around low tide.

The Gunnamatta Surf Life Saving Club was founded in 1966. This is a very hazardous beach, with an average of 113 rescues a year, second only to its neighbouring Portsea Beach.

SWIMMING

This is a potentially hazardous beach for swimming, with usually high waves and strong rips close to shore. Definitely stay between the flags, on the bar and away from the rips, rocks and reefs.

SURFING

Gunnamatta offers the best beach breaks on the Mornington Peninsula, with consistency guaranteed by the high swell and reefs. Good breaks are found down the beach past the surf club, in front of the first and second car parks, and up the beach at the Pumping Station, which is, however, polluted by the sewerage outfall. Best conditions are with a low to moderate swell and north-easterly winds.

FISHING
Deep rip holes and gutters, together with rocks and reefs, are a permanent feature of this beach and make it a popular spot for beach and rock fishing.

GENERAL
A high energy, hazardous beach backed by extensive sand dunes. Best suited for experienced bathers and surfers.

The outdoor environment as a gymnasium

Shutterstock.com/Ammit Jack

Figure 2.5 The challenges inherent to an activity are often what attract people to them.

When the environment is portrayed primarily as a place for humans to participate in recreational activities, it is being regarded as a gymnasium. Here, it is the venue for human outdoor experience – whether it be to run, walk, jump, climb, swim or paddle. It provides terrain, gradients, rapids, holds or waves that are the playground in which to enjoy outdoor experiences. The challenges that these different elements of the environment present are what is attractive to the participant. They can be seen as problems that need to be solved or a series of potential achievements. Portrayals of these types can often incorporate an element of thrill, as the challenges that the environment provides include an element of risk. There has been a rise in popularity of 'extreme' activities, whether through actual participation or to promote a product, and these rely heavily on this type of portrayal.

There are many ways to portray a river environment. Someone who is interested in using a river to test their paddling skills and to experience the thrill of shooting the rapids while successfully completing a set course would probably describe the river in terms of its degree of difficulty. The components of the river, such as flow rate, obstacles, drops and eddies, are dissected and analysed to determine the amount of physical skill required to safely participate in recreation in this environment. The river, to them, is the gymnasium.

The outdoor environment as a resource

Alamy Stock Photo/Image Professionals GmbH

Figure 2.6 The ancient forest as a museum

Portraying the outdoor environment as a resource reflects the attitude that it should be used to support our needs. This can include our survival needs (the things that we need to live on), such as foods and other materials. These are harnessed from the environment by taking them directly or by utilising and modifying the environment to produce them.

A cattle grazier or dairy farmer will provide pasture on land cleared specifically for the raising of their herd. This industry is required to support the human population's need for food. Although the purpose of the land here is primarily for resource production, there is an increasing awareness of conservation and sustainability issues associated with their interactions with their land. Many farmers are today involved in groups such as Landcare, Land for Wildlife and the Victorian Farmers Federation. Although the focus of such groups is often on **sustainability** of their industry, the way farmers regard their land and the understanding of the benefits of a healthy environment is leading to increasingly positive impacts.

sustainability
the ongoing capacity of Earth to maintain all life

The production of resources can also be driven by human desires for improvement and enhancement. This may include aspects such as making money from the environment or utilising the environment to advance human standards of living. Power companies are increasingly searching for the expansion of existing power production or new power sources. Our reliance on power such as electricity for domestic and industrial purposes provides the demand for this resource and makes it a rich commodity. The potential to make money will encourage continual exploration of the environment, and industries such as these will put pressure on specific environments in the pursuit of resources.

The outdoor environment as a museum

When the environment is portrayed as a museum, it is presented as a place of history that can be used to help us understand human relationships with the outdoor environment. When we think of a museum, we may think of a storehouse of items from history. Not only do these things help us with our understanding of the past, but they also help us with our understanding of where we are going in the future. The environment can be thought of in this way too.

Due to the rapid rate of vanishing wilderness in both the past and present, the areas that remain intact can help us understand the effect humans are having on certain venues. The use of reference zones in the management of national parks involves demarcated areas ranging from minimal human access to no human access. These zones can be used to compare with other areas that are subject to a higher level of human impact (such as recreation zones) to determine what effect we are having and to formulate management strategies to mitigate these effects.

The preservation of the natural components of the environment does not only have to be for research purposes. Someone might simply enjoy visiting a natural outdoor environment to escape the human-dominated landscapes that surround them in their day-to-day lives. With an ever-increasing separation of humans from nature and intact environments becoming more and more rare, the environment becomes a small example of what was once all around us and where we came from. To highlight this, we might portray it as something that is continually threatened by the impacts of our lifestyles.

The outdoor environment as a cathedral

This portrayal involves regarding the outdoor environment in a spiritual way. Often, such a portrayal will highlight the 'specialness' of an environment or an aspect of an environment, and emphasise how its existence and health is essential to human wellbeing (see Chapter 1). Here, we are not necessarily talking about formal religious doctrines, but rather a connection that one makes with the outdoors that is fundamental to their existence. This type of connection is becoming increasingly more popular in developed cultures such as ours and can spring from a range of factors.

Indigenous rights and awareness movements are having an increasing influence on our perceptions of the outdoor environment. They bring with them integral human–nature connections, which have been the basis for successful, long-lasting, sustainable relationships with the environment.

Social movements, including those centred on the environment, have endured and grown since the 1960s and have provided alternative belief structures to traditional religions. With a decline of formal religion among young people today, many are in search of a spiritual meaning for life, and the environment can sometimes provide this.

Andrew Watson / Alamy Stock Photo

Figure 2.7 A spectacular view from a summit as a cathedral

The notion of Kinship in relation to outdoor environments (see Chapter 1) also has at its core a fundamental link between humans and nature. The environment is portrayed as something that is subjectively identified with self. In this instance, humans are not separate or superior to nature and there is little distinction between the two. This portrayal of outdoor environments is consistent with Indigenous Australian perspectives and more recent perceptions such as **deep ecology**.

deep ecology a conservationist philosophy that regards humans as one of many equal components of a global ecosystem

WHAT IS DEEP ECOLOGY?

Chris Johnstone from the Institute for Deep Ecology (UK) says: 'Deep Ecology is a holistic approach to facing world problems that brings together thinking, feeling, spirituality and action. It involves moving beyond the individualism of Western culture towards also seeing ourselves as part of the Earth. This leads to a deeper connection with life, where ecology is not just seen as something "out there", but something we are part of and have a role to play in'.

2.1.3 REASONS FOR DEPICTIONS OF EXPERIENCING OUTDOOR ENVIRONMENTS

There is a range of reasons for depicting outdoor experiences and outdoor environments in a specific way, including the following:

- To celebrate nature and the outdoors – an amazing film or television documentary about a particular experience or a particular place will often do this, as will many artworks. We might also find a celebration of the outdoors in a media work that is focusing on a human protagonist, such as a film about surfing or mountain climbing where the focus is on the outdoor experience or the adventurer.
- To sell a place or an experience – this is the role of advertising and marketing, and we find it throughout the media. People are perhaps literally trying to sell a place, or they may be focusing on tourists visiting a particular place or seeking a particular experience. We can even find depictions of outdoor experiences and environments in the selling of other things, such as cars, iPads or ice cream.
- To inform – in newspapers and magazines, and across the internet, outdoor experiences and the environment are depicted in order to inform us in some way, such as articles about coral bleaching of the Great Barrier Reef or discoveries of new animal specifies deep in our oceans.

- To understand and develop our knowledge – related to informing, this is the depiction of outdoor experiences to educate us. Wildlife documentaries are an example of this.
- To challenge – film and television often challenge the ways we see things, and this is one of the main purposes of artworks. Music can challenge us in a powerful way – consider songs with environmental or other social messages, such as those by Midnight Oil, Billie Eilish, Lady Gaga and Joni Mitchell.
- To scare – creating fear is something that media can do very well. Depictions of outdoor environments that scare us aren't hard to find, particularly in film and television. The film *Wolf Creek* is an example of where the depiction of outdoor experiences can scare us. Indeed, in many of these types of depictions, the environment itself is said to almost become another character in the story that we are watching, reading or hearing.

2.1.4 INFLUENCE OF DEPICTIONS ON PERSONAL RESPONSES

How do depictions of experiences in outdoor environments influence our own personal responses? Since our personal responses are emotional responses, depictions can influence our responses by tapping into our emotions or influencing them in some way.

There are many ways depictions can influence our responses, including:

- motivating – such as inspiring people to care for or about something, or motivating them to go to a particular environment or participate in a particular outdoor experience
- changing behaviour – such as moderating the way someone acts in a particular place
- informing – such as educating people about an issue related to a particular environment or outdoor experience
- influencing actions – such as influencing people to buy and use the latest equipment and technology.

Depictions of people experiencing outdoor environments can be found in the second-hand information that is being provided to us through written, oral and visual methods (secondary sources). These depictions can influence how we respond and relate to outdoor environments. We can explore the influence of depictions of experiencing outdoor environments on personal responses from the following sources:

- the mainstream media
- social media
- music
- art
- writing
- advertising.

mainstream media traditional forms of mass media, as television, film, radio, magazines and newspapers

AJ Pics / Alamy Stock Photo

Figure 2.8 Paul Hogan in *Crocodile Dundee* (1986)

Mainstream media

The **mainstream media** refers to the traditional forms of mass media, television, radio, magazines and newspapers. Images of outdoor experiences and outdoor environments figure prominently in the media. There are two significant uses of images:

- to add emphasis, clarity or explanation to a discussion relating to the outdoor environment
- to add impact to an unrelated issue (often the case in advertising).

Television shows such as the late 1960s Australian classic *Skippy the Bush Kangaroo* and the popular *Home and Away* have showcased unique experiences and idyllic Australian outdoor environments. The 1986 film *Crocodile Dundee* introduced the world to the Australian outback and

Kakadu National Park. People around the world have responded positively by wanting to personally experience these iconic Australian outdoor environments. These media depictions have led to significant increases in tourism in these areas.

Often, personal responses are shaped by the way a pursuit or endeavour, and any associated incidents, are depicted by the media. They can include exaggerated reports that sensationalise an account of what has occurred by over-emphasising or under-emphasising particular aspects. For example, if someone requires rescuing from an outdoor environment, reports might concentrate on the costs incurred by taxpayers through the use of emergency services, emphasise the amount of harm or trauma suffered by participants, allege negligence on the part of operators or instructors or suggest a lack of responsibility from participants by putting themselves in harm's way. Although the intention of the media here is not necessarily to provide an inaccurate report of what has happened, the language used and details provided may be chosen to spark the interest of potential readers by capturing their imagination or courting controversy. Media reports of misadventure in outdoor environments may cause negative responses such as fear and dislike of the outdoors, leading to destructive behaviours.

LEARNING ACTIVITY 2.1A

Media, motivations and responses

Collect at least six articles, headlines or photos from a variety of websites, newspapers and magazines that depict adventure in various ways. Apply the following points and questions to the articles you have collected:

1 Explain the image that is being depicted. For example, is it a sensationalised report on misadventure? A glamorised description of adventure? Does it convey the value and rewards of the outdoor experience?

2 Explain the impact this image is likely to have on participation. For example, will it discourage people? Inspire them? Attract participants?

3 Describe the possible motivations of people who choose to participate in this outdoor experience. For example, is it the challenge? The fame and glory?

Worksheet
2.1a Media, motivations and responses

Social media

Social media are forms of electronic communication through which users create and share information, or participate in social networking. As of January 2023, there were approximately 5 billion social media users around the world. This is more than 75% of the total global population over the age of 13, an increase from less than 20% just 10 years ago. The typical user spends an average of two-and-a-half hours per day on their social media accounts.

social media forms of electronic communication through which users create and share information, or participate in social networking

As social creatures, we are compelled to share our experiences. Instagram analytics show that more than 95 million selfies, snaps of sunsets and staged photos are posted every day. Social media apps such as TikTok, Instagram, Facebook and Snapchat enable people to plan, capture and share their experiences in outdoor environments. This has led to an increase in awareness of the natural wonders of the world around us. Ironically, the outdoors is often seen as one of the few places where we can disconnect from the busyness of everyday life and the technologies that come with it.

In some cases, social media has been responsible for a surge of people visiting Insta-worthy landscapes and locations, such as The Pillars in Mount Martha, as discussed at the start of this chapter. Many social media users will take a selfie as proof they were in attendance, reflecting a contemporary 'I've been there' attitude.

Figure 2.9 Using the infamous 'selfie stick'. See, I've been there!

Figure 2.10 Capturing memories or experiencing nature via a 13 cm screen?

Figure 2.11 Images of native animals like these puggles (baby platypuses) are used by conservation organisations in social media posts as part of their campaigns to raise awareness of endangered species.

However, social media can also influence people to act sustainably and protect and preserve our outdoor environments. It has been used as a valuable tool by organisations such as Australian Conservation Foundation to encourage people to stand up for nature.

Social media posts have the power to educate the masses about the beauty, fragility and importance of outdoor environments. It is critical that people remember the adage, 'Take nothing but photographs and leave nothing but footprints'.

Worksheet 2.1b Media

LEARNING ACTIVITY 2.1B

Media

1 Describe the scene in Michael Leunig's 'Understandascope' cartoon.
2 What is the specific message being suggested?
3 How might it influence responses to outdoor environments?

Figure 2.12 Michael Leunig's 'Understandascope'

Music

The Midnight Oil song 'River Runs Red' describes some human attitudes to outdoor environments as destructive, such as a lack of sustainable practices.

Many of Midnight Oil's songs focus on the plight of outdoor environments. Similarly, Australian musician John Williamson's song 'Rip Rip Woodchip' presents images of logging of ancient forests. Common responses to these songs are ones of outrage against the companies causing the destruction and support for conservation and sustainability practices. However, in the case of John Williamson's song, there was also a sense of outrage from logging workers who felt betrayed by a country singer siding with 'Greenies'.

Farifax Photo/Gelnn Hunt

Figure 2.13 John Williamson's 'Rip Rip Woodchip' highlights the loss of ancient forests.

Musicians from many genres have referenced environmental themes, from acknowledging and appreciating its beauty to advocating for its protection and conservation. Joni Mitchell explained, 'I wrote "Big Yellow Taxi" on my first trip to Hawaii. I took a taxi to the hotel and when I woke up the next morning, I threw back the curtains and saw these beautiful green mountains in the distance. Then, I looked down and there was a parking lot as far as the eye could see, and it broke my heart … this blight on paradise. That's when I sat down and wrote the song.'

LEARNING ACTIVITY 2.1C

Songs

Search online for Midnight Oil's 'River Runs Red' or other songs with an environmental theme (or use one that you already know).

1 Provide the lyrics to the song.
2 How is the environment depicted in the song?
3 How effective is this depiction in providing the intended message?
4 How could this influence how a listener might interact with the outdoor environment?

Worksheet
2.1c Songs

Art

Artistic depictions of outdoor environments have provided a rich insight into the changing nature of the nation's identity over time. Relationships with the Australian outdoor environment have been portrayed in many artworks, including:

- rock paintings, many thousands of years old, by Indigenous Australian communities
- the early European oil paintings of John Glover
- paintings by members of the Heidelberg School (Australia's first significant art movement), such as Tom Roberts, Frederick McCubbin and Arthur Streeton.

Australian Indigenous art is the oldest ongoing artistic tradition in the world. There is evidence of Indigenous rock art dating back at least 60000 years in the Northern Territory, depicting extinct

Gippsland, Sunday night, February 20th, 1898 John Longstaff

Figure 2.14 John Longstaff, Gippsland, *Sunday Night, February 20th, 1898*

megafauna. Common subjects of the artwork include animals of the surrounding environment, and the Dreaming – when ancestral spirits came to the land and created rivers, plants, people, animals and tribal laws (see Chapter 1). Common positive responses to the artwork include appreciation, inspiration and connection: feelings that we are all connected due to these special places.

John Longstaff's painting *Gippsland, Sunday night, February 20th, 1898* presents the perception of fire within the outdoor environment as destructive and frightening. This is in direct contrast to Indigenous communities' understanding of fire as a useful environmental management tool.

Frederick McCubbin's three-panel painting *The Pioneer* (1904) is considered to be one of the masterpieces of Australian art. The painting tells the story of a free selector and his family establishing a home in the Australian bush over three time periods. This painting demonstrates the artist's understanding that the Australian outdoor environment is harsh and untamed, but through hard work and persistence it is possible to progress and achieve.

The Pioneer, 1904, Frederick McCubbin

Figure 2.15 *The Pioneer* (1904) by Frederick McCubbin

LEARNING ACTIVITY 2.1D

Worksheet
2.1d Art

Art

Conduct an online image search to find four Australian artworks that depict the outdoor environment in different ways.

1 Describe how the environment is depicted in each.
2 How does the artist achieve this imagery?
3 How might each artwork influence the responses of those who view them?

Writing

Outdoor environments have been depicted in writing as both places of wondrous beauty in need of protection, and harsh and hostile locations where death awaits at every turn. A good travel book can transport you to snow-capped mountains, tropical islands or enchanting forests.

Then there are the almost unbelievable tales of survival from harsh and unforgiving outdoor environments. Aron Ralston's book *Between a Rock and a Hard Place* documents his extraordinary story of survival after a freak accident caused him to be trapped by a 363-kilogram boulder pinning his arm for six days in a remote canyon in Utah, USA. The book depicts Aron's outdoor experience as understandably traumatic, prompting a range of responses from how dangerous and unpredictable outdoor environments can be to how amazing the human spirit is to overcome challenging obstacles.

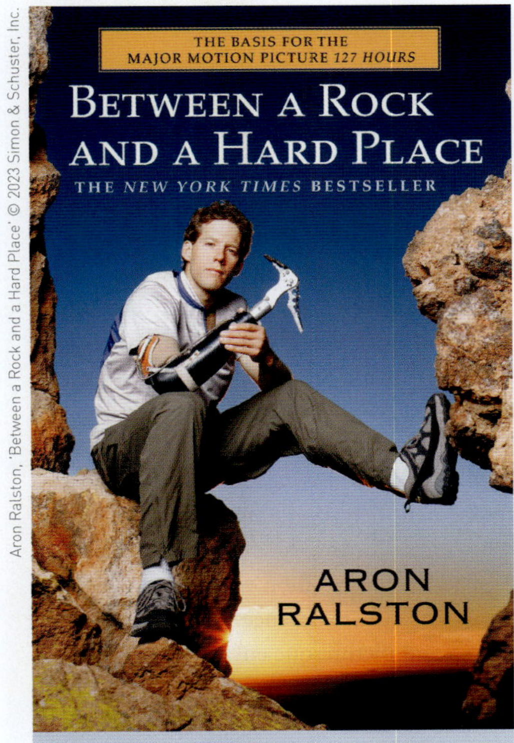

Figure 2.16 Aron Ralston's book *Between a Rock and a Hard Place* demonstrates how unpredictable outdoor environments can be.

Figure 2.17 'I am the Lorax. I speak for the trees.'

Worksheet
2.1e A fight for survival

LEARNING ACTIVITY 2.1E

A fight for survival

Research an example of where an individual's experiences in an outdoor environment turned into a fight for survival.

Complete the following questions based on your example 'fight for survival'.

1 Provide a background of the individual (include prior experience and skills relating to the outdoors).
2 Where did the fight for survival take place? Include the characteristics of the location.
3 How is the outdoor experience depicted in the story?
4 How was the outdoor environment portrayed?
5 What was your response to the story?
6 Were your responses influenced by how the outdoor experience was depicted in the story?

The destruction of outdoor environments has inspired many to express their emotions in the writing of novels, poems and stories. Examples include *The Rabbits* (John Marsden and Shaun Tan, 1998) and *The Lorax* (Dr Seuss, 1971). In *The Lorax*, outdoor environments are portrayed as resources. The Once-ler's unchecked corporate greed via 'biggering' his manufacturing of 'Thneeds' from the tuff of the Truffula Tree ultimately leads to total environmental destruction. Given its popularity as a children's book, it is possible that *The Lorax* influenced the perceptions of many young environmentalists. Interestingly, in parts of America *The Lorax* was banned from schools and libraries near logging communities. The logging industry sponsored a rebuttal book, called *The Truax*, to help children understand the necessity of harvesting timber.

Adobe Stock/Roman

Figure 2.18 Advertisements use images to depict nature as a place to be conquered in a four-wheel drive.

Advertising

Outdoor experiences and outdoor environments have been used extensively to promote the sale of products. It is rare to see an advertisement for a four-wheel drive vehicle that does not involve images of vehicles scaling steep mountains and splashing their way through raging rivers.

Tourism Australia advertisements are famous worldwide for showcasing the unique and spectacular outdoor experiences and environments Australia has to offer. Tourism Australia's 2018 *Crocodile Dundee*–inspired advertisement featuring Chris Hemsworth was first broadcast during the Super Bowl to a TV audience of more than 100 million Americans. It has since had over one billion views.

The main goal of **advertising** is to create awareness and influence their audience. By associating their brands, products and services with exciting outdoor experiences and picturesque environments, advertisers hope to create positive responses.

advertising the practice and techniques employed to bring attention to a product or service

2.1 KEY CONCEPTS

- Outdoor experiences and environments can be depicted in certain ways depending on the message that a person or group is trying to make.
- Depictions are how someone or something is represented in words or images.
- Responses refer to the feelings or emotions that an outdoor environment or outdoor experience creates in your mind.
- The way you respond to or feel about an environment is likely to influence the way you act in and with that environment.
- Positive responses such as appreciation, awe and exhilaration will probably result in positive behaviours towards that place – behaviours that might protect, preserve or enhance that place in some way.
- Negative responses such as fear or revulsion may lead to damaging behaviours or dangerous actions, such as the desire to clear it, change it or remove things from it.
- In Outdoor and Environmental Studies, we group perceptions, interactions and impacts together as a way of understanding our relationships with outdoor environments.
- The outdoor environment can be portrayed as an adversary, a gymnasium, a resource, a museum or a cathedral.
- There is a range of reasons for depicting outdoor experiences and outdoor environments in a specific way. These include to celebrate nature and the outdoors, to sell a place or an experience, to inform, to challenge or to scare.
- There are many ways depictions can influence our responses, including to motivate, to change behaviour, to inform and to influence actions.
- Social media are forms of electronic communication through which users create and share information, or participate in social networking.
- Musicians from many genres have referenced environmental themes such as acknowledging and appreciating its beauty to advocating for its protection and conservation.
- Artistic depictions of outdoor environments have provided a rich insight into the changing nature of the nation's identity over time.
- The destruction of outdoor environments has inspired many to express their emotions in the writing of novels, poems and stories, such as *The Lorax*.
- By associating their brands, products and services with exciting outdoor experiences and picturesque environments, advertisers are hoping to create positive responses.

2.1 CONCEPT QUESTIONS

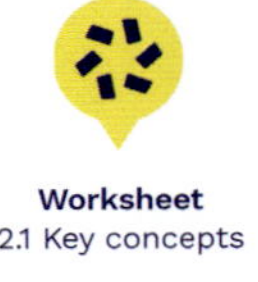

Worksheet
2.1 Key concepts

REMEMBERING

1 What is meant by depictions of outdoor experiences?
2 Provide examples of positive and negative responses to experiences in outdoor environments.
3 Identify and describe the three components that make up relationships with outdoor environments.

UNDERSTANDING

4 Explain how an environment may be portrayed as both an adversary and a gymnasium.
5 Compare three different ways the word 'nature' may be used.
6 Discuss two different ways depictions can influence a person's responses.

APPLYING

7 Research a social media account that includes depictions of people experiencing outdoor environments. Present these depictions on a poster with an analysis of the influence of the depictions on people's relationships with outdoor environments.

EXTENSION CHALLENGE

8 Research three media examples (newspapers, magazines, social media, etc.) that include depictions of people experiencing outdoor environments. Complete the following questions for each depiction:
   a What is the purpose of the depiction of the ourdour experience?
   b Do you think this purpose has been achieved? Why or why not?
   c How has the outdoor environment been portrayed in the depiction?
   d What might be the influence of the depiction on people's responses?
   e Do you think this depiction is positive, negative or neither? Why or why not?

2.2 FACTORS THAT AFFECT ACCESS TO EXPERIENCING OUTDOOR ENVIRONMENTS

KEY KNOWLEDGE

- factors that affect access to experiencing outdoor environments, including socioeconomic status, cultural background, age, gender and physical ability

KEY SKILLS

- analyse factors that affect access to experiencing outdoor environments

2.2.1 FACTORS THAT AFFECT ACCESS TO OUTDOOR EXPERIENCES

If a person views a social media post depicting their friends chilling out on a beach and enjoying themselves, they most likely will respond positively and be motivated to participate in a similar outdoor experience. However, there are a number of reasons that may stop them accessing the same experience, such as not having the time due to work commitments, not being old enough to drive or not having enough money. We explore the factors that may affect how easy or difficult it may be to participate in an outdoor experience. We will focus on the following five factors:

1 socioeconomic status
2 cultural background
3 age
4 gender
5 physical ability.

Socioeconomic status

Many people in Australia face diverse challenges in relation to their desire to access outdoor environments. For example, walking in a local park is an outdoor experience that's relatively easy to access, compared with an expensive activity or one that requires a degree of specialised skill or equipment, such as downhill skiing.

Socioeconomic status (SES) can be described as an individual's or family's economic and social position in relation to others, based on income, education and occupation. It can be loosely categorised as low, middle or high. Those of low SES are characterised by lower income, lower levels of education and fewer employment options. In contrast, people of high SES are associated with higher income, higher levels of education and more employment options. Most people in Australia fall into the middle SES category, and therefore most of you will have an understanding about how socioeconomic status can affect access to the outdoors, as you may have experienced an instance when you wanted to do some type of adventure activity but were restricted by the costs associated with it.

Shutterstock.com/Monkey Business Images

Figure 2.19 Some outdoor experiences are free.

socioeconomic status an individual's or family's economic and social position in relation to others based upon income, education and occupation

Australia is ranked as one of the most expensive places to participate in adventure activities such as rock climbing, surfing and skiing, even though there are many locations around the country where outdoor experiences can be enjoyed at no cost. SES can influence the type of outdoor experience a person may be able to access, and also the kind of experience they are likely to have. Take surfing as a simple example. A person of high SES may have a brand-new surfboard and full-length wetsuit, and could surf all year round. Compare this situation with a person of low SES, who may have to borrow an old surfboard from a friend and only has board shorts to wear, so can only enjoy surfing in warm weather and when their friend is able to lend them the board. However, this doesn't necessarily mean that one person's experience is better than another's. For instance, someone who swims in the ocean with their friends may have a more enjoyable time than someone who rides alone on a jet ski.

LEARNING ACTIVITY 2.2A

Accessible or inaccessible?

1 Select an outdoor experience (other than the ones mentioned previously) and complete a table similar to the one shown, detailing how different levels of SES might influence the accessibility of your chosen outdoor experience.

Low SES	High SES

2 Analyse how accessible or inaccessible this particular outdoor experience is based on socioeconomic status.

Worksheet 2.2a Accessible or Inaccessible?

Cultural background

Cultural background is a broad and complex topic that is presented in a simplified manner here to suit the purposes of Outdoor and Environmental Studies. Culture affects who we are, how we think, how we behave, how we learn and how we respond to the environment. It includes patterns of thinking, feeling and acting that stem from the social context of one's life experience, such as membership of various

cultural background patterns of thinking, feeling and acting that stem from the social context of one's life experience, such as ethnicity, race, socioeconomic status, gender, language, religion, sexual orientation and geographical area

iStockhoto/Vladimir Vladimirov

Figure 2.20 Hijab combined with a wetsuit for surfing

groups based on ethnicity, race, socioeconomic status, gender, language, religion, sexual orientation and geographical area.

Each country in the world has a unique culture with its own set of rules, regulations and expectations. Often, if someone moves from one country to another, they will take their culture (or aspects of it) with them. Australia is described as being multicultural, as close to 30% of our population was born overseas. Nearly half of Australians have at least one parent born overseas. Our culture is very diverse and can mean many different things to many different people.

Often the effect of culture on our access to outdoor environments will depend on the circumstances of a given scenario. Some people say that being outdoors is the Australian way, suggesting that Australians have greater access to a variety of environments and, therefore, a greater level of associated experiences.

By using simple examples, we can highlight how access might be influenced by culture. Many Middle Eastern cultures believe traditional outer garments that cover much of the body should be worn, which limits participation in outdoor activities where this type of clothing may be restrictive. There is now a clothing range of specifically designed garments such as head-covering wetsuits to address this.

Some outdoor education participants may experience a form of 'culture shock'. With the new and unfamiliar surroundings of a unique outdoor environment, students from overseas may experience feelings of disorientation and anxiety, with the additional separation from their own cultural norms. However, being of a particular culture or carrying with you selected beliefs doesn't automatically preclude you from being involved in outdoor activities.

Shutterstock.com/PhotoSky

Figure 2.21 Sometimes a person's age can affect their outdoor experience. This older climber is not wearing the required safety helmet. This may be because, in his youth, safety helmets were not part of the safety precautions that casual climbers took.

Age

Age is possibly one of the most influential factors that can affect access to and the types of outdoor experiences undertaken. It is accepted that we reach the peak of our physical fitness at approximately 27–30 years of age; beyond this, our fitness declines and we become weaker. It is common to hear people say, 'I'm too old to do that', as they may feel their body would not be able to cope with the demands placed on it. Of course, there are also the old sayings, 'Age is only a number' and 'You're only as old as you feel', suggesting that just because someone may be 70 years of age doesn't necessarily mean they are restricted by certain outdoor pursuits.

If age doesn't affect the type of outdoor experience someone can do, it could affect the intensity with which they participate. For example, the distance a 20-year-old could cover bushwalking in one day may be significantly further than the distance a 60-year-old could cover in the same time period.

Conversely, being young can also limit access to the outdoors. An example is someone who is 17 years of age living in central Victoria who loves to surf – they would need to rely on family members or friends to drive them to the beach or they might need to catch a bus. However, as soon as they turn 18 (provided they can obtain a vehicle and have a licence), they can drive themselves to the beach and their access to the coastal environment and surfing will increase dramatically.

Newspix/Russell Brown

Figure 2.22 Some outdoor experiences are more suited to younger than older people.

Gender

gender
the state of being on the male–female continuum, but also often used with reference to social and cultural differences

Gender is the state of being on the male–female continuum, but the term is often used with reference to social and cultural differences as well as biological differences. The concept of gender often includes our perception of physical, mental and behavioural characteristics. In the past, outdoor activities were generally more accessible to males than females, as boys and men were often motivated and encouraged to participate, male adventurers predominantly appeared in the media and images of males were used throughout advertising campaigns. However, the profile of female adventurers has grown in recent years and the number of women participating in outdoor activities has increased.

The perception that the majority of outdoor activities are considered 'manly' has been dissolved, as there is nothing men can or have done that women cannot or have not done – such as climb Mount Everest, surf 15-metre waves or complete a non-stop circumnavigation of the world. It can be suggested that gender does not impact access to outdoor experiences as much as in the past as gender stereotypes associated with specific activities are less prevalent in our society.

AAP Image/Dean Lewis

Figure 2.23 Jessica Watson (OAM) become the youngest person to sail solo, non-stop, around the world at aged 16. In response to the prime minister declaring her an Australian hero, Jessica said she didn't consider herself a hero, just an 'ordinary person, who had a dream, worked hard at it and proved that anything really is possible'.

Physical ability

physical ability
the quality of being able to perform some type of physical action

Physical ability can be described as the quality of being able to perform some type of physical action. In relation to outdoor environments and

Shutterstock.com/Andrea Izzotti

Figure 2.24 Participation in some outdoor activities, such as diving, are dependent on accessibility and physical ability

their associated activities, physical abilities are specific to particular locations and actions. For example, someone who excels at hiking in steep terrain may not excel at surfing. It is more likely that a low level of (or a limit to) someone's physical ability may restrict their access to an outdoor experience or environment.

Limits to physical ability may include those associated with mobility, vision or hearing impairments. People with mobility impairments, such as those who require the use of a wheelchair, may find it difficult or impossible to move along bush tracks. People who are sight impaired may find it difficult to participate in rock climbing, and those with hearing impairments could find activities that rely on verbal communication, such as rafting or canoeing, difficult. People with a disability may have less access to the outdoors than able-bodied people. However, there are many ways to overcome the limits to physical ability, such as modified equipment, extra support and instructors with specific qualifications.

ACCESSIBILITY FACTORS

Accessibility can also be determined by:

- travel distance and remoteness of location
- time required to plan, prepare and participate
- costs associated with travel, equipment, preparation, permits, food, supplies, staff and taking time off work
- scale or size of the group
- fitness, training and experience or qualifications required
- motivation
- weather
- regulations
- physical environment such as terrain, water availability and emergency access
- risk and required safety precautions and plans
- history and popularity of certain experiences and activities.

By taking all of these factors into consideration, you can determine how easy or difficult it may be to participate in an outdoor experience.

Worksheet
2.2b Watch a documentary

LEARNING ACTIVITY 2.2B

Watch a documentary

Watch either of the documentaries *Touching the Void* (2003) or *Everest* (2015) online. During viewing, make note of all the factors that make these particular experiences low access. At the conclusion of the documentary, compile everyone's notes on the board in a brainstorm session or mind map.

LEARNING ACTIVITY 2.2C

Factors affecting access

1 Select an outdoor experience and create a fishbone diagram similar to the one shown below.

Activity: ______________________

a Place the name of the outdoor experience at the head of the fish (the red box on the sample diagram).

b Write the factors that can affect access to this outdoor experience (SES, cultural background, age, gender and physical ability) along the top and bottom of the spine (the blue boxes on the sample diagram).

c Detail how each factor could potentially affect someone's access to the outdoor experience you have chosen (you could include both restrictions to access or improved access).

2 Display your diagram in the classroom and discuss similarities and differences of factors affecting access to outdoor experiences.

Worksheet
2.2c Factors affecting access

2.2 KEY CONCEPTS

- There are a number of factors that may affect how easy or difficult it may be to participate in an outdoor experience. These include socioeconomic status, cultural background, age, gender and physical ability.
- Socioeconomic status (SES) can be described as an individual's or family's economic and social position in relation to others based on income, education and occupation. This may restrict a person from an outdoor experience due to the costs associated with it.
- Expectations in some cultures for females to wear traditional outer garments that cover much of the body may limit a person's access to some outdoor experiences.
- Age is one of the most influential factors that can affect access to and the types of outdoor experiences undertaken. For example, not being old enough to drive will limit access to some outdoor environments.
- It can be suggested that gender does not impact access to outdoor experiences as much as in the past because gender stereotypes associated with specific activities are less prevalent in our society.
- People with a disability may have less access to the outdoors than able-bodied people. However, there are many ways to overcome the limits to physical ability, such as modified equipment, extra support and instructors with specific qualifications.

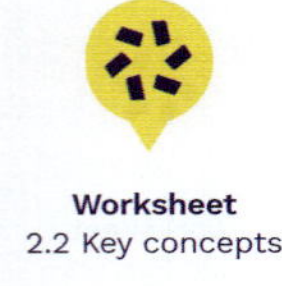

Worksheet
2.2 Key concepts

2.2 CONCEPT QUESTIONS

REMEMBERING

1 What is meant by the term socioeconomic status (SES)? How might it affect a person's access to an outdoor experience?
2 Describe how the weather might influence a person's access to an outdoor experience.

UNDERSTANDING

3 Explain how an environment might be portrayed as both an adversary and a gymnasium.
4 Compare two different outdoor experiences that would have participants from different cultures.
5 Using a specific example, discuss two factors that have affected your access to an outdoor experience.

APPLYING

6 Create a timeline of outdoor experiences that are suited to different age groups. Include activities suitable for an infant, toddler, primary school-aged person, teenager, adult and senior person. Compare your response with other members of the class. Are there any differences? Why?
7 Investigate and produce a PowerPoint presentation about one device that can affect a person's access to outdoor experiences, such as the amphibious Mobichair.

EXTENSION CHALLENGE

8 Research one of the following Australian female adventurers (or select another with your teacher's permission):
- Allie Pepper
- Belinda Ritchie
- Brigitte Muir
- Danielle Murdoch
- Jessica Watson
- Kay Cottee
- Linda Beilharz
- Nancy Bird Walton
- Sorrel Wilby
- Michelle Bloomcamp, Noelene Weightman and Sandra Floate (the 'Ice maidens').

Produce a multimedia presentation that tracks their particular adventure(s). Your presentation may be relatively short, but it should be entertaining and feature images, videos, media grabs and music.

2.3 RELEVANT TECHNOLOGIES AND THEIR INFLUENCES ON OUTDOOR EXPERIENCES

KEY KNOWLEDGE

- relevant technologies and their influences on outdoor experiences

KEY SKILLS

- explain the influence of relevant technologies on experiencing outdoor environments

In the last section we explored the range of factors that can influence a person's access to some outdoor experiences. As an old saying goes, 'necessity is the mother of all invention', meaning that over time we have used our creativity to overcome many of the reasons our choices have been restricted. One example of this is the use of technology to improve access to outdoor experiences, such as the amphibious Mobichair designed to improve beach access for those with a physical disability.

Technological developments such as advancements and modifications made to clothing and equipment have enhanced our participation in the ways we experience outdoor environments. Technologies have made experiences easier and safer, and have improved our access to them. However, some say that technologies have caused a disconnection between people and their environment, making challenging activities too easy, de-skilling individuals and causing an over-reliance on devices that can fail in the outdoors. In this section, we explore technologies and their influences on how we experience outdoor environments.

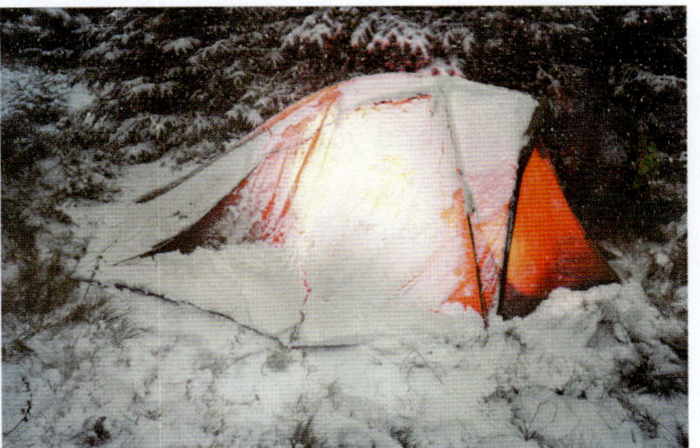

Shutterstock.com/Roman Mikhailiuk

Figure 2.25 Modern tent design enables access to otherwise inhospitable outdoor environments.

2.3.1 TECHNOLOGIES USED IN OUTDOOR EXPERIENCES

The term '**technology**' refers to the application of scientific knowledge for practical purposes to extend our human abilities and to manipulate nature to satisfy our wants and needs. Ongoing technological advancement continues to influence almost every aspect of our society. The function of technology is basically to make our lives easier by providing tools that enable us to solve everyday problems. The advent of such tools gives us an increasing level of control over our environments. Innovations in a number of areas have changed the way we experience outdoor environments, including:

- communication devices, such as radios, smartphones and EPIRBs (emergency position-indicating radio beacons)
- navigational devices, such as GPS (global positioning system)
- synthetic materials and clothing, such as fleece, Gore-Tex, polypropylene, Dri-FIT, nylon, plastics and Smartwool
- specialised equipment, such as fibreglass canoes, lightweight mountain bike frames, kernmantle climbing ropes and autofocus cameras.

These innovations allow us access to outdoor environments that would be too difficult to reach without technological assistance. They can provide us with the means to participate in a wide range of environmental conditions or enable us to enjoy year-round participation, despite the climatic variation.

technology
the application of scientific knowledge for practical purposes to extend our human abilities and to manipulate nature to satisfy our wants and needs

Getty Images / Gualter Fatia

Figure 2.26 Surfing has seen many advances in technology.

Shutterstock.com/maxpro

Figure 2.27 Mountain bike technology makes difficult terrain more accessible.

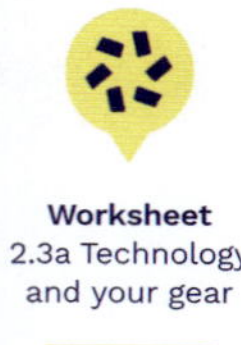

Worksheet
2.3a Technology and your gear

LEARNING ACTIVITY 2.3A

Technology and your gear

When preparing for (or unpacking after) a practical experience, spread your clothing and equipment out and think about how advances in technology have contributed to the amount of gear you take into the outdoors, the weight of this gear and its influence on your outdoor experience.

Changes in technologies over time

Technology has had an enormous impact on the way in which adventure activities have been undertaken by relatively unskilled participants over the past 60 years. The creation of synthetic fibres paved the way for improvements in adventure clothing and equipment; this was coupled with more efficient production techniques and a decrease in the overall cost. Participants in the outdoors are now confronted with a seemingly overwhelming range of technologically advanced equipment and clothing.

Table 2.1 Examples of technological changes over time

Surfing	Camping	Rock climbing
• Use of different materials in the construction of boards, from various woods to foam and fibreglass, and plastic (epoxy) and carbon fibre • Changes in the shape, size and curvature of boards • Changes in the number, design and placement of fins • Different mechanisms for fin attachment on boards • Changes in wetsuit materials • Use of heated wetsuits • Use of web cams for viewing breaks offsite • Use of the internet for weather analysis and surf break prediction • Use of wax and deck grips	• Changes in design and materials of tents, tent poles and pegs • Development of lightweight gear, including fuel stoves • Development of sleeping bags and sleeping mats • Development of packs and walking poles • Changes in the design and materials of shoes and footwear • Development of navigation supplements from compasses to GPS units • Development of breathable waterproof and water-resistant fabrics	• Design of new types of ropes • Development of specialist climbing shoes and footwear • Development of climbing harnesses • Development of climbing aids, such as carabiners, belay plates and camming devices • The use of helmets and other safety equipment • Changes in clothing used • Development of portable ledges and other devices to extend climbs • Introduction of simulated climbing environments

Worksheet
2.3b Old versus new

LEARNING ACTIVITY 2.3B

Old versus new

Compare one change in equipment technology associated with experiencing outdoor environments. Examples include:

- closed-cell foam mats/self-inflating mats
- leather ski boots/plastic ski boots
- handheld radios/mobile phones
- compass and paper maps/handheld GPS devices.

Use the following table to present your comparison.

	Old technology	New technology
General information		
Positive influences on outdoor experiences		
Negative influences on outdoor experiences		

LEARNING ACTIVITY 2.3C

Research task

1 Visit the National Geographic website and read the article, 'Everest Climbing Gear – Then and Now', which compares the equipment that was carried to the summit of Mount Everest in 1953 by Sir Edmund Hillary and Tenzing Norgay with that used by modern mountaineers.

2 Produce a visual presentation similar to that on the website comparing the technological changes for one piece of equipment used in a different outdoor adventure activity. Suggestions for equipment include surfboards, skis, outer shell jackets, thermals, canoes, navigation devices, cameras or rock-climbing equipment.

Worksheet 2.3c Research task

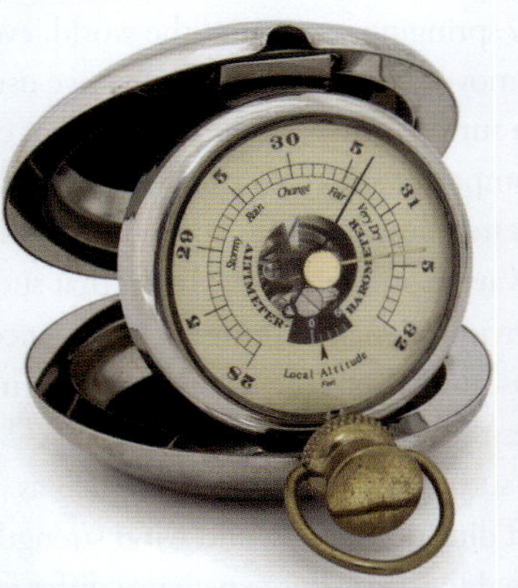

Figure 2.28 Antique pocket barometer altimeter and a modern sports watch with altimeter

Shutterstock.COM/Zysko Sergii, Shutterstock/ Martina I. Meye

The impact of new household technological devices

Household technological devices have drastically altered the way we live. They provide us with the means to perform a majority of what we need to do to survive within the confines of our own homes. For example, information and communication technologies allow us to maintain social contact and entertain ourselves without having to leave the house. Thus, there are fewer reasons to venture into the outdoors than ever before. A child today will spend less time exploring the neighbourhood and playing with other children than in the past, as they are likely to be indoors consumed by gaming and interacting through social media. Adults are increasingly finding it hard to escape the ease and comfort of a more sedentary indoor lifestyle.

The impact of new technology on how we experience the outdoors

Even when we do venture into the outdoors, technology can still alienate us from the environments we visit. Can the use of Gore-Tex clothing, hi-tech tents, sleeping bags and other camping equipment detract from the outdoor experience? Does it remove us from the environmental conditions that are present? Consider the rock-climbing venue that includes bolts and hangers strategically placed across a cliff face to provide the climber with anchor points to attach to during an ascent. Does modifying the environment to assist us in our participation actually diminish the connection we make with that environment? If a climb is too easy for us, do we not fully respect the challenges of that environment?

Shutterstock.com/Dragon Images

Figure 2.29 Indoor rock climbing is an artificial environment that provides the physical activity of climbing, yet diminishes our connection with the outdoors.

We are now creating our own artificial environments designed to mimic the outdoors, but with 'improvements' that provide more controlled levels of participation. Indoor rock climbing will provide you with the physical activity of climbing, but with a multitude of graded routes all in one contained venue, with a roof over your head to keep out the weather. There will be no actual rock in sight.

Artificial wave pools are considered the next step in surfing participation. At one time, this concept was thought of as far-fetched and unachievable, but wave pools are now springing up around the world, even in areas far removed from the beaches that we usually associate with surfing. They produce waves that can be used by competent surfers. Surfing legend Kelly Slater has invested heavily in this concept with his Kelly Slater Wave Company in the belief that such an artificial environment can bring the experience of surfing to those away from the coast. The structure also eliminates the reliance on environmental variables that surfing conditions rely on, such as swell sizes and directions, tides, and wind strengths and directions. How does this experience differ from actually travelling to the coast looking for waves?

Alamy/Stuart Westmorland / DanitaDelimont

Figure 2.30 Surfing in an artificial wave pool

Worksheet
2.3d Technology paradox

LEARNING ACTIVITY 2.3D

Technology paradox

> The paradox is that each new layer of invention and innovation simultaneously enhances and separates us from the outdoor experience.
>
> Mike Bartle, Technology and Outdoor Education, 2000

1 Discuss this quote using an example of outdoor adventure that you have participated in.
2 How has technology influenced your outdoor experience in this example?
3 Was it enhanced? Did you feel that you had been separated from the outdoors?

The following table outlines some of the ways technology has an influence on our relationships with outdoor environments:

- the way outdoor environments are perceived
- how we interact with outdoor environments
- the impact that we have on outdoor environments.

Table 2.2 Influence of technology on relationships with outdoor environments

Influence	Perception	Interaction	Impact
Alienation	• Separation from outdoor environments • Removal from outdoor environments • Outdoor environments are uncomfortable	• Less participation in the outdoors • Participation in a wider range of indoor pursuits: social media, game consoles, information technology • Greater use of creature comforts: air conditioning, heating, lighting	• Increased urbanisation causing loss of habitat • Fewer people travelling to outdoor environments less often has less impact on certain environments
Easier participation	• Outdoor environment is more attractive as a venue for participation • Can decrease 'wilderness experience' as people are not as exposed to environmental conditions • Can decrease the thrill of participation	• Increased participation by a greater range of ages and cultures • Participation in a greater range of activities (e.g. soft surfboards, full body wetsuits)	• Increased numbers participating can have a negative impact on specific environments (e.g. campsite overuse, beach/dune erosion)
Activities are safer	• Environment is a safe place to participate • Participants have a sense of security in the outdoors	• Greater participation in the outdoors is possible through: - more venues available - people being able to participate in a wider range of weather conditions - people being able to participate even if they are less skilled/experienced	• Greater number of people will put stress on specific environments
Performance is improved	• Environment is a venue for physical activity • Environment is a series of problems to solve through recreational activities	• Participation encouraged in 'sensation seekers' • Higher participation rates (e.g. mountaineering equipment, mountain bikes, thrusters, surfboards, jet skis for tow surfing)	• Equipment costs can reduce numbers participating • Increased participation can put stress on specific environments
Better access	• The outdoor environment is an accessible resource for leisure participation	• Increased participation through improved transport (e.g. better alpine roads) • Access to remote areas • Snowmaking facilities to enable more skiing	• Increased participation can put stress on specific environments • Widespread impacts due to increased accessibility to a greater range of venues

iStockphoto/mato750

Figure 2.31 With a GPS unit, people can find their location with the touch of a button.

DE-SKILLING AS A RESULT OF NEW TECHNOLOGIES

While the use of technologies is great for helping unskilled individuals to partake in outdoor experiences with more confidence and efficiency, technological advancement has also been associated with the de-skilling or loss of self-reliance among **contemporary** outdoor adventurers. In many cases, technological advances that have made outdoor activities easier have also removed the need for adventurers to develop their own skills through personal experience.

contemporary events or actions that have occurred within the past 15 years

Navigation is an example of where de-skilling has occurred. In the past, a bushwalker would need to ensure that their skills of map reading and compass use were proficient to a high degree. In gaining information on the area they were planning to bushwalk in, they would most likely have spoken to locals and people who had previously visited the area. Today, this is simply not required. Instead, bushwalkers have a range of navigational devices at their disposal, such as high-quality topographical maps, guidebooks, handheld GPS devices and smartphones. These technologies allow bushwalkers to undertake journeys that they may not have been capable of in the past. In this example, technologies have replaced the need for personal navigational skills, and perhaps caused an over-reliance on devices that can potentially fail at any time in the outdoors. Many people argue that drastically reducing the difficulty of tasks weakens the environmental experience of those involved.

AAP Image/Dean Lewins

Figure 2.32 A personal locator beacon (PLB) uses satellite technology to relay a message from your location (on land or in water) to search and rescue crews in case of emergency.

THE UPSIDE OF NEW TECHNOLOGIES

On the positive side, more people can experience the environment as a result of technological advances, including people with a disability, and in many ways the need for expensive rescues and medical attention has been reduced. Technologies have enhanced adventures by creating safer experiences and by providing greater options within outdoor environments, resulting in increased participation rates in many outdoor activities. For example, synthetic fibres are water-repellent, quick-drying, durable, lightweight and breathable, and this means people can spend longer in the outdoors with less clothing and equipment, and can endure more extreme conditions in comparative comfort. Likewise, a Personal Locator Beacon (PLB) is a compact, self-contained radio transmitter that can be activated in life-threatening situations – carrying this device enhances the feeling of security people have in the outdoors, making their experience less stressful and more enjoyable. However, is this also creating a false sense of security?

LEARNING ACTIVITY 2.3E

Worksheet 2.3e Class debate

Class debate

Your teacher will select one of the debate topics from the following list (or provide another for you to research). The class will be divided into two groups – one in favour of the selected topic and one against it.

- Technological advances have made outdoor experiences safer and alleviated many dangers.
- Technological advances can only enhance our experience with the outdoors.
- Technological advances have only increased the separation between people and nature.
- Technological advances have created a false sense of security for many venturing into the outdoors.

Prepare for the debate according to your teacher's instructions.

LEARNING ACTIVITY 2.3F

Worksheet 2.3f Surf websites

Surf websites

Visit a surf forecasting website such as Swellnet, Magic Seaweed, Surfline or Willi Weather, then complete the following:

1 Describe the information available on your selected website.
2 How can the use of this technology influence:
   a the ease of participation in surfing?
   b safety during surfing participation?
3 How can this technology influence our relationship with the environment? Think about:
   a how we interact with an environment:
      - understanding conditions
      - desire to participate at a venue
      - personal connection to a venue
4 How could this technology impact on specific surfing environments (positively and negatively)?
5 Visit a surf forecasting website and choose a venue to go surfing at tomorrow. Include reasons why you chose this venue.

2.3 KEY CONCEPTS

- The term 'technology' refers to the application of scientific knowledge for practical purposes to extend our human abilities and to manipulate nature to satisfy our wants and needs.
- Technological developments such as advancements and modifications made to clothing and equipment have enhanced our participation in the ways we experience outdoor environments.
- Some say that technologies have caused a disconnection between people and their environment, making challenging activities too easy, de-skilling individuals and causing an over-reliance on devices that can fail in the outdoors.
- Participants in the outdoors are now confronted with a seemingly overwhelming range of technologically advanced equipment and clothing.

- We are now creating our own artificial environments designed to mimic the outdoors, but with 'improvements' that provide more controlled levels of participation.
- Increased urbanisation and people living in suburban environments make it more difficult to reach the outdoors for leisure.
- Technological advances have made outdoor activities easier by removing the need for adventurers to develop their own skills through personal experience.
- Technologies have enhanced adventures by creating safer experiences and by suggesting that more is now possible within outdoor environments, resulting in increased participation rates in many outdoor activities.

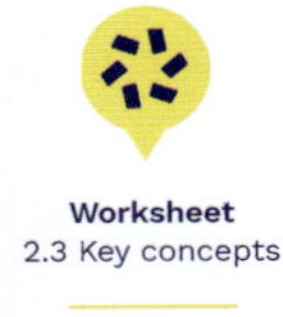

Worksheet
2.3 Key concepts

2.3 CONCEPT QUESTIONS

REMEMBERING

1 Define the term 'technology'.
2 Describe the technological development of one piece of equipment used within outdoor environments.

UNDERSTANDING

3 Discuss how one piece of technology you used influenced your experiences in an outdoor environment.
4 Using examples, explain why some people say that technologies have caused a disconnection between people and their environment.
5 Outline the ways technology may influence our relationships with outdoor environments.
6 Explain how technology may be used to promote sustainable interactions with outdoor environments.

APPLYING

7 Create a two-minute oral presentation for your class about the latest technological advancements associated with a specific recreational activity.
8 Develop a list of the advantages and disadvantages of utilising technologies when participating in outdoor experiences.

EXTENSION CHALLENGE

Research task

9 Investigate the changes, developments and advancements that the equipment or the outdoor experience has undergone in relation to technologies. You may like to address the following questions in your research:
   a What changes have occurred (such as shape, materials, price or durability)?
   b Have the changes altered any aspect(s) of the experience?
   c Have the changes in equipment affected participation rates?
   d Have the changes in equipment brought participants closer to the environment or separated them from it?

   Present your research in a basic report format (including images) to highlight the changes in technologies.

2.4 PERSONAL RESPONSES TO RISK WHEN EXPERIENCING OUTDOOR ENVIRONMENTS

KEY KNOWLEDGE

- the variety of personal responses to risk when experiencing outdoor environments, including the interplay between competence, perceived risk and real risk

KEY SKILLS

- compare a range of personal responses to risk when experiencing outdoor environments

2.4.1 RISK WHEN EXPERIENCING OUTDOOR ENVIRONMENTS

> Only those who will risk going too far can possibly find out how far they can go.
>
> T.S. Eliot, 1931

Risk is sometimes seen as an inherently bad thing. We often try to avoid risk wherever we can and we certainly take steps, both individually and as a society, to reduce the risks associated with many parts of our daily lives. However, there are risks that can be good for us as humans.

Shutterstock.com/Greg Epperson

Figure 2.33 Rock climbers usually manage the risk of the activity with a variety of safety technologies. How is this climber managing risk? Are there things he could do better?

Stepping outside of our comfort zone can be challenging, but it's also been shown to improve our confidence and sense of self-worth, and provide other psychological benefits. We feel better about ourselves when we can face a risky, challenging situation and succeed at it. And if we don't succeed? Learning about failure can also help us. If the risky situation is managed well, initial failure may encourage us to try again.

Before we explore risk in more detail, a note of caution: risk in outdoor experiences should be managed or controlled as much as we possibly can, without, of course, taking away the risk itself. We'll explore this in the next chapter by investigating risk management of outdoor experiences.

In this section we explore the concept of risk, how to identify risks, factors contributing to risk, the different types of risk and the variety of personal responses to risk when experiencing outdoor environment.

What is risk?

risk
the potential to lose something that you value measured against the possibility of gaining something you value

Risk is the potential to lose something that you value measured against the possibility of gaining something you value. The thing we might lose may be physical (such as an injury to your body), psychological (such as embarrassment at failure) or financial (such as damage to expensive equipment). The thing we might gain could also be physical (such as developing physical fitness, or some sort of biochemical change such as an adrenaline boost), psychological (such as a rise in self-esteem) or financial (such as some sort of monetary reward).

Risk involves uncertainty. When we take a risk, we're not exactly sure what will happen. In some outdoor activities, the risk may be in potentially facing injury or even death measured against the physical, psychological and emotional benefits of succeeding.

Identifying risks

There is a large number of possible risks inherent in outdoor experiences. We can break these risks into three main categories:

1. **Environmental risks** – those that originate from the outdoor environments themselves and can have an impact on the experience. They include such factors as weather, terrain, the availability of shelter and the remote aspect of many outdoor environments, as well as the inherent dangers associated with some Australian flora and fauna.
2. **Risks associated with people** – those that can be connected to the people involved in outdoor experiences, such as leaders and participants, as well as other people that groups may encounter. They include such factors as the skills, knowledge, experience, health and fitness, age, fear and other emotions that participants bring to an outdoor experience.
3. **Risks associated with equipment** – those associated with the specialised equipment that we use while participating in outdoor experiences, and the equipment used to get to the places that we visit. They include such things as clothing, buoyancy aids, kayaks, surfboards, bikes, tents, climbing ropes, helmets, motor vehicles and fuel stoves.

absolute risk
the uppermost limit of risk in a particular situation or activity, assuming safety has not been considered

perceived risk
the subjective assessment that a person makes about the risk they are about to face in a particular situation

real risk
the risk that actually exists for a particular situation or activity, given that safety has been considered and controls put in place

competence
the ability of someone to be able to deal with the situation they are in, which comes from the skills and experience that they have

Types of risk

How we perceive risk is subjective; that is, how a person views the risks involved in an activity will be influenced by their level of competence, past experience and emotions. In the context of outdoor activities, we can categorise three different types of risks:

- **Absolute risk** – the uppermost limit of risk in a particular situation or activity, assuming safety has not been considered. Think of it as the 'worst-case scenario' type of risk; for example, the risk of death associated in climbing a rock wall with no safety devices.
- **Perceived risk** – the subjective assessment that a person makes about the risk they are about to face in a particular situation. The perceived risk can vary dramatically, and could be much higher or lower than the actual risk. Factors that may influence perceived risk include confidence in the activity or environment, familiarity with equipment, experience levels of group members, emotional state, awareness of limitations and fear of the unknown.
- **Real risk** – the risk that actually exists for a particular situation or activity, given that safety has been considered and controls put in place. For example, the risk associated in climbing a rock wall using ropes, harnesses, belay devices, helmets and other equipment.

Given that we would probably never undertake an activity without a consideration of safety issues, perceived risk and real risk are of more importance to us in outdoor experiences than absolute risk.

Managing risk

A useful way to help think about managing risks related to outdoor activities is sometimes called the competence–difficulty model.

Competence is the ability of someone to be able to deal with the situation they are in, and comes from the skills and experience that they have. Someone with more skills and greater knowledge and experience is going to be more competent to deal with a particular situation than someone with fewer skills and less experience.

Psychologist Abraham Maslow first described the term 'peak experience' in the 1960s: 'Think of the most wonderful experience of your life: the happiest moments, ecstatic moments, moments of rapture, perhaps from being in love, or from listening to music or suddenly "being hit" by a book or painting, or from some creative moment.' For Maslow, the peak experience was an event or moment that was incredibly joyous and exciting. It stood out from everyday events and involved sudden feelings of intense happiness and wellbeing, wonder and awe. The memory of such events was long-lasting, and Maslow likened them to intense spiritual or religious experiences. For people undertaking outdoor activities, achieving a peak experience is often the goal.

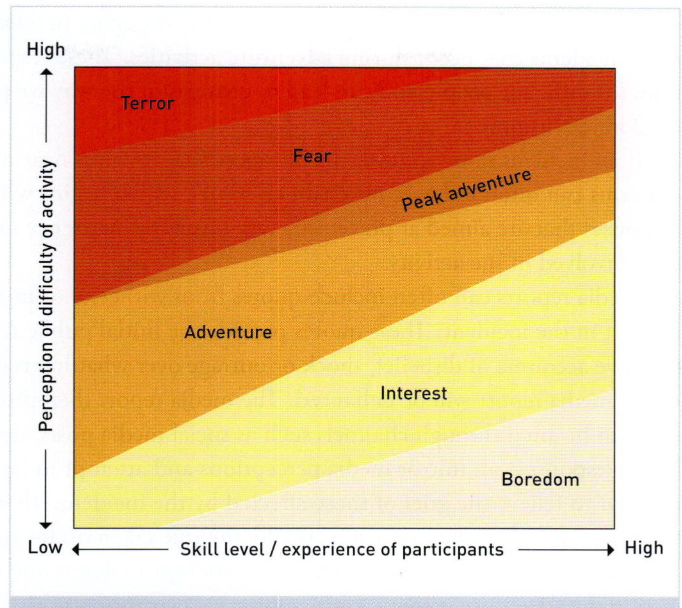

Figure 2.34 Competence–difficulty model

As can be seen in the competence–difficulty model, the peak adventure experience is wedged between adventure and fear. According to this model, it requires just the right mix of skill/experience and perception of difficulty. Not enough skill or too high a perception of difficulty and we are terrified; too much skill and not enough difficulty and we are bored.

You might notice the similarity between the diagram that represents the competence–difficulty model for risk and the diagram for the idea of flow (which we looked at earlier when considering the motivations that people have for participating in outdoor activities). The notion of the peak experience synchronises very closely with the idea of flow – it seems that when someone is having a peak experience, they are probably in this psychological state called flow.

Andrew Mannion

Figure 2.35 How would you assess the risk of these conditions?

Responses to risk

For some people in society, adventure and risk-taking are important; for others, these activities are perceived to be senseless and dangerous. The ways in which people respond to risks in outdoor experiences vary and they are generally related to perceived risk rather than real risk; that is, what a person believes the dangers are versus the actual or real risk.

People respond in a variety of ways to others engaging in what they perceive as risk-taking behaviour or to incidents that occur during adventure activities. These responses relate to the way society defines roles for different people and can lead to changes in the way we perceive, interact with and impact on the outdoor environment.

If an incident does occur in the outdoors that involves near miss, injury or death, a fairly typical chain of events can often follow. This chain of events, which begins with media coverage, will usually result in responses that are aimed at preventing any future reoccurrence of the incident by reducing the specific risks involved in the activity.

Media reports can often include quotes from witnesses or members of the public with a specific interest in the incident. These quotes provide the initial public response that can often lean towards negative accounts of disbelief, shock or outrage over what has transpired. Again, the sensational aspect of the media report will be enhanced. The media report then provokes further responses from the public that can be aired through channels such as social media posts, newspaper editorials or talk-back radio. These responses can mirror media perceptions and attempt to lay blame and search for a guilty party, or attempt to relieve the grief of those affected by the incident. There also exists the opportunity to respond with reason and balance, which is usually the role taken on by authorities or the adventure industry.

The investigation into an incident will attempt to determine the actual events that have occurred during the adventure activity. Here, subjective perception of risk is less important than the real risks involved, as those in charge of the investigation will be more interested in what has gone wrong and how it can be avoided in the future. Who the investigation is conducted by will depend on the severity of the incident. If a death is involved, the coroner will usually conduct the investigation, which provides a high level of formality in a legally sound setting. People involved will be interviewed along with others with expert knowledge and experience in the particular activity or outdoor environment. A report containing the findings of the investigation, including those liable and recommendations for future participation, will then be presented to authorities. Similar proceedings will follow less severe incidents and might be conducted by other authorities such as police or by the outdoor adventure industry itself.

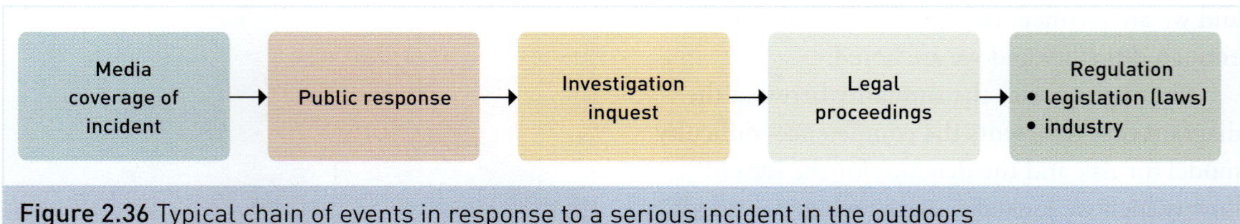

Figure 2.36 Typical chain of events in response to a serious incident in the outdoors

Legal proceedings may follow if an investigation suggests that negligence has actually occurred. Although a vast majority of outdoor adventure activities are conducted by suitably qualified staff or instructors in appropriate environments adhering to relevant guidelines, there are rare circumstances where this is not the case and a legal precedent needs to be set to ensure future participant safety. In doing this, however, the level of inherent real risk of outdoor adventure is not completely ignored and is considered in the context of the legal decisions.

Mitigating risk with regulations

Response to tragedy by authorities such as local, state or federal governments, and statutory bodies such as Parks Victoria, usually involves the imposition of regulations. Often, such regulations take the form of training (as has occurred in rock climbing and canoeing) and registration requirements. Another response is to restrict access. For example, at Hanging Rock in Central Victoria, rock climbing has been banned in part because of the risk to other users of the area from dislodged rocks and debris. Sometimes sites are closed altogether.

One way of implementing these measures is to introduce or redefine zoning restrictions on public land such as national parks. A direct effect of such restrictions may be to reduce the sense of freedom experienced by users and, in some cases, reduce the opportunity for friction to occur between groups with different values, such as bush walkers and trail-bike riders. Such restrictions can also increase the safety of both users of the natural environment and its inhabitants.

Introducing and enhancing safety regulations is also a response that can be initiated by other participants in a particular activity following an incident. They may work together with authorities in order to introduce a code of conduct or to implement a certification system so participants are knowledgeable about the skills and safety techniques required to safeguard themselves and/or the public. An example of this type of certification is in surf lifesaving, where members must earn a Bronze Medallion in order to patrol beaches and coastlines. While such measures certainly lead to safer activities, for some they can also reduce their sense of freedom of choice and escape from everyday situations. They also increase the cost of pursuits due to new equipment requirements and training. Increases in insurance premiums based on the calculated risk of injury can further contribute to increasing operator costs, which will be ultimately worn by the paying participant.

Alamy Stock Photo/Antonio Siwiak

Figure 2.37 Many outdoor experience industries regulate how many participants can be in a guided group (the guide–participant ratio). Usually, the smaller the group, the safer the activity. What does this mean for the costs of running activities?

Shelters, huts, snow pole lines, signs, fences and barriers at lookouts are other tangible signs of the ways in which society can respond to risk-taking activities. Such infrastructure improves safety, sometimes even providing a false sense of security, but can also diminish one's sense of wilderness and adventure due to its visual impact.

CASE STUDY

School climbing trip goes horribly wrong as boy, 16, plunges from a mountain while his mates watch on helplessly before becoming trapped on the cliff face

Three teenage boys have been rescued from a cliff face after a rock-climbing adventure went horribly wrong. The students from Melbourne were retrieved from Mount Arapiles, west Victoria, about 3 pm on Tuesday after one boy fell and two others became stuck. The 16-year-old was taken by ambulance to Horsham Public Hospital where he is in a stable condition and undergoing treatment for lower body injuries.

'We're unsure of the distance he's actually fallen but he's landed on a rock ledge approximately five to seven metres above the ground level,' a rescuer told Seven News. 'He was complaining of the injuries, obviously with a fall like that.' The rescue teams initially tried to reach the teenager with a helicopter and winch but had to resort to using ropes to lower him to the ground.

Resource
Case study: School climbing trip goes horribly wrong

The news comes after four other rock climbers had to be rescued from the mountain in late February. A 67-year-old climber suffered a head injury and became stuck – along with three other climbers – on the mountain's cliff face. A high angle rescue team was able to reach the group two hours later. SES Victoria has warned Mount Arapiles is a very challenging climb, even for experienced climbers.

Shutterstock.com/Alizada Studios

Figure 2.38 Mount Arapiles is very challenging, even for experienced climbers.

QUESTIONS

1 Using examples from the article, discuss the three different types of risks.
2 Describe your personal response to the risks involved in the activity of rock climbing.
3 How may personal responses to the risks involved in rock climbing differ?
4 What role did level of competence in rock climbing play in the described incident?

'School climbing trip goes horribly wrong as boy, 16, plunges from a mountain while his mates watch on helplessly before becoming trapped on the cliff face', Ashley Nickel, *Daily Mail*, 5 April 2022

It is important for participants in any outdoor activity to strike a balance between their competence and awareness of risks. This balance promotes safety while still allowing for an enjoyable outdoor experience. Education, training and experience can help participants develop greater competence in outdoor activities. This enables more informed and accurate considerationsof the real risks when venturing into outdoor environments.

Worksheet
2.4a Safety audit

LEARNING ACTIVITY 2.4A

Safety audit

1 While participating in an outdoor activity, observe and record any evidence of regulation or attempts to increase safety and reduce risk. This may include fences at cliffs, 'beware of ...' signs, snow pole lines, shelter huts and steps.
2 Why are these items in place?
3 Do you feel they are necessary? Give reasons for your answer.

2.4 KEY CONCEPTS

- Risk is the potential to lose something that you value measured against the possibility of gaining something you value.
- Stepping outside of our comfort zone can be challenging, but it's also been shown to improve our confidence and sense of self-worth and provide other psychological benefits.
- In some outdoor activities, the risk may be in potentially facing injury or even death measured against the physical, psychological and emotional benefits of succeeding.

- Environmental risks are those that originate from the outdoor environments themselves and can have an impact on the experience.
- Risks can be connected to the people involved in outdoor experiences, such as leaders and participants, as well as other people that groups may encounter.
- Risks can be associated with the specialised equipment that we use while participating in outdoor experiences, and the equipment used to get to the places that we visit.
- Absolute risk is the uppermost limit of risk in a particular situation or activity, assuming safety has not been considered.
- Perceived risk is the subjective assessment that a person makes about the risk they are about to face in a particular situation.
- Real risk is the risk that actually exists for a particular situation or activity, given that safety has been considered and controls put in place.
- Competence is the ability of someone to be able to deal with the situation they are in, and comes from the skills and experience that they have.
- Someone with more skills and greater knowledge and experience is going to be more competent to deal with a particular situation than someone with fewer skills and less experience.
- The ways in which people respond to risks in outdoor experiences vary and they are generally related to perceived risk rather than real risk.
- Often, society's responses are shaped by the way a pursuit or endeavour, and any associated incidents, are portrayed by the media.
- These reports can influence what we believe is appropriate in terms of the level of risk we or others pursue, or are subjected to, in the outdoors.
- If an incident does occur in the outdoors that involves near miss, injury or death, a fairly typical chain of events can often follow.
- Introducing and enhancing safety regulations is also a response that can be initiated by other participants in a particular pursuit following an incident.

2.4 CONCEPT QUESTIONS

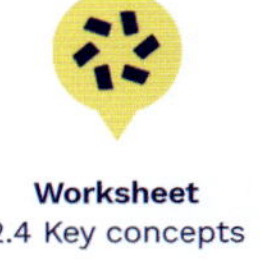

Worksheet
2.4 Key concepts

REMEMBERING

1 Define the term 'risk'.
2 Describe three benefits of participating in activities that involve a degree of risk.

UNDERSTANDING

3 Using your own examples from an outdoor experience this year, distinguish between the three types of risk.
4 Explain the interplay between competence, perceived risk and real risk.
5 Compare two different responses you have experienced when participating in an outdoor recreational activity.
6 What is meant by the concept of peak experience?

APPLYING

7 Copy the competence–difficulty model into your workbook. For each section of the model provide an example of when a person may experience this response to an outdoor activity. For example, a highly skilled mountaineer may experience boredom when asked to sit through a beginner's rock-climbing lesson.

EXTENSION CHALLENGE

8 Visit a local adventure playground or make use of equipment in your gym. Work in pairs and complete a course laid out over, under and through a variety of objects. When you are familiar with the course, complete it wearing a blindfold while your partner acts as a spotter. Swap roles. Change the circuit around. Discuss how you felt after completing the circuit, first with sight and then without.
   a What would have made you feel safer?
   b Provide a definition of perceived risk and discuss how it relates to this situation.

Resource
Glossary – Chapter 2

Assessment
Exam-style questions

Glossary test

EXAM-STYLE QUESTIONS

1 Analyse two factors that have affected your access to experiencing a specific outdoor environment. (6 marks)

2 a Describe one example of how an outdoor experience is depicted in one of the following: the mainstream media, social media, music, art, writing, or advertising. (2 marks)

b Describe two ways you responded to the depiction described in part a. (2 marks)

c Analyse the influence of your described depiction on your personal responses. (2 marks)

3 a Discuss the technological developments of one piece of equipment you have used on an outdoor experience this year. (3 marks)

b Explain the influence of that piece of equipment on your experiences in a specific outdoor environment. (3 marks)

4 a Using examples from your experiences in outdoor environments this year, explain the difference between perceived risk and real risk. (6 marks)

b Describe how you responded to the risks involved in the outdoor experience. (2 marks)

c Discuss how a person's experience in an outdoor activity may influence their perception of the risk involved in the activity. (2 marks)

CHECK YOUR KNOWLEDGE AND SKILLS

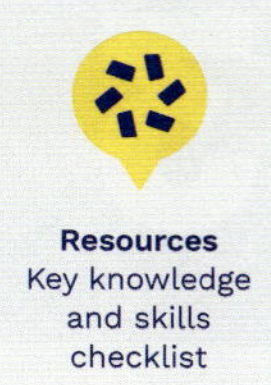

Resources
Key knowledge and skills checklist

Checklist for students to mark and date making judgments of their knowledge and understanding.
(**D** = developing **C** = consolidating **ER** = exam ready)

D	C	ER	Check your knowledge and skills
☐	☐	☐	the influence of depictions of experiencing outdoor environments on personal responses, such as in the mainstream media, social media, music, art, writing and advertising
☐	☐	☐	analyse the depictions of experiencing outdoor environments on personal responses
☐	☐	☐	factors that affect access to experiencing outdoor environments, including socioeconomic status, cultural background, age, gender and physical ability
☐	☐	☐	analyse factors that affect access to experiencing outdoor environments
☐	☐	☐	relevant technologies and their influences on outdoor experiences
☐	☐	☐	explain the influence of relevant technologies on experiencing outdoor environments
☐	☐	☐	the variety of personal responses to risk when experiencing outdoor environments, including the interplay between competence, perceived risk and real risk
☐	☐	☐	compare a range of personal responses to risk when experiencing outdoor environments

CHAPTER 3

UNIT 1 – AREA OF STUDY 3

Safe and sustainable participation in outdoor experiences

KEY KNOWLEDGE

- safe and sustainable interactions with outdoor environments, involving minimum impact strategies for individuals, route planning, tent-site selection, fuel stove usage, navigation and packing a pack
- basic first aid skills, including blister management, small wounds, snake bite, severe bleed treatment, Cardio-Pulmonary Resuscitation (CPR) and immobilisation techniques
- equipment required to safely explore outdoor environments
- risk management of outdoor experiences.

KEY SKILLS

- plan for participation in a range of sustainable outdoor experiences
- demonstrate and apply basic first aid knowledge and skills
- use appropriate skills for safe and sustainable interactions with outdoor environments
- explain the use of equipment designed to improve the safety of participants in outdoor experiences
- identify hazards, analyse risks and suggest controls for an outdoor experience in a chosen outdoor environment
- analyse relevant information collected during outdoor experiences.

VCE Outdoor and Environmental Studies Study Design 2024–2028, pp. 16.

- safe and sustainable interactions with outdoor environments, involving minimum impact strategies for individuals, route planning, tent-site selection, fuel stove usage, navigation and packing a pack

- basic first aid skills, including blister management, small wounds, snake bite, severe bleed treatment, Cardio-Pulmonary Resuscitation (CPR) and immobilisation techniques

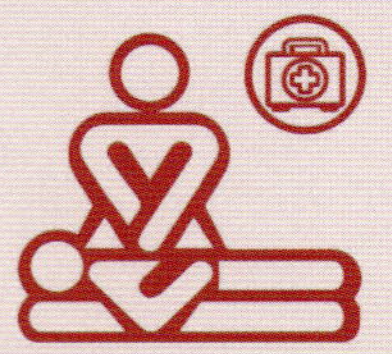

- equipment required to safely explore outdoor environments

- risk management of outdoor experiences

KEY TERMS

code of conduct
global positioning system (GPS)
hazard
likelihood
minimal impact strategies
mud maps
risk
risk assessment
risk rating
stakeholder
topography
topographic maps

Worksheets

3.1a Campfires **p. 105**
3.1b Old codes of conduct **p. 108**
3.1c Create your own code of conduct **p. 108**
3.1d Planning a route **p. 112**
3.1e OES Masterchef **p. 116**
3.1f Fuel stove rankings **p. 117**
3.1 Key concept **p. 124**
3.2a First aid **p. 136**
3.2b Emergency case studies and role-plays **p. 137**
3.2 Key concepts **p. 137**
3.3a Emergency information card **p. 140**
3.3b Gear checklist **p. 146**
3.3 Key concept **p. 146**
3.4a Common-sense approaches to dangers in the outdoors **p. 148**
3.4 Key concept **p. 151**
3.4b Outdoor experience: design and reflection **p.152**

Weblinks

- Australian Alps National Park codes of conduct **p. 108**
- National Outdoor Leadership School case studies **p. 137**

Resources and templates

- Glossary – Chapter 3 **p. 154**
- Key knowledge and skills checklist **p. 155**

Assessments

- End of chapter exam **p. 154**
- Glossary test **p. 154**

Nelson MindTap

To access resources above, visit **cengage.com.au/nelsonmindtap**

SPOTLIGHT

Safe and sustainable participation in outdoor experiences

The rescue of a man who set out for a bushwalk with only potatoes, naan bread and matches is likely to have cost the state $10 000.

A 29-year-old Victorian man planned to cross the Blue Mountains in just three days but came unstuck after injuring his ankle. A search operation was launched on Wednesday after he failed to meet friends at a pre-designated rendezvous point.

'He endeavoured to walk what was, for a person equipped with ample provisions, [a] seven-day plus [hike],' said Inspector Max Wallace from Katoomba police. 'He lacked all those essential preparations.'

Police found the man with one kilogram of potatoes and naan bread in his backpack after he lit a fire to attract the attention of two helicopters passing overhead.

Police fined the man $500 for risking his own safety and that of others. A police spokesman said it was only the second such fine issued this year.

iStockphoto/Ryan Fletcher

Figure 3.1 A lack of appropriate preparation for outdoor experiences can lead to dangerous and expensive search and rescue operations.

Police refused to comment on the cost of the rescue, but typically charge between $2500 and $3300 per hour for use of helicopters for non-police matters, according to a schedule of fees. Similar missions in other states have been estimated to cost between $10 000 and $12 000 an hour.

The helicopters – understood to have spent hours in the air – were supported by State Emergency Service volunteers and NSW Volunteer Rescue Association members, who battled fading light in a three-hour search.

Previous calls to bill unprepared or negligent bushwalkers for the full cost of their rescues have been dismissed as impractical because the costs are often in the tens of thousands of dollars. A former president of Bushwalking Australia, David Reid, said charging hikers for their rescue could prove harmful.

'To introduce a broad-based penalty for people who get into trouble could well deter them from seeking assistance when it's important to do so,' he said.

The NSW Volunteer Rescue Association commissioner, Mark Gibson, said many searches would be prevented if all bushwalkers carried locator beacons.

Last year, about 1700 beacons were triggered.

'Bushwalker's rescue likely to come with $10,000 price tag', Bevan Shields and James Robertson, *The Sydney Morning Herald*, 1 January 2013

Outdoor environments are beautiful and unique; however, as the above media report highlights, they can also be unpredictable. Not being appropriately prepared with the correct food, equipment and knowledge can result in expensive and dangerous search and rescue operations, where the lives of others may be put at risk. Of course, lack of preparation can lead to more serious consequences for those involved. Before embarking on an experience in the outdoors, it is crucial to carefully consider appropriate planning and preparation.

As covered in Chapter 1, Indigenous Australians' connections to Country, both land and sea, are integral to cultural values, identity, traditions and heritage. Participants in adventure activities have a responsibility to recognise and respect the cultural and natural values of these areas. Planning an outdoor experience should acknowledge the traditional custodians of the land, ensure awareness of culturally significant sites and modify the activity to avoid or reduce potential conflict and impact.

This area of study focuses on planning and participating in outdoor experiences. Experiencing outdoor environments safely requires an understanding of how to plan and conduct sustainable outdoor experiences in outdoor environments.

However, just because you are prepared does not mean things can't go awry. In Chapter 2 we discussed the traumatic events that occurred to experienced mountaineer Aron Ralston. Aron failed to notify anyone of his plans to explore the narrow canyons of Utah alone. The consequence of failing to 'let someone know before you go' were realised once Aron found himself trapped by a massive boulder in a very isolated location. Aron was able to draw upon his prior knowledge, skills and self-determination to survive.

The first two chapters of this textbook have focused on developing your knowledge and understanding of how humans connect with outdoor environments and why these connections are important. We now move on to preparing you and your classmates to plan for safe and sustainable participation in outdoor experiences. We explore a range of practical skills and knowledge including route planning, navigation, basic first aid, minimal impact strategies and risk management.

3.1 SAFE AND SUSTAINABLE INTERACTIONS WITH OUTDOOR ENVIRONMENTS

KEY KNOWLEDGE

- safe and sustainable interactions with outdoor environments, involving minimum impact strategies for individuals, route planning, tent-site selection, fuel stove usage, navigation and packing a pack

KEY SKILLS

- plan for participation in a range of sustainable outdoor experiences

3.1.1 SAFE AND SUSTAINABLE INTERACTIONS WITH OUTDOOR ENVIRONMENTS

By failing to plan, you are preparing to fail.

Unknown

Some of the outcomes we might want from experiences in outdoor environments include fun, enjoyment and challenge. But we also want to make sure we are safe and we do not negatively impact the environment – we want our experience to be sustainable. Living and travelling in remote outdoor environments brings with it a whole set of demands and issues we don't normally face in our everyday lives.

Experiencing outdoor environments safely requires an understanding of how to plan and conduct sustainable outdoor experiences in outdoor environments.

If we are to sustain healthy outdoor environments, both in Australia and globally, then individuals and society need to be environmentally responsible. Through your studies in Outdoor and Environmental Studies and your practical experiences, you will develop an understanding of both the beauty and the plight of outdoor environments.

In this section, we will explore the knowledge and skills for safe and sustainable interactions with outdoor environments, including minimum impact strategies for individuals, route planning, tent-site selection, fuel stove usage, navigation and packing a pack.

Minimum impact strategies

We are incredibly fortunate to have a wide variety of outdoor environments at our doorstep. Unfortunately, the vast majority of these have been negatively impacted by humans to some degree. Through the development of monitoring technologies, we have been able to establish clearer links between our interactions and their resultant impacts on outdoor environments. Therefore, there has been a greater focus on the need to adopt minimal impact strategies to promote sustainable and healthy outdoor environments to ensure future generations are able to enjoy the same outdoor experiences that we do today. **Minimal impact strategies** are practices that aim to have as little environmental impact as possible.

minimal impact strategies practices that aim to have as little environmental impact as possible

There are some classic phrases used to encourage and remind people about minimal impact:

- Leave no trace.
- Tread softly.
- Tread lightly.
- Take nothing but photographs, leave nothing but footprints.
- Pack it in, pack it out.
- Kill nothing but time.

You can use these as a simple mantra while out on your trip to help remind you about the idea of minimal impact.

Minimal impact strategies for safe and sustainable interactions with outdoor environments include the following:

- Plan ahead and prepare (the whole point of this section).
- Educate yourself on the specific outdoor environment you are visiting.
- Travel and camp on durable surfaces – avoid walking or pitching your tent on areas that are potentially sensitive.
- Dispose of waste properly – in most cases this will mean taking your rubbish with you. Even when rubbish bins are around, remember the effort required to have someone empty and maintain them. You can help out simply by keeping your rubbish with you and disposing of it at home.
- Leave what you find.
- Collect only fallen dead wood for the fire.
- Do not use soaps in water sources.
- Respect wildlife by not feeding or touching them and not camping on animal trails or near breeding sites.
- Use established toilets where possible.
- Be considerate of other users in an environment.
- Respect Traditional Owners and their Country. Be aware of sacred sites and practices.

CAMPFIRES

There's nothing quite like sitting around a campfire after a hard day of outdoor activities. But from the scars created by fires to the collecting of wood for fuel, campfires can have a major impact on fragile environments. Assuming fires are allowed in a particular place (they aren't allowed everywhere) and assuming it's not a Total Fire Ban day, building a safe fire is a great idea.

Andrew Mannion

Figure 3.2 Many protected places in Victoria are designated as 'fuel stove only' environments, meaning campfires are not allowed – find out what you can and can't do, and take the right gear with you when you go.

LEARNING ACTIVITY 3.1A

Campfires

Make a list of things you could do to reduce the impact of a campfire on an outdoor trip.

Worksheet
3.1a Campfires

TRANSPORT TO AND FROM A TRIP

When thinking about minimal impact, people often focus only on the actual activity part of a trip, such as the walking, canoeing or skiing. But a trip begins at home or school, and ends at home or school. In other words, there's the travel component to and from the trip venue itself, which could be considered to have a greater impact on the overall environment than anything you might do while in the bush. If you're going on a trip to a particular place, you obviously have to get there somehow. Try to think about ways of reducing your travel impacts:

- Use as few vehicles as possible or public transport if it's an option.
- Consider extending the trip to include further walking, cycling or canoeing to help you get to the venue.
- Avoid using eroded and damaged roads.
- Always use made roads and designated car parks where available.
- Think about ways to offset the impact of your travel, perhaps through tree-planting programs at school, at home or in your local community.

TOILETS

Assuming you want to stay healthy and not cause damage to your internal organs, at some point on a trip you'll need to go to the toilet. Toilets, including drop toilets, are often provided at campsites (especially the more popular sites). While some of these can be very smelly, you should use these if provided.

Always bury your waste. If digging a toilet, keep it away from water sources, including creeks and rivers. Your toilet hole, which should be at least 15 centimetres deep, needs to be far enough away from a water source that it doesn't contaminate the water. How far is that? Some people suggest about 15–50 metres, or even 100 metres, from the water source. This will depend on the environment you're travelling and living in. In some places, it won't be possible to get 100 metres away from the water. A simple rule of thumb is: get as far away from water as is reasonable and possible for your group.

CODES OF CONDUCT

code of conduct
a set of rules outlining the responsibilities of, or proper practices for, an individual, group or organisation undertaking a particular activity in the outdoors

stakeholder
a person, group or organisation that has interest or concern in an issue

A formal way of encouraging the use of minimal impact strategies to promote safe and sustainable interactions with outdoor environment is through the establishment of what is known as codes of conduct. A **code of conduct** in relation to the outdoor environment is a set of rules outlining the responsibilities of, or proper practices for, an individual, group or organisation undertaking a particular activity in the outdoors.

The need for a code of conduct often arises from the recognition that unregulated use is leading to significant negative impacts on an environment. The creation of a code of conduct in this case aims to reduce the negative impacts and promote sustainability. Codes of conduct are developed by different **stakeholders**, such as government, commercial and community-based groups working together. Some of these include:

- Department of Energy, Environment and Climate Action (DEECA)
- Department of Primary Industries (DPI)
- Environmental Protection Authority (EPA)
- Four Wheel Drive Victoria and other recreational activity peak bodies or associations
- Outdoor education companies such as Bindaree Outdoor Education and The Outdoor Education Group
- Parks Victoria
- Victorian Government and its associated bodies.

Considerations in relation to codes of conduct include:

- ideal group sizes
- seasonal restrictions
- vehicle restrictions
- best practices for the conduct of the activity
- practices that should be avoided.

Codes of conduct are often associated with minimal impact strategies and are used for both prevention and education. This means there can be specific reasons for their development, as they are used by individuals, commercial groups, outdoor educational organisations and schools. Abiding by a code of conduct is a question of ethics. It is everyone's responsibility to follow the guidelines that have been provided to them. Some codes of conduct include penalties. For example, it may be a condition of membership to a certain group that you adhere to a particular code of conduct. However, in most cases, codes of conduct related to recreational activities are recommendations that cannot be strictly enforced by law.

Leigh Park

Figure 3.3 Code of conduct signs can be displayed in prominent places to promote safe and sustainable interactions with outdoor environments.

Camping codes of conduct

Camping is a recreational activity often associated with other pursuits and is highly relevant to school groups. Therefore, it is important for you to have an understanding of a typical camping code of conduct. For example:

- Use established campsites if they are available.
- Observe and listen to wildlife, but don't touch or disturb it.

- Leave no trace – leave the environment as you found it, or in a better condition (for example, pick up any rubbish you find).
- Use toilet facilities if they are available, or dig a hole at least 15 centimetres deep and as far away from natural water courses as practicable. Bury all faecal waste and paper.
- Use bins if they are available; otherwise take all rubbish and leftover food home with you.
- Have campfires in designated fireplaces and use only fallen dead timber; completely extinguish fires before leaving the area.
- Don't wash dishes in natural water courses; instead wash them at least 50 metres away.

Shutterstock.com/noomcm

Figure 3.4 Typical designated fireplace found at a campsite

Codes of conduct related to specific recreational activities vary depending on the type of activity, location and source. Some typical code of conduct extracts follow.

Alamy Stock Photo/Bjorn

Figure 3.5 Following a boardwalk in an area of sensitive vegetation

Bushwalking code of conduct

When bushwalking:

- stay on the tracks provided; if tracks do not exist, then spread out rather than walking in a single file (as this can create a new track)
- be aware of sensitive areas
- avoid marking the track with tape or rock cairns (as this can confuse other walkers)
- observe and enjoy the wildlife, but do not disturb it
- follow appropriate toileting procedures
- use boot-cleaning stations if they are provided
- carry out what you carry in and pick up any rubbish you see along the way.

Alpine travelling code of conduct

When alpine travelling:

- use campsites provided or choose one away from regular ski trails
- dismantle snow walls and/or shelters you have built
- use a fuel stove rather than burning firewood (as this can leave fire scars on fragile alpine grasses)
- minimise the use of fires in huts during the winter season and restock any wood you have used
- only use huts in emergency situations (as others may need to use these)
- use the ski trails provided and avoid sensitive vegetation.

Worksheet
3.1b Old codes of conduct

Weblink
Australian Alps National Park codes of conduct

LEARNING ACTIVITY 3.1B

Old codes of conduct

Visit the Australian Alps National Park website and search for their codes of conduct. Read some of the scanned versions of old code of conduct brochures for bushwalking, car-based camping, mountain biking, horse riding, river users, snow camping and huts. (These were previously published by the Australian Alps Liaison Committee, but are no longer in print.) Select one of these codes of conduct that relates to a recent practical experience you have been on and answer the following:

1 Were you aware of this code of conduct before your practical experience? If yes, then how did you find out about it?
2 While on the practical experience, did you abide by the code of conduct? Provide examples of how you may and/or may not have followed the code.
3 From your observations, did other people (from your class or members of the public) abide by the code of conduct?
4 Evaluate the effectiveness of this particular code of conduct, based on your experience and observations.

Worksheet
3.1c Create your own code of conduct

LEARNING ACTIVITY 3.1C

Create your own code of conduct

Create your own code of conduct for a recreational activity of your choice. Your code of conduct can be inclusive of some of the information provided in this book, but should also include some updated or new information. Describe the stakeholders or groups of people you would involve in the creation process (if they were available). Provide illustrations, diagrams or a sample of how you would produce and market your code of conduct to ensure it was used effectively.

Route planning

The uniqueness of Australian outdoor environments offers visitors a vast range of places to see and experiences to be enjoyed. Therefore, to make the most of any visit it is important to carefully plan the route to be taken to provide sufficient opportunities to challenge, to enjoy and to learn about our diverse environments. Planning your own route can be a fun and rewarding experience. Whether it's a short activity over several hours or a multi-day journey, there are many factors to consider when planning a route to promote safe and sustainable interactions with the outdoor environment. These include the following:

- Nature of the terrain: Is it likely to be flat or mountainous? Open grasslands or thick forest? Sandy beach or muddy swamp? The sort of terrain you face will affect how easy (or otherwise) it is for a group to move through the environment.
- Possible weather conditions: Although you can't necessarily know beforehand exactly what the weather will be like on a trip, you can make some good deductions about likely conditions based on where and when you're going, and you can be prepared for likely extremes. Some routes may be inaccessible, such as over creeks, during or after wet weather.
- Capabilities of the group: The size, skills, fitness and the experience of the group will determine the planned route, equipment and activities you can include, and the time you'll need to allow.
- Time of year: This will determine the sort of clothing, equipment and food required for the experience, and may affect the availability of water. Some routes are closed at certain times of the year due to weather or proximity to sensitive breeding areas.

Marcia Cross

Figure 3.6 Day trip or many days? Walking on tracks or off track? In summer or winter? Asking questions of you and your group is a key part of trip planning.

- Location of campsites (if needed): Suitable places for campsites are a requirement for a multi-day trip, not just for the comfort and convenience of the group but for the likely sustainability of your journey. It is easy for a large group setting up tents to trample, disturb and otherwise impact sensitive environments – this should be factored in when planning the route.
- Location of water: Water is an absolute requirement. Whether the group carries all their water requirements (very difficult for multi-day trips), relies on water drops (needing careful planning beforehand) or sources water from the environment as the group travels through it (needing a decision about whether or not to treat the water before drinking), this is a major consideration for the planned route.
- Availability of maps: Although maps have become less essential with modern GPS (global positioning system) technology, a suitably prepared leader will have a good map in the case of emergencies or flat batteries in devices. Not all environments have good maps available. This doesn't necessarily mean you shouldn't go to those places, but it does mean you should treat that place with caution.
- Possible escape routes when facing emergencies: The most obvious emergency facing many groups will be bushfire. A group would be foolish to attempt a journey in an environment under bushfire threat, but many fires are sudden and unexpected. A planner should consider ways a group might escape from a fire depending on their location while on a trip. Other emergencies that may require a rapid change of route include floods and injury to one or more members of the group.

TOPOGRAPHY

When route planning, it is necessary to consider the **topography** of the environment. The physical features of a region will offer a range of interesting and picturesque locations to visit, but these may also have a significant influence on the difficulty of a planned route. Map reading and navigation (these are discussed later in this chapter) play a crucial role in route planning. The following topographic features of a region may influence route planning.

topography the landforms or surface features of a region

Spurs

A spur is sometimes described as a tongue of land that descends from the high point of a ridge or mountain top. Spurs are usually easier to travel up than down, since it can be easy to lose the spur (especially near the bottom), and vegetation tends to be less dense on the crest of a spur.

Ridges

A ridge is usually an extended high prominence that connects two (or more) high points or peaks. Ridges are usually excellent features to walk along, since they're generally easy to stay on and their elevation will often make for great views of surrounding areas.

Creeks and rivers

Creeks and rivers are great for helping you to locate your position and for travelling, although it's usually easier to travel downstream with the flow rather than upstream, where smaller creeks and tributaries can become confusing

Steep climbs

If lots of climbing or steep ascents are involved in a route, it's probably better to include them at the start of the trip, or the start of the day, when everyone is still fresh.

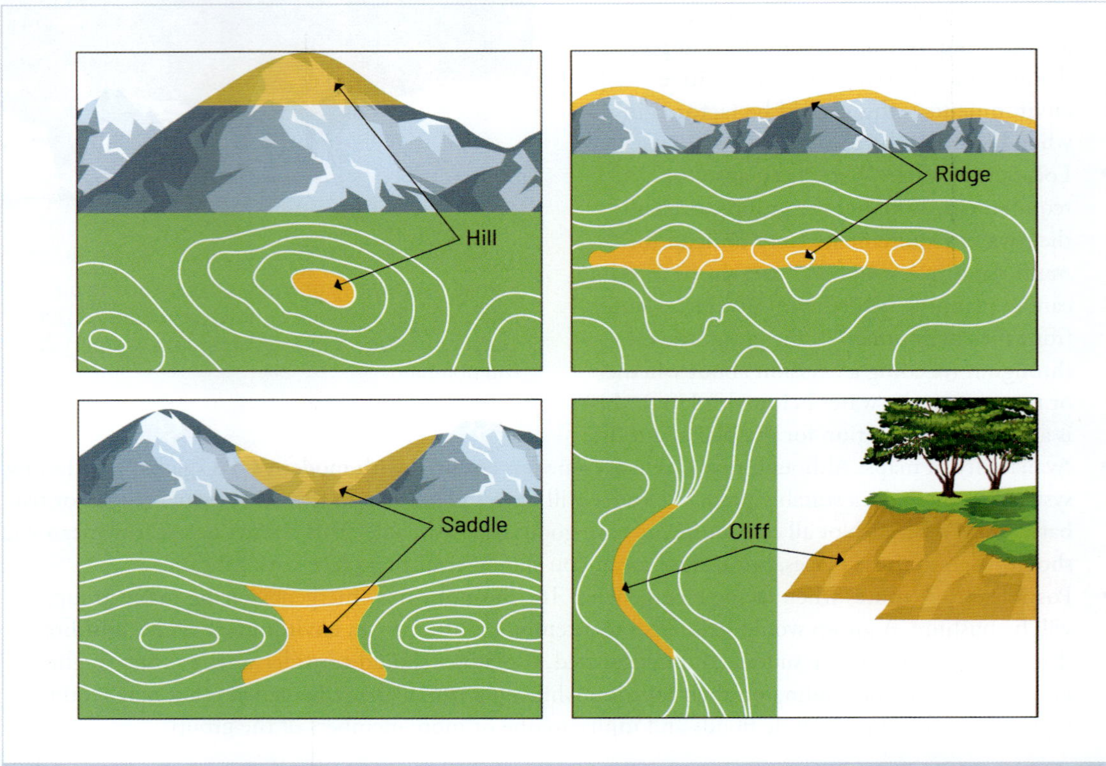

Figure 3.7 Being able to identify landscape features on a map is an important skill in route planning.

PREDICTING WALKING TIME

If planning a multi-day walking journey, you'll need to figure out how long it will take to get from campsite to campsite. Most people tend to underestimate the amount of time it will take them to travel a particular route, assuming that they will be able to move faster than they actually do. This is especially true for larger groups. A good rule of thumb is: the larger the group, the slower they will move.

One way of predicting walking time is to use Naismith's rule, which was developed by Scottish mountaineer William Naismith in 1892. His basic rule breaks into three simple points:

1 Allow one hour of travel time for every five kilometres forward.
2 Allow one hour for every 600 metres of ascent.
3 When you're in a group, calculate all times based on the slowest person.

Other people have subsequently added to Naismith's original idea, adding time for descents (often figured at one hour for every 1000 metres of descent), as well as decreasing the forward distance achieved in one hour if carrying a heavy pack or walking through more challenging bush.

The specifics probably don't matter too much, as you'll need to be flexible in terms of how your group manages a particular route. Remember that as a day progresses, people get tired, so you will need to factor in rest breaks and time for food. Also, remember that an outdoor experience is intended to be fun – you won't want to be slogging it out all day, so it's worth factoring in time to 'chill out' and do other things.

If the group is moving through an environment in a way other than walking (such as canoeing, cycling or skiing), then speeds will probably be faster and travel times will be shorter.

ROUTE PLANNING WEBSITES AND APPS

There are many online resources to help you plan a route through an outdoor environment. Planning your route online allows you to use tools to easily create, review and change your route with a few clicks of a mouse.

Some mapping applications will calculate distances, elevations and estimated travel time, and provide a 3D visualisation of your route. Viewing in 3D brings your route to life. You can fly through the virtual landscape to check out the steepness of the terrain, locations of water sources and possible campsites, and discover alternate routes. You may be able to print maps to use in the field or send the route to a handheld GPS device.

Figure 3.8 3D visualisations using online mapping software

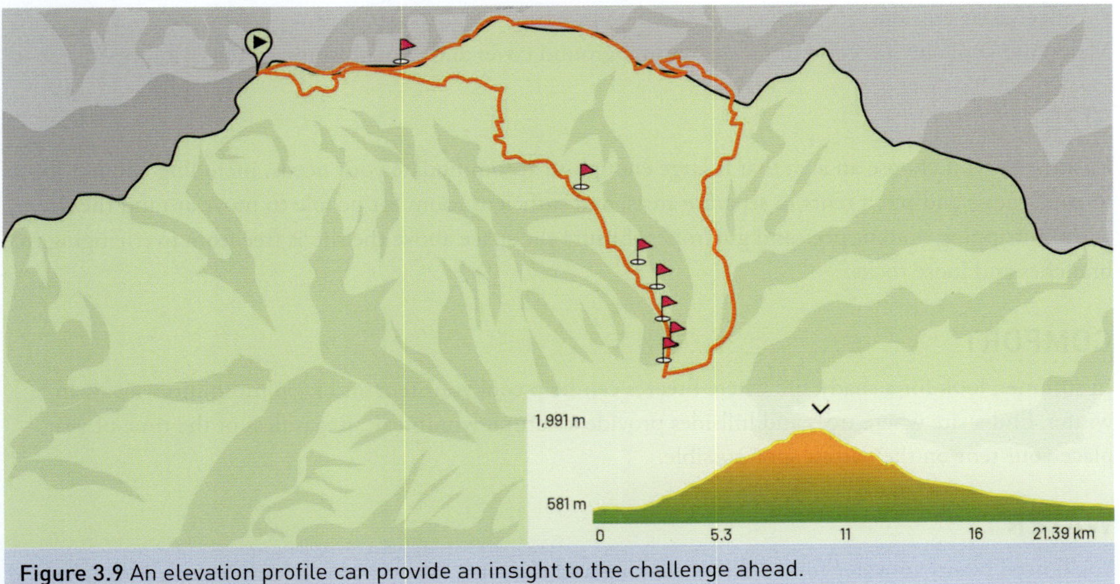

Figure 3.9 An elevation profile can provide an insight to the challenge ahead.

Worksheet
3.1d Planning a route

LEARNING ACTIVITY 3.1D

Planning a route

Use one of the following web-based mapping software or apps to plan a route for an upcoming outdoor experience.

- Avenza Maps
- AllTrails
- CalTopo
- GaiaGPS
- Google Earth
- Maps.me
- Routie

Tent-site selection

When venturing out into an outdoor environment for an extended period of time, getting a good night's sleep is crucial to your enjoyment and, most of all, your safety. Where you choose to set up your tent can make or break a good night's sleep. Consider the following points when selecting your home away from home.

iStockphoto/tsvibrav

Figure 3.10 A good campsite is found, not created. This one is at Johanna Beach on the Great Ocean Walk.

SAFETY

Before establishing your campsite, you should always assess trees in the area and ensure that you pitch your tent away from any trees that are dead or that have limbs that might fall in a storm. Also, avoid camping in areas that could be:

- targets for lightning, such as near lone trees or on mountaintops and high ridges
- affected by flash flooding, such as low-lying areas like valleys, canyons and banks of rivers
- affected by erosion.

Make sure your site has good drainage, natural ground cover and enough sunlight for visibility.

SIZE

Ensure that you choose an area that is large enough to accommodate your needs, including enough space to cook meals and pitch tents in separate areas, and sufficient room for people to move around the site without tripping on tent pegs and guy ropes. Ensure the space above the site is free from overhanging tree branches and loose rocks.

COMFORT

In summer, look for a shady site where breezes can help cool your tent and keep mosquitoes away. In winter, find a site where trees and hillsides provide a natural windbreak. Regardless of the time of year, place your tent on the flattest spot possible.

WATER

You'll need plenty of water for drinking, cooking and cleaning. Water from streams, rivers or lakes should be properly treated before use. In some dry places, you might need to carry all your water to camp.

FIRE

Where fires are not allowed or where wood is scarce, plan on using a camp stove to heat water and cook food. Where fires are permitted, look for a campsite with an existing fire ring. Use firewood that is dead and down; never cut live trees.

PRIVACY

Respect the privacy of others by selecting campsites away from theirs. Trees, bushes and the shape of the terrain can screen your camp from trails and neighbouring campsites. Keep noise down so you won't disturb nearby campers, and respect quiet hours at public campgrounds.

PERMISSION

Check ahead of time with national parks and reserves. They can issue any permits you will need. Never camp on private property without permission.

ENVIRONMENTAL IMPACT

To minimise your impact on the environment, set up your camp on an established campsite wherever possible. If this is not possible, ensure that your campsite is set in an area that won't be easily damaged and avoid clearing objects, such as rocks and logs, or return them to their original position before leaving.

> **FAST FACT**
>
> Many outdoor enthusiasts and adventurers make their own gear, from portable stoves to tents and bivvy bags. For example, a lightweight methylated spirits fuel stove can be made from a recycled soft-drink can.

Fuel stove usage

One of the many advantages of studying Outdoor and Environmental Studies is the range of practical skills that build your independence. One such practical skill is that of cooking your own meal. Whether you are already an avid chef or have never cooked a meal for yourself, throughout your experiences in this subject you will have the opportunity to prepare and cook a range of meals. Cooking on a fuel stove is a skill that takes practice. Once mastered, not only will you be able to prepare delicious meals, but you will also have developed an important skill for life. Using a fuel stove for cooking a meal is a more sustainable choice than an open fire as they emit less carbon into the atmosphere and any wood used for a fire, even if it has already fallen from a tree, naturally provides habitat for wildlife.

There is a sense of accomplishment felt when you have cooked your own nutritious meal on a fuel stove in an outdoor environment without the modern conveniences of running water, electricity and, at times, tables and chairs. The technological development of portable fuel stoves has allowed people to travel further and stay longer in remote and harsh outdoor environments. The remoteness and distance from emergency services highlights the need to follow strict safety and sustainable practices to protect yourself, others and the environment from possible harm caused by the misuse of a fuel stove. Used correctly, fuel stoves are safe and convenient, and provide independence while experiencing outdoor environments.

iStockphoto/O Kemppainen

Figure 3.11 Fuel stoves are capable of cooking any number of delicious meals.

SAFETY AND SUSTAINABILITY GUIDELINES FOR THE USE OF FUEL STOVES

Some common considerations for the safe and sustainable use of fuel stoves include the following.

Always

- Set up the stove on clear, level ground with all non-essential items put away, especially anything that could burn or melt.
- Ensure highly flammable tent fabric is well away and secured to prevent it flapping or blowing against the stove if cooking.
- Ensure there is adequate ventilation.
- Keep pot/pan handles out of the flame to avoid burns when picking them up.
- In a hut use specific cooking areas.
- Refuel a stove outside and away from the cooking area.
- Keep movements near a lit stove to a minimum.
- Keep children well away from lit stoves and liquid fuel bottles.
- Keep unused fuel in a cool place out of direct sunlight.
- Keep fuel bottles separate from water bottles.

Never

- Stoop or bend over a stove while lighting it.
- Cook inside a tent.
- Leave a lit stove unattended.
- Use a stove in a poorly ventilated space (e.g. in a snow cave or igloo). Carbon monoxide is a deadly poisonous, colourless and odourless gas and therefore dangerous. Ensure there is adequate ventilation.
- Refuel a stove near any flame (e.g. another lit stove, candle or campfire).

Source: https://bushwalkingmanual.org.au/food-and-water/stove-safety/

Fuel stoves come in a variety of sizes and shapes, with differing features to suit different applications. Most stoves designed for use in outdoor environments fall into four categories:

- solid fuel stoves
- methylated spirits stoves
- gas stoves
- liquid fuel stoves.

SOLID FUEL STOVES

Solid fuel stoves are compact, lightweight, inexpensive and very quiet. A well-known solid fuel stove is the German manufactured Esbit, which was used extensively throughout World War II. These stoves use a solid fuel tablet that burns a low-medium flame within a stand to place a cooking pot on. However, they have slow cooking times, are very susceptible to wind and give off an unpleasant odour when in use.

Shutterstock.com/MarioJTurco

Figure 3.12 Compact and lightweight, the Esbit stove is a great choice for ultra-lightweight travel.

METHYLATED SPIRITS STOVES

Methylated spirits stoves (or alcohol stoves) are sturdy, lightweight and reliable stoves that are simple to use. This makes them a popular option for less experienced users and school groups. A reservoir of methylated spirits is lit, which in turn causes a vapour to escape from a ring of small holes to create a safe, consistent flame to cook on. A main advantage of methylated spirits stoves is that they are commonly supplied as a self-contained package with a wind shield, set of bowls, fry pan, pot/pan grips (also known as 'spondonicles') and sometimes even a kettle.

Disadvantages of methylated spirits stoves include that due to the low calorific value (heating value), they require more fuel than that of a gas stove, which also means they are not suitable to be used at low temperatures or at high altitudes. Methylated spirits stoves also hold a relatively small amount of fuel, and can only be safely refilled once the stove has cooled down.

One of the most popular methylated spirits stoves is the Swedish-manufactured Trangia. The Trangia is popular due to its light weight, durability and easy-to-use design. The following set of instructions is from the Trangia website. A video on how to use the Trangia stove is also available.

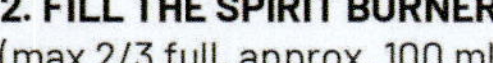
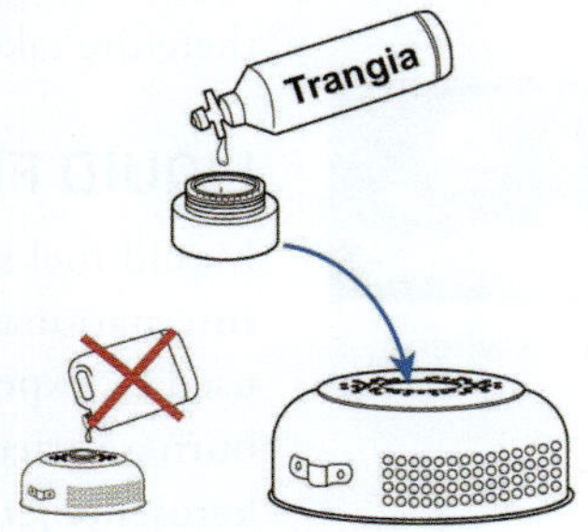
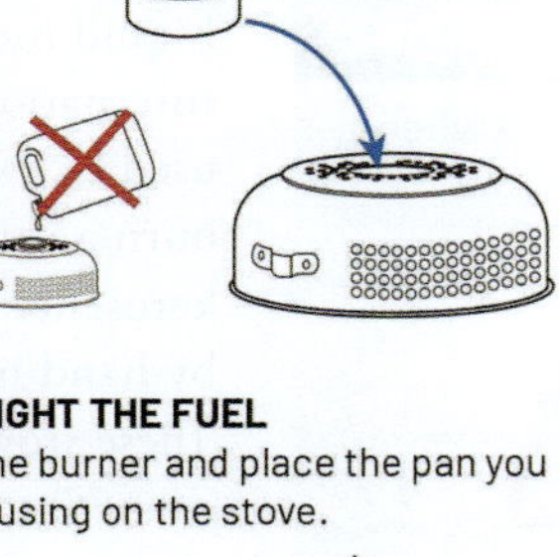
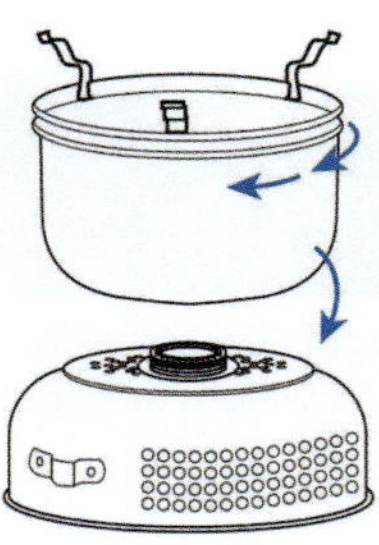
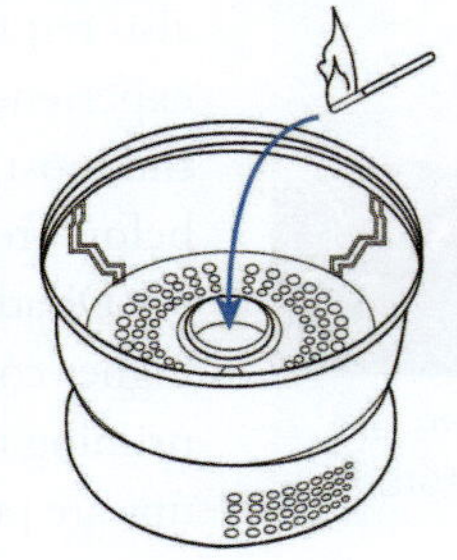
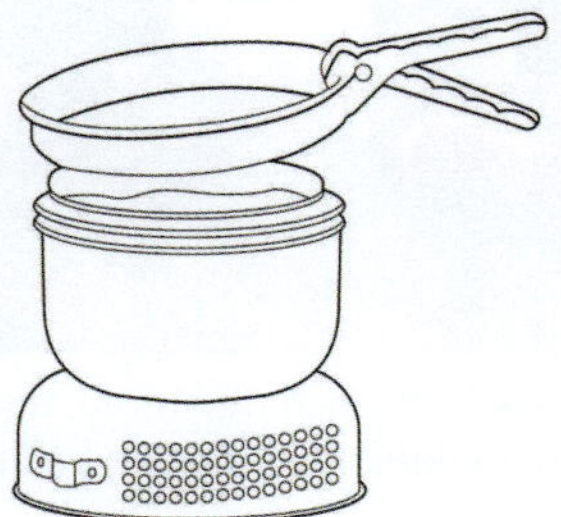
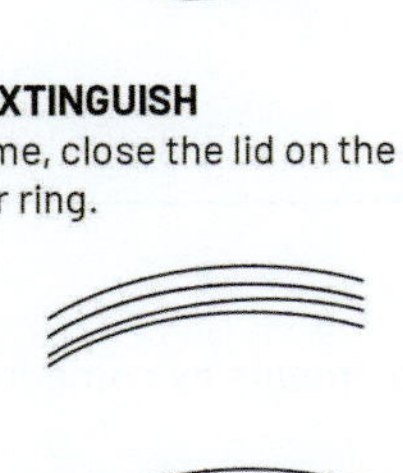

Figure 3.13 User guide for the Trangia stove

iStock.com/gregepperson

Figure 3.14 Gas stoves are a great option when space and weight are limited.

iStockphoto.com/kotangens

Figure 3.15 Once mastered, liquid fuel stoves are fast, fuel efficient and can be used at any attitude.

GAS STOVES

Gas stoves are a commonly used type of fuel stove due to their convenience, ease of use and relatively low cost. Gas stoves use gas canisters full of pressurised gases such as butane and propane to provide an easy-to-control, consistent and fuel efficient flame to cook on.

Most canisters use some mix of butane and propane or isobutane and propane. The screw on/off canister provides a ready-made stable platform to cook on, or the stove may be hung from above if required without the hazard of spilling fuel. Unfortunately, once all the gas is used the canister needs to be disposed of, therefore adding to landfill. Gas stoves don't work as efficiently in cold weather as the canister loses pressure, and therefore takes longer to cook a meal.

LIQUID FUEL STOVES

Liquid fuel stoves such as the MSR International and WhisperLite are commonly used by experienced outdoor enthusiasts. They burn a variety of liquid fuels (shellite, gasoline, kerosene, jet fuel or diesel) that are pressurised by hand into a vapor from a refillable bottle. These stoves typically have a high heat output so are fast and fuel efficient; however, they also require higher technical knowledge and experience to use safely. Liquid fuel stoves are the most efficient for trips at high altitude or below freezing temperatures.

Disadvantages of liquid fuel stoves include higher cost, regular maintenance and, because priming the stove is required, fuel spills and flare-ups are possible.

Worksheet
3.1e OES Master Chef

LEARNING ACTIVITY 3.1E

OES Master Chef

Develop your skills in preparing nutritious meals by competing in a MasterChef-inspired OES challenge.

Get out the fuel stoves and compete against your classmates to cook a range of delicious meals. Challenges could include blind taste test, mystery box, pressure test, and team or teacher challenge.

LEARNING ACTIVITY 3.1F

Fuel stove rankings

Investigate different types of fuels stoves and create rankings based on a set of criteria, which could include:

- ease of use
- safety features
- portability
- time taken to boil 100 mL of water
- amount of fuel used to boil 100 mL of water
- price.

Write up your findings as a report. Include a final ranking and your preferred choice, with justification of the stove that is best suited to your camping requirements.

Worksheet
3.1f Fuel stove rankings

Navigation

 Not all those who wander are lost.

JRR Tolkien 

A key skill when participating in experiences in outdoor environments is the ability to know where you have been, where you are and where you are going – in other words, knowing how to navigate. Being 'geographically embarrassed' (otherwise known as being lost!) is a common fear that stops people venturing into the great outdoors. The skill of navigation through an unfamiliar environment may seem initially complex; however, with practice it becomes second nature and will enhance the safety and sustainability of any outdoor experience.

The level of navigation skills needed will vary depending on the complexity of the outdoor experience. Even a well-defined track may be impacted by a flooded creek or landslide that calls for an alternative cross-country route, requiring skills in the use of a map and compass. With the technological developments of GPS-enabled mobile phone apps that provide easy access to exact locations plus directions to follow, the skill of using a map and compass may seem redundant. However, when a tech fail strikes – that is, no mobile coverage and/or a dead battery – it is necessary to have a Plan B. In these cases, knowing how to navigate using a map and compass can literally be a lifesaver.

To be confident in navigation requires the knowledge and understanding of how to read a map, use a compass and use a GPS device.

Alamy/Cavan Images

Figure 3.16 Caro Ryan, leader with the Sydney Bush Walkers Club: 'One of the most fundamental skills needed for a hike or bushwalk is knowing how to navigate.'

MAPS

A map is a pictorial representation of a specific area. Maps come in all sorts of types, many of which are useful for groups planning an outdoor experience. The two most common maps used on trips are topographic maps and mud maps (sometimes called sketch maps).

Topographic maps

topographic maps maps showing detailed graphical representations by contour and lines of features that appear on the Earth's surface

Topographic maps are designed to represent vertical relief (the third dimension) on a flat two-dimensional sheet. When combined with a compass, these are a powerful navigational tool. All of Victoria is represented by the Vicmap Topographic 1:25 000 map series, and also by the newer online 1:30 000 map series by Land Victoria.

The map scale shows the way the map has reduced the real world in making the representation. For example, 1:25 000 means that 1 centimetre on the map corresponds to 25 000 centimetres in the real world (or 1 centimetre = 250 metres). Most maps will contain a legend that defines the features and symbols used on the map. One important piece of navigation information on a map is the north arrow. This will help you to orient the map so the landscape features line up with their representation on the map. This will help you to identify the direction of travel to your destination.

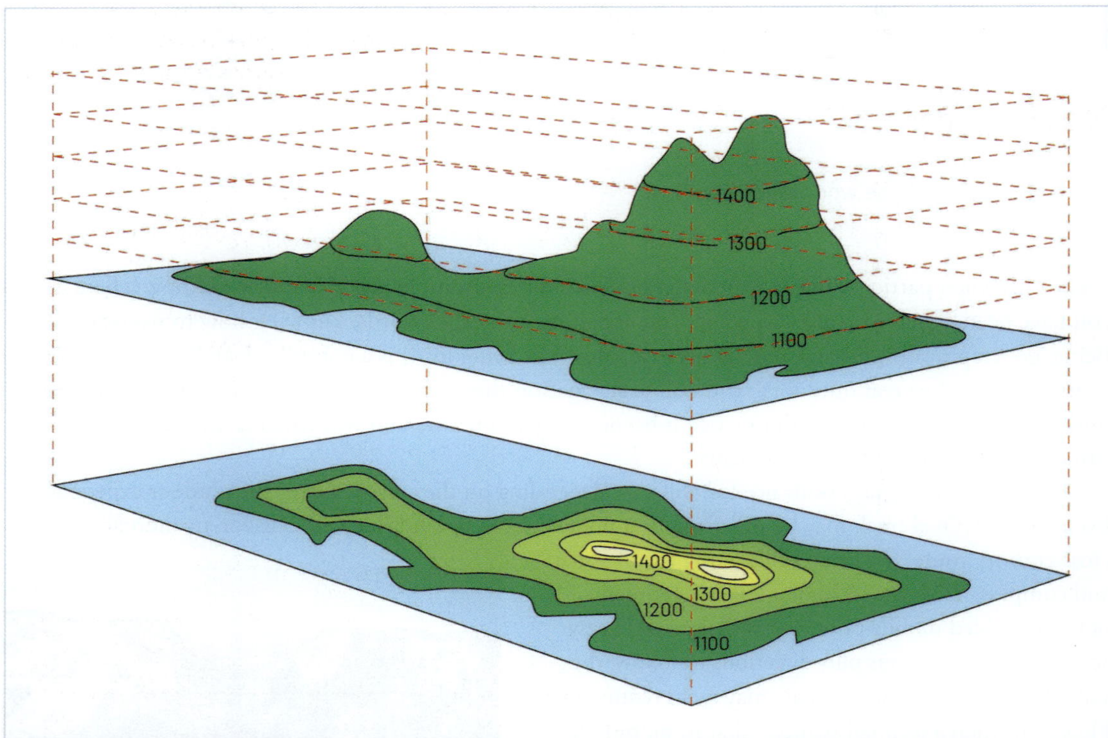

Figure 3.17 Contour lines join points of equal height on the map to illustrate the shape of the land surface or its topography.

Mud maps or sketch maps

mud maps rough sketches of a place or journey that show key features and likely routes, but are not drawn to scale

Literally once drawn with sticks in mud, **mud maps** are rough sketches of a place or journey. They show key features and likely routes, but are not drawn to scale. Mud maps are very useful as a rough plan, but they should not be used for complex navigational purposes.

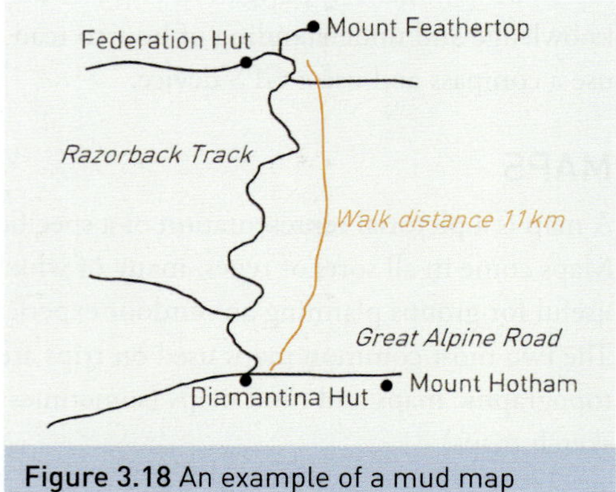

Figure 3.18 An example of a mud map

COMPASS

A compass is a valuable tool that assists you in determining your location and navigating your surroundings. It can be an incredibly useful tool, and if you ever find yourself lost, it can even be critical in finding your way back to safety. Although there are many types and brands of compasses available, most of them have the same key features, including:

- a magnetic needle suspended on a central pivot. The red end of the needle points towards the north pole of the Earth. It is important to note that the direction of magnetic north, where the compass needle points, is not identical to north on a map grid. This is because magnetic north varies across different regions of the world and changes over time.
- a compass wheel or dial that is inscribed with the cardinal points (N-S-E-W) and calibrated every 2 degrees. These markings are used to determine your bearings; that is, the direction from your current location to your destination.
- a base plate or casing of the compass, which is marked with an orienting arrow. The plate is transparent so it can be placed upon the map for orienting purposes. By rotating the wheel so that the red end of the needle aligns with the orienting arrow on the base plate, you can establish your bearing on the compass wheel.

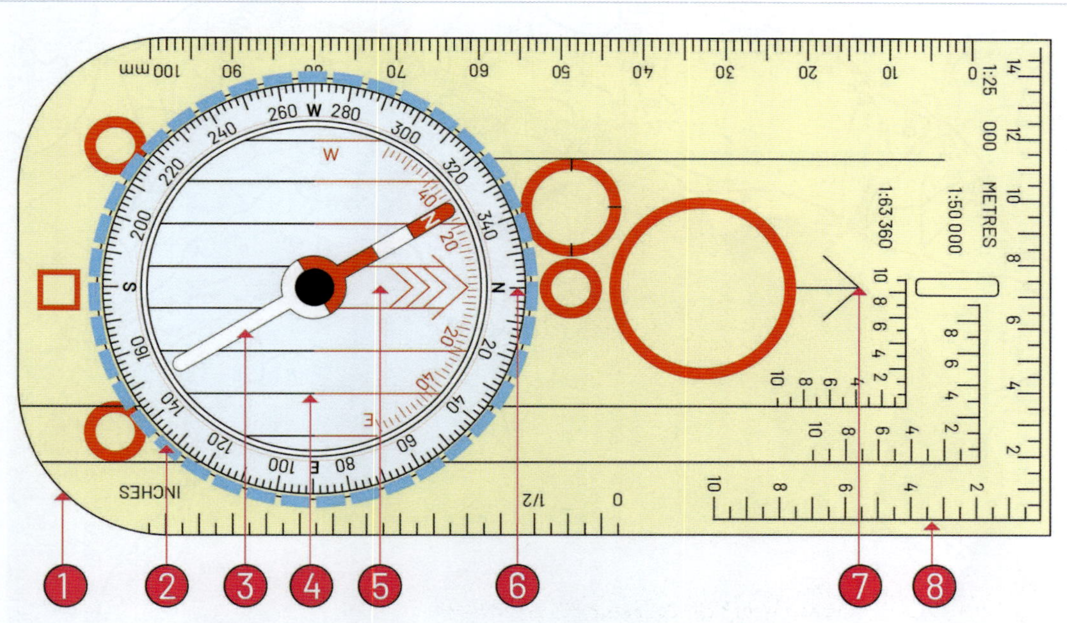

www.snowys.com.au/blog

1. Baseplate - usually clear plastic base.
2. Compass housing - also known as the compass wheel or dial, with a mark preferably every 2 degrees covering 360 degrees, and N-S-E-W (the 'cardinal points').
3. Magnetic needle - red end for north and white for south (this can be black end for north and white for south in some cases).
4. Compass lines - on the bottom of the baseplate (also called 'orienting lines').
5. Orienting arrow - fixed and aligned to north within the compass housing.
6. Index line - extension of the direction of travel arrow.
7. Direction of travel arrow - the big arrow at the end of the baseplate.
8. Map scales 1:25 000, 1:50 000 and metric measurer (known as Romer scales).

Figure 3.19 Elements of a compass

HOW TO USE YOUR COMPASS

This guide will make much more sense if you have the following equipment at hand as you go through each step:

- topographic map
- ruler
- pencil (ideally) or highlighter
- protractor
- eraser

Step 1

Lay your map down somewhere flat and place your compass on top. Draw a line between your starting point and your destination to show the direction of travel. Now, line up the base plate edge with the direction in which you want to go, represented by the highlighted line on the map.

Figure 3.20 Line up the base plate edge with your direction of travel.

www.snowys.com.au/blog o

Step 2

Keeping the base plate edge of your compass in line with your direction of travel, carefully rotate the graduated dial until the N, orienting arrow (5) and compass lines (4) are all pointing in the direction of north on your map. On most maps north is straight up, but make sure you check with the legend on the map that you are using.

Figure 3.21 Rotate the graduated dial to line up with grid lines.

/blo

Step 3

Remove the compass from the map and hold it level out in front of you with the direction of travel arrow (7) pointing straight ahead. Turn your body until the north end of the magnetic needle is directly over the orienting arrow (5), pointing to the 'N' on the dial.

The direction of travel arrow is now pointing in precisely the direction you want to travel in order to reach your destination. The easiest way to use your compass now is by using the 'snap or sight a line' method. While holding your compass in the direction of travel, look up and sight a landmark or object that is not too far away and is in the direction you want to travel.

Put your compass away or hang it around your neck and start walking towards the landmark or object that you spotted/sighted. Once you reach it, repeat the process by holding your compass as before making sure it is still set according to your map, sighting another landmark and walking to it.

Continue doing this until you reach your destination.

Figure 3.22 Line up the needle with north on the graduated dial.

Yury Sevryuk / Alamy Stock Photo

www.snowys.com.au/blog

GPS

global positioning system (GPS) a satellite-based navigation system that can determine accurate and precise locations and give directions to other destinations

Global positioning system (GPS) technology was developed in the 1970s by the US Department of Defense. These days it can be found in mobiles, cars, watches and dog collars for mapping and navigation purposes.

It works by utilising a network of 24 satellites that orbit 20200 kilometres above the earth. A GPS device needs to receive signals from at least four of these satellites in order to determine your position.

The GPS device requires a clear view of the sky, so being stuck in a canyon with high cliff faces on either side or lost in a heavily forested area may render the device useless, which highlights the importance of the back-up of a map and compass.

Mobile apps that utilise the phone's built-in GPS have become powerful navigation devices; however, traditional handheld GPS units offer the advantages of being more rugged and having superior satellite reception, and many offer easily replaceable batteries that last for days.

Handheld GPS devices come in a range of different sizes and offer features from basic navigation to advanced features such as cameras, barometers, altimeters, wireless file transfer, direct messaging and direct social media uploading.

As each GPS model has different settings and features, it is necessary to consult the user guide for specific instructions. However, all GPS devices will have basic functions such as displaying your location, displaying trip data such as distance covered and distance to destination, and the ability to navigate from point to point.

COMPASS OR GPS?

Which is better for navigation: a compass or a GPS? Mobile phones use orbiting satellites to precisely determine your location on the Earth. Many groups will have access to purpose-designed GPS devices for navigation. Table 3.1 compares the advantages and disadvantages of each approach.

Table 3.1 Comparing a GPS with a compass

	GPS	Compass
Advantages	• Will accurately locate your position and able to track your movements • Can be used to enter a route before a trip, which can then be followed while on a trip • Will determine altitude, which can be useful when on a trip in the Alps	• Very tough and durable • Doesn't need power to operate
Disadvantages	• Satellites can take some time to be found, and in thick vegetation may not be able to be located • Usually needs batteries (although solar chargers can be used) • Might be damaged by water, cold or shock	• Requires knowledge about how to use them – they're not as intuitive as a GPS

It is always worth taking a compass (assuming you have someone in your group who can use it), but a GPS is a good idea as well.

Do you even need a navigation device? While many trips will follow tracks and signs, it's always a good idea to be prepared for many possibilities. Remember, you're planning for safe and sustainable interactions with outdoor environments.

Packing a pack

Hiking packs have evolved significantly since the early days of the swaggie rolling up his bedroll, tying it with rope and slinging it over his shoulder. Modern-day hiking packs are lightweight, have an internal frame, are made from waterproof materials and have an adjustable harness for support and comfort.

The hiking pack is one of the most crucial pieces of equipment to get right. When choosing a hiking pack, ensure you try it on with some weight to ensure it suits your body type. Eighty per cent of the pack's weight should rest on your hips for maximum support and comfort. Speak to the experts – outdoor shop staff usually have a wealth of knowledge to suggest a pack that is suitable for you and for your upcoming outdoor experience.

Learning to be self-sufficient in terms of food and equipment comes with practice and experience. Packing has become an art form, with many books written and websites dedicated to the art of packing a pack. Many an adventurer has also studied packing techniques of other people, and will probably refine their packing strategy each time they pack. While learning from others is a great start, packing becomes a very personal task as our preferences for different types of food and equipment will influence how a bag is packed. There is no one right way to pack a pack, but there are common strategies to help distribute weight, protect fragile items and allow easy access to water bottles, snacks, rainwear and first aid kits.

iStockphoto/MaximFesenko

Figure 3.23 A well-packed pack allows more time to enjoy the journey.

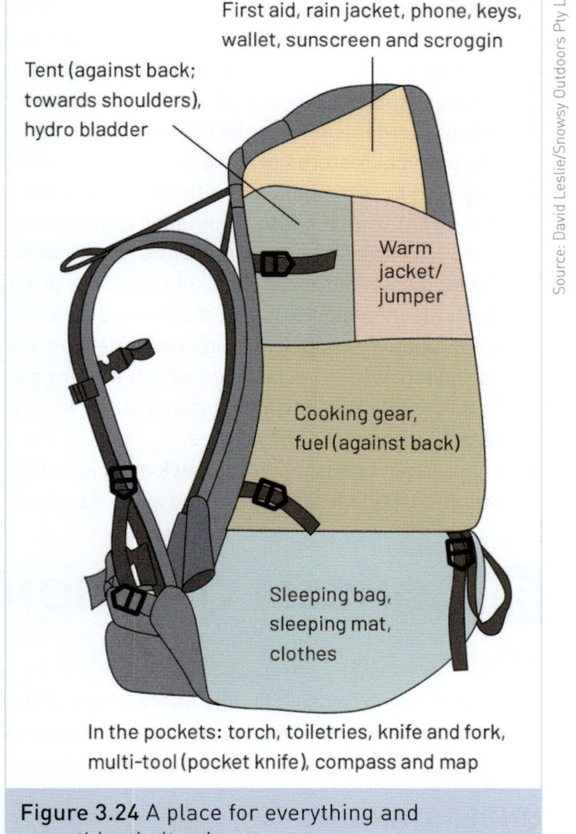

Source: David Leslie/Snowsy Outdoors Pty Ltd

Figure 3.24 A place for everything and everything in its place

Considerations when packing a pack

- Waterproof your equipment – sleeping bag, sleeping mat and clothes should be completely waterproof. You can do this with a pack liner, dry bags or garbage bags.
- Pack your sleeping bag first – first in, last out. This eliminates the possibility of having to get your sleeping bag out in the rain when trying to locate other equipment or food.
- Pack heavy items close to your back – to keep a low centre of gravity to improve the stability of your pack, pack heavy gear such as stoves low and close to your back. Depending on the size of your tent, it should also be packed close to your back to enable your waist belt to take most of the weight.
- Fill the gaps – gaps between equipment in your back can be filled with rolled up clothes and food items.
- Water – this is best carried in a hydro bladder with a tube enabling access to the water without having to take your pack off. Pack water close to your back, which allows your shoulders to take the weight and improve your balance.
- Easy access – pack your rain jacket, jumper, first aid kit, medications and snacks close to the top of your pack to allow easy access. Alternatively, these items can be packed in external storage pockets, along with a torch, mobile, GPS device, pocket knife, sunscreen and toilet paper.

Packing a pack is not an exact art. Expect to pack and repack a few times to ensure a well-balanced pack. Usually a neat pack is a well-packed pack. Because the unexpected can happen, it's best to know exactly where everything is in your pack. Once packed, take time to correctly adjust your pack to suit your body shape and know how to adjust 'on the fly' so you are able to change the weight distribution between your hips and shoulders. Carrying a pack takes time to get used to, but it also enables you to access more remote locations, challenge yourself and develop personal attributes such as resilience and independence.

3.1 KEY CONCEPTS

- Before embarking on an experience in the outdoors, appropriate planning and preparation is crucial.
- Not being appropriately prepared with the correct food, equipment and knowledge can result in expensive and dangerous search-and-rescue operations.
- Experiencing outdoor environments safely requires an understanding of how to plan and conduct sustainable outdoor experiences in outdoor environments.
- Through the development of monitoring technologies, we have been able to establish clearer links between our interactions and their resultant impacts on outdoor environments.
- Minimal impact strategies are practices that aim to have as little environmental impact as possible.
- Codes of conduct are sets of rules outlining the responsibilities of, or proper practices for, an individual, group or organisation undertaking a particular activity in the outdoors.
- The physical features of a region will offer a range of interesting and picturesque locations to visit, but may also have a significant influence on the difficulty of a planned route.
- When venturing out into an outdoor environment for an extended period of time, getting a good night's sleep is crucial to your enjoyment and – most of all – your safety.
- Cooking on a fuel stove is a skill that takes practice. Once mastered, not only will you be able to prepare delicious meals but you will also have developed an important skill for life.
- A key skill when participating in experiences in outdoor environments is the ability to know where you have been, where you are and where you are going – in other words, knowing how to navigate.
- To be confident in navigation requires the knowledge and understanding of how to read a map, use a compass and use a GPS device.
- There is not one right way to pack a pack, but there are common packing strategies to help distribute weight, protect fragile items and allow easy access to water bottles, snacks, rainwear and first aid kits.

3.1 CONCEPT QUESTIONS

Worksheet
3.1 Key concepts

REMEMBERING

1. What is meant by sustainable interactions with outdoor environments?
2. Annotate a diagram of a hiking pack with advice on the placement of specific equipment.
3. Discuss three reasons for learning how to use a map and compass for navigation.

UNDERSTANDING

4. Why is it necessary to implement minimal impact strategies when participating in experiences in outdoor environments?
5. Explain what is meant by the saying, 'Take nothing but pictures, leave nothing but footprints and kill nothing but time'.

APPLYING

6. Working together with the students in your class, create a cookbook of your favourite recipes that are nutritious, delicious and suitable for overnight experiences in outdoor environments.
7. Develop your own instructional booklet on how to safely and sustainably use a fuel stove. In addition to a step-by-step set of instructions, you may wish to include your own annotated photos.

EXTENSION CHALLENGE

8 Create a brochure outlining the minimal impact strategies for individuals to promote safe and sustainable interactions with a specific outdoor environment. Use the following headings in your brochure:
- Preparation for outdoor experience
- Travel to outdoor environment
- Camping and campfire
- Hygiene and waste
- Wildlife
- Respect for Traditional Owners

3.2 BASIC FIRST AID SKILLS

KEY KNOWLEDGE

- basic first aid skills, including blister management, small wounds, snake bite, severe bleed treatment, Cardio-Pulmonary Resuscitation (CPR) and immobilisation techniques

KEY SKILLS

- demonstrate and apply basic first aid knowledge and skills
- use appropriate skills for safe and sustainable interactions with outdoor environments

3.2.1 FIRST AID

For all medical emergencies, dial triple zero (000) immediately and ask for an ambulance.

You may have heard of the saying 'Plan for the worst and hope for the best'. Considering situations where you may need to apply basic first aid is not pleasant; however, the reality is accidents and incidents occur and it may be necessary to support a person who has an injury. Outdoor environments can be unpredictable; therefore, it is important we prepare for a range of possibilities. Knowing what to do if someone injures themselves is crucial to the success of any outdoor experience.

This section covers a range of basic first aid skills, including blister management, small wounds, snake bite, severe bleed treatment, Cardio-Pulmonary Resuscitation (CPR) and immobilisation techniques. A first aid course will explain and provide essential training and practice of the actions to be taken for each of the listed basic first aid skills.

iStockphoto/warrengoldswain

Figure 3.25 Would you know what to do?

FIRST AID TRAINING

This information is NOT a substitute for first aid training. The following organisations offer comprehensive training to learn how to recognise an emergency and to provide basic first aid until professional help arrives.

- Australian Red Cross (www.redcross.org.au)
- Emergency First Aid (www.emergency.com.au)
- Royal Life Saving Society Australia (www.lifesavingvictoria.com.au)
- St John Ambulance Victoria (www.stjohnvic.com.au)
- Wilderness First Aid (https://www.wildernessmedicine.com.au/)

Basic life support

BASIC LIFE SUPPORT ROUTINE

The basic life support actions to take when providing first aid are referred to as DRSABCD. These steps are crucial for a patient who has a serious injury or illness.

- DANGER – Check for Dangers
- RESPONSE – Check Response
- SEND – Send for help
- AIRWAY – Clear and open the Airway
- BREATHING – Check for normal Breathing
- COMPRESSION – If trained, give required chest Compressions and rescue breaths
- DEFIBRILLATE – Automatic External Defibrillator (AED). Most unlikely to be available on a bushwalk.

Figure 3.26 The first aid kit location should be clearly visible and known to all group members.

Providing First Aid

When first aid is needed for an ill or injured group member, a person with first aid training should take control of providing first aid to them until they are either recovered or medical aid has been obtained. Other support tasks can be delegated to other group members (e.g. calling emergency services or keeping a history [log] of the patient's condition).

Do not attempt to move or carry an injured person, except when necessary to get them to a safe location and where sufficient people to assist are available. Moving an injured person is very difficult and can cause further injuries.

Calling for Assistance

If the injury is serious:

- Call 000 and ask for an ambulance if there is mobile or satellite phone reception.
- Activate emergency communication device if you have one (e.g. PLB, SPOT or InReach).

Recording First Aid Care

The first aid care provided to a sick or injured person should be recorded to:

- monitor and track the patient's condition
- provide medical personnel with relevant information about the patient and treatment provided. Written records should be clear, concise and factual observations. This information should be provided to medical professionals when requested and otherwise be kept confidential.

Caring For an Ill or Injured Person

Care provided for an ill or injured person until help arrives should include:

- prevention of further injury or aggravation of existing injuries
- protection from weather
- providing maximum comfort possible – regulate their temperature and provide ground insulation if needed
- providing adequate food and water as appropriate
- providing assistance with personal hygiene
- providing companionship and reassurance
- explaining what actions are being taken
- aiming to care for the whole person, not just the injury.

Group Actions

The leader can delegate roles and tasks to other experienced members and should avoid trying to do everything. Allow time to think. 'Hasten slowly' and consult other experienced members. Group members can provide the following assistance:

- caring for the ill or injured person (as above)
- recording first aid care
- ensuring other group members are safe and together
- putting up tents/shelter for warmth or shade
- getting a stove or fire going for a hot drink in cooler weather
- collecting water
- contacting emergency services. Call for help sooner rather than later.

Blister management

Blisters rarely need medical attention, unless they are severe, recurrent, caused by burns or are due to an underlying infection.

Figure 3.27 Blister prevention: if you become aware of a localised 'hot' area on your foot, stop and tape the area immediately.

If you have a blister, it's best to avoid bursting it if possible, as this can lead to infection and slow down the healing process. However, if you do need to pop a blister, start by cleaning the area thoroughly with soap or disinfectant. Then, use a sterilised needle (heated over a flame) to gently puncture the blister. Allow the fluid to drain out slowly, and let the roof of the blister collapse down onto the base. This will act as a natural dressing, which you can cover with a bandage or adhesive plaster. Do not remove the roof of the blister, as this can delay healing and increase the risk of infection. If the blister fills up again over the next day or two, you may need to repeat the process.

The following advice is from the Victorian Government's Better Health Channel website. Suggestions for treating a simple friction blister are listed in the following extract.

BLISTER MANAGEMENT

- If the blister has burst, don't peel off the baggy skin pocket – let your body heal the area in its own way and in its own time.
- Apply antiseptic and a dressing or sticking plaster to the area to protect it and keep it free from dirt or irritants.
- Don't use tape alone for the dressing, as removing the tape may rip the roof skin off the blister.
- Change the dressing daily and re-apply antiseptic.
- Avoid 'folk remedies' like applying butter or vinegar. These don't work.
- See your doctor or other health professional for treatment if:
- the blister is caused by a burn, scald or severe sunburn
- the blister starts weeping pus (yellow or green, sometimes smelly, fluid)
- the area becomes increasingly swollen or inflamed
- you suspect the blister is infected
- you develop multiple blisters without any preceding skin injury.
- Blister prevention strategies include the following:
- Wear properly fitted shoes.
- Choose moisture-wicking socks (socks that draw sweat away from your feet) or change socks twice daily if you have sweaty feet, as wet socks cause friction and rubbing.
- Wear 'sports socks' when exercising or playing sports.
- If you become aware of a localised 'hot' area on your foot, stop your sport and tape the area immediately.
- Apply a foot spray deodorant to reduce sweating and the risk of fungal infection.
- Change damp socks promptly, as wet socks can drag against the skin.
- Wear heavy-duty work gloves when using tools such as shovels or picks.
- Protect yourself against sunburn with clothing, hats and sunscreen lotions.
- Avoid unnecessary skin contact with chemicals.
- Be careful when dealing with steam, flames or objects that radiate heat (such as electric stovetops).

Small wounds

The two broad categories of small wounds include abrasions, when the skin is scraped or broken, and incised wounds, when the skin is cut.

ABRASIONS

An abrasion refers to a break or scrape in the surface layers of the skin that can often contain particles of dirt. Abrasions occur more commonly on areas of the body that are bony and thin-skinned, such as knees, elbows and ankles.

The following advice is from the Victorian Government's Better Health Channel website. First aid treatments for an abrasion are listed in the following extract.

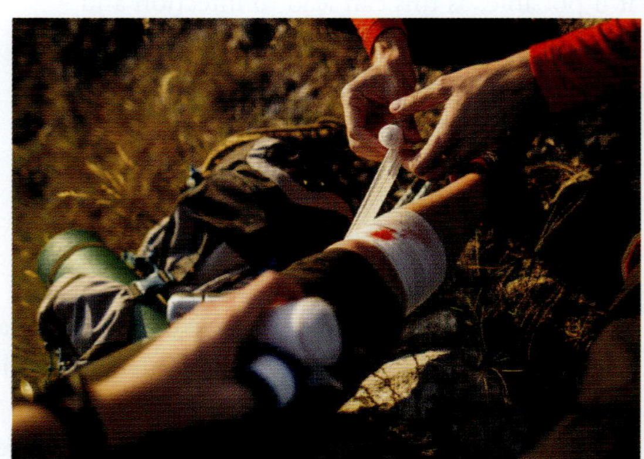

Figure 3.28 Cover the cleaned wound with an appropriate non-stick sterile dressing.

TREATING AN ABRASION

- Clean the wound with a non-fibre shedding material or sterile gauze, and use an antiseptic such as Betadine. If there is embedded dirt, Savlon may be used as it contains an antiseptic and a surfactant to help remove debris. Rinse the wound after five minutes with sterile saline or flowing tap water [if available].
- Don't scrub at embedded dirt, as this can traumatise the site even more.
- Cover the cleaned wound with an appropriate non-stick sterile dressing.
- Change the dressing according to the manufacturer's instructions (some may be left in place for several days to a week). If you reapply antiseptic, wash it off after five minutes and then redress the wound.

INCISED WOUNDS

Incised wounds are cuts that occur when a sharp object, like a knife or glass, slices into the skin. The severity of the wound depends on the depth and location of the injury, and if any underlying blood vessels are affected, significant blood loss may occur. If an artery is severed, it becomes a medical emergency due to the muscular action of the blood vessel, which can pump out the entire blood supply from the wound in a matter of minutes.

The following advice is from the Victorian Government's Better Health Channel website (betterhealth.vic.gov.au). First aid treatment for severe bleeding is shown in the following extract.

TREATING AN INCISED WOUND

- Remove clothing around the site for easier access.
- Apply pressure directly to the wound with your hands to stem the blood flow.
- Cover the wound with a sterile dressing, if possible, and continue to apply direct pressure (bandage firmly).
- Try to raise the injured area above the level of the person's heart.
- Don't remove existing dressings if they become saturated with blood, but instead add fresh dressings over the top.
- Seek urgent medical attention. You may need to call an ambulance if you cannot stop the bleeding, are feeling faint, sweaty or dizzy.

Snake bite

Approximately 3000 people are bitten by snakes each year in Australia. You need to be particularly vigilant for any signs of snakes in the warmer months (spring and summer) and particularly around wetlands and waterways. If you see a snake, keep calm and move away from them. The tiger snake is the source of the most bites and cause of envenoming in Victoria. Other dangerous snakes to watch out for when outdoors in Victoria include the copperhead snake, red-bellied black snake and common or eastern brown snake (also known as the common brown snake).

The following advice for recognising the symptoms and treating a snake bite is from the Victorian Government's Better Health Channel website.

(Left to right, Top to bottom) iStockphoto/Ken Griffiths; iStockphoto/Ken Griffiths; iStockphoto/Ken Griffiths; Shutterstock.com/Sylvie Lebchek; Shutterstock.com/Ken Griffiths

Figure 3.29 The four most common dangerous snakes in Victoria are: (A) copperhead snake (B) tiger snake (C) eastern brown snake (D) red-bellied black snake.

TREATING A SNAKE BITE

Signs of a snake bite are not always visible. In some cases, a person may not even feel a snake biting them. Symptoms may not be obvious for an hour or more after being bitten. It is important to act quickly if you suspect someone has had a snake bite. Depending on the type of snake, signs and symptoms may include:

- immediate or delayed pain at the bite site
- swelling, bruising or local bleeding
- bite marks (usually on a limb) that may vary from obvious puncture wounds to scratches that may be almost invisible
- swollen and tender glands in the groin or armpit of the bitten limb
- faintness and dizziness
- nausea and vomiting
- headache
- abdominal pain
- oozing of blood from the bite site or gums
- double or blurred vision
- drooping eyelids
- difficulty in speaking or swallowing
- limb weakness or paralysis
- difficulty in breathing
- occasionally, initial collapse or confusion followed by partial or complete recovery.

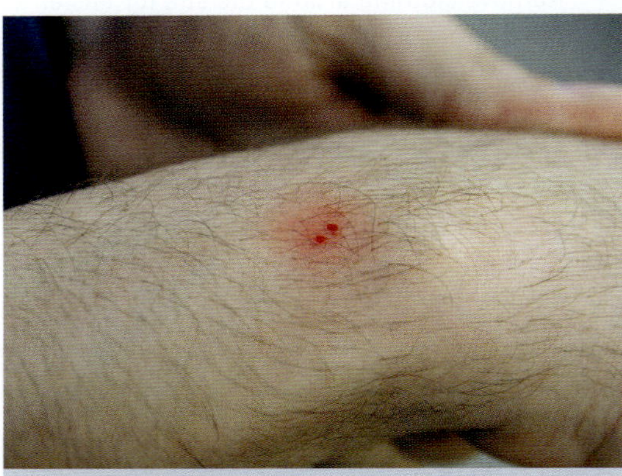

Shutterstock.com/marketlan

Figure 3.30 Snake bite marks on the skin — these might be obvious puncture wounds or almost invisible small scratches.

Not all Australian snakes are venomous, but you should follow basic first aid techniques, just in case.

1 Follow DRSABCD.
2 Call triple zero (000) and ask for an ambulance.
3 Lie the person down, ask them to keep still and reassure them.
4 Remove jewellery on the limb (if you are able to do so).
5 If the bite or sting is on a limb, apply a broad pressure bandage over the site. Mark the site where the bite is on the bandage with an X.
6 Apply a further elasticised roller bandage (10–15 cm wide), starting just above the fingers or toes and moving upwards on the bitten limb as far as can be reached. Apply the bandage as firmly as possible to the limb. You should be unable to easily slide a finger between the bandage and the skin.
7 Immobilise the bandaged limb using a splint.
8 Keep the person lying down and completely still (immobilised).
9 Continue to check the colour, temperature and feeling in their fingers or toes.
10 Write down the time of the bite and when the bandage was applied. Try to mark the location of the bite site (if known) on the skin with a pen, or photograph the site. Do not wash venom off the skin or clothes because it may assist identification.
11 Stay with the person until medical aid arrives.
12 Do not make the person walk to the rescue vehicle – bring the vehicle as close to them as you can.
13 Do not cut a bite to release the venom or try to suck the venom out of the wound.

Severe bleed treatment

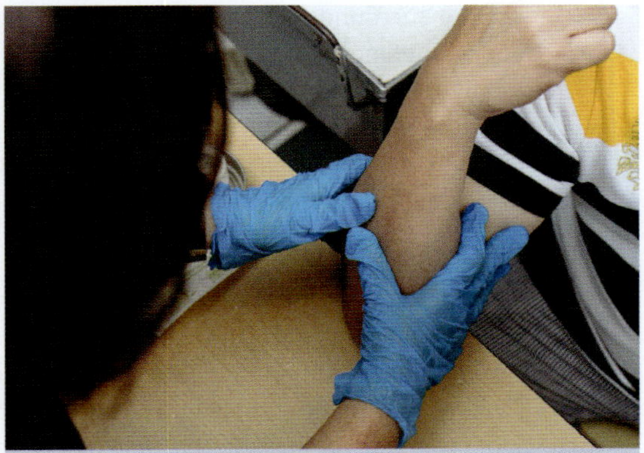

Bleeding is the loss of blood caused by injuries to the skin, which range from abrasions and small incisions to deep cuts and the loss of limbs. Internal bleeding, which is not visible, is caused by injuries to the body that can also range from minor bruising of the skin to massive bleeds that can be life-threatening.

Figure 3.31 Apply direct pressure to the bleeding wound and raise the injured area above the level of the person's heart

When treating external bleeding, apply pressure to the wound and, if possible, raise the injured limb above the level of their heart. For internal bleeding, the first step is to call for emergency medical assistance immediately. While waiting for help to arrive, the person should be kept warm and still. If the person is unconscious, they should be placed in the recovery position. It is important not to give the person anything to eat or drink, as they may require surgery and anaesthesia. In all cases of severe bleeding, it is essential to monitor the person's condition closely and be prepared to perform CPR if necessary.

The following advice for treating bleeding is from the Victorian Government's Better Health Channel website.

TREATING BLEEDING

Severe External Bleeding

Even a small injury can result in severe external bleeding, depending on where it is on the body. This can lead to shock. In medical terms, shock means the injured person no longer has enough blood circulating around their body. Shock is a life-threatening medical emergency. First aid management for severe external bleeding includes the following:

- Check for danger before approaching the injured person. Put on a pair of gloves (nitrile ones, if available).
- If possible, send someone else to call triple zero (000) for an ambulance.
- Lie the person down. If a limb is injured, raise the injured area above the level of the person's heart (if possible).
- Get the person to apply direct pressure to the wound with their hand or hands to stem the blood flow. If the person can't do it, apply direct pressure yourself
- You may need to pull the edges of the wound together before applying a dressing or pad. Secure it firmly with a bandage.
- If an object is embedded in the wound, do not remove it. Apply pressure around the object.
- Do not apply a tourniquet.
- If blood saturates the initial dressing, do not remove it. Add fresh padding over the top and secure with a bandage.

Internal Bleeding – Visible

The most common type of visible internal bleed is a bruise, when blood from damaged blood vessels leaks into the surrounding skin. Some types of internal injury can cause visible bleeding from an orifice (body opening). For example:

- bowel injury – bleeding from the anus
- head injury – bleeding from the ears or nose
- lung injury – coughing up frothy, bloodied sputum (spit)
- urinary tract injury – blood in the urine.

Internal Bleeding – Not Visible

It is important to remember that an injured person may be bleeding internally even if you can't see any blood. An internal injury can sometimes cause bleeding that remains contained within the body; for example, within the skull or abdominal cavity. Listen carefully to what the person tells you about their injury – where they felt the impact, for example. They may display the signs and symptoms of shock (pale skin, sweating, nausea). In the case of a head injury, they may display the signs and symptoms of concussion. Therefore, it is important to ask the right questions to collect the relevant information.

Symptoms of Concealed Internal Bleeding

The signs and symptoms that suggest concealed internal bleeding depend on where the bleeding is inside the body, but may include:

- pain at the injured site
- swollen, tight abdomen
- nausea and vomiting
- pale, clammy, sweaty skin
- breathlessness
- extreme thirst
- unconsciousness.

Signs and symptoms specific to concussion (caused by trauma to the head) include:

- headache or dizziness
- loss of memory, particularly of the event
- confusion

- altered state of consciousness
- wounds on the head (face and scalp)
- nausea and vomiting.

Internal Bleeding is a Medical Emergency

First aid cannot manage or treat any kind of internal bleeding. Prompt medical help is vital.
Suggestions include the following:

- Check for danger before approaching the person.
- If possible, send someone else to call triple zero (000) for an ambulance.
- Check that the person is conscious.
- Lie the person down.
- Cover them with a blanket or something to keep them warm.
- If possible, raise the person's legs above the level of their heart.
- Don't give the person anything to eat or drink.
- Offer reassurance. Manage any other injuries, if possible.
- If the person becomes unconscious, place them on their side. Check breathing frequently. Begin cardiopulmonary resuscitation (CPR) if necessary.

Spread of Disease Through Bleeding

Some diseases can be spread through open wounds. Remember:

- If possible, wash your hands with soap and water before and especially after administering first aid. Dry your hands thoroughly before putting on gloves.
- First aid kits contain gloves. Always put on gloves beforehand if available. If not, improvise.
- Do not cough or sneeze over the wound.

Better Health 'Conditions and Treatments' © Copyright State of Victoria 2021

Cardio-Pulmonary Resuscitation (CPR)

Cardiopulmonary resuscitation (CPR) is a critical emergency technique used when someone is not breathing normally or has experienced cardiac arrest. It involves a combination of chest compressions and rescue breathing (mouth-to-mouth) to keep the blood circulating and provide oxygen to the body until specialist medical treatment is available. CPR can be lifesaving because there is typically enough oxygen in the blood to keep the brain and organs alive for a few minutes, but without circulation from CPR, the oxygen is not delivered to these vital areas.

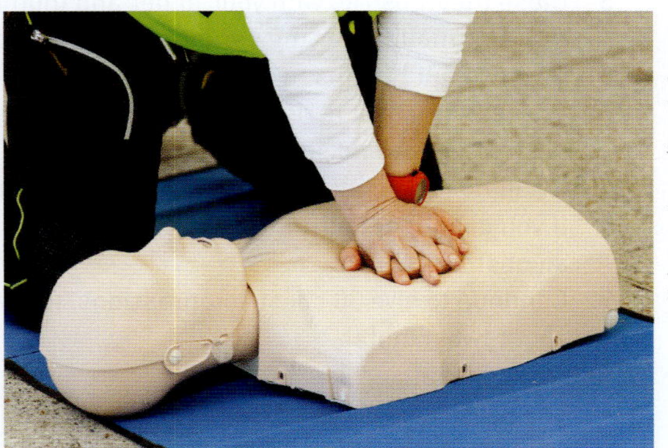

Shutterstock.com/Antonio Gravante

Figure 3.32 Aim for five sets of 30 chest compressions and two breaths in about two minutes.

Basic emergency first aid and CPR knowledge is important as it can give someone a chance of survival when otherwise there would have been none. While there is no guarantee that someone will survive from CPR, it can make a significant difference. Without CPR, the brain can become injured due to a lack of oxygen in just a few minutes.

While the steps for CPR are the same for adults and older children, the technique for babies and young children (0–5 years) is slightly different.

The following advice on performing CPR is from the Victorian Government's Better Health Channel website.

GIVING CPR

How To Give CPR To Adults And Older Children

Aim for five sets of 30 chest compressions and two breaths in about two minutes. If you can't do mouth-to-mouth, stick with continuous compressions at a rate of approximately 100 per minute. To perform CPR on adults and older children:

1. **A** = Airways – open the person's airways (nose, mouth and throat) and check they are clear. Remove any blockage (such as vomit, blood, food or loose teeth). Don't spend too much time doing this – CPR is your main priority.
2. Make sure the person is in a neutral position (such as on their back). Gently tilt their head back and lift their chin.
3. **B** = Breathing – are they breathing normally? If so, gently roll them onto their side (known as the recovery position).
4. If they are not breathing or breathing abnormally (such as grunting or gasping for air) you will need to give them CPR.
5. **C** = CPR consists of two techniques – 30 chest compressions and two breaths of mouth-to-mouth.

Chest Compression Steps

Place the heel of one hand on the lower half of the person's breastbone (in the middle of their chest).

1. Place your other hand on top of your bottom hand and grasp your wrist. Or you may like to interlock your fingers – depending on what feels comfortable.
2. Keep your arms straight and press down on their chest by one-third of their chest depth.
3. Release the pressure – this counts as one compression.

Mouth-To-Mouth On Adults And Older Children

1. If the person is not breathing normally, make sure they are lying on their back on a firm surface.
2. Open the airway by tilting the head back and lifting their chin.
3. Close their nostrils with your finger and thumb.
4. Put your mouth over the person's mouth and blow into their mouth. Make sure no air is leaking.
5. Give two full breaths to the person (this is called 'rescue breathing').
6. Check their chest is rising and falling. If this is not happening, tilt their head back, pinch their nostrils tightly and seal your mouth to theirs.
7. If still no luck, check their airway for any obstruction.
8. If you cannot get air into their lungs, go back to chest compressions – this may help shift an obstruction.
9. Continue to give 30 chest compressions, followed by two breaths (30:2). Aim for five sets of 30:2 in about two minutes.

Keep going until:

- The person becomes responsive. They may begin to move, breathe normally, cough or talk. Then put them in the recovery position (onto their side).
- The ambulance arrives and the paramedics take over.

CPR can be tiring. If you need a break, ask someone else to assist with minimal disruption. Rotate the person performing compressions every two minutes. If you find mouth-to-mouth difficult, continue with chest compressions until medical help arrives. They can still save someone's life.

Can It Be Dangerous Doing Chest Compressions?

Sometimes, people will have their ribs broken by chest compressions. This is still better than the alternative of not receiving CPR. If this occurs, pause and reposition your hands before continuing or get someone else to take over.

Automated External Defibrillators (AED)

A defibrillator is a lifesaving device that treats someone who is having a cardiac arrest. It can analyse abnormal heart rhythms and send an electric shock or pulse to get the heart to return to its normal pumping rhythm. CPR must be continued until an automated external defibrillator (AED) becomes available. The pads must be attached to the skin and the machine turned on. There are different types of AEDs and some are available in public places (such as shopping centres and schools). It is important to follow the prompts on the AED. Do not touch the person during analysis or shock delivery.

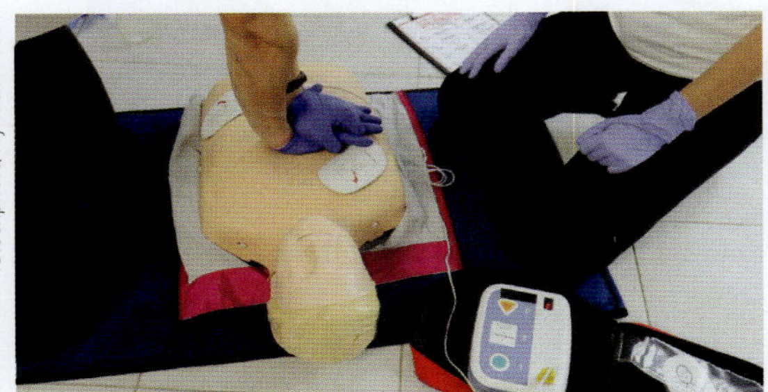

iStockphoto/MyndziakVideo

Figure 3.33 AEDs are easy to use – voice prompts tell you what to do.

Immobilisation techniques

Immobilisation techniques may be required for a range of injuries such as for fractures, possible neck and spinal injuries, ankle injuries and snake bite. In the case of a snake bite, pressure bandaging and immobilisation slows the movement of venom through the lymphatic system. The lymphatic system is a network of tubes that drains fluid (lymph) from the body's tissues and empties it back into the bloodstream.

IMMOBILISATION TECHNIQUES

Bandaging the wound firmly tends to squash the nearby lymph vessels, which helps to prevent the venom from leaving the puncture site. If you don't have any bandages on hand, use whatever is available such as:

- clothing
- stockings
- towels.

Immobilising the limb is another way to slow the spread of venom, sometimes delaying it for hours at a time. This is because the lymphatic system relies on muscle movement to squeeze lymph through its vessels.

Pressure-immobilisation is recommended for bites from all Australian snakes, funnel web spider bites and blue-ringed octopus stings.

Do not use pressure-immobilisation first aid for jellyfish stings, stonefish stings or bee, wasp and ant stings.

How to apply a pressure-immobilisation bandage:

1 Keep the person who has been bitten as still as possible. Lie the patient down.
2 Apply a firm wide elasticised bandage over the bitten area.
3 Bandage the entire limb (fingers to shoulder or toes to the hip) – the bandage should firm.
4 Apply a rigid splint to the limb (piece of wood, branch or rolled-up paper).
5 Keep the patient still until the arrival of the ambulance for transport.

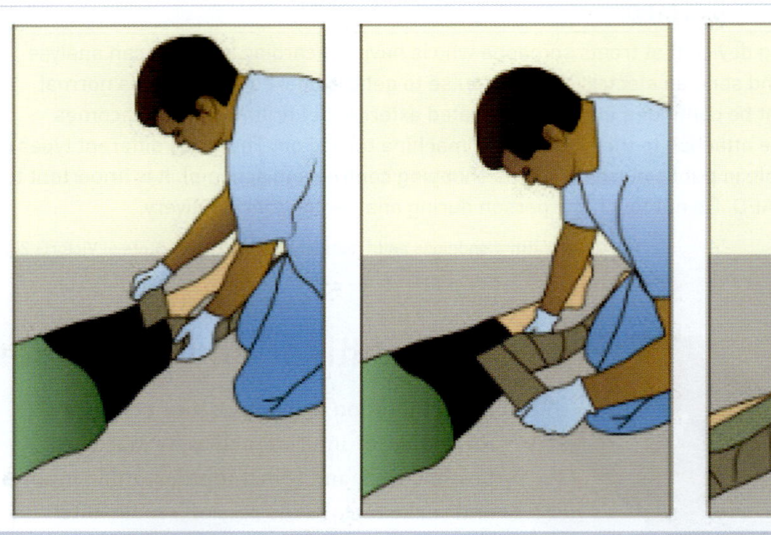

Figure 3.34 Applying a pressure bandage and immobilising with a splint

Don't tourniquet or cut bites or stings

In the past, a tight tourniquet was recommended as the best method to cut off blood flow and prevent the circulation of venom through the body. This is no longer advised.

Worksheet
3.2a First aid

LEARNING ACTIVITY 3.2A

First aid

1 Choose one aspect of first aid to research from the following list:
   - anaphylaxis
   - asthma
   - bites and stings
   - bleeding
   - burns
   - care of an unconscious patient
   - diabetes
   - ear and eye injuries
   - epilepsy
   - head injuries
   - heart attack and stroke
   - hyperthermia
   - hypothermia
   - lower limb fractures
   - management of an accident scene
   - poisons
   - shock
   - sprains and strains
   - upper limb fractures.

2 For your chosen aspect of first aid, investigate and produce a classroom display that includes the following:
   a the nature of the condition and possible causes, where appropriate
   b appropriate emergency treatment
   c any first aid or other equipment that should be carried as a precaution or which could be used in an emergency
   d any preventative measures (short- or long-term) that could be taken to try to prevent the condition occurring.

LEARNING ACTIVITY 3.2B

Emergency case studies and role-plays

As a class, look at some examples of wilderness and outdoor emergency case studies such as those found on the National Outdoor Leadership School website. Discuss some of the ways you might respond in these scenarios. Use some of these scenarios in role-plays about dealing with emergencies in the outdoors.

Worksheet
3.2b Emergency case studies and role-plays

Weblink
National Outdoor Leadership School case studies

3.2 KEY CONCEPTS

- **For all medical emergencies dial triple zero (000) immediately and ask for an ambulance.**
- Outdoor environments can be unpredictable, so it is important we prepare for a range of possibilities. Knowing what to do if someone injures themselves is crucial to the success of any outdoor experience.
- First aid courses provide comprehensive training to learn how to recognise an emergency and to provide basic first aid until professional help arrives.
- When first aid is needed for an ill or injured group member, a person with first aid training should take control of providing first aid to them until they are either recovered or medical aid has been obtained.
- Written records should be clear, concise and factual observations. This information should be provided to medical professionals when requested and otherwise be kept confidential.
- The basic life support flow chart specifies actions to take when providing first aid and is referred to as DRSABCD.
- When treating a blister, resist the temptation to burst it. You could cause an infection or hinder your body's healing process.
- For small wounds, clean the wound with a sterile gauze, use an antiseptic and cover with a non-stick sterile dressing.
- For snake bite, follow DRSABCD, lie the person down, ask them to keep still and apply a broad pressure bandage over the site and upwards on the limb as far as possible.
- To manage severe bleeding, apply direct pressure to the wound, maintaining the pressure using pads and bandages, and raise the injured limb above the level of the heart if possible.
- CPR helps keep the blood circulating and delivers oxygen to the body until specialist treatment is available.
- When providing CPR, aim for five sets of 30 chest compressions and two breaths in about two minutes.

Pressure bandaging and immobilisation slows the movement of venom through the lymphatic system. To splint a limb, you can use any hard objects (such as sticks) gently tied or bandaged to the limb.

3.2 CONCEPT QUESTIONS

Worksheet
3.2 Key concepts

REMEMBERING

1 Discuss three ways that outdoor environments can be unpredictable.
2 Name and describe the actions to take when providing first aid as a part of the basic life support referred to as DRSABCD.
3 List five actions that should be followed when caring for an ill or injured person until help arrives.

UNDERSTANDING

4 Why is it important not to scrub at embedded dirt when treating an abrasion?
5 Explain what is meant by the saying, 'Plan for the worst and hope for the best'.
6 Compare the signs, symptoms and treatments of severe external and internal bleeding.

APPLYING

7 Annotate a diagram of a foot with descriptions of strategies to prevent blisters.

8 As a class, create a list of possible accidents that could occur during an outdoor experience. In groups, respond to a range of mock accidents to demonstrate and apply your basic first aid knowledge and skills. Critique each other on the effectiveness of the response.

EXTENSION CHALLENGE

9 Working together with the students in your class, create a pocket-sized pictorial first aid book that outlines basic first aid skills, including blister management, small wounds, snake bite, severe bleed treatment, Cardio-Pulmonary Resuscitation (CPR) and immobilisation techniques. Alternatively, you could create a personal first aid kit for use when out exploring the outdoors on your own or with friends and family.

3.3 EQUIPMENT REQUIRED TO SAFELY EXPLORE OUTDOOR ENVIRONMENTS

KEY KNOWLEDGE

- equipment required to safely explore outdoor environments

KEY SKILLS

- explain the use of equipment designed to improve the safety of participants in outdoor experiences
- identify hazards, analyse risks and suggest controls for an outdoor experience in a chosen outdoor environment

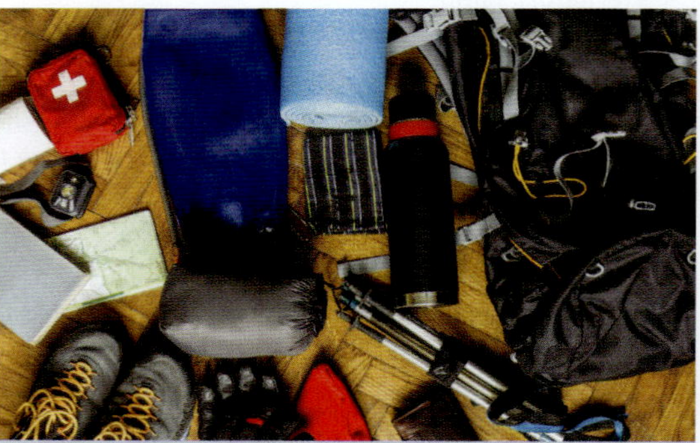

iStockphoto/gubernat

Figure 3.35 Today's preparation promotes tomorrow's success.

You may have heard of the saying about the 6Ps: 'Proper Prior Planning Prevents Poor Performance', or a similar variation. The success of any outdoor adventure is very much reliant on how well prepared all participants are for the experience. In particular, planning pays off when (ironically) things don't go to plan. The unforeseen injury or break in equipment can be attended to with greater confidence if planning has addressed these possibilities.

When planning for safe and sustainable interactions in outdoor environments, there are many factors to consider. In this section we focus on the use of equipment designed to improve the safety of participants in outdoor experiences. We also discuss the need for adequate food and water supplies.

The aim of this section is not to provide an all-inclusive equipment list for your next adventure but rather to focus on equipment designed specifically to improve the safety of participants. The equipment discussed falls within the following categories: first aid, communication, food and water, activity-specific and personal protective equipment.

3.3.1 FIRST AID

For all medical emergencies dial triple zero (000) immediately and ask for an ambulance.

In the previous section, we outlined a range of basic first aid skills, and many of these treatments required the use of specific equipment. All groups exploring outdoor environments should carry an appropriate number of first aid kits. The contents of any first aid kit should be determined based on the size and needs of the group, and the location and the nature of the activities undertaken on the experience.

FIRST-AID KIT

The Royal Flying Doctor recommends the following items be included in a first aid kit:

- first aid booklet
- triangular bandages – non-elastic bandages used for slings, to hold splints in place and to restrict movement
- crepe ('conforming' or elastic) bandages of varying widths – these elastic bandages are used to create pressure, hold dressings in place, reduce swelling and provide some support
- non-adhesive (non-stick) dressings of varying sizes – best used for covering burnt or abraded (scraped or grazed) skin
- disposable gloves (medium and large), preferably made of non-latex material
- thermal blanket – can be used when a patient is suffering from shock
- adhesive tape (2.5 cm wide – preferably a permeable tape such as Micropore)
- resuscitation mask or face shield
- medium combine dressing pads (9 × 20 cm) – used to help control bleeding and reduce the risk of infection
- large combine dressing pads (20 × 20 cm) – used to help control bleeding and reduce the risk of infection
- adhesive dressing strips (bandaids) – used for minor cuts and skin injuries; never use adhesive dressings on burnt or abraded skin
- medium gauze dressing (7.5 × 7.5 cm)
- four sterile tubes of saline solution (minimum 10 mL) – to flush out eyelashes, dust, sand or similar particles from the eye; never attempt to remove an object that is embedded in or has penetrated an eye – in such an instance, seek urgent medical attention
- one pair of scissors
- one pair of tweezers
- plastic bags of varying sizes
- notepad and pencil.

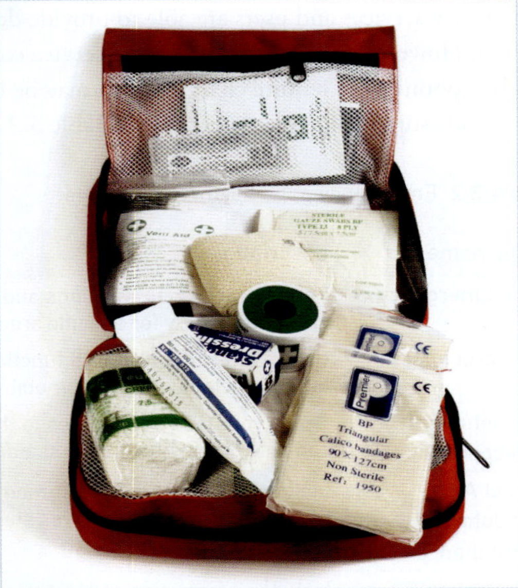

It is not recommended to store medications in your first aid kit. Store your first aid kit in a cool dry spot, and ensure everyone in the group knows where it is. Some items in the first aid kit may have use-by dates, so make sure you're checking these items and replace them when necessary.

The Royal Flying Doctor Service of Australia

Alamy Stock Photo/Trevor Chriss

Figure 3.36 Ensure all members of the group know the location, contents and how to use the first aid kit.

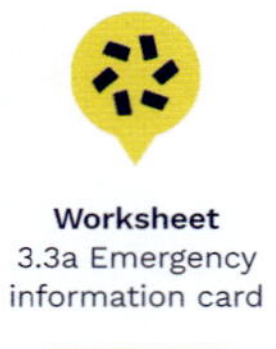

Worksheet
3.3a Emergency information card

LEARNING ACTIVITY 3.3A

Emergency information card

Create a wallet sized, waterproof information card that can be located easily in case of an emergency. Include the following information:

- Name:
- Date of birth:
- Contact details:
- Emergency contact details:
- Medical information (allergies/asthma etc.):
- Doctor contact details:

3.3.2 COMMUNICATION

Communication technology is rapidly improving, but there is no one perfect device that can ensure your safety in outdoor environments. However, there is a range of communication devices that can be used to contact help in case of an emergency. These include mobile phones, satellite phones and personal locator beacons (PLBs).

Mobile phones

Mobile phones are highly effective means of communication as messages are instantly received by emergency services and users are able to provide details of the incident and receive advice until help arrives. However, mobile phones require service coverage that may not be available in remote areas. Mobile phones also require charging, and may be lost or damaged. There is a large range of apps available to provide support in an emergency; see Table 3.2 for some examples.

Table 3.2 Emergency apps for mobile phones

App name	Description
Vic Emergency	Provides access to warnings and incidents for fires, floods, storms, earthquake, tsunamis and water safety for specific areas
Emergency+	Uses a smartphone's GPS functionality to help a triple zero (000) caller provide the critical location details needed to mobilise emergency services
St John First Responder	Sends your GPS coordinates to the operator when you call triple 000 for an ambulance. It can also link in qualified first aiders to deliver critical first aid to patients in need
First Aid – St John's Ambulance	Includes 13 of the most common and critical conditions that may require immediate lifesaving first aid; it features step-by-step emergency first aid information with large clear images for each step
Australian Bites and Stings	Includes information and pictures of venomous spiders, aquatic creatures, jellyfish and creepy crawlies, and first aid instructions on what to do if bitten or stung by a venomous creature
Fires Near Me Australia	Provides national information on current bush fires and other incidents, as well as warnings about fires that may affect where you live
Get Prepared	Helps you connect with your key support people, create an emergency plan and create checklists to follow in the case of an emergency
AirRater	Helps people with asthma, hay fever and other lung conditions to better manage their symptoms and improve their quality of life; it allows people to prepare and plan travel activities, taking into account air quality in the area
Vic Traffic	The official traffic app of VicRoads Victoria, which provides information about traffic incidents and alerts

Healthdirect Australia

Satellite phones

A satellite phone operates using a network of orbiting satellites instead of traditional mobile towers to send and receive calls or messages. As long as the user has a clear view of the sky, no matter how remote an area is, they will still stay connected. Satellite phones are expensive to purchase, are heavier and bulkier than a mobile phone, and must stay charged to work effectively.

Figure 3.37 A satellite phone enables emergency communication in remote areas.

Personal locator beacons

A personal locator beacon (PLB) is a dependable and efficient method of notifying emergency services about a life-threatening situation. Once activated, a PLB transmits its location and device identification to emergency communication satellites and commercial aircraft. However, the devices are limited to one-way communication, and do not convey any details about the nature of the emergency. PLBs work in a similar way to an emergency position indicating radio beacon (EPIRB), a waterproof buoyant device that when activated emits a radio distress signal via satellites to help search and rescue authorities find a vessel at sea in distress. EPIRBs are registered to a boat, while PLBs are designed for use by an individual.

An alternative to a PLB is a satellite communication device such as the Garmin InReach or Spot. These palm-size devices use the satellite network to send and receive text messages, track and share your journey and, if necessary, trigger an SOS alert to contact emergency services.

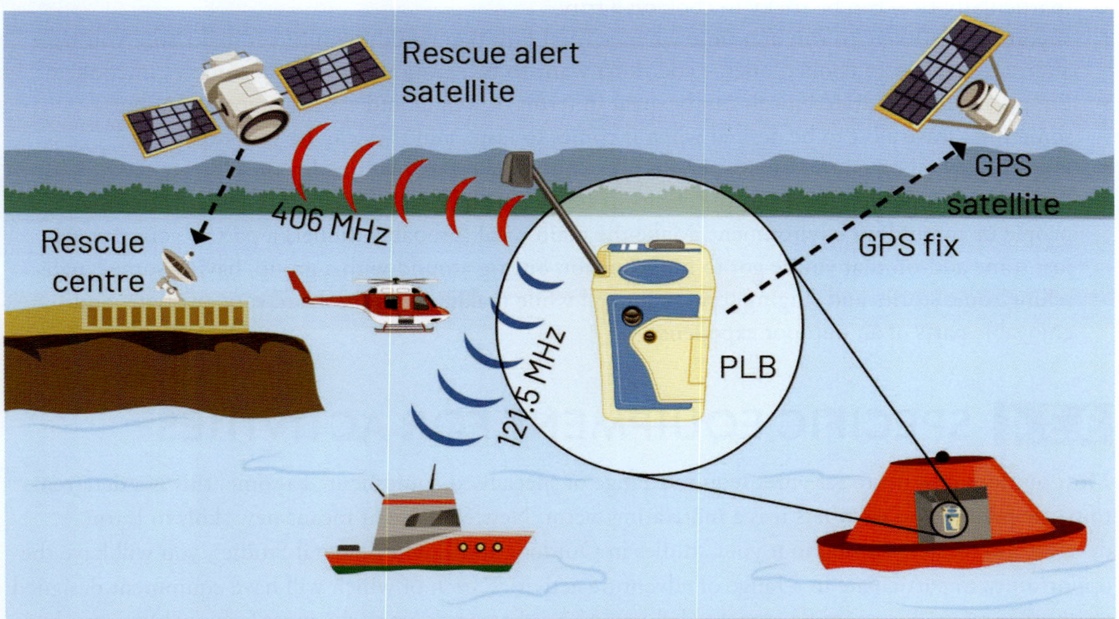

Figure 3.38 PLBs should only be activated in a situation of grave and imminent danger – when faced with a life-threatening situation.

FAST FACT

Two hours of walking with a backpack can use between 800 and 1000 extra calories per hour, so be sure to include more food than you would normally eat when you plan your menu.

3.3.3 FOOD AND WATER

Figure 3.39 Staying hydrated by drinking small amounts every 10 to 15 minutes keeps you focused and improves stamina.

Shutterstock.com/Daxiao Productions

Eating sufficient amounts of nutritious food is necessary to ensure your body is capable of producing the energy required to safely participate in the planned outdoor experience. Many students have made the mistake of not adjusting for the required increase in food and water intake (compared to a normal day) due to the greater energy demands of outdoor activities. Running out of energy while mountain biking or surfing could have significant consequences.

There are a range of considerations when selecting appropriate food and sufficient water for an outdoor experience. These include:

- You need food that is nutritious, light to carry, tasty, easy to prepare and easy to clean up afterwards. This can be a tall order, so it is often worth experimenting in the ease and comfort of home rather than trying out your ideas for the first time on a trip, only to find it doesn't work or that you (or your group) don't like it.
- Stay hydrated by drinking water slowly in the hours before participating in physical activity. You should consume about one litre of water for every hour of activity, and more if it's hotter and/or participating in more intense activity. Think about having both water and a sports drink available to ensure an adequate intake of electrolytes.
- Food that requires water to cook it (such as noodles, pasta and rice) can be very light and is useful to take on trips, but make sure you will have access to enough water, and note that some of these (particularly rice) can be tricky to cook on a trip.
- If your trip is likely to run at a time when a Total Fire Ban could be in force, make sure you have alternative meals to cooked ones – you don't want to be eating your pasta or noodles uncooked.
- Pre-packaged dehydrated meals can be great in terms of weight and ease of preparation, but sometimes leave a bit to be desired when it comes to taste, and they can be very expensive.
- Hygiene is vital on an outdoor experience, so don't forget to allow time for the clean-up.
- The preparation, the eating and the clean-up are all part of the social aspect of being with other people in an outdoor environment. Make the main meal (probably dinner) a part of your day, not just some add-on that you've got to find time for. Sitting around with a group, having some laughs, telling some stories and singing some songs, all while making your meal, are very satisfying and enjoyable parts of an outdoor experience.

3.3.4 SPECIFIC EQUIPMENT FOR ACTIVITIES

Most outdoor adventure activities require a range of specialised equipment. To some, this is a deterrent to participation, and to others it is a motivating factor. New equipment means new skills to learn, develop and master. Throughout your studies in Outdoor and Environmental Studies, you will have the opportunity to participate in a range of adventure activities, each of which will have equipment designed to improve safety. For example, goggles, helmets and wrist guards aren't the most fashionable items, but they do reduce the risk of injury when snowboarding. Similarly, the use of a G-Board in a beginner's surf lesson isn't the easiest way to carve a wave, but it is a far safer option than the pointy-nosed fibreglass alternative. The following presents some of the equipment designed to improve the safety of participants in a range of outdoor experiences.

Figure 3.40 Safety equipment for a range of outdoor activities

Personal protective equipment

Protecting yourself against the elements – whether it be heat, wind or cold – is crucial to your safety when exploring outdoor environments. Ever-changing technological developments in materials and design mean new types of equipment that boast ways to increase our safety, comfort and enjoyment in outdoor activities.

Clothing

In the early 1900s, out of necessity, the puffer jacket was invented exclusively for hard-core mountaineers to survive the extreme conditions of Mt Everest. Then, in the 1970s, along came Gore-Tex, a waterproof, breathable fabric that had many outdoor enthusiasts rushing to the specialist shops to spend hundreds on a new jacket. Nowadays, however, you are just as likely to meet someone wearing a Patagonia Gore-Tex hardshell or North Face puffer jacket at your local shopping centre as you are hiking the snow-capped peaks of Victoria's High Country. Thanks to clever marketing, hiking gear has become high fashion, and unfortunately it has also attracted the high price tag to match.

When exploring outdoor environments, we may experience extremes in temperature. It is important that we carefully consider our clothing to best protect ourselves from the serious medical conditions of hyperthermia and hypothermia.

Hyperthermia is an abnormally high body temperature, and heat stroke is a life-threatening form of this. It occurs when the body is overwhelmed by heat and unable to control its temperature. Symptoms include rapid pulse, lack of sweating, dry flushed skin, feeling faint and confusion. If you suspect someone has heat stroke, call triple zero (000) for an ambulance immediately. Treatment for hyperthermia includes getting the person to a cool place such as shade, a cool shower and applying a cool wet cloth to the wrists, neck, armpits and/or groin to help cool the blood.

Hypothermia occurs when your body's temperature drops from a healthy 37°C to below 35°C. It is caused by exposure to prolonged cold conditions. Symptoms include shivering, cool pale skin, confusion and drowsiness. People with severe hypothermia may stop shivering, breathe slowly, have a slow heart rate and could fall into a coma. If you suspect someone has hypothermia, call triple zero (000) for an ambulance immediately. Treatment for hypothermia includes removing wet clothes, placing the person somewhere warm, keeping them still and providing warm drinks. One of the best strategies to regulate your body temperature and protect yourself from weather extremes is to layer your clothing. Layers can be added or taken off as required. Each layer plays an important role in protecting you from the elements:

- base layer (underwear layer): wicks sweat off your skin; the best materials include synthetics like polyester and nylon, or natural fibres like merino wool and silk
- mid layer (insulating layer): retains body heat to protect you from the cold such as fleece or puffer jackets; usually thicker means warmer
- shell (outer layer): shields you from wind and rain; waterproof and durable are the best but the most expensive options.

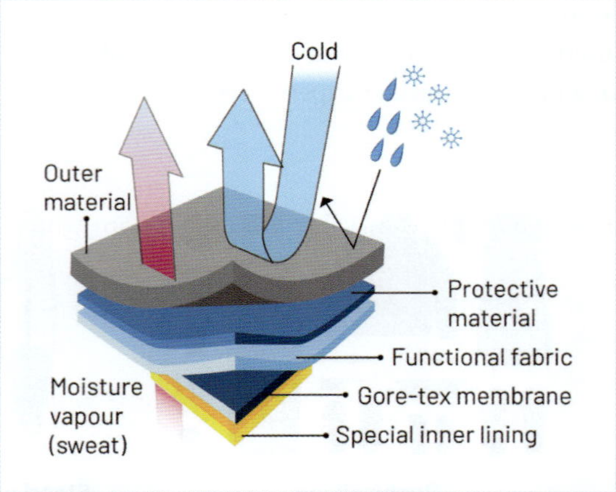

Cold
Outer material
Protective material
Functional fabric
Moisture vapour (sweat)
Gore-tex membrane
Special inner lining

Gore-Tex®

Figure 3.41 Gore-Tex® material allows your body to breathe at the same time it stops rain, hail and snow from penetrating, allowing a more comfortable and safer experience.

iStockphoto.com/swissmediavision

Figure 3.42 Layers can be added or taken off as required, such as when taking a break or stopping for lunch on the trail.

Other items of clothing that play an important role in safety when exploring outdoor environments include sunglasses, gloves and footwear.

Sleeping

A good night's sleep is important for many physical and psychological reasons, including our ability to pay attention, stay focused, reduce stress and improve mood. These factors can have a significant influence on our ability to participate safely in an outdoor experience. For example, a sleep-deprived momentary lapse of concentration while mountain biking or rock climbing can have serious consequences. Your choice of equipment has a direct impact in your ability to have a good night's sleep during an outdoor experience.

SHELTER

Depending on your adventure, your outdoor accommodation may range from a mattress and sleeping bag on the ground, with an uninterrupted view of the night sky, to a tarp strung between two trees or a sophisticated single-skin ultralight tent.

iStockphoto/pixdeluxe

Figure 3.43 Who needs five-star accommodation when camping offers a thousand-star rating?

SLEEPING BAGS

Your sleeping bag is one of the most important pieces of equipment to ensure a good night's sleep. They come in a variety of sizes, shapes and materials, and can vary significantly in cost; therefore, it is important to consider the individual, time of year, location of trip and budget when selecting a suitable bag. Most sleeping bags use a temperature rating system to aid selection.

Shutterstock.com/Ashley-Belle Burns

Figure 3.44 Getting your sleeping system right plays an important role in the success of any outdoor adventure.

SLEEPING MATS

Sleeping mats provide insulation and comfort when sleeping on cold or hard ground. They also vary in size, shape and materials to suit the adventure. Old-style closed cell foam mats are cheaper, and bulkier, than the more expensive lighter self-inflating mats that offer greater warmth at a fraction of the size.

To ensure a restful night's sleep when spending time outdoors, here are some tips to consider:

- Stay hydrated throughout the day to avoid going to bed thirsty. Keep a water bottle nearby in case you wake up thirsty.
- If there is extra space in your sleeping bag, stuff extra clothing along the sides to prevent heating unnecessary space. Also, keep extra clothing nearby to add layers as needed when the body's metabolism slows and temperatures drop.
- Wear loose-fitting clothes for sleeping, mimicking the comfort of sleeping at home. And put on warm, dry socks just before getting into bed (unless you are trying to use your body heat to dry out damp socks).
- Wait until it gets dark before going to bed and ensure you are warm before settling in. Stay active until you are ready for bed.
- Opt for sleep-promoting, non-caffeinated drinks like herbal tea in the evening, and avoid stimulants like caffeine and chocolate.
- Consume high-calorie foods, particularly those rich in fats and carbohydrates, just before bedtime to provide extra body heat a few hours later.
- Practise relaxation exercises or meditation techniques and listen to soothing music to help your body and mind relax.

Worksheet
3.3b Gear checklist

LEARNING ACTIVITY 3.3B

Gear checklist

1 Create a checklist of gear required for a specific outdoor experience.
2 Compare your checklist with those from other students. Use this comparison to refine your checklist.

3.3 KEY CONCEPTS

- **For all medical emergencies dial triple zero (000) immediately and ask for an ambulance.**
- The success of any outdoor adventure is very much reliant on how well prepared all participants are for the experience.
- All groups exploring outdoor environments should carry a first aid kit suited to the size and needs of the group, and the location and the nature of the activities undertaken.
- Have a range of charged communication devices, such as mobiles and personal locator beacons, available in case of emergency.
- There is a large range of apps available to provide support in an emergency, such as Vic Emergency and First Aid – St John's Ambulance.
- Eating sufficient amounts of nutritious food is necessary to ensure your body is capable of producing the energy required to safely participate in the planned outdoor experience.
- Stay hydrated by drinking small amounts every 10 to 15 minutes; this keeps you focused and improves stamina.
- If your trip is likely to run at a time when a Total Fire Ban could be in force, make sure you have alternative meals to cooked ones.
- Throughout your studies in Outdoor and Environmental Studies, you will have the opportunity to participate in a range of adventure activities, each of which will have equipment designed to improve safety. Take the time to learn how to correctly use this equipment.
- Protecting yourself against the elements – whether it be heat, wind or cold – is crucial to your safety when exploring outdoor environments.
- Gore-Tex® material allows your body to breathe at the same time as stopping rain, hail and snow from penetrating, allowing a more comfortable and safer experience.
- One of the best strategies to regulate your body temperature and protect yourself from weather extremes is to layer your clothing. Layers can be added or taken off as required.
- A good night's sleep is important for many physical and psychological reasons, including our ability to pay attention, stay focused, reduce stress and improve mood.
- There is a massive range of tents, sleeping bags and sleeping mats to suit individual needs. Seeking expert advice is well worth the effort to aid you in making the right choice.

Worksheet
3.3 Key concepts

3.3 CONCEPT QUESTIONS

REMEMBERING

1 What factors need to be considered when deciding on the first aid kit requirements for a school group?
2 Name and describe the purpose of the equipment associated with one outdoor recreational activity you will participate in as a part of your studies in this subject.

UNDERSTANDING

3 What is meant by the saying 'Proper Prior Planning Prevents Poor Performance'?
4 Describe the clothing strategy of layering to aid in protection from weather extremes.
5 Why is it important to get a good night's sleep when experiencing outdoor environments?

APPLYING

6 Create a table summarising the advantages and disadvantages of the different communication devices designed to improve the safety of participants in outdoor experiences.

7 Create a menu for an upcoming outdoor experience (or three-day camp) that includes a range of suitable meals to provide sufficient nutrition and hydration to meet your energy requirements.

EXTENSION CHALLENGE

8 Working together with the students in your class, investigate and produce A3 information posters on a range of mobile apps designed to support people in emergencies.

3.4 RISK MANAGEMENT OF OUTDOOR EXPERIENCES

KEY KNOWLEDGE

- risk management of outdoor experiences

KEY SKILLS

- identify hazards, analyse risks and suggest controls for an outdoor experience in a chosen outdoor environment
- analyse relevant information collected during outdoor experiences

3.4.1 MANAGING RISK IN OUTDOOR EXPERIENCES

There are many things in our everyday lives that have the potential to harm us. Two examples that come to mind quickly are cars and electricity, but we use cars and electricity all the time, despite the dangers. Why do we do this? Because we prepare for, and try to minimise, the dangers of potentially harmful actions. We wear seatbelts, drive using socially agreed and legally binding rules, and cross roads carefully. We don't stick knives into toasters, and we hire qualified electricians to work on our household circuits and lights. These sorts of precautions and preparations often don't even cross our minds – they are just a part of life in 21st-century Australia.

Shutterstock.com/Soloviova Liudmyla

Figure 3.45 Participating in risky outdoor experiences can have significant benefits, including exercising personal responsibility, building self-confidence and gaining a sense of accomplishment.

In the last chapter, we explored the variety of personal responses to risk and the interplay between competence, perceived risk and real risk. We now move on to using this information to identify hazards, analyse risks and suggest controls for experiences in outdoor environments.

When we live and travel in the outdoors – such as when bushwalking, mountain biking, camping in tents, canoeing or climbing – there are life-threatening dangers we need to be aware of. In the same way we prepare for and are cautious about dangers in our everyday lives at home, we can also prepare for and be cautious of these outdoor dangers too. While this is not intended to be a comprehensive list, any collection of dangers on outdoor experiences will include:

- drowning
- impacts with solid things (which either fall onto you or you fall onto)
- exposure (including hypothermia and heat stroke)
- burns from bushfires and fuel stoves
- lightning strikes
- poisonous bites.

Worksheet 3.4a Common-sense approaches to dangers in the outdoors

LEARNING ACTIVITY 3.4A

Common-sense approaches to dangers in the outdoors

As a class, divide into six groups. Each group will be allocated one of the dangers listed above. As a group, develop some simple commonsense guidelines that could be followed to minimise the potential threat from this danger. Present your group's guidelines to the class, and listen to the guidelines for other dangers.

Types of risks in the outdoors

Risk management is a process that all teachers and outdoor leaders must include in their planning of outdoor experiences with students and other groups. There are a large number of possible risks inherent in outdoor experiences. We can break these risks into three main categories:

- Environmental risks – these originate from the outdoor environments themselves and can have an impact on the experience. They include such factors as weather, terrain, the availability of shelter and the remote aspect of many outdoor environments, as well as the inherent dangers associated with some Australian flora and fauna.
- Risks associated with people – these can be connected to the people involved in outdoor experiences, such as leaders and participants, as well as other people that groups may encounter. They include such factors as the skills, knowledge, experience, health and fitness, age, fears and other emotions that participants bring to an outdoor experience.
- Risks associated with equipment – these can be associated with the specialised equipment that we use while participating in outdoor experiences, and the equipment used to get to the places that we visit. They include such things as clothing, buoyancy aids, kayaks, surfboards, bikes, tents, climbing ropes, helmets, motor vehicles and fuel stoves.

Once risks have been identified, the next step is to implement control measures to minimise the likelihood of an injury/illness occurring. This process is known as a risk assessment.

iStock.com/Christie Cooper

Figure 3.46 Risky outdoor experiences require careful and considered planning.

Risk assessments

risk assessment a systematic process of analysing the potential risks that may be involved in an activity

A **risk assessment** is a systematic process of analysing the potential risks that may be involved in an activity. Risk assessments are completed by many industries (e.g. construction) to create safer and healthier workplaces. The purpose of a risk assessment in outdoor education is to ensure everyone involved in an activity is as safe as possible when participating in outdoor experiences. The removal of hazards and their associated risks is always the best option; however, this is not always possible. Safety procedures (known as controls) are implemented to reduce risks to an acceptable level.

Key questions to be answered when completing a risk assessment are as follows:

1 What could cause harm (hazard)?
2 What could go wrong (risk)?
3 How likely is it to happen (likelihood)?
4 How bad will it be (consequences)?

Risk assessments should raise awareness about hazards and risks, identify who may be at risk and determine whether there are existing and adequate controls in place. It is important that a risk assessment is shared and discussed with all participants as a part of the planning for an outdoor experience.

RISK ASSESSMENT STEPS

There are many different ways to consider the risks involved in an outdoor experience; however, generally a risk assessment will have four steps:

Step 1: Identify hazards

Assess the outdoor experience to identify existing and potential hazards that may cause harm. A hazard is something with the potential to cause loss or injury. Hazards can be identified by surveying the outdoor experience location, activities and travel.

Step 2: Analyse risks

A risk is the possibility of loss or injury. It is necessary to identify who may be harmed and the severity of the loss or injury cause by the hazard by determining a risk rating.

Step 3: Suggest controls

Consider controls to either remove the **hazard** or reduce the **risk** as much as possible and minimise the likelihood of a loss or injury occurring. Controls may include facilitating a safety briefing, using specific equipment such as life jackets and ratios of instructors to participants. An additional risk rating is determined based on the controls being implemented.

hazard
something with the potential to cause loss or injury

risk
the potential to lose something that you value measured against the possibility of gaining something you value

Step 4: Review risks

Risk assessments should be reviewed regularly to account for changing conditions and to assess the effectiveness of controls in removing hazards and reducing risks.

iStockphoto/coolendelkid

Figure 3.47 What would be the risk involved in an activity such as bushwalking in a remote area?

RISK ASSESSMENT TEMPLATE

There are many different formats that a risk assessment may take. Table 3.3 provides an example of the common components of a risk assessment for an outdoor experience.

Table 3.3 Assessment table template

Identified hazard	Who may be harmed?	Risk rating before controls	Controls	Risk rating after controls
Dislodgement of rocks when climbing	All participants	High	• Review site for loose rock prior to climbing • Always wear helmets • Teach calls to warn of rock fall and appropriate protective response activity	Low

risk rating
the result of using a likelihood and consequence matrix to determine the severity or rating of a risk

likelihood
the possibilities, high or low, that someone will come into contact with the hazard

To determine the **risk rating**, a risk rating matrix is used such as the one presented in Figure 3.49. A risk rating is determined by making a judgement of the **likelihood** of a risk occurring (Rare, Unlikely, Possible, Likely, Almost certain) and a judgement of the consequence of the risk (Negligible, Minor, Medium, High, Catastrophic). Once the risk rating (Very low, Low, Moderate, High or Extreme) is determined, a judgement of whether it is safe or not to participate in the outdoor experience is made, or whether additional controls need to be implemented to promote participant health and safety.

iStockphoto/Philip Thurston

Figure 3.48 Carefully manage risks and reap the rewards

Likelihood	Consequences				
	Negligible	**Minor**	**Medium**	**High**	**Catastrophic**
Almost certain	Moderate	Moderate	High	Extreme	Extreme
Likely	Low	Moderate	High	High	Extreme
Possible	Low	Low	Moderate	High	High
Unlikely	Very Low	Low	Moderate	Moderate	High
Rare	Very Low	Very Low	Low	Moderate	Moderate

Figure 3.49 Risk rating matrix

Risk assessments are not 'one size fits all'. It is crucial that a risk assessment is completed based on the components of the specific upcoming outdoor activity, as elements will differ between experiences (e.g. location, group size, forecast weather and level of experience of participants). Risk assessments should be completed, reviewed and updated on a regular basis.

3.4 KEY CONCEPTS

- When we live and travel in the outdoors – such as when bushwalking, mountain biking, camping in tents, canoeing or climbing – there are life-threatening dangers we need to be aware of.
- In the same way we prepare for and are cautious about dangers in our everyday lives at home, we can prepare for and be cautious of these outdoor dangers too.
- Risk management is a process that all teachers and outdoor leaders must include in their planning of outdoor experiences with students and other groups.
- Environmental risks are those that originate from the outdoor environments themselves and can have an impact on the activity, such as weather, terrain, flora and fauna.
- Risks can be connected to the people involved in outdoor activities, such as leaders and participants, as well as other people that groups may encounter.
- Risks can be associated with the specialised equipment that we use while on outdoor trips, and the equipment used to get to the places that we visit (e.g. clothing, buoyancy aids, kayaks, surfboards and bikes).
- A risk assessment is a systematic process of analysing the potential risks that may be involved in an activity.
- The purpose of a risk assessment in outdoor education is to ensure everyone involved in an activity is as safe as possible when participating in outdoor experiences.
- The removal of hazards and their associated risks is always the best option; however, this is not always possible. Safety procedures (known as controls) are implemented to reduce risks to an acceptable level.
- It is important that a risk assessment is shared and discussed with all staff and students as part of the planning for an outdoor experience.
- It is crucial that a risk assessment is completed based on the components of the specific upcoming outdoor activity, as elements will differ between experiences (e.g. location, group size, forecast weather and level of experience of participants).
- A hazard is something with the potential to cause loss or injury.
- A risk is the possibility of loss or injury.
- A risk rating is the result of using a likelihood and consequence matrix to determine the severity or rating of a risk.
- Risk assessments should be completed, reviewed and updated on a regular basis.

3.4 CONCEPT QUESTIONS

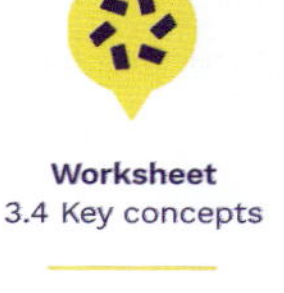

Worksheet
3.4 Key concepts

REMEMBERING

1 What is meant by risk management?
2 Identify, describe and provide an example of the three main categories of risk in outdoor experiences.
3 What is a risk assessment?

UNDERSTANDING

4 Explain the purpose of a risk assessment.
5 List the four key questions to be answered when completing a risk assessment.
6 Why is it import that a risk assessment is shared with all participants as part of the planning for an outdoor experience?
7 Describe each of the four steps involved in the creation of a risk assessment for an outdoor experience.

APPLYING

8 Create a risk assessment using the textbook template and risk rating matrix for one outdoor experience in a specific outdoor environment you have participated in or will be participating in as a part of this subject.

EXTENSION CHALLENGE

9 Prepare a risk management analysis of an outdoor experience:
   a Compile a list of possible risks that you and your group may face as a part of that experience.
   b Divide these risks into the three categories: environmental, people and equipment.
   c Analyse and describe ways of preparing for and reducing these risks.
   d Work as a class to prepare, share and discuss a risk assessment for an upcoming outdoor experience.

Worksheet 3.4b Outdoor experience - design and reflection

EXTENDED LEARNING ACTIVITY 3.4B

Outdoor experience – design and reflection

An expectation for Unit 1 is that students contribute to designing an outdoor experience that enables them to appropriately demonstrate knowledge and skills, as well as undertake the outdoor experience, and reflect on its success, suggesting changes for the future. The following activity is provided to support you in meeting this expectation. Your teacher may ask you to contribute to the design and reflection of a specific outdoor experience. Provide the details of your outdoor experience under the following headings.

OUTDOOR EXPERIENCE - Design and reflection

Part 1 – Preparation for outdoor experience	
Name of outdoor experience	
Transport to and from arrangements	
Itinerary for trip	
Traditional Owners' rights and interests in the area	
Listing of cultural significance of places and sites	
Weather forecast	
Strategies for safe and sustainable participation	
Route information (distance, time, elevation change, etc.)	
Campsite location information	
First aid requirements	
Equipment list	
Risk assessment	
'Let someone know before you go'	
Emergency contacts	

Part 2 – During outdoor experience	
Group safety check	
Group morale check	
New knowledge acquired	
New skills acquired	
Weather (impact on experience)	
Observations of impacts on outdoor environment	
Updates of risk assessment	

Part 3 – After outdoor experience	
Evaluation of the effectiveness of the strategies for safe and sustainable participation	
Reflection on success of outdoor experience	
Aspects of outdoor experience needing changes	
Suggested changes for future experiences	
Final risk assessment	

9780170477123

Resource
Glossary – Chapter 3

Assessments
End of chapter exam

Glossary test

EXAM-STYLE QUESTIONS

1 Why it is important to implement strategies for safe and sustainable participation in outdoor experiences? (3 marks)

2 Explain how two minimal impact strategies you or your school group implemented were used to promote safe and sustainable interactions with outdoor environments. (4 marks)

3 Describe three factors to consider as part of route planning in order to promote safe and sustainable interactions with the outdoor environment. (3 marks)

4 List four safety precautions to be considered when selecting a tent site. (4 marks)

5 Describe the actions to take when providing first aid as a part of the basic life support when presented with an injured person. (6 marks)

6 Discuss the first aid treatment for two of the following injuries: blisters, small wounds, snake bite and severe bleed treatment. (6 marks)

7 Explain the use of equipment designed to improve the safety of participants in an outdoor experience you have participated in this year. (3 marks)

8 Using examples from your experiences in outdoor environments this year, explain the difference between a hazard and a risk. (4 marks)

9 Identify and describe the four steps involved in the creation of a risk assessment for an outdoor experience. (8 marks)

CHECK YOUR KNOWLEDGE AND SKILLS

Resources
Key knowledge and skills checklist

Checklist for students to mark and date, making judgements of their knowledge and understanding.
(**D** = developing **C** = consolidating **ER** = Exam Ready)

D	C	ER	Check Your Knowledge
			I know how to use appropriate skills for safe and sustainable interactions with outdoor environments, involving:
☐	☐	☐	• minimum impact strategies for individuals
☐	☐	☐	• codes of conduct
☐	☐	☐	• route planning
☐	☐	☐	• tent-site selection
☐	☐	☐	• navigation
☐	☐	☐	• packing a pack
			I know how to demonstrate and apply basic first aid skills, including:
☐	☐	☐	• sbasic life support
☐	☐	☐	• blister management
☐	☐	☐	• small wounds
☐	☐	☐	• snake bite
☐	☐	☐	• severe bleed treatment
☐	☐	☐	• Cardio-Pulmonary Resuscitation (CPR)
☐	☐	☐	• immobilisation techniques.
			I know how to explain the use of equipment designed to improve the safety of participants in outdoor experiences including:
☐	☐	☐	• first aid equipment
☐	☐	☐	• communication devices
☐	☐	☐	• food and water
☐	☐	☐	• activity-specific equipment
☐	☐	☐	• personal protective equipment.
			I know the process of risk management of outdoor experiences involving the following steps
☐	☐	☐	• identify hazards
☐	☐	☐	• analyse risks
☐	☐	☐	• suggest controls for an outdoor experience in a chosen outdoor environment.

9780170477123

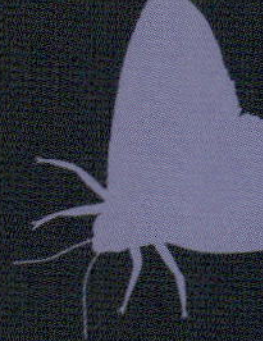

Unit 2
Discovering outdoor environments

Unit 2 – Introduction

Unit 2 of the VCE Outdoor and Environmental Studies study design titled 'Discovering outdoor environments' focuses on the different ways to understand outdoor environments and the impact of humans on outdoor environments.

We study the effects of natural changes and impacts of land management practices on the sustainability of outdoor environments, develop the practical skills required to minimise the impact of humans on outdoor environments, and build knowledge of the vocational perspectives that inform human use of outdoor environments.

Chapter 4 (Understanding outdoor environments) introduces students to a range of understandings and types of outdoor environments from several perspectives, and how these environments are managed.

Chapter 5 (Observing impacts on outdoor environments) focuses on human activities undertaken in outdoor environments and their impacts on those environments, both positive and negative.

Chapter 6 (Independent participation in outdoor environments) directs students to plan for their outdoor experience and peer lead that experience with their peers. During the experience they will analyse the impacts of other users and themselves on the outdoor environment and investigate ways to reduce this and promote sustainable interactions.

Through their studies of Unit 2, students will develop the practical skills required to minimise the impact of humans on outdoor environments while discovering a range of vocational perspectives that inform human use of outdoor environments. On reflection upon their experiences of outdoor environments, students make comparisons between outdoor environments, as well as develop theoretical knowledge about natural environments.

CHAPTER 4

UNIT 2 – AREA OF STUDY 1

Understanding outdoor environments

KEY KNOWLEDGE

- scientific understandings of a range of outdoor environments including:
  - the interrelationships between biotic and abiotic components
  - the effects of natural and human-induced changes on a range of outdoor environments, such as day to night, seasons, tides, flood, drought, fire, migration and climate change
- Indigenous peoples' land management understandings and perspectives of an outdoor environment
- understandings of vocational perspectives of outdoor environments, including at least two of the following:
  - natural resource management
  - nature-based tourism
  - outdoor leading and guiding
  - environmental research and policy
  - land management roles including Traditional Owner groups and National Indigenous Australians Agency programs
  - education
  - agriculture.

KEY SKILLS

- compare abiotic and biotic interrelationships in an outdoor environment
- describe the effect of natural and human-induced changes and apply these to a range of environments
- explain Indigenous peoples' perspectives and different forms of land management within an outdoor environment
- compare a range of vocational perspectives on outdoor environments.

VCE Outdoor and Environmental Studies Study Design 2024–2028, pp. 18-19.

Scientific understandings of outdoor environments

Scientific investigations into nature have provided some of the most important discoveries of our world. This section examines how we can understand outdoor environments from a scientific perspective.

Indigenous peoples land management and perspectives

Indigenous peoples of Australia have lived on this country for thousands of years and their perspective is informed by this time-tested relationship and high regard for country as their 'mother'.

Vocational perspectives of outdoor environments

When comparing **vocational** perspectives, we are comparing the different ways people can regard outdoor environments through the point of view of their occupation.

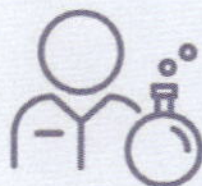

KEY TERMS

abiotic
aquaculture
atmosphere
biogeochemical
biorhythms
biotic
bush food
carbon dioxide (CO_2)
carnivore
circadian rhythms
climax vegetation
customary laws
diurnal
El Niño
fire intensity
fossil fuel
fuel load
geochemical cycles
greenhouse gas
herbivore
immigration
impervious surfaces
intertidal zone
kinetic
La Niña
land management
matter
metaphor
middens
migration
net sink
neutral phase
nocturnal
other special rights
outdoor environments
perspective
photosynthesis
precipitation
renewable energy
respiration
sacred places
sustainable tourism
totemism
transpiration
trophic level
vocational

To access resources above, visit **cengage.com.au/nelsonmindtap**

PERSPECTIVES OF OUTDOOR ENVIRONMENTS

Outdoor environments are wondrous places that provide much value to everyone who interacts with them. Our individual perspectives on different outdoor environments will vary depending on our previous experiences, our beliefs and our knowledge of each environment. This forms our point of view of the value or use of either a specific outdoor environment or nature in general. If we selected 10 people at random from different backgrounds and placed them in a line in front of the same section of bushland, those 10 people would likely each have a different **perspective** about the location.

perspective
a particular attitude towards or way of regarding something; a point of view

When considering different human–nature relationships, it is important for us to understand that different perspectives are not right or wrong, and can be helpful in expanding our thoughts and ideas on a topic. A common trap, especially when it comes to opposing perspectives, is to assume that everyone will see things the way that you do. This can lead to conflict and often comes from a struggle to understand different perspectives.

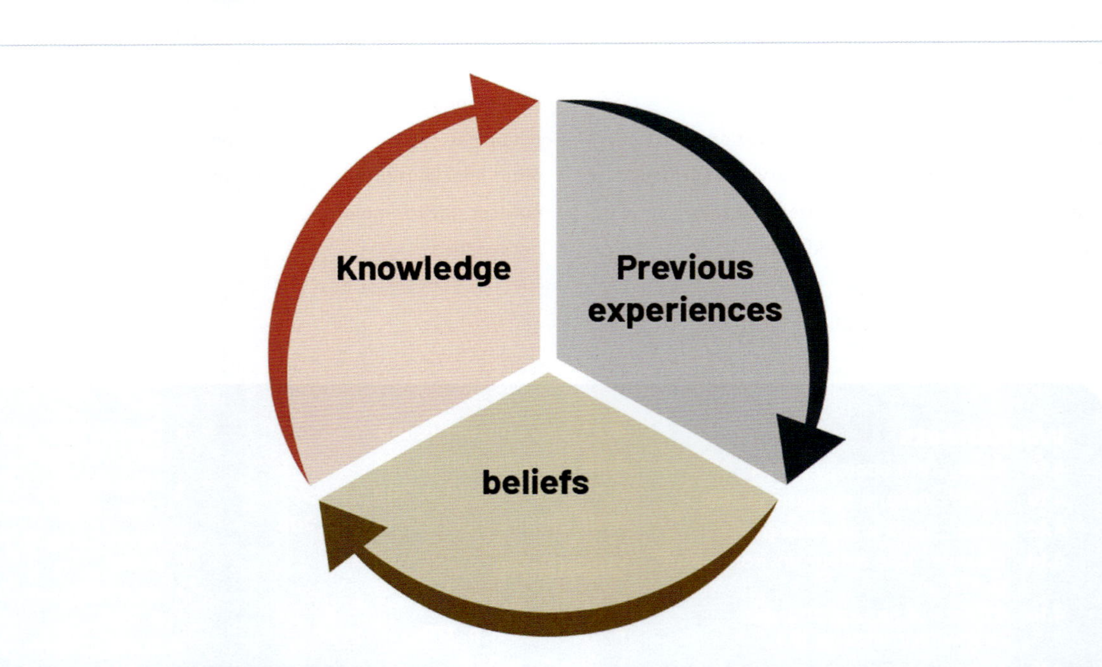

Figure 4.1 There are three main factors that influence people's perspectives.

CASE STUDY

PERSPECTIVES

Resource
Case study: Perspectives

Our perspectives about everything in life begin developing very early on, and between the ages of 2 and 6 especially we build strong connections that inform these perspectives. All of the things that we are exposed to and experience inform who we are and our outlook on life.

Read the following excerpt from Robert Fulhgum's book *All I Really Need to Know I Learned in Kindergarten*. If everyone had the same experiences in their early years of life, would we all see the world in the same way?

These are the things I learned (in kindergarten):

1. Share everything.
2. Play fair.
3. Don't hit people.
4. Put things back where you found them.
5. CLEAN UP YOUR OWN MESS.
6. Don't take things that aren't yours.
7. Say you're SORRY when you HURT somebody.
8. Wash your hands before you eat.
9. Flush.
10. Warm cookies and cold milk are good for you.
11. Live a balanced life – learn some and drink some and draw some and paint some and sing and dance and play and work every day some.
12. Take a nap every afternoon.
13. When you go out into the world, watch out for traffic, hold hands and stick together.
14. Be aware of wonder. Remember the little seed in the Styrofoam cup: The roots go down and the plant goes up and nobody really knows how or why, but we are all like that.
15. Goldfish and hamster and white mice and even the little seed in the Styrofoam cup – they all die. So do we.
16. And then remember the Dick-and-Jane books and the first word you learned – the biggest word of all – LOOK.

Robert Fulhgum, *All I Really Need to Know I Learned in Kindergarten*, Ballantine Books, Penguin Random House 2003

QUESTIONS

1 From the list above, which of the points describe aspects of kindergarten that could influence people's perspectives on outdoor environments?

2 Can you remember experiences you had at kindergarten age that influenced your perspective on outdoor environments?

3 Is your perspective the same now as it was in kindergarten?

4.1 SCIENTIFIC UNDERSTANDINGS OF OUTDOOR ENVIRONMENTS

KEY KNOWLEDGE

- scientific understandings of a range of outdoor environments including:
  - the interrelationships between biotic and abiotic components
  - the effects of natural and human-induced changes on a range of outdoor environments such as day to night, seasons, tides, flood, drought, fire, migration and climate change

KEY SKILLS

- compare abiotic and biotic interrelationships in an outdoor environment
- analyse the effect of natural and human induced changes and apply these to a range of environments

Scientific investigations into nature have provided some of the most important discoveries of our world, such as the theory of evolution by British naturalist Charles Darwin (1809–1882). These investigations inform our perspectives on outdoor environments from a scientific point of view. This section examines how we can understand outdoor environments from a scientific perspective. This includes the interrelationships between components, and the effects of natural changes and fire on the environment.

4.1.1 THE BIOSPHERE

A shell approximately 20 kilometres wide surrounds the Earth and stretches from the deepest oceans to the highest mountains. This shell is called the biosphere, and includes all forms of life on Earth and all their interactions. Within the biosphere, there are three major regions of the Earth that interact:

- the **atmosphere** – a gaseous envelope of air surrounding the Earth
- the hydrosphere – all water on the Earth, including oceans, lakes, rivers, ice caps and the water vapour in the atmosphere
- the lithosphere – the soil and rocky crust of the Earth, and it's upper mantle

atmosphere
the thin layer of gases that surrounds the Earth

Components of an ecosystem

The interactions occurring between organisms and their environment make up an ecosystem. The concept of an ecosystem was developed to make the study of different parts of the biosphere easier. The components of an ecosystem can be grouped in various ways, including the following:

- An organism is an individual living thing.
- A species is a group of organisms of the same type, which are capable of interbreeding and producing fertile offspring.
- A population is a group of organisms of the same species living together in the one area.
- A community consists of a number of populations of different species living together in a particular environment.

In a community, organisms interact with one another and with their surroundings. It is these surroundings – chemical and physical, living and non-living – that make up an **outdoor environment**. 'Habitat' refers to the specific place within an ecosystem that is occupied by an organism or population. Habitat size can vary from an ocean to a decomposing leaf. A broader concept is that of 'ecological niche', which describes the place and role of an organism in a community. This concept includes information about what an organism consumes, what consumes it, where it lives and its interaction with the biotic and abiotic components of its environment.

outdoor environments
environments that have minimal influence from humans, but may also include those that have been subjected to human intervention

biotic
a living feature in an environment

abiotic
a non-living feature in an environment

Interrelationships between biotic and abiotic components

In order to understand outdoor environments, it is crucial that we appreciate the complex interrelationships between the **biotic** components (living plants and animals) and **abiotic** components (non-living structures such as rocks, soils, sunshine and water).

The interactions between components of an ecosystem are complex. Plants compete with one another for water, light and soil nutrients, while animals and birds compete for food and habitat. Non-living elements such as soil, water and wind also affect the species within an ecosystem.

All ecosystems have inputs and outputs, together with processes that occur along the way to change the inputs in some way. Inputs and outputs can be categorised as **matter** or energy (in the form of thermal, solar, wind, **kinetic**, sound or potential energy). For example, a plant requires the input of carbon dioxide (CO_2), water (H_2O) and sunlight (solar energy) to produce glucose and oxygen through the process of photosynthesis. The materials leaving an ecosystem can be quite different from those that entered, as can be seen in Figure 4.2.

matter
in physics, that which has both a mass and volume, which occupies space and which possesses a rest mass (as distinct from energy)

kinetic
relating to or resulting from movement

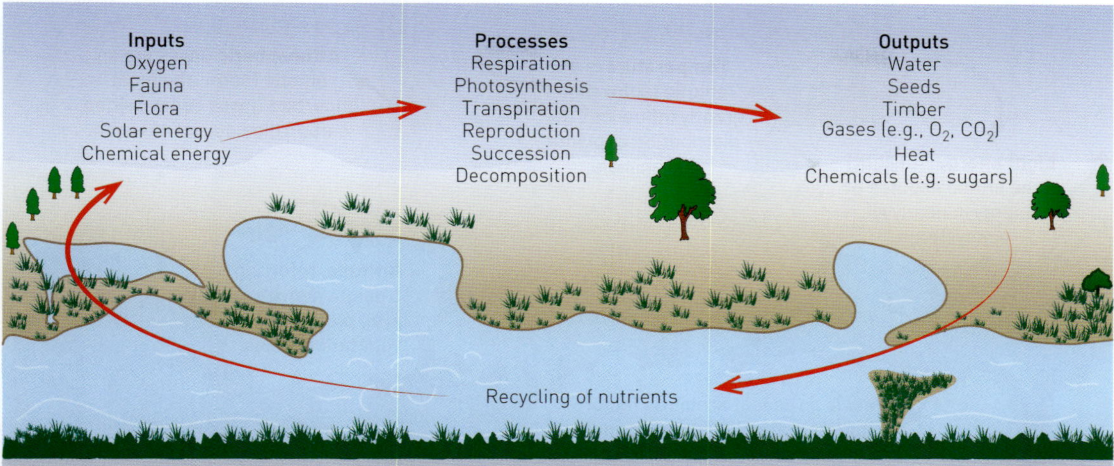

Figure 4.2 Natural systems are complex and involve many interrelated processes.

Figure 4.3 The biotic components of an environment are essentially the living things that make up that environment.

Figure 4.4 The abiotic components are non-living chemical and physical factors in the environment.

biogeochemical the cycle in which simple substances and chemical elements are transferred between living elements and the environment

Worksheet 4.1a Ecosystem ecology

Weblink Video: Ecosystem Ecology: Links in the Chain – Crash Course Ecology #7

Biogeochemical cycles

The Earth's environments have many varied cyclic processes that enable nutrients to be reused. These include the **biogeochemical** cycles that allow elements to move throughout ecosystems, such as the carbon, oxygen, nitrogen and water cycles. Each cycle relies upon both biotic and abiotic components of the ecosystem to function effectively.

The cycling of these elements and water can be disrupted by human activity, such as overgrazing and land clearing. Nitrogen and phosphorus levels in the soil can quickly diminish, while burning fossil fuels contributes to increased levels of carbon dioxide in the atmosphere.

LEARNING ACTIVITY 4.1A

Ecosystem ecology

Watch a short video on ecosystem ecology 'Ecosystem Ecology: Links in the Chain – Crash Course Ecology #7' on the WatchKnowLearn.org website. This video describes the components and processes of ecosystems. Take notes identifying the key components and processes of ecosystems.

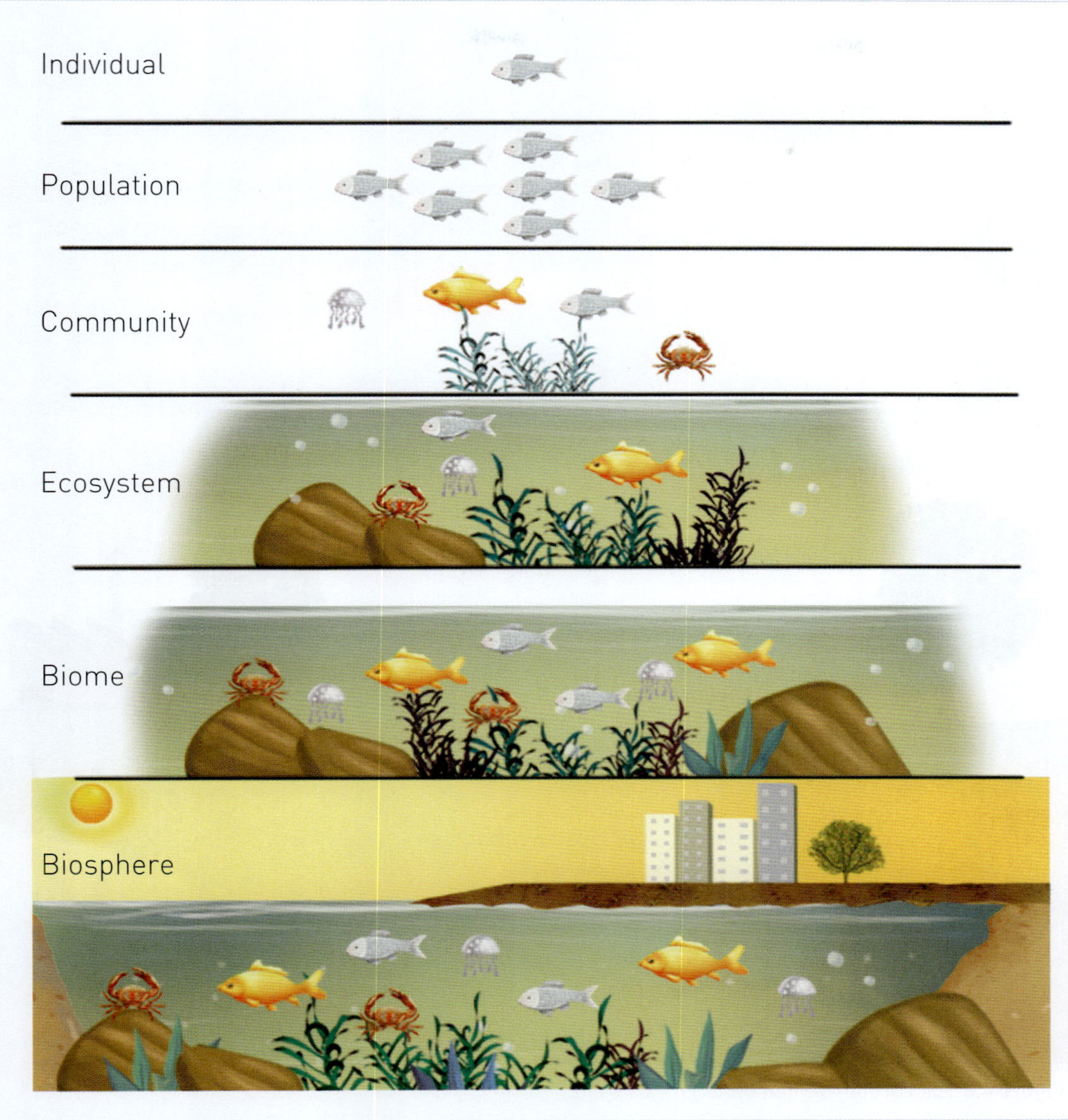

Figure 4.5 The levels of organisation of material

photosynthesis
a process used by plants to convert light energy into chemical energy to grow

carbon dioxide (CO_2)
a colourless, odourless gas that is the fourth-most abundant in the atmosphere (approximately 0.04%); it is produced from the burning of fossil fuels and acts as a greenhouse gas

respiration
the physiological process that enables animals to exchange carbon dioxide

fossil fuels
a deposit, such as petroleum, coal or natural gas, derived from the accumulated remains of ancient plants and animals and used as fuel

Carbon–oxygen cycle

During **photosynthesis**, plants absorb **carbon dioxide (CO_2)** into their leaves from the atmosphere. The carbon from the carbon dioxide is then integrated into materials such as glucose and starch, which allow plants to function effectively and maintain and build their structure. Carbon then passes along the food chain, as animals acquire nutrients from plants and each other. The processes of **respiration** and decomposition eventually release carbon back into the atmosphere. As carbon is also found in **fossil fuels** (such as oil, gas and coal) that have derived from decomposed plant matter, it is released into the atmosphere as carbon dioxide when these substances are burned in industrial processes, cars or domestic use. Some natural processes, such as volcanic eruptions and erosion, can also contribute to carbon levels in the atmosphere.

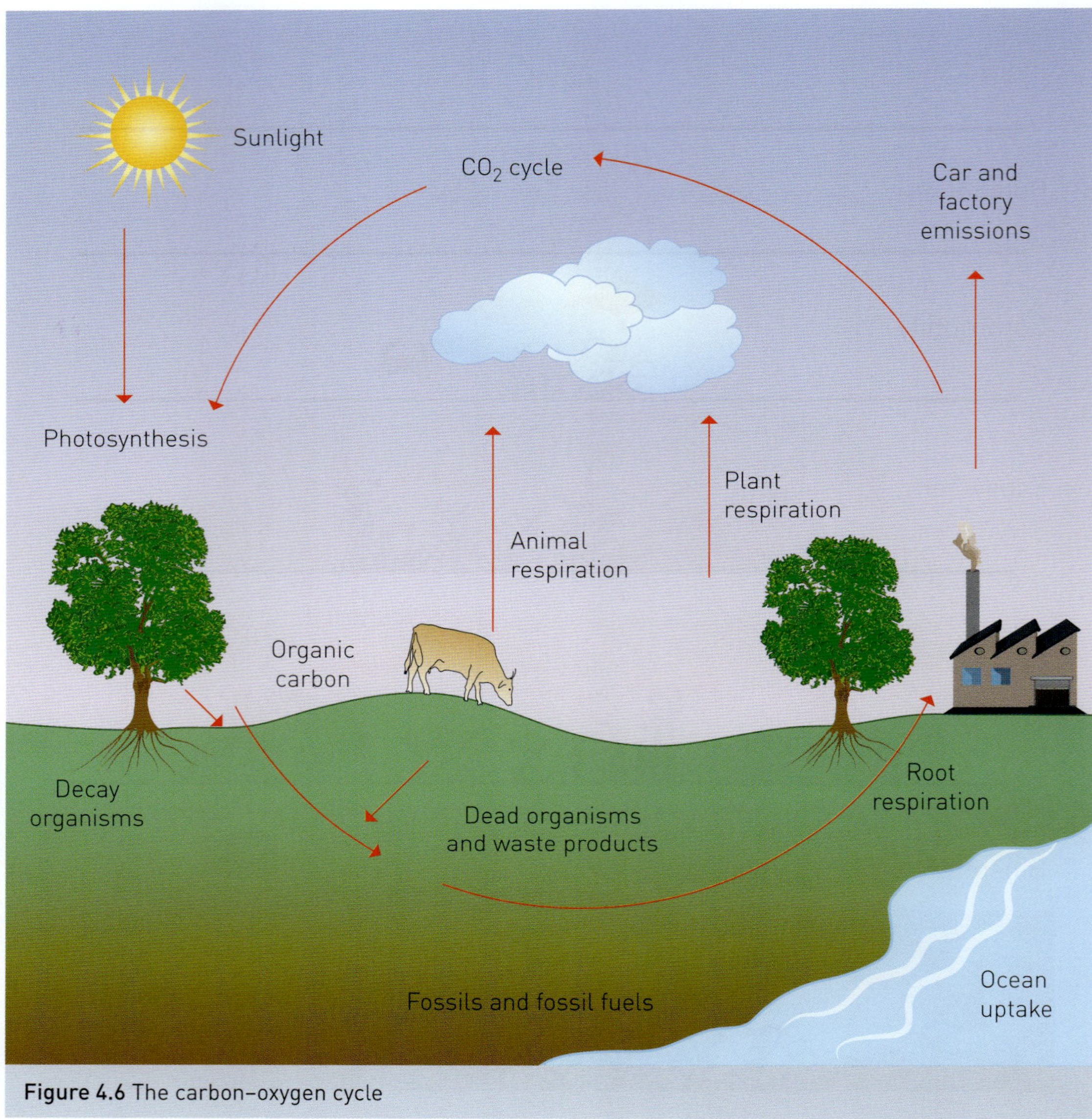

Figure 4.6 The carbon–oxygen cycle

Nitrogen cycle

The nitrogen cycle is the process by which nitrogen is converted between its various chemical forms. Approximately 80% of the atmosphere consists of nitrogen. It is an essential element required by organisms to manufacture protein and vitamins and is a vital component of DNA. Plants cannot absorb nitrogen as a gas and so rely on soil bacteria that convert the nitrogen (gas) to nitrates (salt), allowing them to absorb it through their roots. Animals obtain nitrogen by eating plants and other animals. Nitrogen is returned to the system via faeces and urine, and the process of decomposition.

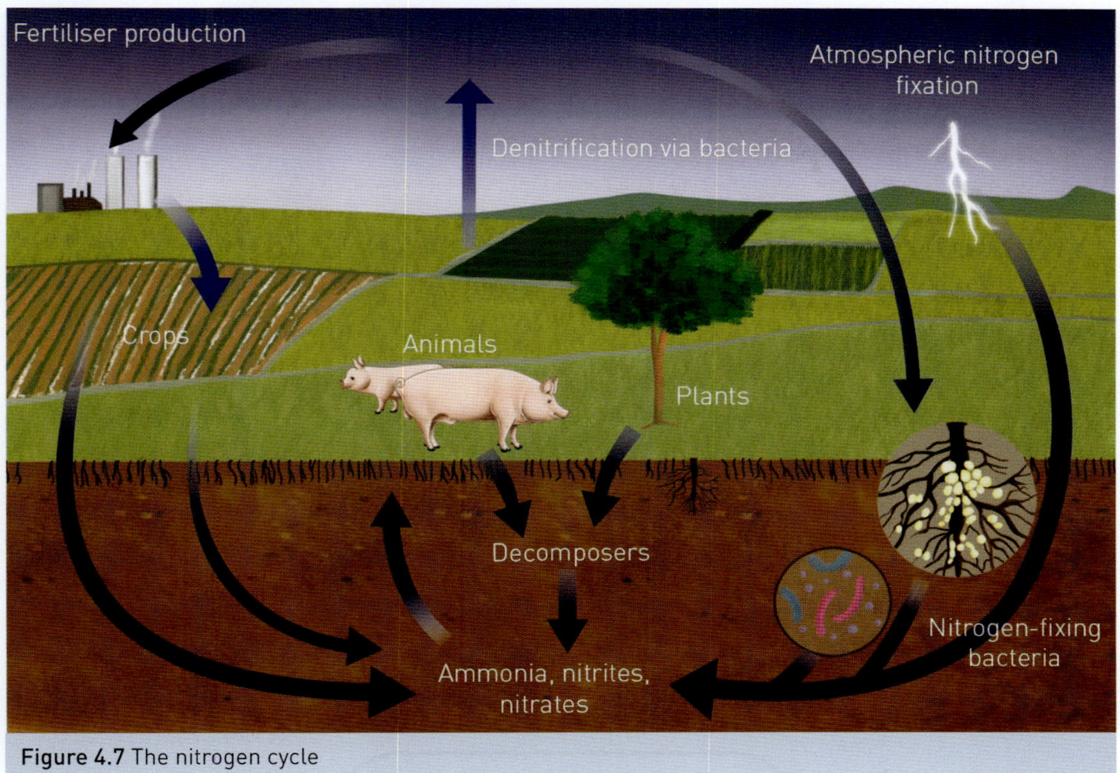

Figure 4.7 The nitrogen cycle

precipitation
when water is released from clouds in the form of rain, freezing rain, sleet, snow or hail

transpiration
the evaporation of water into the atmosphere from the leaves and stems of plants

Water cycle

All life depends on water for survival. The sun's energy powers the water cycle through evaporation and **precipitation**. Plants absorb water from soils and use it during processes such as photosynthesis, and release excess water back into the atmosphere via **transpiration**. Animals (including humans) release water into the atmosphere through respiration, sweat and urine. In addition, the water contained in dead animals and plants is quickly evaporated into the system.

> **FAST FACT**
> There are approximately 1 397 918 500 trillion litres of water on Earth.

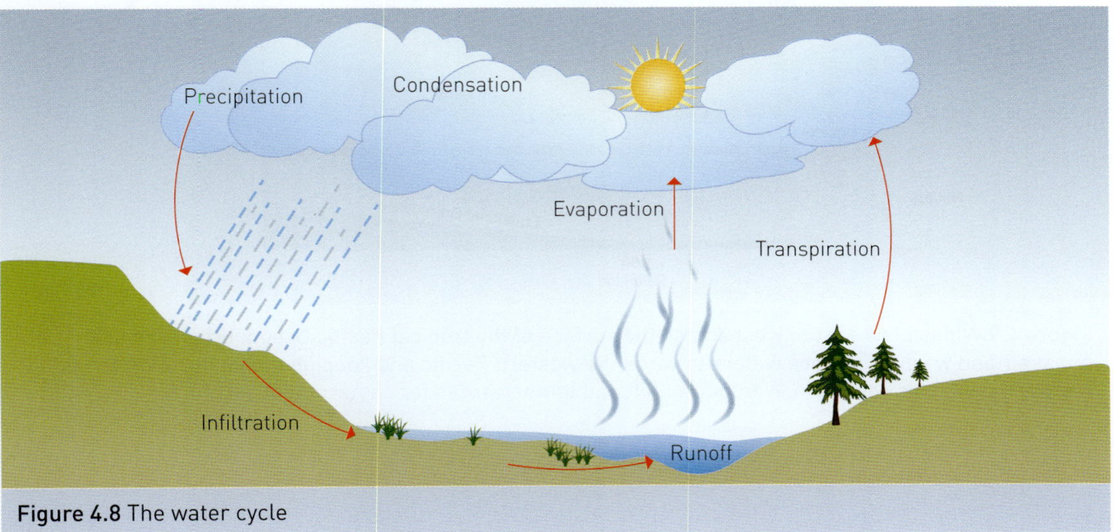

Figure 4.8 The water cycle

Worksheet
4.1b Earth's natural systems

LEARNING ACTIVITY 4.1B

Earth's natural systems

1 Identify and describe each of the components of the biosphere.
2 Using a diagram, name and describe the components of an ecosystem.
3 Create a classroom display illustrating the natural processes involved in the carbon, nitrogen or water cycle.

neutral phase
where warming winds towards the western Pacific keep the central Pacific Ocean relatively cool, associated with average rainfall and temperatures for eastern Australia

La Niña
extensive cooling of the central and eastern tropical Pacific Ocean, associated with increased probability of wetter conditions in Australia

El Niño
extensive warming of the central and eastern tropical Pacific, associated with an increased probability of drier conditions in Australia

Climatic cycle

Our eastern Australian climate is influenced heavily by the natural cycle of the El Niño–Southern Oscillation (ENSO), centred on the tropical Pacific Ocean. This is the natural change that brings us dry or wet spring and summer seasons, depending on which part of the cycle we are in. It heavily impacts outdoor environments, with the incidence of flood or bushfire linking closely with this cycle.

EL NIÑO–SOUTHERN OSCILLATION

The ENSO cycle loosely operates over timescales from one to eight years, shifting from the **neutral phase** to **El Niño** (think dry and hot) to **La Niña** (think wet and cool).

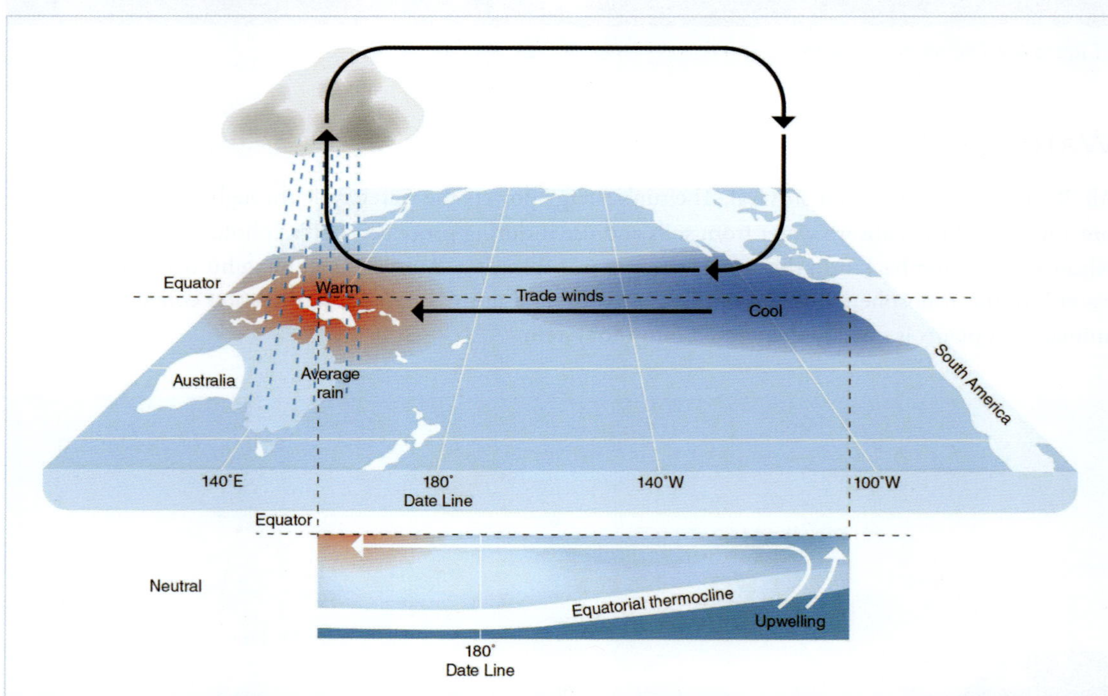

Figure 4.9 Winds blow east to west across the surface of the tropical Pacific Ocean, bringing warm moist air and warmer surface waters towards the western Pacific and keeping the central Pacific Ocean relatively cool. This results in average rainfall and temperatures for eastern Australia.

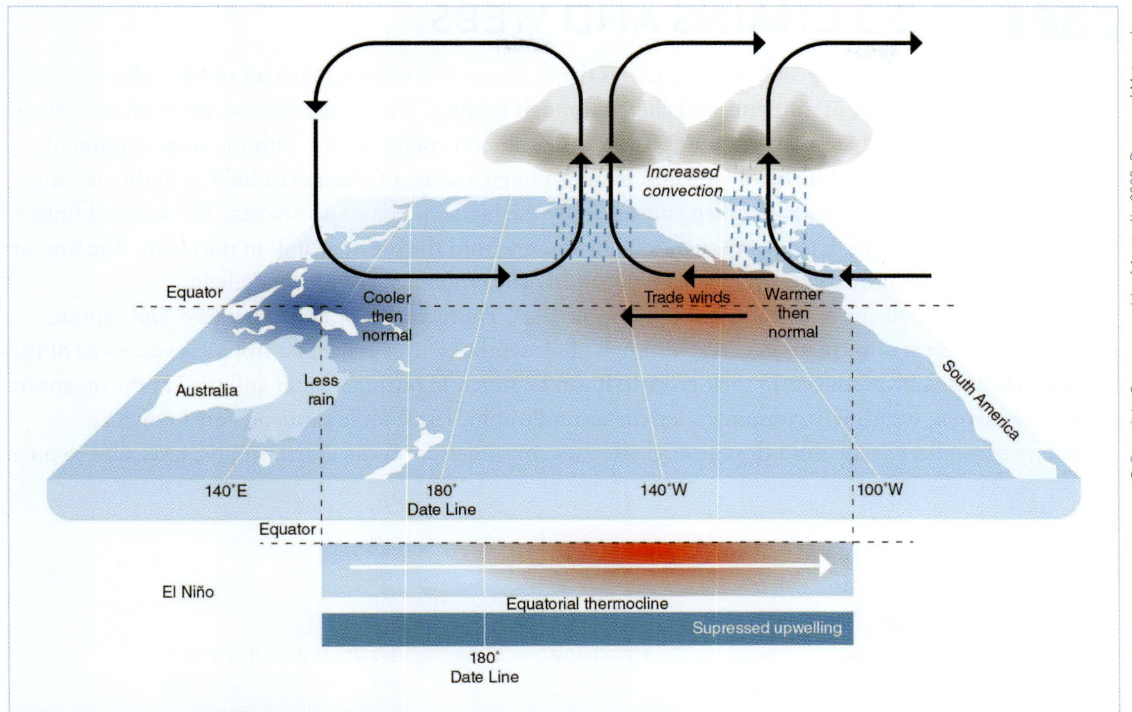

Figure 4.10 The warming of the central and eastern tropical Pacific Ocean results in decreased rainfall and hotter weather for eastern Australia.

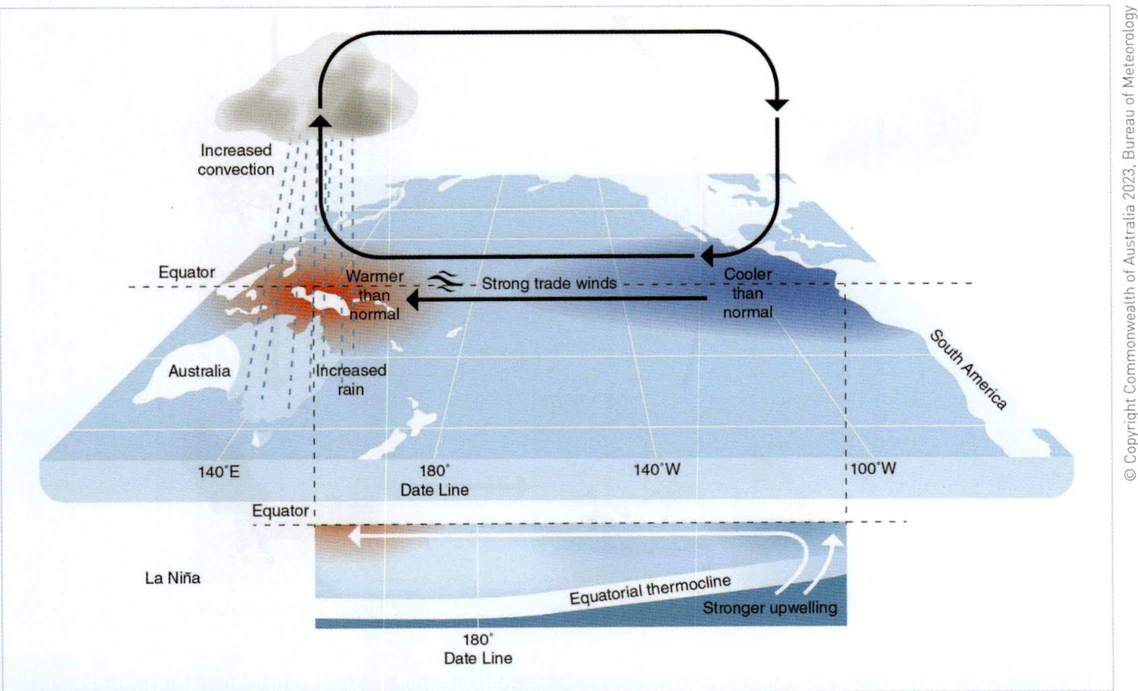

Figure 4.11 A cooling of the water in the equatorial Pacific is associated with widespread changes in weather patterns for eastern Australia, including mainly increased rainfalls and cooler daytime temperatures.

geochemical cycles
the circulation of biological, geological and chemical substances

herbivores
an animal that feeds on grass and other plants

carnivores
an animal or plant that feeds on animals

trophic level
a feeding level; organisms that form one link in a food chain – producers, consumers or decomposers

4.1.2 FOOD CHAINS AND WEBS

The connection between the **geochemical cycles** is the exchange of the basic materials of life – the consumption and transfer of nutrients and energy through feeding. The living components of an ecosystem cannot maintain themselves without energy. The transfer of food energy begins with the consumption of the producers (green plants) by herbivores and passes through a series of changes or links as **herbivores** are eaten by **carnivores** that, in turn, are then consumed by the higher predator carnivores. This series of links is referred to as a food chain. Each component obtains energy from the previous link in the chain. The sun, or solar radiation, is the primary source of energy for nearly all components of the food chain.

Each link in a food chain is referred to as a **trophic level**. Producers occur at the first level and capture sunlight to convert it into chemical energy through photosynthesis. This is the first trophic level. Some of this energy is then used by producers in their growth. It can be stored in organic matter and used as the organism grows. Herbivores, or primary consumers, are the second trophic level, while carnivores are known as secondary consumers or the third trophic level. Tertiary consumers form the fourth trophic level, and so on.

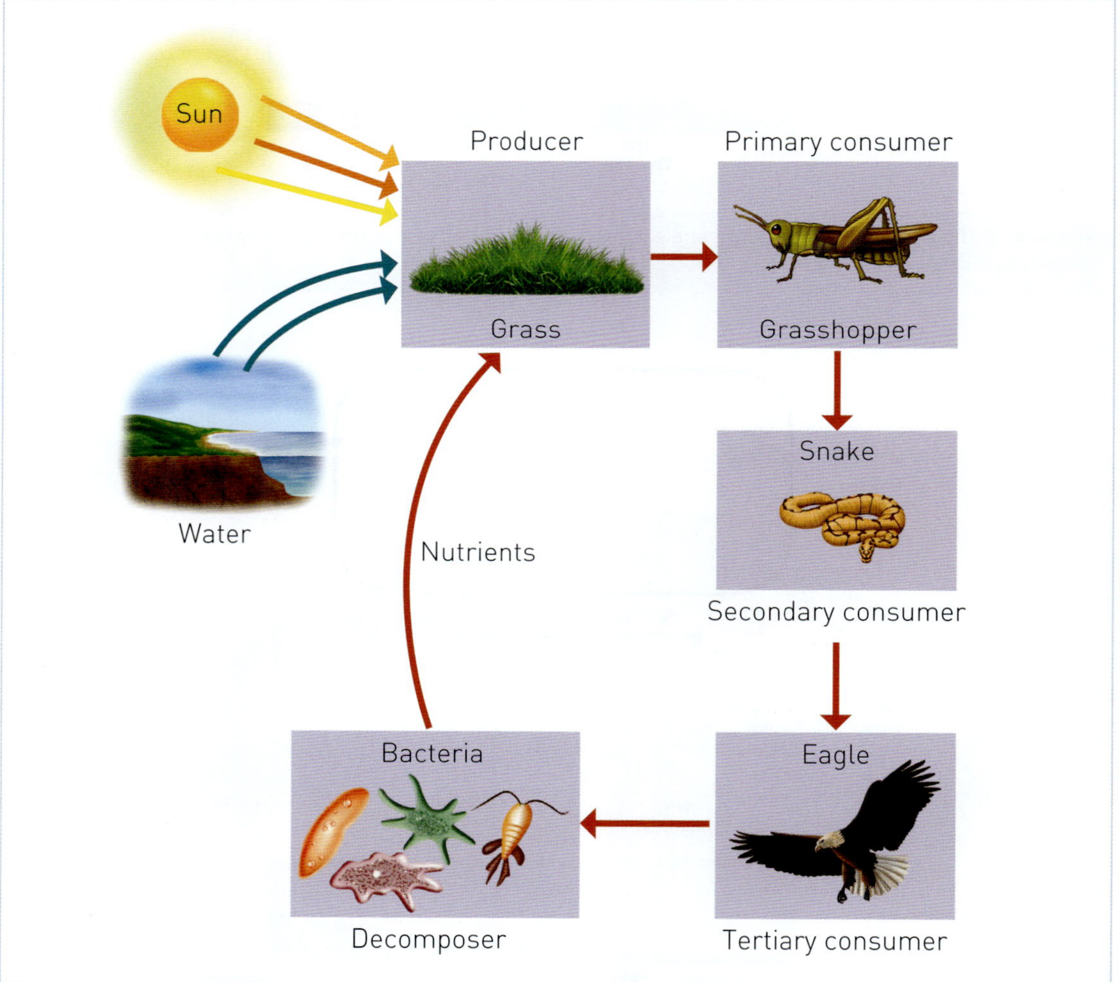

Figure 4.12 A simple food chain in which energy is passed from the sun to the producer (here, the grass) and then to the consumers (in order of hierarchy, the grasshopper, snake and eagle).

Most animals consume or are consumed by several types of organism, so each one is linked into several different food chains. These complex relationships can be described using a food web.

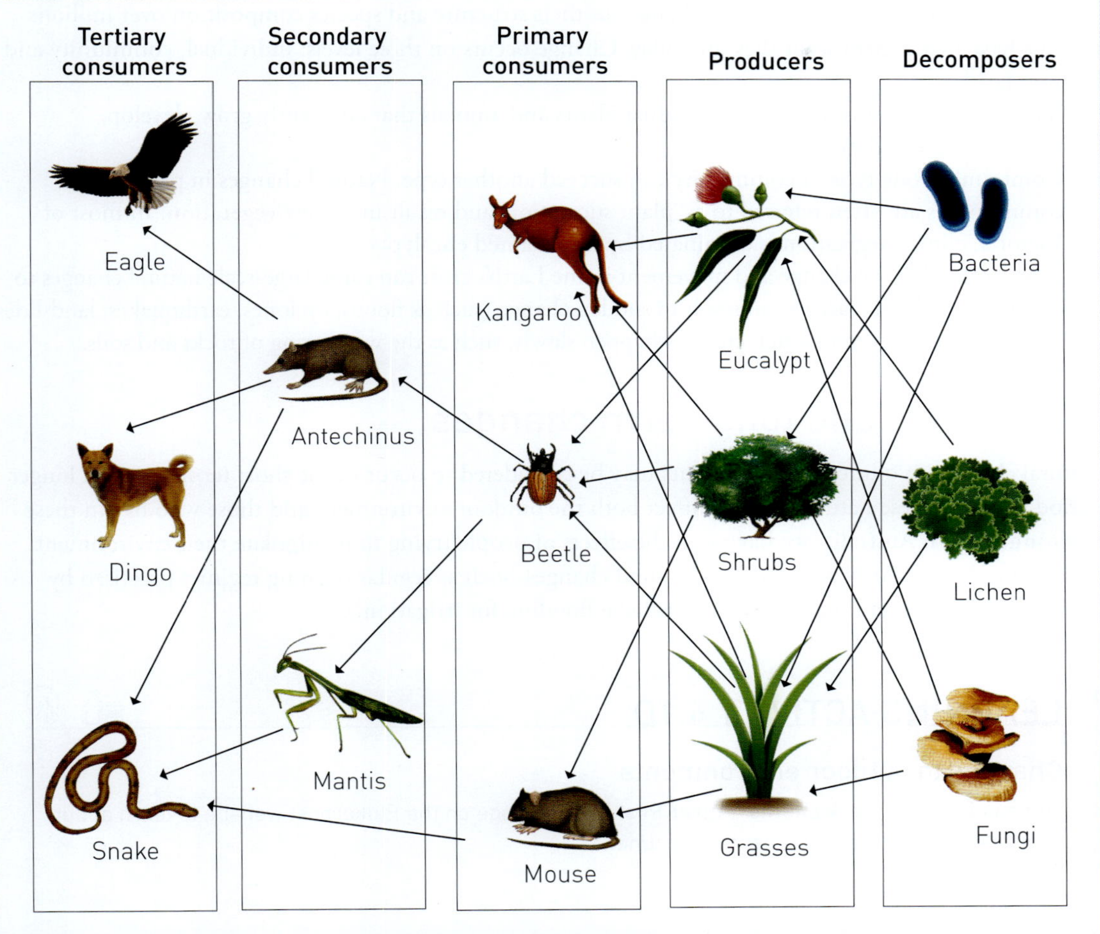

Figure 4.13 Food web for a forest ecosystem

Note: decomposers are included in this diagram showing their place in the food web although they are not a trophic level.

Weblink
Ecolink

LEARNING ACTIVITY 4.1C

Food chains and webs

Learn about food chains and food webs at the EcoLink website and then answer the following questions.

1. Describe the difference between a food chain and a food web.
2. How is energy passed from one organism to another within a food web?
3. Why is the role of decomposers important in a food web?
4. Construct your own food web for an environment you have visited or investigated this year.

Worksheet
4.1c Food chains and webs

4.1.3 EFFECTS OF NATURAL AND HUMAN-INDUCED CHANGES ON OUTDOOR ENVIRONMENTS

Outdoor environments are dynamic. Changes in their structure and species composition over millions of years have made them what they are today. Change occurs on three levels: individual, community and large-scale:

- Individual organisms: these are individual plants and animals that constantly grow, develop, reproduce, die and decompose.
- Community: one type of community can succeed another type. Natural changes in plant communities are often referred to as 'plant succession' and result in **climax vegetation**. In most of Victoria, climax vegetation is dominated by fire-adapted eucalypts.
- Large-scale: climatic change and movement in the Earth's crust can cause large-scale natural changes to occur. Some abiotic factors can result in sudden change (such as floods, cyclones, earthquakes, landslides and tsunamis), but more often changes happen slowly, such as the weathering of rocks and soils.

climax vegetation vegetation that establishes itself in an area over a long time in the absence of any major disturbances

Short-term and long-term changes

Natural changes to outdoor environments may be considered to occur in the short term or over a longer period of time. These natural changes affect both the outdoor environment and those who live in these environments. In Australia, we have seen the effects of people trying to manipulate their environment by altering the regularity and degree of natural changes, such as regular burning regimes practised by Indigenous peoples and, more recently, artificial flooding for irrigation.

Worksheet 4.1d Changes to outdoor environments

Weblink Fluker Post

LEARNING ACTIVITY 4.1D

Changes to outdoor environments

Jump online and check out the 'Fluker Post Project' page on the Flukerpost website to learn about changes to outdoor environments over time.

Day to night

The regular change from day to night has some complex and profound effects on both humans and outdoor environments. Once the sun has set, temperatures decrease on both land and in bodies of water such as oceans, lakes and rivers. This can lead to changes in the speed and direction of wind. **Nocturnal** animals become active and **diurnal** species, unable to function as well in darker conditions, may seek out nesting sites. This includes humans who, without the aid of artificial light and heating, would traditionally seek shelter and rest. In our highly urbanised society, the effects of the change from day to night are not felt as keenly. Insulated houses and warm clothes, along with artificial light and heat sources, allow us to ignore the effects of darkness.

nocturnal animals and plants that are active at night

diurnal animals and plants that are active during the day

Shutterstock.com/deb talan

Figure 4.14 The quoll (genus Dasyurus), a native Australian marsupial, is nocturnal.

During the day, plants use sunlight for the process of photosynthesis, where they take in carbon dioxide and produce oxygen. Plants are also continuously respiring to release energy for their functions. This means that, at night, plants can be net carbon dioxide producers. Vegetation that relies on animal species active during daylight hours may close their petals and their leaves may droop during the night. Other plant species attract night-time pollinators through strong perfume or light-coloured flowers.

Native animals, including most mammals, are active from dusk to dawn to avoid predators, seek out food and avoid extremes in temperature. With the exception of species such as owls, tawny frogmouths and owlet nightjars, most birds roost at night in hollows or nests, although some (such as plovers and magpies) will come out during moonlit nights when there is more light available.

The theory of **biorhythms** claims that the daily lives of humans are affected by rhythmic cycles such as day to night. One such cycle involves **circadian rhythms**, which involve sensitivity to light and darkness, and sleep/wakefulness patterns. In the absence of cues such as sunlight, aspects of metabolism, physiology and behaviour continue to be regulated by a 'biological clock'. This is one explanation for the effects of long-distance plane travel, known as jetlag.

biorhythms
cyclic pattern of changes in activity of living organisms

circadian rhythms
a 24-hour cycle in the physiological processes of living organisms

LEARNING ACTIVITY 4.1E

Effects of natural changes to environments

During a practical outdoor experience, take time out by yourself to note the changes in the environment from day to night and record how these changes affect you.

Worksheet
4.1e Effects of natural changes to environments

Seasons

Indigenous Australians understand the seasons and the natural changes that occur throughout them and utilise them to their advantage. The movement of people across landscapes was often dictated by seasonal changes in particular regions. For example, prior to European settlers dominating the landscape, the Indigenous peoples of Victoria's north-east would move to the river plain areas to take advantage of flowing water caused by snowmelt. This was a time of breeding for birds and other animals, and a range of plant foods was readily available.

At the end of the floods, people would return to the riverbanks and then travel to the high plains for both the cooler weather and the bogong moths. This was a time of ceremonies, initiations, trading, dispute settlement, marriage and renewing of acquaintances. When the weather cooled in the mountains, people would once again descend to the rivers and plains areas. Prior to departing, the dry grasses would be set on fire to create regeneration for the following year.

Prior to European settlement and the introduction of dams and locks on the river, during the dry period the Murray River was often a string of waterholes, filled with fish, animal and bird life attracted to the limited water supply. This made gathering food quite easy. After several heavy rains, the creeks and rivers in the foothills would be flowing again and people would move to these areas to seek shelter from the onset of the colder months.

Gwangal moronn
Season of honey bees – autumn (late-March to June)

Sunrises, bees and flocking birds
Autumn (the season of native honey bees or gwangal moronn) is when the country starts to cool down after the summer heat.

Chinnup
Season of cockatoos – winter (June to late-July)

Cold, cockatoos and early wildflowers
Morning frosts, bleak mists and freezing winds make winter (season of cockatoos or chinnup) the coldest time of year.

Larneuk
Season of nesting birds – early spring (late-July to late-August)

Nesting birds and changeable weather
Early spring (season of nesting birds or larneuk) is usually the wettest time of the year with rivers running high. It is a time of dramatic weather changes, with up to six seasons in one day.

Petyan
Season of wildflowers – late spring (late-August to mid-November)

Wildlife and wildflowers
In spring (season of wildflowers or petyan), the bush bursts into life. Nature's rock gardens amaze. The days are warmer, although the weather can still be tempestuous.

Ballambar
Season of butterflies – early summer (mid-November to late-January)

Warmth, butterflies and wetland plants
With the onset of summer heat, the land dries and the weather stabilises. This is the season of butterflies or ballambar.

Kooyang
Season of eels – late summer (late-January to late-March)

A parched landscape
Late summer (season of eels or kooyang) is the hottest and driest time of the year. The risk of bushfire (Piikorda) is high. Streams dry up.

Figure 4.15 The traditional peoples of Gariwerd (the Grampians), the Jardwadjali and Djab Wurrung, recognised six distinct weather periods that influenced their lives and the surrounding flora and fauna.

Worksheet
4.1f Natural changes report

LEARNING ACTIVITY 4.1F

Natural changes report

1 How do seasonal variations influence and affect your participation in outdoor activities?
2 Investigate the relationship between the sun, the Earth's position on its axis and the seasons. Produce a report that explains how these elements interact, utilising diagrams.
3 Describe the effects of the changing seasons on people and the outdoor environment.

Tides

Tides are the rise and fall of sea levels caused by the combined effects of the gravitational forces exerted by the moon and the sun, and the rotation of the Earth. As the moon circles the Earth, its gravity gently pulls the waters of our seas and oceans toward it. In Australia, most coastal areas (with some exceptions) experience two high tides and two low tides every 24 hours and 50 minutes; therefore, high and low tide times vary every day. The seashore can be divided into four separate zones based on the tides:

- splash zone, which is rarely covered by water
- upper zone, which is covered by water during some high tides
- middle shore, which is covered at every high tide and exposed every low tide
- lower shore, which is almost always covered by water.

Different plants and animals live in different seashore zones.

Getty Images/Jason Edwards

Figure 4.16 There are many treasures to be found on the shore at low tide along the Victorian coastline. However, even small children should be taught that, as responsible global citizens, we do not take treasures home but leave them on the beach.

THE INTERTIDAL ZONE

The area exposed to the air between the high and low tide marks is known as the **intertidal zone**. This area provides insight into the mysterious lives within marine environments.

With increases in the urbanisation of coastal areas in Australia, more people are visiting our beaches on a regular basis, placing greater pressure on the health of these areas. The intertidal zone has been a popular place to collect bait-like worms, clams, yabbies and crabs, leading to loss of biodiversity. Beachside developments such as piers, seawalls, groynes and marinas have influenced the natural tidal processes and marine life. People collecting shells from intertidal zones could reduce the availability of habitat for animals such as hermit crabs.

KING TIDES

King tides can lead to the loss of sand dunes and coastal vegetation, and can also lead to flooding of developed areas. It is possible to harness the power of tides to produce clean **renewable energy**.

intertidal zone
the area of foreshore and seabed exposed to air at low tide and submerged at high tide (i.e. the area between low-tide and high-tide marks)

renewable energy
energy that can be obtained from natural resources that can be constantly replenished

FAST FACT

The highest tides occur with the coming of the full and new moons, and many believe these natural events coincide with a rise in mental illness, hence the term 'lunacy'.

LEARNING ACTIVITY 4.1G

Tides report

1 Investigate the relationship between the moon and tidal flow.
2 Produce a report that explains how these two elements interact and why. Include diagrams.
3 List the effect of the tides on people and on a specific outdoor environment.

Worksheet
4.1g Tides report

Floods

Shutterstock.com/Ahmed Alkyat

Figure 4.17 Flooding in regional Victoria, Shepparton, October 2022.

The effects of floods can sometimes be devastating to those who live and work in low-lying areas, with loss or damage to life and property. However, it is important to remember that flooding is part of the natural water cycle. Land that borders waterways, such as rivers, creeks, lakes and swamps, form natural floodplains. The size of floodplains can vary from a few hundred metres to hundreds of kilometres, such as the lower Goulburn Valley around Shepparton and the Riverina area of the Murray River. Both of these areas experienced severe flooding as recently as the end of 2022 during a La Niña–influenced climate cycle.

Prior to European settlement, the Yarra River had experienced floods up to 10 metres above the normal level of the river. Throughout the 19th century, there were several large floods that inundated the central business district of Melbourne and other low-lying areas such as South Melbourne and Collingwood. Because of natural valley storage areas installed at Eltham, Yering Flats and Fairfield, the incidence of major flooding along the river has been greatly reduced, but it can still occur.

FLOODPLAIN MANAGEMENT

> **FAST FACT**
> In recent years, efforts have begun towards rehabilitating some waterways to return them to their natural state, including sections of the Moonee Ponds Creek.

Fairfax Photo/Justin Mcmanus

Figure 4.18 Concrete channels along Moonee Ponds Creek

The need to avoid future problems for both humans and the outdoor environment is crucial, but planning and development processes in the past have been known to overlook such needs. Effective floodplain management can be achieved through introducing measures that concentrate on preventing new problems caused by urbanisation and development, and alleviating existing problems. Management strategies that can have an impact on the frequency and volume of flooding include flood warning and emergency procedures; the construction of retarding basins, storage areas and levee banks; excavation of waterways; and concreting of channels (such as along the Moonee Ponds Creek). Many of these measures can have extremely negative effects on the natural environment through the removal of natural vegetation and associated habitats.

Stream erosion is a natural process, particularly around bends. However, changes in land adjacent to a stream can lead to instability and result in continuous erosion along its channel. Such changes include increased runoff from **impervious surfaces** (such as paved roads and car parks) and piped tributaries, uncontrolled access by stock or direct interference, such as straightening and channelling. When a stream has been straightened or channelled using concrete banks, the stream may be stable with little erosion, but it will have little vegetation cover and provide a minimal amount of habitat.

impervious surfaces areas that have been covered by any material that impedes the infiltration of water into the soil

BENEFITS OF FLOOD

Flooding can be essential in the reproductive cycle of some plants. For example, the river red gum, *Eucalyptus camaldulensis*, needs to be inundated by flood waters for its seeds to germinate. The lack of flooding of rivers in some areas, such as in the Barmah National Park, because of river damming for irrigation, has led to the absence of young river red gum trees in these and other areas.

Drought

According to the Australian Bureau of Meteorology, drought is 'a prolonged, abnormally dry period when the amount of available water is insufficient to meet our normal use'. Drought does not mean low rainfall – it is measured on the availability of water based on our ever-increasing needs. For this reason, human interactions with the environment can impact greatly on whether an area is considered to be drought affected. Droughts are usually associated with extended dry periods of lower-than-expected rainfall and in eastern Australia are highly influenced by the El Niño– Southern Oscillation (ENSO) climate cycle.

Australia is the driest inhabited continent in the world. Our climate is highly variable – drier periods and wetter periods are natural cycles, and their extent varies from year to year. Therefore, we must learn to live and adapt to naturally occurring drought conditions. Based on climate records for the past 100 years, on average in Australia for every 10 years there will be three years of drought.

Shutterstock.com/Lauren Cameo

Figure 4.19 A river red gum (*Eucalyptus camaldulensis*) at Barmah National Park relies on regular flooding to thrive.

Impacts of drought

Drought can have far-reaching effects on outdoor environments and the people who live and rely on the services provided by these areas. Some of the impacts of drought on the outdoor environment not only include the initial loss of crops and livestock, toxic algae outbreaks and increased threat of bushfires, but also erosion and the loss of valuable topsoil, leaving land infertile for years to come.

Ultimately, drought impacts upon the lives of all Australians. Drought can have a significant impact on the economy due to the reduction in agriculture production and associated industries such as transport and manufacturing, leading to food shortages and rising food prices. In some countries, drought can lead to widespread famine, causing many deaths.

Alamy Stock Photo/Raymond Warren

Figure 4.20 Drought affects the lives of all Australians.

Droughts can lead to water restrictions, such as limiting the watering of gardens and lawns, and the washing of motor vehicles. These limitations encourage people to adopt water-saving practices in their homes and businesses. Major droughts also have social implications, such as an increased incidence of depression and suicide among farmers affected by a prolonged lack of rainfall.

Fire

Fire occurs both through natural and human-induced triggers in Australia; can be deliberate, accidental or natural (i.e. as a result of lightning strike); and can vary greatly in **fire intensity**. Indeed, our outdoor environments have developed over time to rely on fire in their natural patterns. For example, many vegetation species (such as the alpine ash) require fire to begin their own reproductive processes.

fire intensity
the amount of heat energy released by a fire; higher intensity equals a hotter fire

LIGHTNING

Lightning is the natural form of fire ignition, often striking on a hot summer day as a cool weather change meets the warm air and begins a storm. Often, fire is started by dry lightning strikes during El Niño weather cycles where no rain falls at the same time. This is an example of a natural ignition and can be the catalyst for a large-scale fire. Emergency service fire personnel and aircraft are often deployed to extinguish fires as quickly as possible.

HUMAN-INDUCED TRIGGERS OF FIRE

Human-induced triggers of fire in outdoor environments include the following:

- Controlled burning – this is where forest management personnel deliberately ignite fires, which are then kept under their control. This management technique occurs in the cooler shoulder months of summer to create a fire that burns only the lower understorey of a forest, usually in locations sharing boundaries with private property or towns. The purpose of the fire is to reduce flammable material build-up to decrease the risk of extreme fires occurring in the summer time in this location.
- Infrastructure – this can occur through a power line falling down, damage to a pole, trees touching the power line, equipment failure or trains throwing out sparks during brake failure.
- Individual human activity – unextinguished cigarettes, burning-off debris, campfires and equipment use (e.g. chainsaws or angle grinders) can all be triggers for fire ignition, along with deliberate fire lighting, which is known as arson.

Figure 4.21 The impact of fire on bushland

IMPACTS OF FIRE

After a fire, the rate of recovery will depend on the length of time that has elapsed since the last event, the fire's intensity and the season in which the fire occurs. As the plants mature in the regrowing forest, they will require light, nutrients and water. Some species will be displaced because of insufficient amounts of these components. This will lead to a decrease in plant density. As the dominant species, eucalypts will grow and reduce the light reaching the forest floor. Species richness will temporarily decline as a result, and the understorey plants will change as shade-tolerant species become more abundant. These changes will suit some faunal species more than others, resulting in further change. Low-intensity fires can cause sterilisation of soils, while higher temperatures can alter the amount and availability of nutrients such as nitrogen and phosphorus. This can result in more fertile soil, but also raises the risk of these water-soluble nutrients being washed away by rainfall.

Worksheet
4.1h Alpine ash

LEARNING ACTIVITY 4.1H

Alpine ash

1. Investigate online the special characteristics of alpine ash and mountain ash trees that make them reliant on fire for their reproduction.
2. Prepare a written response that describes the necessity for fire as part of any management of alpine ash and mountain ash forests.
3. Describe the risk of frequent fires to alpine ash and mountain ash forests.

Migration

Migration occurs when species move from one location to another in response to changes in habitat. When the variation in habitat is predictable, so is the pattern of migration. The significance of migration is that animals move between habitats as each is only suitable for part of their lives. Migration can include dispersal to new areas, nomadism (or wandering) and **immigration**. Some migration patterns are linked to climate and food availability, while others relate to breeding seasons and habitat change. The large range of migratory patterns within Australia is a reflection of the range of habitats available.

migration
the movement of a species from one location to another in response to changes in habitat

immigration
the movement of a species into another country or region to which they are not native in order to settle there; species may be seeking better resources or mating partners, or escaping danger and environmental changes

BOGONG MOTH MIGRATION

In spring, millions of bogong moths migrate from the inland plains of eastern Australia to the Victorian Alps and Snowy Mountains to escape the summer heat. The moths provided a regular high-protein and fat food source for Indigenous peoples for thousands of years; these days the local mammals and birds rely on the annual migration for the same reasons.

MUTTON-BIRD MIGRATION

The short-tailed shearwater (commonly known as the mutton bird) breeds in burrows along the shores of Tasmania in summer. It undertakes one of the longest migrations in the world to the north Pacific (via Japan) and along the west coast of America before returning to Australia. Chicks emerge from their burrows at about 97 days old and follow their parents (who have departed up to two weeks earlier) along the migratory path. This route seems to take advantage of available food, while utilising prevailing winds.

Migrating birds and the outdoor environments they seasonally rely on for habitat and resources can be recognised and protected internationally (such as being identified as a Ramsar site). This can have a positive effect on the outdoor environment – more funds and stricter regulations in place may increase the health and sustainability of these areas. Stricter planning regulations to protect the habitat of migrating birds may place limits on how people can use outdoor environments, such as limiting development in protected areas.

iStockphoto/Imogen WarrenT

Figure 4.22 The *Puffinus tenuirostris* (short-tailed shearwater) is commonly known as the mutton bird.

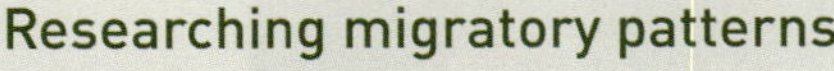

LEARNING ACTIVITY 4.1I

Researching migratory patterns

1. Investigate the migratory patterns of a species in an area you have visited this year or the area where you live.
2. Prepare a written response, including a map, along with a report to be presented to the class.

Worksheet
4.1i Researching migratory patterns

Climate change

> Long-term changes in the climate, particularly the changed incidence of extreme events such as cyclones and bushfires, place pressure on many aspects of the Australian environment. Further changes in the climate, driven by past and future emissions of greenhouse gases, will continue to make climate a major pressure on the Australian environment and communities for the foreseeable future.
>
> Department of Climate Change, Energy, the Environment and Water, 2021

Introduction | Australia state of the environment 2021 (dcceew.gov.au)

Figure 4.23 The profound effects of climate change on our environment and society

Australian climate has and is changing. Following rising temperatures over the past 100 years, average temperatures are now at unprecedented levels, and are consistent with global findings that recent years have been warmer than any multiyear period in at least the past 2000 years. Average temperatures over Australia have increased 1.4 °C since 1910. Human activity, particularly global emissions of **greenhouse gases**, has been the main driver of global temperature increases.

However, Australian greenhouse gas emissions have fallen substantially since 2007, although the rate of change has decreased since 2013 (except for large falls in 2020 that are partly due to reductions resulting from the COVID-19 pandemic, which are likely to be temporary). Changes in our land use – particularly in the forestry sector, which has moved from being a net source to a **net sink** – has made the biggest difference. The increasing proportion of Australian electricity generated from renewable sources has also contributed to the decrease.

greenhouse gas
a gas in an atmosphere that absorbs and emits radiation; examples include carbon dioxide, methane, nitrous oxide and ozone

net sink
where a process such as forestry absorbs more carbon than it emits

We have already observed major changes to climate in Australia, including:

- increased frequency and intensity of heat extremes (on land and in the ocean)
- rising sea levels
- rainfall increases in northern and eastern regions during summer periods and decreases in the south-east.
- increased occurrence of dangerous fire weather in southern and eastern Australia that's well outside the range of previous historical record.

Adaptation to climate change is very important to reduce catastrophic long-term changes to society and the environment, and is being pursued across all levels of government and industry. However, climate is changing at a speed that is not allowing natural adaptation to keep pace and is presenting massive challenges to human societies.

Climate change is also having a profound effect on Indigenous peoples as their use and relationship with the environment is different from that of non-Indigenous people. Their cultural knowledge and traditional practices, as well as understanding of climate and weather, are all connected to the slow cycles and climatic variations Australia has experienced over tens of thousands of years. With climate now changing more rapidly, traditional knowledge and Indigenous methodologies are being forced to adapt, and can be used to complement non-Indigenous Australians' developing knowledge of Country.

LEARNING ACTIVITY 4.1J

Climate change investigation

1 Investigate an outdoor environment you have visited or studied this year. Describe the long-term changes that have shaped this environment, including climatic, geological and vegetation changes. Use illustrations to explain your findings.
2 Create a concept map illustrating some of the long-term effects that have occurred as a result of human impact on the specific environment.
3 Prepare a response to the statement: 'Extinction is a naturally occurring long-term change, and as such humans should not interfere with this process.'

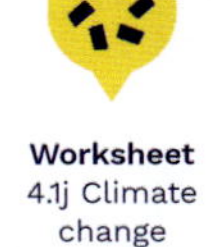

Worksheet
4.1j Climate change investigation

NOTES FOR THE EXAM

For the exam, you should be able to:

- describe the effects of natural and human-induced changes on a range of outdoor environments such as day to night, seasons, tides, flood, drought, fire, migration and climate change
- compare abiotic and biotic interrelationships in an outdoor environment
- analyse the effect of natural and human induced changes and apply these to a range of environments.

4.1 KEY CONCEPTS

- Perspective is a particular attitude towards or way of regarding something; a point of view and is influenced by a person's:
  - previous experiences
  - beliefs
  - knowledge of the situation/topic.
- Biogeochemical cycles allow elements to move throughout ecosystems. These include water, carbon, nitrogen and climatic cycles.
- Food webs involve links called trophic levels, from decomposers to tertiary consumers.
- Biotic components of an ecosystem are living, like plants and animals, and abiotic components of an ecosystem are non-living, like rocks, water and the atmosphere.
- Biotic and abiotic components interact through ecosystems and natural cycles.
- Natural and human-induced changes on outdoor environments occur at three levels: individual organisms, community and large-scale.
- Natural and human-induced changes can affect:
  - day to night (nocturnal and diurnal species)
  - seasons (including Indigenous seasons)
  - tides (low and high plus intertidal zone of life)
  - floods (exacerbated by La Niña climate cycle)
  - drought (exacerbated by El Niño climate cycle)
  - fire (lightning and human-made)
  - migration (immigration and emigration)
  - climate change (temperature averages in Australia have increased 1.4°C since 1910).

4.1 CONCEPT QUESTIONS

Worksheet
4.1 Key concept

REMEMBERING

1 Define the following terms:
   a ecosystem
   b biotic
   c abiotic.
2 Give four examples of natural systems.
3 Describe the steps of one of these natural cycles.

UNDERSTANDING

4 Identify and describe the inputs, components, processes and outputs of a specific environment you have visited or investigated this year.
5 Identify which of the following are biotic or abiotic components of the environment:
   a eucalypts
   b granite
   c kangaroo grass
   d nitrogen
   e oxygen
   f red-back spider
   g sunshine
   h wallaby
   i water
6 Describe five biotic and five abiotic requirements for human survival.

APPLYING

7 What would happen without the sun's energy to a natural system such as the water cycle?

8 Evaluate the impact of the collapse of ocean currents bringing warm air from the Pacific Ocean to eastern Australia.

EXTENSION CHALLENGE

9 Analyse the impact of climate change on outdoor environments. Describe what will happen to outdoor environments if average temperatures increase by more than 1.5°C worldwide.

10 Describe strategies for reducing climate change and adapting to climate change currently being implemented or developed.

4.2 INDIGENOUS PEOPLES' LAND MANAGEMENT AND PERSPECTIVES

KEY KNOWLEDGE

- Indigenous peoples' land management understandings and perspectives of an outdoor environment

KEY SKILLS

- explain Indigenous peoples' perspectives and different forms of land management within an outdoor environment

4.2.1 AN ANCIENT CIVILISATION

Indigenous peoples of Australia have lived on this country for at least 60 000 years, according to confirmed evidence of rock art paintings in the Northern Territory, and active archaeological research is constantly being undertaken to further explore this timeline of human interaction.

Understanding the timescale of Aboriginal and Torres Strait Islander peoples' habitation of the Australian continent highlights their rich and profound knowledge of this land and their deep understanding and experience in managing it. This understanding also serves to highlight how much non-Indigenous Australians can learn from this knowledge and experience.

middens
mounds of discarded shells built up over time; some of the most visible archaeological evidence of Indigenous peoples' campsites and diets

Our knowledge of when the earliest people arrived in Australia is based on archaeology, a science that involves unearthing the physical remains of human activity. In Australia, these remains are mainly in the form of stone tools, rock art and ochre, shell **middens**, charcoal deposits and human skeletal remains.

The following timeline of the first human settlement of Australia is based on archaeological discoveries to date.

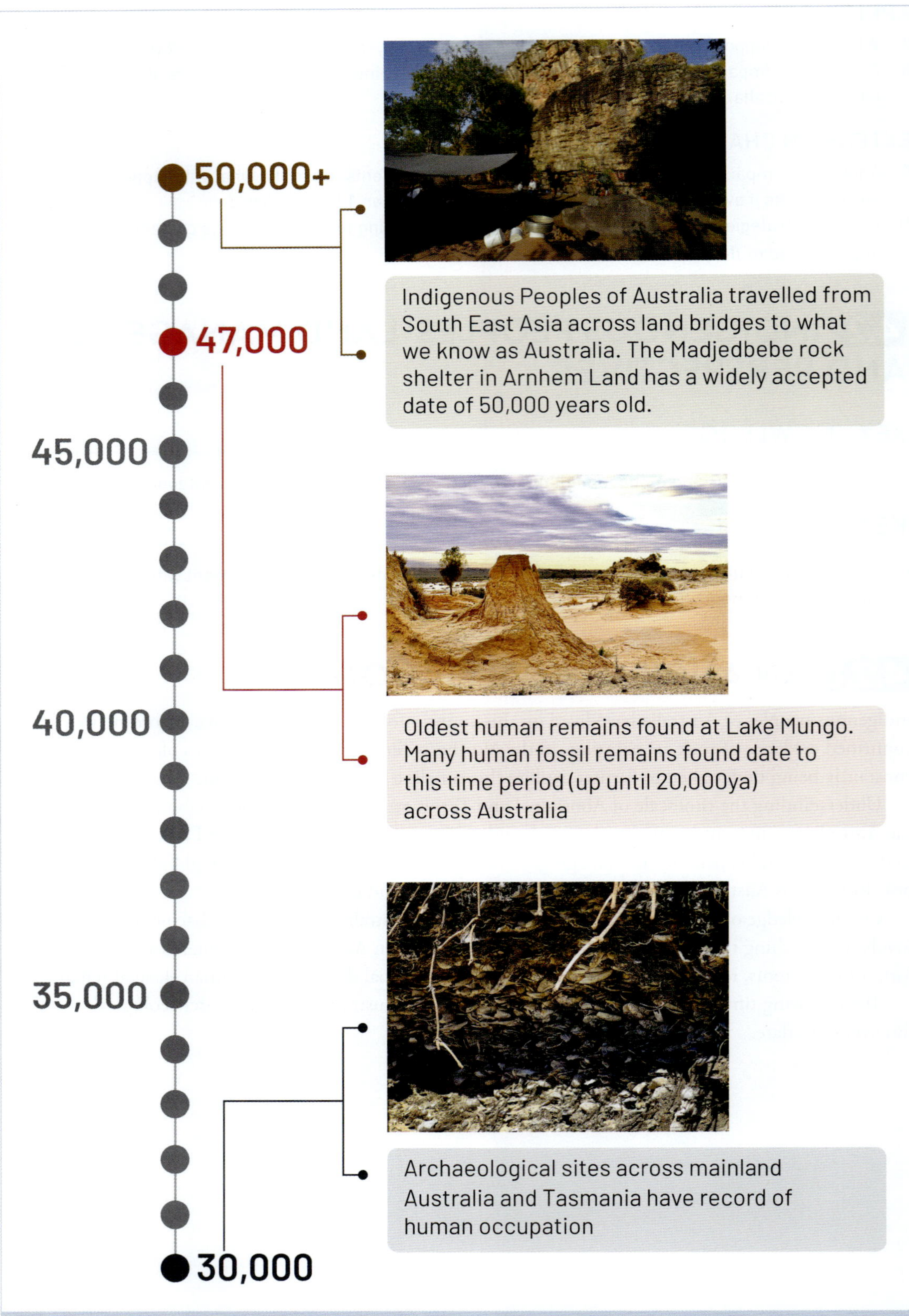

Figure 4.24 Timeline of Aboriginal and Torres Strait Islander peoples living in Australia

4.2.2 INDIGENOUS PEOPLES' PERSPECTIVES ON LAND MANAGEMENT

Aboriginal and Torres Strait Islander peoples' perspectives on how best to manage land is informed by this time-tested relationship. Their perspective is that the land is their mother, and is to be respected as the giver of life. All daily requirements come directly from the land, which requires proficient knowledge, as a lack of understanding can have dire consequences. Thousands of years of experiencing the natural and climatic variations of Australia has informed Indigenous Australians about how best to care for their Country and continue to thrive as communities.

> For Aboriginal peoples, country is much more than a place. Rock, tree, river, hill, animal, human – all were formed of the same substance by the Ancestors who continue to live in land, water, sky. Country is filled with relations speaking language and following Law, no matter whether the shape of that relation is human, rock, crow, wattle. Country is loved, needed and cared for, and country loves, needs and cares for her peoples in turn. Country is family, culture, identity. Country is self.
>
> 'Ambelin Kwaymullina, 'Seeing the Light: Aboriginal Law, Learning and Sustainable Living in Country', *Indigenous Law Bulletin* May/June 2005, Volume 6, Issue 11

Alamy Stock Photo/Chris Putnam

Figure 4.25 Gulgurn Manja Shelter Walk in the Northern Gariwerd (Grampians) region of Victoria, Australia

FAST FACT

Australian Indigenous art is the oldest ongoing tradition of art in the world. Indigenous rock art dating back at least 60 000 years can be found in the Northern Territory, depicting extinct megafauna, animals of the surrounding environment and the Dreaming – when ancestral spirits came to the land and created rivers, plants, people, animals and tribal laws.

LEARNING ACTIVITY 4.2A

Analysis of an outdoor environment

Select a Victorian outdoor environment you have visited or investigated this year.

1 Name the Indigenous people or peoples who are the traditional owners of this land.
2 Describe how they interacted with their Country.

Worksheet 4.2a Analysis of an outdoor environment

4.2.3 INDIGENOUS PEOPLES' LAND MANAGEMENT

Indigenous land management practices are most readily observed in areas in northern Australia where traditional lifestyles have been less interrupted by European colonisation. In Victoria, local peoples were often displaced by colonisers, which interrupted the practice of traditional land care, leading us to rely on longstanding oral histories shared by Elders, past and present. Today, these histories are being recorded and being used to continue or reinstate traditional land practices by Aboriginal peoples (see the later section, 'Contemporary Indigenous land management practices').

Our other source of observation of Indigenous land management is early European explorers and their diaries. For example, as William Hovell navigated the mountainous terrain of north-east Victoria in 1824, he wrote about 'encounters with the Indigenous peoples of the region noting land management techniques, such as grass burning and damming of rivers to catch fish'. Many more explorer accounts that support Indigenous peoples' land management are highlighted in the following sections.

Traditional burning

Fire was an important way in which Indigenous peoples cared for the land, providing a sustainable form of land management that kept it thriving and healthy. Evidence of this can be found in

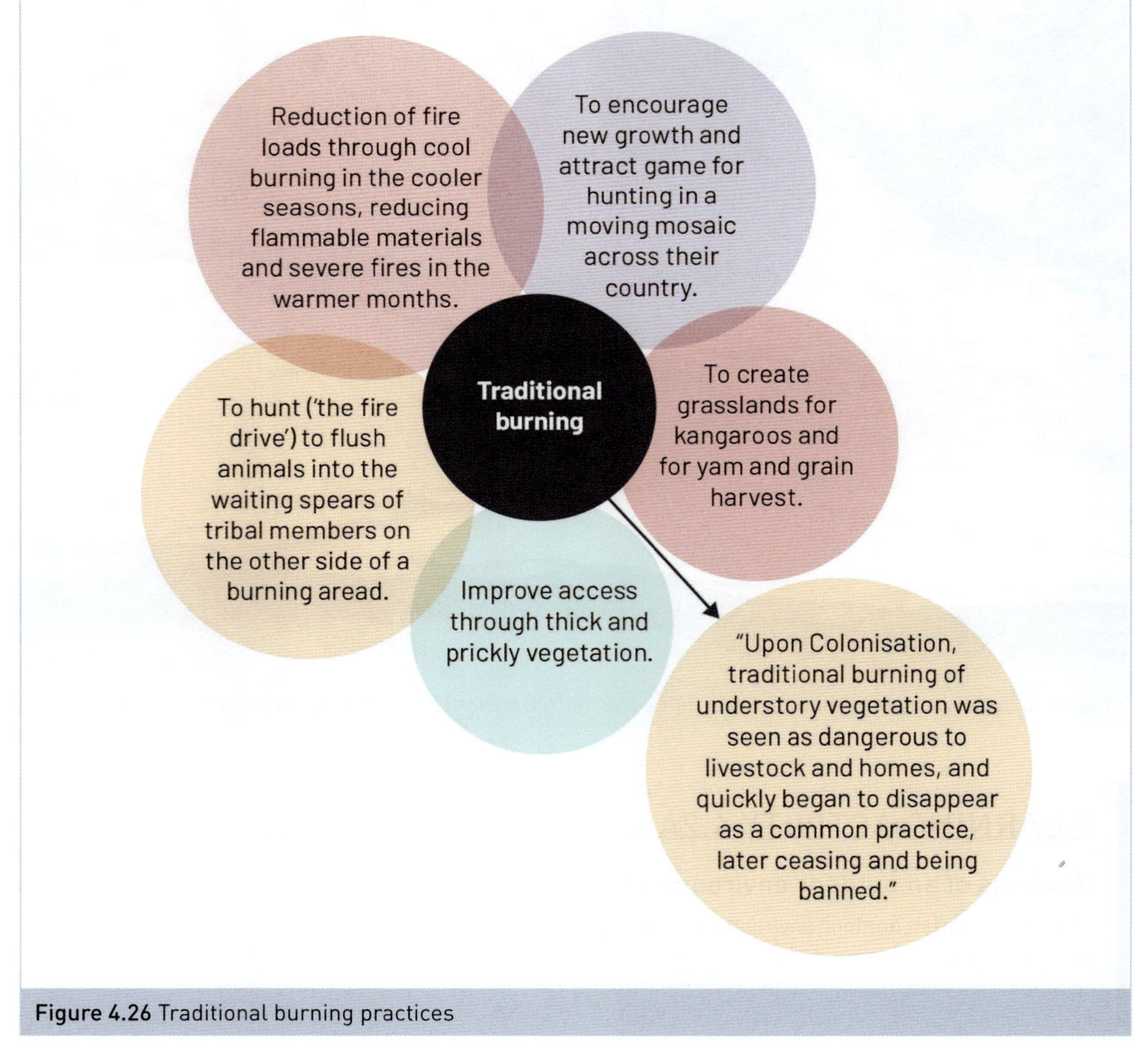

Figure 4.26 Traditional burning practices

Sir Thomas Mitchell's 1848 journal, in which he observed that fire was used by local Aboriginal peoples as a method of farming and creating open forests to ensure food for both people and wildlife. Aboriginal peoples have used fire intentionally and frequently for thousands of years, limiting burns to low intensities contained within relatively small areas. This made the impacts of fire a natural part of the landscape and aided life on the land in several ways

4.2.4 AGRICULTURE

Historical and archaeological findings in recent years have supported the development of a new perspective on the practices of many pre-European Indigenous communities, including those spread throughout Victoria. *The Biggest Estate on Earth* by historian Bill Gammage, *Dark Emu* by Indigenous writer Bruce Pascoe, along with anthropologist and architect Paul Memmott's 2022 publication *Gunyah Goondie + Wurley*, all describe and discuss the growing evidence that Indigenous communities practised a large-scale and long-term form of agriculture.

Pascoe, citing much of his evidence from the eyewitness accounts of early European settlers and explorers, describes communities that were practising plant domestication, sowing, irrigation, harvesting and storing of foodstuffs. In their journals, some early explorers of Australia described scenes of large-scale agriculture by Aboriginal people, including Sir Thomas Mitchell, who observed '9 miles of harvested grain', and Sir George Grey, who described a large area of 'deeply tilled gardens used for growing vegetables'. For Gammage, this was evidence of a sophisticated form of land management developed and refined over the many thousands of years of Indigenous communities' engagement with Australian environments.

J. H. Wedge (1835)

Figure 4.27 LEFT: This drawing by J. H. Wedge (1835) shows women digging roots of the Murnong (yam daisy) using a digging stick (RIGHT). According to Bruce Pascoe, 'This practice was observed in the 1830s by explorer Sir Thomas Mitchell and again by George Augustus Robinson in 1941. Land was managed through fire to encourage grassland and the proliferation [or spread] of this food source.'

4.2.5 HUNTING AND GATHERING

'Hunting and gathering' is a term used to describe a way of directly living off the land to source food. For Aboriginal and Torres Strait Islander peoples, this was a way of life that supported and sustained both themselves and the land.

Alamy/ Trung Nguyen

Figure 4.28 Bush food from Central Australia

HUNTING

Hunting for undomesticated animals (such as kangaroos, emus and wallabies) provided a nourishing source of protein. The hunting of animals was not indiscriminate (meaning that a person could not hunt and kill any or as many animals as they wanted) but rather was guided by **customary laws** that governed not only the killing of certain species but also the consumption and distribution of food within the group. These laws described sacred areas where no hunting was allowed and included the concept of totemism, which protected certain animals that were sacred to the group.

customary laws
a series of laws or rules informed by Kinship, enforced by sanctions and used as a way of resolving disputes

> Hunting ties the past to the present, but is not simply a survival of some prior subsistence gambit ... Most importantly, it is an aspect of the law. As such it offers a venue through which certain men can and do display concern for the belief system ... Just like ritual, hunting affords men the opportunity of making claims regarding their position and right to authority in the group ... To hunt, then, is, as with ritual participation, to follow the Law, demonstrate its great potency and guarantee its continuance.

SACKETT, L,'The Pursuit of Prominence: Hunting in an Australian Aboriginal Community' (1979) 21 Anthropologica 197.

As mentioned, customary laws around totemism and sacred places shaped what was hunted, and how, where and when. This resulted in deeper care for the land and the protection of certain species and places from hunting. Today, we describe this approach as sustainable **land management**.

land management
the responsibility of managing the use and development of land resources

Totemism

Totemism refers to a spiritual and physical connection between humans, animals and the natural world, which is governed by a complex belief and values system. Individuals and groups in Indigenous cultures are given an animal or totem. This animal is seen as a close relative and needs to be carefully protected and taken care of, and therefore cannot be hunted.

totemism
an Indigenous term referring to spiritual and physical connection between humans, animals and the natural world that is governed by a complex belief and values system

Sacred places

Sacred places fulfil many functions for Indigenous peoples. In relation to land management, sacred places are regarded as prohibited areas, within which all hunting and food gathering is forbidden. Even wounded animals cannot be pursued into such forbidden places. This also allowed space for the safety of native plants and animals, further supporting the sustainable management of the land.

LEARNING ACTIVITY 4.2B 

Totems

Watch the video, 'Totems' by Lynette Riley from the University of Sydney to learn about totems. Riley, a Wiradjuri and Gamilaroi woman from NSW, describes how totems link a person to land, air, water and geographical features and are about creating a balance.

Worksheet
4.2b Totems

Weblink
Totems

Aquaculture

The Gunditjmara people of south-western Victoria provide the best example of **aquaculture** in Victoria. Evidence of thousands of years of aquaculture can be seen where the people used volcanic rock created by the Budj Bim (Mt Eccles) lava flow to create a complex of eel traps, weirs, dams and channels. This practice ensured year-round food and encouraged the Gunditjmara people to settle, building stone houses and developing more sophisticated fish traps. These eel traps are now part of a UNESCO World Heritage site.

Fairfax Photo/Jason South

Figure 4.29 Gunditjmarra Aboriginal site, Budj Bim

sacred places
areas or natural features in the landscape that are significant under Aboriginal tradition, including rocks, waterholes, trees, plains and billabongs

aquaculture
the farming of aquatic organisms, including fish, molluscs, crustaceans and aquatic plants, with some sort of intervention in the rearing process to enhance production

Natural resource use

You can find many examples of the ways Indigenous peoples use, and have used, the resources from their environments. A local example from Victorian Indigenous communities is the construction and use of stone axes, which were used particularly for stripping bark from trees to make containers and canoes.

CANOE BUILDING
1. Greenstone was collected to make axe blades from a site near Lancefield, north of Melbourne, known today as the Mount William quarry. Greenstone outcrops at the quarry were superheated with large fires set close to the rock faces. The heat from the fires created cracks in the rock into which branches were wedged. Greenstone pieces were then levered off the main rock faces.
2. The axe heads were sharpened (requiring immense skill) through long, slow grinding on wet rocks by rivers and creeks. There are many places around the state today where these grinding holes can be seen.
3. Black wattle branches were used to create handles. The wattle branch was bent back on itself to form a loop into which the axe head was placed. Sap collected from the fronds of grass trees was used to make a glue to help bind the axe head to the wattle handle. Finally, kangaroo sinew was used to bind the handle tightly so that the wattle loop couldn't spring free.
4. The axe was used to create an outline of the canoe on the tree. Toeholds were cut into the tree to help the canoe builder get up high enough. The outline was gradually and carefully deepened, and then stone wedges were placed along the length of the outline. By carefully hitting the wedges, the bark outline could gradually be peeled off the tree in one complete piece.
5. If the bark was removed at the right time of year when the tree was rich in sap, then it would be fired directly – laid over a fire to help soften the bark. If not, it would have to be soaked in water before firing. The firing helped make it easier to bend and shape the canoe, particularly the ends that could be bunched up and tied. Just the right amount of fire needed to be applied – too much and the bark would burn up.
6. The canoe was ready to assist its user to travel by river for trade or hunting.

4.2.6 CONTEMPORARY LAND MANAGEMENT PRACTICES

Land managers in a contemporary context are responsible for ensuring the long-term health and sustainability of outdoor environments, including the protection of water resources, control of pest animals, eradication of noxious weeds and conservation of soil. It is important that all land managers have an intimate knowledge and understanding of the land they are responsible for – their decisions can have a dramatic impact on the environment.

In Victoria, the Department of Energy, Environment and Climate Action (DEECA) is responsible for the management of Victoria's public land. Covering more than 4 million hectares, this includes parks and reserves, alpine resorts and catchments. More than 60% of Victoria is private land. Landowners, including farmers, have many obligations in terms of managing their land.

Auscape/Jean-Paul Ferrero

Figure 4.30 A canoe tree

Contemporary Indigenous land management practices

Land management by Indigenous peoples, also referred to as 'caring for Country', has been steadily increasing in recent times, and includes a range of environmental, natural resource and cultural heritage management activities, including water management, the harvesting of food and fibre, and the undertaking of controlled burns. Indigenous land management has a clear purpose and is focused on protecting, maintaining, healing and enhancing the health and ecological diversity of the land and broader ecosystems.

In 2020, the Department of Agriculture, Fisheries and Forestry mapped the land in which Indigenous peoples and communities have ownership, management and co-management, or **other special rights**. Their report offers the following breakdown:

- 134 million hectares of land in Australia (17%) is Indigenous owned, including 22 million hectares of forest.
- 174 million hectares of land in Australia (22%) is under some form of Indigenous management, comprising:
  - 141 million hectares that is Indigenous managed, including 18 million hectares of forest
  - 33 million hectares that is Indigenous co-managed, including 10 million hectares of forest.
- 337 million hectares of land in Australia (44%) is subject to other special rights for Indigenous people, including 51 million hectares of forest.

other special rights
land or forest subject to Native Title determinations, registered Indigenous Land Use Agreements and legislated special cultural use provisions

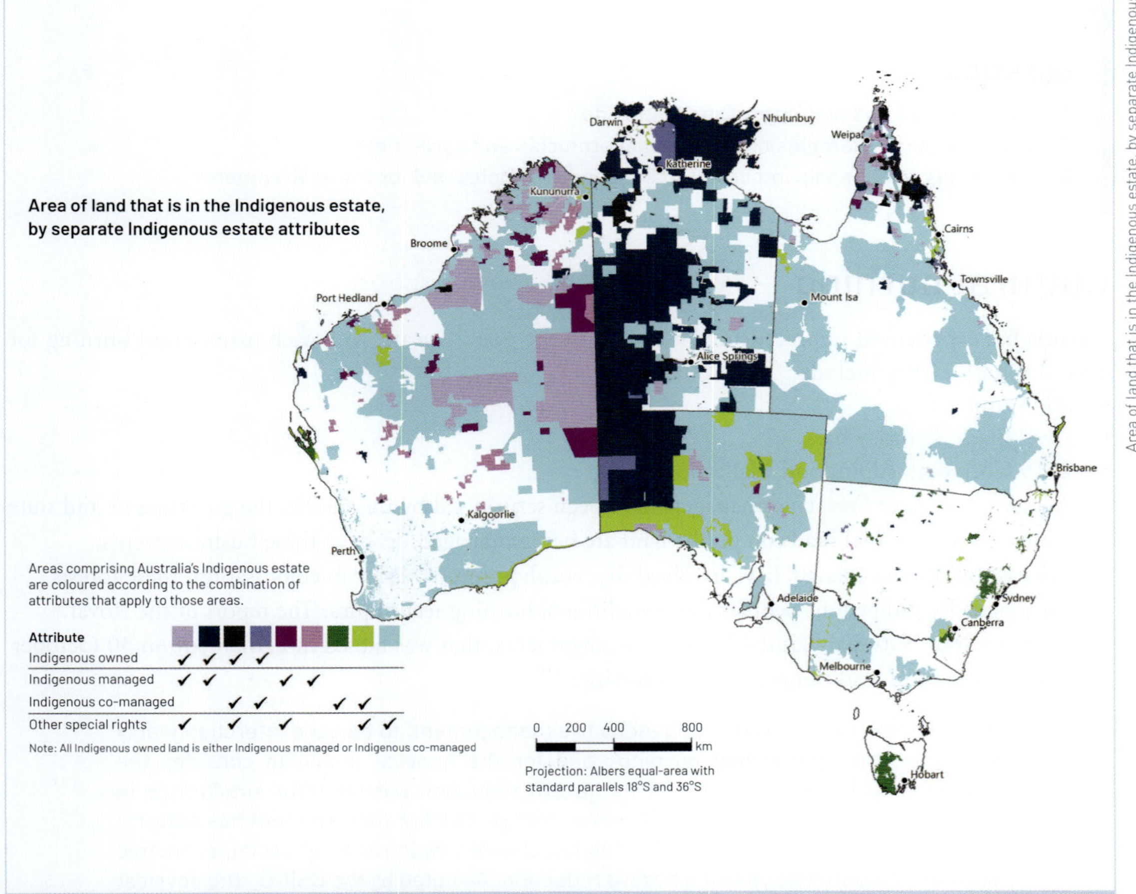

Figure 4.31 Area of land that is in the Indigenous estate

Resource
Case study: Gunaikurnai Land and Waters Aboriginal Corporation

CASE STUDY: CULTURAL BURNING – A LINK TO COUNTRY

Gunaikurnai Land and Waters Aboriginal Corporation

Gunaikurnai Land and Waters Aboriginal Corporation (GLaWAC) represents Traditional Owners in the Gippsland region of Victoria from the Brataualung, Brayakaulung, Brabralung, Krauatungalung and Tatungalung family clans, who were recognised in a Native Title Consent Determination, made under the Traditional Owner Settlement Act 2010 (Vic).

GLaWAC has a partnership with the Victorian Government to jointly manage 10 parks and reserves in Gippsland. These environments include forests, rivers, beaches, plains and animals and are all part of 'Country' and the cultural identity of the Gunaikurnai. This formal partnership arrangement brings together the combined skills, expertise and cultural knowledge of the Gunaikurnai people and the Victorian Government in a way that respects and values the culture and traditions of the Traditional Owners.

GLaWAC believes that collaboration and knowledge sharing between Traditional Owners and government agencies is integral in generating an adaptive fire management practice. The Corporation has formal and informal protocols and agreements with land management agencies on when and how they should be consulted and involved.

QUESTIONS

1 What does a Land and Water Corporation do?
2 What are some examples of the informal protocols and agreements?
3 Why is this relationship important to Indigenous peoples and for the environment?

Cultural burning

Victoria has experienced unprecedented bushfires in the past 20 years, with each major event burning for days or months. They include:

- 2003 Alpine fires
- 2009 Black Saturday fires
- 2019 Corryong, Alpine and Mallacoota fires.

As a result of these fires, land managers have been scrutinised by the public, the government and state parliament as to how **fuel loads** on public land are being managed. Each of these bushfire events, exacerbated by climate change, have involved dry, eucalypt forests. Prior to colonisation, these forests were managed by Indigenous peoples using traditional burning techniques. The report of the Royal Commission into National Natural Disaster Arrangements, that was tabled in Parliament on 30 October 2020, recognised this traditional practice, asserting:

fuel loads
a measure of how much fuel is present and available to burn in the form of grass, undergrowth or forest litter

> It is common for the term 'Indigenous fire management' to be used interchangeably with the term 'prescribed burning', and for the general public to consider the approach exclusively in terms of its hazard reduction outcomes or similarities in technique (e.g mosaic burning). However, Indigenous fire management has cultural origins and broader objectives. It aims to achieve a wide range of social, economic and cultural outcomes beyond hazard reduction. As noted by the CSIRO, 'the physical impact of Indigenous cultural burning is complemented by a cultural and symbolic significance that is passed from generation to generation'.

The Royal Commission went on further to note that 'cultural burning' is an example of how local knowledge has successfully informed land management for tens of thousands of years.

Today, Indigenous land management maintains its traditional and cultural importance, while also leveraging technologies such as helicopters and satellites. Cultural burning, already on the increase across Victoria, supported by Landcare and the Department of Energy, Environment and Climate Action, has had the spotlight further placed on it and we expect to see the Indigenous peoples' land management practice continue to increase for the benefit of all the lands and its peoples.

LEARNING ACTIVITY 4.2C

Traditional Aboriginal burning in modern-day management

Read the article, 'Traditional Aboriginal burning in modern day land management', published by Landcare Australia in 2017, to further grow your knowledge and understanding of Landcare's support of Indigenous peoples' management practices.

Worksheet
4.2c Traditional Aboriginal burning in modern-day management

Weblink
Traditional Aboriginal burning in modern day land management

Resource
Case study: Dja Dja Warrung people and cultural burning

CASE STUDY

DJA DJA WARRUNG PEOPLE AND CULTURAL BURNING

The Dja Dja Wurrung people of Central Victoria, combined with other Elders and Traditional Owner groups, have strong aspirations to increase collaboration with land and fire management agencies to facilitate the reintroduction of cultural burning in Victoria's natural landscapes.

Partly as a result of their input into the Victorian Traditional Owner Cultural Fire Strategy of 2019, they have secured funding (as the Dja Dja Wurrung Clans Aboriginal Corporation) to employ Forestry and Fire Team Leaders to lead a team of fire crew to support fire and forestry works on Djandak (meaning Country in Dja Dja Wurrung language) through the delivery of culturally informed fire activities.

View the short video, 'Djandak Wi – Traditional Burning Returns' by the Department of Energy, Environment and Climate Action.

Weblink
Video: Djandak Wi — Tradtional Burning Returns

Hunting and gathering

The traditional Indigenous practices of hunting and gathering continue in contemporary Australia, including through legally protected hunting of '**bush food**', which adheres to both customary law and Australian law. However, the land itself has changed, including the destruction of native flora and fauna, and has become heavily regulated. Although Indigenous hunting is permitted on Crown land, the space for this has shrunk dramatically.

bush food
the term for all foods native to Australia, such as seeds, starchy roots, fruits, vegetables, spices and nuts, as well as meat, insects, fish and seafood

Resource
Case study: Indigenous land and sea corporation

CASE STUDY

INDIGENOUS LAND AND SEA CORPORATION

Indigenous Land and Sea Corporation

The Indigenous Land and Sea Corporation (ILSC) was established in 1995 to provide for the contemporary and future land needs of Indigenous Australians, particularly those unlikely to benefit from Native Title or Land Rights. As the custodian of funds held in trust for Aboriginal and Torres Strait Islander people, we assist Indigenous Australians to acquire and manage land- and water-related rights so that they can enjoy the rightful entitlements, opportunities and benefits that the return and management of country brings. In redressing dispossession, the ILSC's acquisition and management functions serve to assist Indigenous Australians to:

1 Maintain and grow the value and productivity of country;
2 Own and manage country sustainably;
3 Influence policy and opportunity for country; and
4 Strengthen culture through reconnection to country.

Indigenous Land and Sea Corporation, 'Welcome to the Indigenous Land and Sea Corporation: Connecting People, Country and Opportunity', p. 5. © Copyright 2021 ILSC https://www.ilsc.gov.au/home/copyright/

The following two case studies exemplify the great collaboration of ILSC and Indigenous peoples' groups.

Iningai People Protect and Preserve Country for Future Generations

In 2018, the Yambangku Aboriginal Cultural Heritage and Tourism Development Aboriginal Corporation (YACHATDAC) approached the ILSC for assistance to protect valuable cultural heritage on Gracevale Station, an 8870-hectare cattle property near Aramac in Queensland. The Iningai Traditional Custodians had limited access to the property and significant cultural sites, including a 200 metre rock art wall that had not been formally recorded or studied.
In April 2019, the ILSC purchased Gracevale Station to support the Iningai Traditional Custodians to reconnect to their country and pursue opportunities to develop tourism and other socioeconomic, cultural and environmental benefits for local Aboriginal people. Since that time, the Iningai people have restored cleared land, excavated disused waterways and reduced cattle numbers on the property to protect culturally and environmentally significant country. In October 2020, the Iningai people reopened the rock art gallery to visitors and renamed the property to 'Turraburra', which reflects the traditional name of the area.

Indigenous Land and Sea Corporation, 'Welcome to the Indigenous Land and Sea Corporation: Connecting People, Country and Opportunity', p. 5. © Copyright 2021 ILSC https://www.ilsc.gov.au/home/copyright/

Nari Nari People Build Property Portfolio and Environmental Sustainability

The traditional home of the Nari Nari people, Gayini (Nimmie Caira) is an environmentally and culturally significant property on the Murrumbidgee floodplains in NSW. Since 2018, Gayini has been the focus of an ambitious and collaborative wetlands restoration program. There have also been over 2000 sites of cultural significance recorded on the property.
A successful partnership between the ILSC, the Nari Nari Tribal Council (NNTC) and the Nature Conservancy (Australia), with support from the John B Foundation, resulted in the official return of country to the NNTC in March 2020. Nari Nari access and reconnection with country has enabled the transfer of knowledge between generations and the development

of new enterprises, with benefits flowing to the local Indigenous community. A further $1.2 million of ILSC funding has assisted the NNTC to purchase vital land and water management equipment, removing the ongoing financial burden of costly equipment and contractor hire, and providing improved employment and training opportunities for local Aboriginal people.

> The Nari Nari people have been using traditional knowledge to sustain our country for thousands of years. Having the property back in Nari Nari hands has allowed us to pursue sustainable sources of income and the intergenerational transfer of knowledge of caring for country.
>
> Ian Woods, Nari Nari Tribal Council (NNTC) Chair.

NOTES FOR THE EXAM

For the exam, you should:

- be able to explain Indigenous peoples' perspectives of the land
- be able to explain different forms of land management techniques before and after colonisation by Indigenous peoples
- be able to provide specific examples of land management techniques in an outdoor environment you have studied.

4.2 KEY CONCEPTS

- Indigenous peoples have managed the landscape of Australia for up to 60 000 years.
- Indigenous peoples' perspective is that the land is their mother and must be cared for.
- Historically, Indigenous peoples managed land through:
  - traditional burning
  - agriculture
  - hunting and gathering governed by totemism and sacred places
  - aquaculture
  - natural resource use.
- Contemporary Indigenous management or 'caring for Country' is increasing and being formalised through collaborative relationships. This is represented in:
  - Indigenous peoples' land and land rights (17% of Australia)
  - cultural burning increasing, supported by Royal Commission report of 2020
  - hunting and gathering continuing on Crown and Indigenous lands.

4.2 CONCEPT QUESTIONS

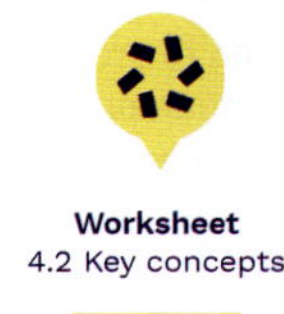

Worksheet
4.2 Key concepts

REMEMBERING

1 How long have Indigenous peoples lived in Australia?
2 What evidence is there for that period of occupation?

UNDERSTANDING

3 Explain what traditional burning is.
4 Compare the practice of traditional burning prior to European colonisation with the approaches used today.
5 Describe how Indigenous peoples practised agriculture prior to European colonisation.

APPLYING

6 Describe what an eel trap is, and where they were built and used extensively in Australia.
7 Describe how much land is formally Indigenous owned in Australia and how that may present challenges for its management and the maintenance of traditional land management practices.

EXTENSION CHALLENGE

8 Find an example not described in this text of an Indigenous group collaborating with land management authorities.

4.3 VOCATIONAL PERSPECTIVES OF OUTDOOR ENVIRONMENTS

KEY KNOWLEDGE

- understandings of vocational perspectives of outdoor environments, including at least two of the following:
  - natural resource management
  - nature-based tourism outdoor leading and guiding
  - environmental research and policy
  - land management roles including Traditional Owner groups and National Indigenous Australian's Agency programs
  - education
  - agriculture

KEY SKILLS

- compare a range of vocational perspectives on outdoor environments

4.3.1 VOCATIONAL PERSPECTIVES OF OUTDOOR ENVIRONMENTS

For this learning section, we are comparing the perspectives of different vocations and how that informs their understanding of outdoor environments.

vocational relating to an occupation or employment

When comparing **vocational** perspectives, we are comparing the different ways people regard outdoor environments through their occupation. Many people pursue jobs or careers in outdoor environments that align with their own perspective of outdoor environments; however, these can change over time. As your job or role, experiences, assumptions and values change, your perspective will also change.

Shutterstock/AuthorLinyt Photography

Figure 4.32 Working in the tourism vocation can be a pathway for a person to see the world through their employment.

Describing perspectives on outdoor environments using metaphors

The next aspect we need to explore is how to describe different perspectives towards outdoor environments. In Outdoor and Environmental Studies, we visit the topic of human relationships with outdoor environments often, and one way to describe these relationships is through **metaphor**.

Basically, a metaphor is when two unlike things are compared with each other because of something they have in common. Metaphors are often used when we are trying to understand something complex, we use a metaphor to compare it with something that we already understand. Here are some examples of common metaphors:

Metaphor: I could *eat a horse.*

Explainer: I am *very hungry* (noted by the size of the meal I could eat).

Metaphor: She was a *lion* in the fight.

Explainer: She was very *strong and brave* when fighting.

Metaphor: The motorbike *flew* down the road.

Explainer: The motorbike moved *quickly* down the road.

Metaphor: He could *sell sand to a desert dweller.*

Explainer: He is *very good at sales* (the metaphor is implying he can sell something to someone that already has the item; in this case *sand*).

Metaphor: Those whales *put on a show.*

Explainer: The whales *performed* for us in an entertaining way.

Here are some examples of metaphors we are going to use to help us to describe vocational perspectives on outdoor environments.

- **Outdoors as a museum:** In a museum, we collect and store valuable things. We protect them and maintain them, and we use them to learn about the world. When describing human–nature relationships that involve conservation, education, protection or preservation, this could be a useful metaphor.
- **Outdoors as a gymnasium:** A gymnasium is a place where we challenge ourselves and look to physically develop our bodies. People go to gyms to work out and 'raise a sweat'. For many people, outdoor environments are places for them to do this (work out, sweat and physically develop), which makes this metaphor useful in this sense.
- **Outdoors as a storehouse:** A storehouse or a warehouse is a place where we find resources: things we need, and we need lots of them. This would be a useful metaphor for people who see the human–nature relationship as being made up of an environment that gives us access to many resources, such as food, air, water, plants, animals, minerals and more.
- **Outdoors as a mother:** A mother is the person who birthed us and who then goes on to protect and care for us. We expect a mother to love her children and to be cared for by her children when she ages. This metaphor describes a belief that nature will always provide for us and care for us as humans, and is often used to describe the relationship of an Indigenous person with nature before colonisation.
- **Outdoors as a performer:** This relates to the performer metaphor – 'Mother Nature never gives us the same show twice' in which humans are perceived as separate from and as an audience of spectacular natural displays.

metaphor
when two unlike things are compared with each other because of something they have in common

Worksheet
4.3a
Perspectives

LEARNING ACTIVITY 4.3A

Perspectives

Compare the following images by considering the body language of each person. What perspectives do these people have of their current situation?

Shutterstock/larisa Stefanjuk

iStockphoto/digitalhallway

Alamy/WILDLIFE GmbH / Alamy Stock Photo

Vocations in outdoor environments

We are now going to explore a number of vocations and develop our understanding of their perspectives towards outdoor environments.

NATURAL RESOURCE MANAGEMENT

Figure 4.33 There are various employment opportunities in natural resource management.

iStockphoto/tdub303

Natural resource management is the management of land, water, soil, plants and animals: the natural resources that make up Australia's landscapes. To manage these resources, a person has to work together with local communities, state/territory governments and the Australian Government.

- Natural resource managers work in:
- universities
- the mining or minerals industries
- environmental organisations
- government agencies that manage community-owned land (such as national and state parks, conservation parks, state forests, local government reserves and Crown land).

Natural resource managers generally have practical experience in conservation and land management, and have completed a VET or university qualification (with the latter being preferred). Degrees in environmental management, environmental science, sustainability, or wildlife and conservation biology are the common pathways to natural resource management.

The role of a natural resource manager is to view the landscape as a resource to be utilised and accessed for the use of humans. At the same time, their role involves balancing this with caring for the environment and employing sustainable approaches to managing resources to ensure supply in the future. The metaphor we can use to describe this perspective of outdoor environments is the 'outdoors as a storehouse'.

PARK RANGER

Park rangers play an integral role in promoting, maintaining and protecting our parks. According to Parks Victoria, 'Working as a ranger is as varied as the environments they manage. On a broad level, the role of a ranger involves extensive planning, researching, strategic thinking and people management to effectively balance conservation and recreational values of each asset.'

Resource
Case study: Kevin Cosgriff, Parks Victoria Ranger

CASE STUDY

KEVIN COSGRIFF, PARKS VICTORIA RANGER

I am a park ranger working with Parks Victoria. I have a Bachelor in Applied Science (Parks, Recreation and Heritage) with Honours from Charles Sturt University. In general, most park rangers have a degree of some sort in the environmental/geography/outdoor field. Prior to getting my current job, I also worked in various areas related to park management, such as fire crew, assisting with visitor surveys in national parks, doing seasonal or summer ranger work, working at a wildlife sanctuary and undertaking flora and fauna surveys for a consulting company. All of these jobs helped in supporting my application when I went for the role as a ranger.

Figure 4.34 Kevin Cosgriff

I grew up in a small town near the beach, so I spent my time after school on my bike, checking out the beach or going to the river with mates. I liked being outdoors. My family would camp a few times a year. As I grew into high school, my family moved to a larger regional Victorian inland town. I still spent time on my bike and went to the river with mates, but probably spent less time outdoors as I became absorbed in high school life – sport, part-time work, hanging at mates' places etc. In Year 10, I was lucky to get a place on a 10-day youth training trip on a sailboat. This sparked my interest in the outdoors again – showing me a different perspective. While I was at university, I spent many weekends with uni friends, camping, exploring, driving to national parks etc. I really enjoyed these days – the freedom, learning about outdoor environments and my newfound knowledge of the places we'd visit.

For me, there are two main purposes of outdoor environments:

1 For the preservation and conservation of natural things – these are places where plants and animals grow, live and survive. The breadth of living things (biodiversity) is key to our life on Earth and outdoor environments provide the place where these living things thrive. This encompasses everything from places where our animals live (including threatened species) to places that contribute to the reversal of climate change (through carbon sequestration and so forth). Without these areas, we all die.
2 For health and wellbeing – so that people have an opportunity to experience our natural world, for recreation or just for latent use (just to see it, to know it exists). This brings with it many benefits, the main being the benefits of individuals' mental health from being in the environment.

Interview by Jarrod Paine

QUESTIONS

1 What training did Kevin need to complete to work in his vocation?
2 Describe Kevin's perspective of outdoor environments using a metaphor.
3 What do you like or dislike about Kevin's vocational perspective of outdoor environments?

NATURE-BASED TOURISM

Nature-based tourism is a broad term that covers all tourism experiences focused on wild or natural environments that have a specific appeal to the tourism market. It can also be the backdrop to an adventure tourism experience.

Tourists visit natural attractions for many reasons, including to:

- enjoy the grandeur of nature
- escape from the pressures of urban life
- escape from humanity
- explore different landscapes
- experience outdoor adventures in a natural setting
- learn about the environment
- participate in conserving the environment.

Employment in the nature-based tourism sector involves standard services-based work in supplying guests' needs, but specifically focuses on expertise in facilitating a guest experience for one of the previously mentioned reasons. Vocationally, this work is guest-facing and a person needs a high level of communication skills and knowledge of nature, relating to both the local and broader environments. The most visible work in the field is in scenic and sightseeing tours: taking bookings, organising tours, answering queries and chaperoning groups.

A person interested in this vocation would often have a keen interest in outdoor environments and an aptitude for environmental education while catering to a customer's needs. Education and training for a person entering this industry involves many entry points, including direct from high school through to a diploma or degree in science, ecotourism or a similar subject, along with first aid qualifications.

With a high level of knowledge of outdoor environments and often a passion for sharing this with guests on nature-based tourism experiences, the vocational perspective of a person in this industry is focused on promoting care for outdoor environments and **sustainable tourism** visiting these places. The metaphor we can use to describe this perspective of outdoor environments is 'Outdoors as a performer', where the outdoors is 'on show' for tourists.

sustainable tourism a type of tourism that reduces the negative impacts of tourism on the environment, society and economy, while maximising positive ones; by conducting responsible travel, sustainable tourism aims to meet the needs of present tourists without compromising those of future generations to enjoy the same natural, cultural and social assets

OUTDOOR LEADING AND GUIDING

Leading and guiding involves taking groups of people (adults, families or school groups) into outdoor environments for varying lengths of time to assist them in achieving an outdoor experience. This may include guided flower identification trips, high-adventure ice climbing and white-water rafting, or leading bushwalking trips. The variation is large but each aspect of the vocation involves:

- customer service: in each instance, a customer is paying for a particular product and the leader or guide's job is to deliver that outdoor experience safely and to a level of high quality.
- education: this includes imparting knowledge about the environment being experienced and the skills required to safely participate in associated activities.
- leadership and decision making: the leader or guide as noted in the title is in charge of leading the group, especially if any incidents occur. They are in charge of the experience in the bush and have to work carefully and collaboratively with guests or students.

Outdoor leading and guiding is a vocational choice in which a person spends lots of time in outdoor environments helping customers to achieve safe experiences. To do this, workers need a high level of skill competency and experience in the outdoors. Typically, a person's skill growth will move from participant to assistant leader to leader, with the time spent in each category dependent on the demands of the outdoor experiences they are training to lead.

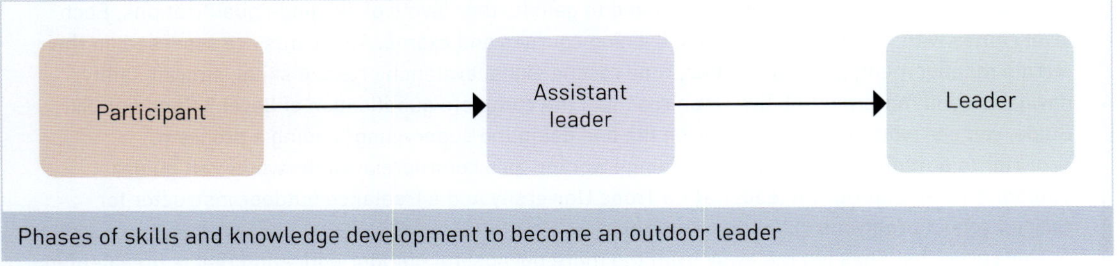

Phases of skills and knowledge development to become an outdoor leader

The participant phase involves a person gaining experience in the skills, knowledge and locations that they are interested in working in. The next step typically involves assisting a leader and may also include further training through higher learning institutions like TAFE or university. That being said, being a leader or guide does not require a university degree or TAFE diploma, but will generally require certificates of competency in first aid and in any specific adventure activity they are leading or guiding. The step to leader involves the culmination of experience gained as participant and assistant leader, coupled with qualifications demonstrating competency in leading groups in that location or activity.

Each of the training steps involves a person spending more and more time outdoors, connecting with experiences and activities. Their relationship with outdoor environments can often be described as 'outdoors as a gymnasium' or 'outdoors as a performer', which keeps a constant stream of customers who need guiding to see the 'show'. It will then often change into 'outdoors as museum', as they begin to be involved in the caring and conservation activities for outdoor environments.

Resource
Case study: Kris Robinson, Backcountry ski guide

CASE STUDY

KRIS ROBINSON, BACKCOUNTRY SKI GUIDE

After high school, Kris went to La Trobe University, Bendigo, to study and complete a Bachelor of Arts in Outdoor Education. While in some ways his current vocation and residence couldn't be further from his beginnings as a guide in Victoria, this pathway began his journey to where he is now. Read on to learn about Kris's story.

Figure 4.35 Kris Robinson at work

I am a professional backcountry ski guide in Canada – I guide backcountry skiing and in the winter mountain environment on both glaciated and non-glaciated terrain. Access is primarily mechanised with the use of either helicopter or snowcat ('cat-skiing' uses an over-snow tracked vehicle like a tank, but for passengers).

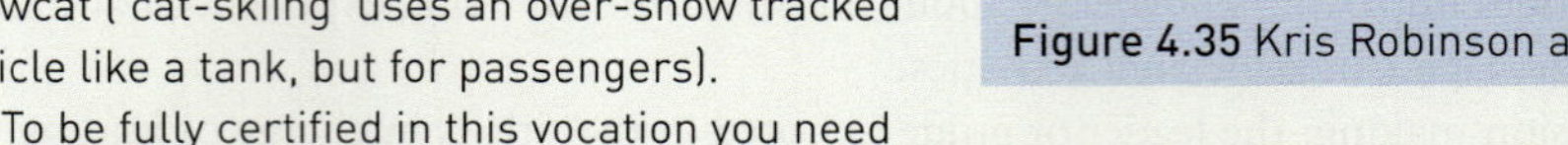

To be fully certified in this vocation you need the following:

- current first aid in a course of minimum length of 80 hours (8–10 days)
- Canadian Avalanche Association (CAA) Operations Level 2 (prerequisites and exams included)
- Canadian Ski Guide Association (CSGA) or Association of Canadian Mountain Guides (ACMG) Ski/Snowboard Guiding certification.

In order to get started in this career, I had to develop extensive backcountry travel experience and proficiency, including some recreational-level Avalanche Skills Training (AST 2). The entry level to this role requires Avalanche Operations Level 1 and many days' ski-touring experience. This qualifies applicants for unpaid or 'practicum' positions or to work as a tail guide (at the rear of the group, mostly in cat-skiing).

In order to start leading groups, you need to gain further levels of ski guide qualifications. Each level comes with its own prerequisites, training courses and exams. Applicants are assessed on their terrain management, guiding abilities, rope rescue skills, avalanche rescue skills, aircraft safety, navigation and movement ability. Upon reaching the complete qualification of Level 3 Canadian Ski Guide, you can work unsupervised or be the one doing the supervising/leading a program.

Prior to pursuing a career in ski guiding I worked as a commercial white-water raft guide, outdoor education program leader at La Trobe University and a freelance outdoor instructor for various school programs in Victoria, Australia.

My career path has led me away from the more purposeful outdoor education that is prevalent in Australian school OE programs. I am now fully embedded in an industry that caters to ski tourism. Our focus is to show our guests to a safe and fun day of skiing in the backcountry of British Columbia. It is typically very fast paced and, while we do have some opportunity to talk about local flora, fauna, indigenous and colonial history, it is not the focus of the typical day ski program.

I believe the purpose of outdoor environments is that moment when things slow down and go quiet, and you look around to just take it all in. Those are the moments I appreciate and try to foster. We still have many of these nature moments just by virtue of being immersed in wild mountainous places, but a ski guide's role is more often laser focused on avalanche safety and group management.

Interview by Jarrod Paine

QUESTIONS

1. What training did Kris need to complete to work in his vocation?
2. Describe Kris' perspective of outdoor environments using a metaphor.
3. Do you agree with or disagree with Kris' vocational perspective of outdoor environments? Justify your response.

ENVIRONMENTAL RESEARCH AND POLICY

iStockphoto/Tenedos

Figure 4.36 Field work is an important aspect of research.

This vocation brings together field science and policy to assist in informing future practice or care for outdoor environments. It can have equal parts outdoor work and indoor work, and can also be split into two separate roles.

In environmental research or science, people measure and record features of the environment. They may study the effects of factors such as terrain, altitude, climatic and environmental change, sources of nutrition, predators, and impacts of humans on animal and plant life. This vocation can also analyse pollution, atmospheric conditions, demographic characteristics, ecology, and mineral, soil and water samples.

The results of scientific research from fieldwork, combined with larger research techniques covering all past information, is then used to develop conservation and management policies to inform private or public practices to manage outdoor environments. Other policy development areas can be for biological resources, such as fish populations and forests, as well as establishing standards and approaches for control of pollution and rehabilitation of areas disturbed by primary industry activities such as mining, forestry and agriculture.

Environmental research and policy workers are employed by federal, state/territory and local governments and organisations, as well as trade unions, political parties, universities and non-profit organisations. To gain employment in the area it is best to complete a degree in environmental science, science or applied science with a major in environmental science, natural resource management, geography, marine science or a related field.

While this vocation is often less interactive with the outdoors, or not as exciting as being an outdoor guide or nature-based tourism operator, it works to contribute to the care for outdoor environments alongside these other vocations. A person working in research and policy will often have a passion for outdoor environments and spend vast time outdoors in their private lives or in the years before working in this role. A job at the level of policy writing has great potential to influence the care for outdoor environments at a national and international level, and commands a great deal of respect. With this in mind, for us to understand the perspective this person has toward outdoor environments we would categorise them metaphorically as seeing the 'outdoors as a museum', working to conserve its wonder.

INDIGENOUS RANGER

The vocation of Indigenous ranger has been created by the Australian Government in partnership with the National Indigenous Australians Agency. Combining traditional knowledge with conservation training, Indigenous rangers work to protect and manage their land, sea and culture. This includes activities such as:

- bushfire mitigation
- protection of threatened species
- biosecurity compliance.

Indigenous ranger groups share skills and knowledge, engage with schools, and generate additional income and jobs in the environmental, biosecurity, heritage and other sectors. Indigenous communities across Australia highly value the work of Indigenous rangers for achieving both environmental and employment outcomes, alongside the wider social, cultural and economic benefits.

Indigenous rangers are reconnecting cultural traditions to land management through this vocation, reintroducing practices that are thousands of years old, including cultural burning, revegetation, weed and pest mapping, and protecting cultural heritage sites.

Figure 4.37 An Indigenous ranger examining rock art in Gariwerd (the Grampians).

Resource
Case Study: Cultural burning - a link to country

CASE STUDY

CULTURAL BURNING – A LINK TO COUNTRY

The Budj Bim Rangers, with Elders and stakeholders, conducted traditional burns on the IPA [Indigenous Protected Area] in May 2020. This began with a 'Welcome to Country' and cleansing ceremony which involved passing through the smoke of the cherry balart. They then created fire in the traditional way to use for their cultural burns.

> Cultural traditional burning has and always will, continue to be intergenerational, a vital link to Country that must endure ...
>
> Budj Bim Rangers

The rangers note that cultural burns are slow, reducing the chance of plants and animals being harmed, and encouraging seed germination.

QUESTIONS

1 What is a 'Welcome to Country' and who performs it?
2 Why is this event of significance from a vocational perspective?
3 What in this short article helps you to understand the vocational perspective of Indigenous rangers toward outdoor environments?

EDUCATION

The vocational perspective of the education industry on outdoor environments might seem at first to be very simple. However, if you probe into the mind of a teacher, you will learn that their motivations or perspectives can be quite complex.

It is the role of an educator to help students to learn new knowledge and to comprehend, challenge, evaluate and analyse that knowledge, while applying it to real-world situations. Therefore, when an educator teaches about open-cut mining for coal, for example, they are teaching about the perspective of a landscape being a resource to be managed, or 'outdoors as a storehouse'. When teachers are educating students about the challenges of nature-based tourism, they are helping the student to understand the point of view of the 'outdoors as a performer', and the importance of caring for the aspects of nature being visited to ensure future employment. Educators may also be presenting the view of the 'outdoors as gymnasium' when preparing students for field trips or outdoor activities.

When you share an outdoor experience with your Outdoor and Environmental Studies teacher, and you are learning about biotic and abiotic components of ecosystems, the teacher is giving the perspective of the 'outdoors as museum' – as something to be studied.

Educational vocational perspectives on outdoor environments involve describing outdoor environments in the myriad ways that they can be perceived, and guiding students to begin to understand all those perspectives – to be able to walk in the shoes of a miner, farmer or guide.

Education is also a massive vocational field. Becoming a university-trained teacher in schools or universities is one option, but educators also exist in nearly every other industry. Education to build skills occurs on offshore gas and oil rigs and within zoos and scuba-diving tours.

iStockphoto/SolStock

Figure 4.38 Teaching the next generation is an important career.

Resource
Case study: Lucy Vardy, Teacher

CASE STUDY

LUCY VARDY, TEACHER

Lucy Vardy

Figure 4.39 Lucy Vardy at work

Lucy is an outdoor education teacher. She grew up on a farm, camped as a child with her family in beautiful outdoor environments, and studied VCE Outdoor and Environmental Studies in high school. Before becoming an outdoor education teacher, Lucy developed her practical outdoor living and outdoor skills experience (camping, bushwalking and skiing) and undertook a 'gap' year as an outdoor education assistant at a school camp.

Lucy went on to complete a Bachelor of Outdoor Education and Graduate Diploma of Education, and has grown her skills in working collaboratively with other people, developing her passion for working with young people and outdoor experiences.

Lucy describes the purpose of outdoor environments in the following ways:

1 They sustain all life on the planet. Humans are a part of them.
2 Time spent in them without distractions from the modern world is hugely important to the wellbeing of people, both mentally and physically.
3 Experiences in them allow us to reflect on who we are as people and also how we interact with those around us.

Positive experiences in them allow us to develop strong connections with outdoor environments, which hopefully then nurtures an ethic of care for outdoor environments.

Interview by Jarrod Paine

QUESTIONS

1 What training did Lucy need to complete to work in her vocation?
2 Describe Lucy's perspective of outdoor environments using a metaphor?
3 What do you like or dislike about Lucy's vocational perspective of outdoor environments?

AGRICULTURE

Shutterstock.com/Taty77

Figure 4.40 Agriculture requires a strong connection to the land and seasons.

Agriculture is often referred to as the backbone of this country. Since European colonisation, Australia has been described as having been 'built on the sheep's back'. Today, however, agriculture is a dynamic and highly diverse industry that has embraced modern technology, and has multiple entry points based on skills and education. These include the following:

- Labour jobs: physical manual tasks such as planting, harvesting, caring for animals and maintaining equipment. No tertiary education is required.
- Administrative jobs: tasks like managing offices, answering correspondence, maintaining filing systems and speaking with clients. Administrative professionals may also handle bookkeeping and accounting tasks, as well as human resources and insurance duties. Administrators often will have completed tertiary education.
- Engineering jobs: tasks such as evaluating, designing, testing and installing a range of equipment and systems. Engineers also oversee manufacturing and maintenance processes. In the agriculture field, most engineers specialise in mechanical, environmental or structural engineering. Tertiary education at a university level is required.
- Sales jobs: tasks include selling materials and products to customers. No tertiary education is required, but often has been completed.
- Science jobs: these use the principles of biology, chemistry, zoology, geology, physics and other areas to research issues, monitor situations, design solutions and develop products and systems. Tertiary education at a university level is required.

The agricultural industry is focused on the metaphor of 'outdoors as a storehouse', as it is the place that gives us access to many resources, such as food, air, water, plants, animals, minerals and more. These resources are then used to grow crops or animals to make a successful 'living off the land'.

NOTES FOR THE EXAM 

For the exam, you should:

- describe a range of vocations that involve outdoor environments
- compare a range of vocational perspectives on outdoor environments.

4.3 KEY CONCEPTS

- A vocation is a person's job within an industry.
- Metaphors help to describe vocational perspectives are:
  - outdoors as a museum
  - outdoors as a gymnasium
  - outdoors as a storehouse
  - outdoors as a mother
  - outdoors as a performer.
- Vocational perspectives of outdoor environments vary depending on the person's relationship with the environment and how they earn their money, but all vocations have a need to establish or maintain sustainable relationships with outdoor environments to ensure their vocation continues.

Worksheet
4.3 Key concepts

4.3 CONCEPT QUESTIONS

REMEMBERING

1 Describe one metaphor for a perspective of outdoor environments.
2 What is a vocation?

UNDERSTANDING

3 Explain the perspective of a natural resource manager and outdoor environments.
4 Compare the perspectives of an agricultural worker and an environmental researcher.

APPLYING

5 Using an example not presented in the text, explain a particular vocation and what its perspective of an outdoor environment could be.
6 Describe a pathway that you could take in order to gain employment in one of the vocations you have studied.

EXTENSION CHALLENGE

7 What would the perspective of a politician be on outdoor environments? Describe why you think that, and what might be the major influences on their perspective.

Worksheet
4.3b Differing perspectives

EXTENDED LEARNING ACTIVITY 4.3B

Differing perspectives

The following two descriptions about wild horses (brumbies) within the Alpine National Park describe the horses from different perspectives. Read the extracts and then answer the following questions:

1 What is the perspective of the writer of article 1?
2 What is the perspective of the writer of article 2?
3 What influences the writers of each article to describe the wild horses from different perspectives?
4 What does the writer of article 1 want as the future of wild horses?
5 What does the writer of article 2 want as the future of wild horses?
6 Using the internet, research the current management plan for horses in the Alpine National Park. What is the current status of the removal of the horses?

ALPINE NATIONAL PARK BRUMBIES

Victoria's Alpine National Park at 646 000 hectares is the state's largest and protects our highest mountains and varied alpine environments. The Alpine National Park has the greatest range of flora and fauna of any national park in Victoria. With adjoining national parks in NSW and the ACT, the Alpine National Park forms a protected area that covers almost all of Australia's high country. It's one of eight Australian Alps national parks that are managed co-operatively. European pastoralists from NSW started moving south into the Alps in the 1830s, with grazing beginning around Omeo and up into the foothills in 1836. Summer grazing soon extended to the higher country, and huts were built there for shelter and storage during stock mustering. From the 1850s to around 1900, gold lured many people to the Alps. The only form of transport and of working the land throughout this time was with horses, so with these settlers and explorers came their horses: Clydesdales, Thoroughbreds, Stockhorses and Timor Ponies.

Working horses and family horses both occasionally ran loose on the massive land holdings and were not always caught again. World War I also played its role in contributing to what we now know as Brumbies. When the men who worked many of the horses went to war, [the horses] ran wild until [the men] returned and not all were caught. When the Depression hit Australia, along with increased industrialisation, many of the locals simply opened their gates and let their horses 'join the wild bush horses…' as Banjo Patterson described it. The Brumbies, having been begun by man, were bred by nature.

Figure 4.41 Brumbies (wild horses) on the Bogong Plains in the Victorian High Country. Australia

Bjorn Svensson / Alamy Stock Photo

'Alpine National Park Brumbies' © 2023 Victorian Brumby Association

FERAL HORSES IN NATIONAL PARKS

Hard-hooved animals are not native to Australia, and many of our native plants and animals are threatened by their impacts.

Feral horses, pigs, goats, camels and deer are now found throughout the land, often in large and growing numbers.

Fortunately, feral horses are only found in two national parks in Victoria, the Alpine National Park in the state's north-east, and Barmah National Park in the state's far north along the Murray River. Nevertheless, their management is difficult.

While the community accepts the control of most feral animals, some methods of feral horse control are not supported by all members of the public.

Doing nothing, however, doesn't help the horses, or the environments they are damaging.

A retired Alpine National Park Ranger has said that if you wanted to design an animal that would do the most damage to alpine ecosystems, a horse would be it.

The alpine region is a very small part of Australia's mainland, and it is particularly vulnerable to hard-hooved grazing animals.

There are no native hard-hooved animals in Australia.

Australia is an old continent, and the mountain ranges of the alpine region of Victoria, NSW and the ACT have worn down over the ages. They are now largely wide open alpine grassy plains, together with areas where accumulated plant material has developed over aeons into extensive peatlands covered in deep blankets of water-holding sphagnum moss and other plants. These peatbeds (also called mossbeds) have built up over tens of thousands of years and are a feature of the high country, acting as huge sponges across the alpine landscape. They hold water at snow melt, slowly releasing it into alpine streams, allowing all season flows into many of our major rivers.

These peatbeds have been eroded, and many destroyed, after a century and a half of horse, sheep and cattle grazing, and now deer invasions. In areas where cattle have been removed for decades and horses have been scarce, such as the northern Bogong High Plains, these peatbeds have been recovering well.

'Feral horses in national parks', Victorian National Parks Association Inc (VNPA)

Resource
Glossary – Chapter 4

Assessments
End of chapter exam

Glossary test

EXAM-STYLE QUESTIONS

1 Describe what biotic and abiotic components of outdoor environments are, using specific examples. (4 marks)

2 Nutrients are reused on earth. Describe the steps of one biogeochemical cycle that allows elements to move throughout ecosystems (4 marks)

3 Explain what human-induced climate change is and describe its impact on a specific environment. (4 marks)

4 a Explain Indigenous peoples' perspectives of outdoor environments. (3 marks)

b Describe one form of traditional land management by Indigenous people. (2 marks)

5 Describe two different perspectives of outdoor environments. (2 marks)

6 Choose two vocational perspectives of outdoor environments:

- natural resource management
- nature-based tourism
- outdoor leading and guiding
- environmental research and policy
- land management roles including Traditional Owner groups and National Indigenous Australians Agency programs
- education
- agriculture.

a Compare your two chosen vocational perspectives of outdoor environments. (4 marks)

b Describe a vocation that you have observed or studied on an outdoor experience and how that person interacted with outdoor environments as a result. (2 marks)

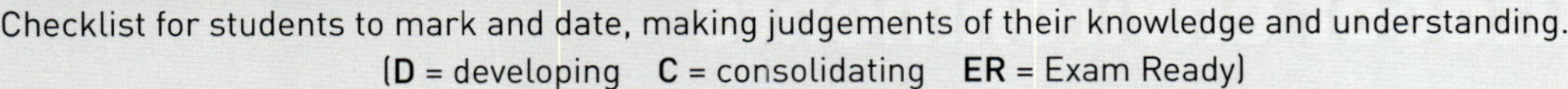

CHECK YOUR KNOWLEDGE AND SKILLS

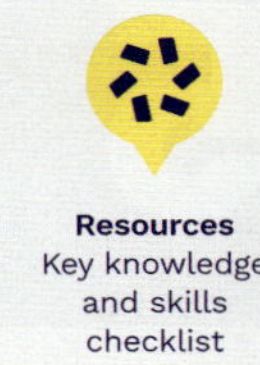

Resources
Key knowledge and skills checklist

Checklist for students to mark and date, making judgements of their knowledge and understanding. (D = developing C = consolidating ER = Exam Ready)

D	C	ER	Check your knowledge and skills
☐	☐	☐	Describe what influences a person's perspective
☐	☐	☐	Draw or describe either the water, carbon, nitrogen or climatic cycle
☐	☐	☐	Demonstrate understanding of short- and long-term changes in natural environments
☐	☐	☐	Compare abiotic and biotic interrelationships in an outdoor environment
☐	☐	☐	Analyse the effect of natural and human induced changes and apply these to a range of environments
☐	☐	☐	Explain the timeline of Indigenous peoples living in this country
☐	☐	☐	Describe historical land management techniques of traditional burning, hunting and gathering, agriculture, aquaculture and natural resource use
☐	☐	☐	Describe examples of contemporary Indigenous land management
☐	☐	☐	Explain Indigenous peoples' perspectives and different forms of land management within an outdoor environment
☐	☐	☐	Describe what is involved in at least two vocations, including the training required
☐	☐	☐	Compare a range of vocational perspectives on outdoor environments using metaphors

CHAPTER 5

UNIT 2 – AREA OF STUDY 2

Observing impacts on outdoor environments

KEY KNOWLEDGE

- impacts of conservation, economic and recreational activities on a range of outdoor environments
- the role of community-based environmental action to promote positive impacts of humans on outdoor environments
- direct and indirect impacts of technologies that support human interactions with outdoor environments, including:
  - equipment manufacture
  - transport
  - snow making
  - recreational vehicles
- the impact of urbanisation on outdoor environments
- identification and management of threatened species and/or ecological communities in an outdoor environment.

KEY SKILLS

- evaluate a range of impacts on contrasting outdoor environments
- compare direct and indirect impacts of technologies on outdoor environments
- describe a local environmental policy that supports management of threatened species or an ecological community
- discuss and predict impacts of urbanisation on outdoor environments
- evaluate the effectiveness of management strategies on an outdoor environment.

VCE Outdoor and Environmental Studies Study Design 2024–2028, pp. 19-20.

Impacts of users on outdoor environments

Did you know humans can also have positive impacts on outdoor environments? We have an impact on all outdoor environments, either through visiting them or through practices and behaviours that create pollution, which then impacts them. This section analyses human impacts on outdoor environments across three modes of interacting: through conservation, economic and recreational-based activities.

Community-based environmental action

Communities play a crucial role in environmental protection by bringing together people with similar interests and goals to create positive change in their local outdoor environment. The focus of this section is on the ecological health of the whole local environment, and in this section we will analyse the impact these groups have on contrasting environments.

Impacts of technology on outdoor environments

Developments in technology have impacted significantly on the ways in which humans relate to and experience outdoor environments. In this section, we compare the direct and indirect impacts of those technologies on outdoor environments, from the ones that help us climb high to the ones that keep us dry.

Impact of urbanisation

Australia is a highly urbanised country – approximately 86% of Australia's population live in urban areas – and this has brought about positive and negative impacts on outdoor environments. In this section we'll pound the footpath of cities, discussing and predicting their impacts. How tall can a building be, after all?

Management of threatened species

Australia has a diverse range of ecological communities important for maintaining biodiversity and ecosystem services, such as clean water and air, nutrient cycling and climate regulation. However, many of these communities are being damaged or are under threat, so the federal, state/territory and local governments are working to care for and protect these outdoor environments. Unless there is a rule about protecting a tree, how will it be protected?

KEY TERMS

carbon tax
community-based environmental group
community engagement
contrasting
conservation
conservation value
direct impacts
ecological community
ecosystem services
emissions
fledgling
green infrastructure
habitat fragmentation
habitat restoration
indirect impacts
land-use planning
non-renewable
population density
protected areas
regenerative farming
remnant vegetation
responsible pet ownership
technology
threatened species management
urban sprawl
urbanisation

To access resources above, visit **cengage.com.au/nelsonmindtap**

As humans, we leave footprints everywhere we go. Even if we are careful, there is evidence of our having been there. It could literally be a footprint in the dirt or one of many other traces of our visitation or habitation.

This chapter focuses on observation of human impacts on outdoor environments from the perspectives of:

- positive impacts
- negative impacts
- direct and indirect technological impacts
- urbanisation impacts.

Observations of these impacts across a range of environments and contrasting user groups are observed, but we don't stop there. We also shift our observation to existing strategies to increase our positive impacts, both as a community and at local and state government policy levels. We also explore positive news stories of people actively working to reduce impacts on outdoor environments.

5.1 SAFE AND SUSTAINABLE INTERACTIONS WITH OUTDOOR ENVIRONMENTS

KEY KNOWLEDGE

- impacts of conservation, economic and recreational activities on a range of outdoor environments

KEY SKILLS

- evaluate a range of impacts on contrasting outdoor environments

5.1.1 HUMAN IMPACTS

Human impacts have changed over time. Indigenous peoples utilised fire to control vegetation, clear out undergrowth to make travel easier and attract feeding animals, both to encourage breeding and to provide prey to hunt. It is also believed that Indigenous peoples introduced the dingo to Australia as an efficient hunting partner.

The arrival of Europeans to Australia signalled the advent of permanent settlements and fenced land, the introduction of exotic animals and plants, and wide-scale timber harvesting, land clearing and farming. (These historical activities and their impacts are explored in detail in Chapter 7.)

Today, humans undertake a variety of activities in outdoor environments, which can be broadly classified as conservation, economic or recreational activities (see Figure 5.1). While many of these activities have had adverse effects on aspects of the environment, some are used for preservation and restoration. Therefore, it is important to recognise that impacts can be both positive and negative, depending on the activity and outdoor environment in question.

Contrasting outdoor environments

When considering human impacts on contrasting outdoor environments, we need to begin by revisiting some environment or biome types. You may recall learning about different biomes in Chapter 1:

- alpine
- coastal
- inland waterways
- grassland
- heathland
- forest
- marine
- arid.

Contrasting environments can be considered as environments that are strikingly different from, or opposite to, one another. Some examples are:

- grassland and forest – an absence of trees versus an abundance of trees
- alpine and coastal – high altitudes and colder average temperatures versus sea-level altitudes and warmer maritime temperatures
- inland waterways and arid – wet versus dry.

Now that we have considered how these environments are in contrast, we can move on to evaluating the range of impacts that conservation, economic and recreational activities have on these environments.

contrasting
of two things that appear as opposites, or strikingly different from each other

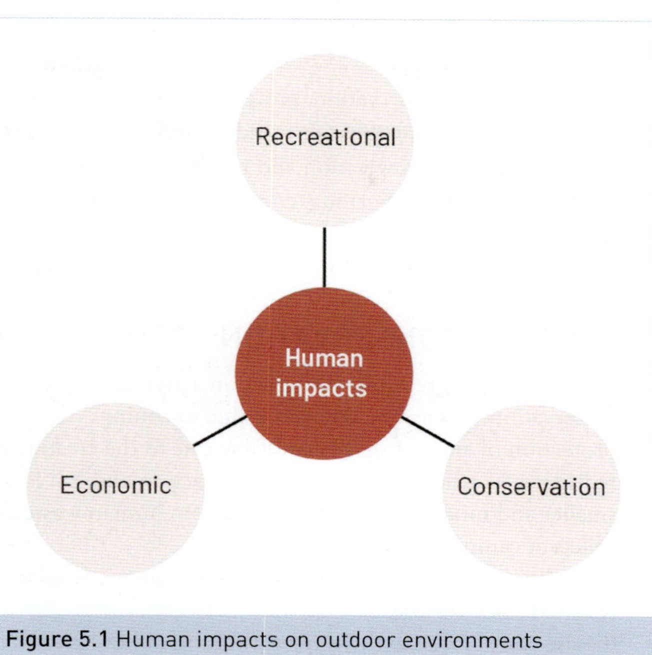

Figure 5.1 Human impacts on outdoor environments

Shutterstock.com/f.ield of vision

Figure 5.2 Surfers catching a wave at Bells Beach beach, Victoria. They can't surf in a contrasting environment of arid western inland Victoria.

Conservation activities

conservation the preservation, protection, management or restoration of the natural environment, inclusive of ecosystems, vegetation, wildlife and natural resources, such as soil and water

Conservation is the preservation, protection, management or restoration of the natural environment, inclusive of ecosystems, vegetation, wildlife and natural resources, such as soil and water. Conservation activities incorporate revegetation and rehabilitation, the establishment of parks and reserves, scientific investigation, controlled burning, community education and the implementation of management strategies. Overall, conservation activities are considered to have positive impacts on outdoor environments, although some negative impacts can also occur, as explored in the following section.

IMPACTS OF CONSERVATION

Positive impacts as a result of conservation activities include:

- preservation of sensitive vegetation due to zoning
- protection of areas of significant value due to the creation of reserves such as state and national parks
- reduction in areas of erosion, soil compaction and soil loss due to the creation of walkways
- smaller and more centralised impacts from human waste due to the introduction of composting toilets at campsites
- rehabilitation and revegetation resulting in increased plant growth and decreased erosion, due to seasonal track closures
- disease control due to the implementation of boot-cleaning stations and quarantine areas
- a reduction in human waste, trail erosion and trampling of vegetation off-track when group size restrictions and permit requirements are mandated
- reintroduction of species into areas they originally inhabited.

Figure 5.3 Tree-planting and revvvegetation are forms of conservation.

Negative impacts as a result of conservation activities include:

- plants and animals being destroyed or otherwise impacted as a result of controlled burning (i.e. the prevention of larger fires through the deliberate burning of fire breaks)
- land clearing due to conservation infrastructure (e.g. walkways, composting toilets)
- environmental intrusion due to the creation of permanent structures in natural areas (e.g. fences, viewing platforms)
- incidental disruption of habitats.

Resource
Case study: Mountain Pygmy possums versus Alpine Resorts

CASE STUDY

Mountain Pygmy Possums versus Alpine Resorts

The mountain pygmy possum is a threatened small marsupial found only in the alpine region of Victoria and NSW. They gorge themselves on mountain plum pine berries and a dwindling annual migration of bogong moths to support their torpor during the winter months, curled up in snow-covered homes the temperature of a fridge.

Figure 5.4 Mt Hotham's 'tunnel of love'

In the past, when Australia's climate was cooler than today, the mountain pygmy possum's range was considerably larger, but since the last ice age female mountain pygmy possums have been restricted to the highest altitudes while the males live further down the high alpine mountains.

Thought to be extinct until a living individual was discovered in 1966, the possums have since been surveyed and found to live within the Alpine National Park, Mt Hotham, Mt Buller and Falls Creek resorts in Victoria. There's estimated to be around 2000 mountain pygmy possums in the wild. Living in such small disconnected populations means that genetic loss is a key threat to the species and work had to be done to protect their habitat.

Since their rediscovery, there have been several measures taken to help the species. In regard to habitat, these include:

- weed control and revegetation
- protection of habitat through fencing and restricted land use
- creation of a boulder field to emulate habitat at Mt Buller
- predator control.

At Mt Hotham, the male and female populations have been separated by the Great Alpine Road – a road built for recreational users to access the alpine resort. This was identified as an issue as long ago as the 1980s and a tunnel was constructed under the road to ensure safe migration of the male pygmy possums to find their female counterparts higher up the mountain. This means they don't have to cross the road and risk car accident or predation in the open.

While numbers stabilised across Victoria due to these strategies, which were implemented by multiple agencies and government departments, threats to the little possum continued.

In 2019, a second 'tunnel of love' was built at Mt Hotham and efforts continue to combat a dwindling food source of bogong moths (due to agricultural pressures in northern NSW), reduced snow cover due to climate change and pressures from the impacts of an alpine resort hustling and bustling side by side an extremely vulnerable species.

This case study demonstrates that humans can impact native animals in significant ways, but it also shows us that we can create and implement strategies to offset our impacts and to care for outdoor environments and their native inhabitants.

QUESTIONS

1 How has the development of the alpine resort at Mt Hotham had a negative impact on mountain pygmy possums?
2 What management strategy has been implemented to reduce the impact of the alpine resort at Mt Hotham on mountain pygmy possums?
3 What other possible solutions can you recognise may be useful to care for this threatened species?

Worksheet
5.1a Conservation in song

Weblink
Video: 'Treat Yo Mama' - John Butler Trio

LEARNING ACTIVITY 5.1A

Conservation in song

Song has long been a tool for expression of the feelings of people. We sing because we are happy and sad, but we also sing to share a message. There are many people around the world who have expressed their views about conservation through song. Listen to the song 'Treat Yo Mama' by the John Butler Trio.

Think about what the lyrics are suggesting about conservation, Mother Earth and decency. Discuss your thoughts with your class members.

EARLY CONSERVATION – VICTORIA'S FIRST NATIONAL PARK

The largest impact of conservation protecting large areas of land in Victoria occurred in 1898 when Wilson's Promontory was named a national park, 19 years after Australia's first national park was created in 1879. These parks are an example of an early attempt at conservation, and as Australians, we can celebrate the fact that the Royal National Park in Sydney was only the second such park to be declared in the world, following the creation of Yellowstone National Park in the USA in 1872.

Wilson's Promontory, as the southernmost tip of mainland Australia, was known as 'Wamoon' or 'Wamoom' by the Indigenous peoples who collected shellfish there over 6000 years ago. The Boonwurrung, Bunurong and Gunaikurnai are the Traditional Owner groups of Wilson's Promontory. After colonisation, non-Indigenous Australians began to exploit all that the 'Prom' had to offer, from sealing and whaling bases to grazing lands and tin mining, only later turning to recreation following the creation of the national park.

Resource
Case study: A Christmas walking tour

CASE STUDY

A Christmas walking tour

During the Christmas holidays of 1884, an event occurred which was to shape the destiny of the Promontory.

Three men, Messrs J. B. Gregory, A.H.S. Lucas and G.W. Robinson, undertook a walking tour from Trafalgar to the Lighthouse by way of the Yanakie Isthmus, the Darby River and Mount Oberon. Lucas collected specimens for subsequent identification by the now-famous Baron von Mueller and later, in collaboration with Gregory, published an account of the trip in the second volume of the *Victorian Naturalist*, the journal and magazine of the Field Naturalists' Club of Victoria. The series of papers makes fascinating reading.

It was they who suggested that the Promontory could be regarded as an ideal resort for tourists and fishermen. They considered that it had little to recommend for commercial development. Indeed, it is in no small measure due to them that Victoria now has Wilson's Promontory as a national park permanently reserved as a refuge and sanctuary for our country's wildlife, as well as a resort where may be seen some of the most entrancing of Victoria's coastal scenery.

Their account of its potentialities as a tourist attraction aroused widespread interest, but the very remoteness of the place from populated centres and the difficulties of access suggested that, for a long time to come, there would be little effort to develop them. In their account of it they wrote: 'We may safely recommend the Promontory as full of interest to naturalists of all persuasions. Practically inaccessible as it is at present (1884), we believe that a future yet awaits it as a summer haunt of lovers of nature and lovers of scenery. In many respects alike, we prophesy that, as the Cornish Peninsula was later to be discovered by tourists, not many generations will pass before means of communication will enable Victorians to find out and do justice to this noble granite promontory – the Cornwall of Victoria.'

A History of Wilson's Promontory, J. Ros. Garnet, Victorian National Parks Association

(attributed to) Albert Charles Cooke, Refuge Cove, Wilson's Promontory, c.1854-72

Figure 5.5 This unsigned watercolour painting of Refuge Cove, attributed to Albert Charles Cooke, is one of the earliest depictions of Wilson's Promontory.

QUESTIONS

1 How did this event shape the destiny of Wilson's Promontory?
2 For what reasons did the three men see Wilson's Promontory as a place of future lovers of nature and scenery?
3 Research the Cornish Peninsula. How would you describe its similarities with Wilson's Promontory both in nature and in tourist visitation?

LEARNING ACTIVITY 5.1B

Threatened species

Visit the Zoos Victoria webpage and navigate to 'Fighting extinction'. Zoos Victoria is involved in conservation programs in six countries, eight grass-roots community conservation campaigns and more than 50 research projects. Read the information on the website, then answer the following questions:

1 Of the 27 priority species, describe three that live in environments you have visited or studied. What are the human-induced impacts that have caused the decline in these species?
2 Describe three actions Zoos Victoria is encouraging we do to help conserve wildlife and wild places. Share your favourite action with the class.

Worksheet
5.1b Threatened species

Weblink
Zoos Victoria - 'Fighting extinction'

Economic activities

As they relate to the environment, economic activities are those that result in a profit or financial gain for an individual or group as a result of utilising the natural environment. They can also be inclusive of large-scale operations that are commonly referred to as industrial or primary industry activities.

Therefore, economic activities include:

- tourism
- fee-for-service recreational activities
- forestry
- farming and agriculture
- the production of energy
- mining
- game hunting
- development and urbanisation.

Economic activities can have a predominantly negative impact on outdoor environments, as they are often associated with the exploitation of natural resources for human use. For this reason, people often give priority to economic activities over the needs of other species or ecosystems.

Adventure companies and tour operators in Victoria who utilise public land need to have relevant licences. As such, the Victorian Department of Energy, Environment and Climate Action (DEECA) suggests that 'indirect benefits provided by tour operators may include providing a platform to learn about ecosystems, biodiversity, flora, fauna, and cultural and historic heritage', which consequently lead to more positive impacts on outdoor environments. However, some economic activities may include the overuse or redevelopment of a particular area, which can lead to increased waste, facilities being created to cater for large groups and forever changing the landscape. The accompanying case study considers the Philip Island penguin parade and ecotourism business that attracts 700 000 visitors each year.

Resource
Case study: Philip Island penguin parade

CASE STUDY

PHILIP ISLAND PENGUIN PARADE

Philip Island is the home of the largest little penguin colony in the world and attracts visitors from around Australia and the world. Managed as an attraction by Philip Island Nature Parks (a not-for-profit body created by the Victorian government), the economic activity of this venture has grown exponentially since 1928, when the first tourism operations visited the colony. Soon after tourism began, efforts commenced to protect the colony, and in 1930 an initial four hectares was gifted to the 'people of Victoria' by landowners, followed by additional land and reserve status being granted by the Philip Island Shire in 1955.

Development of mass tourist infrastructure started in 1961 with fences and concrete viewing stands being added, although the Summerland housing estate was also being developed, and soon boasted 177 houses.

By 1985, it was noted that penguin numbers were declining. Predation by foxes and dogs, deaths from being run over by cars and penguins becoming separated by the Summerland housing estate were all identified as issues. This led to a decision by the then state government to buy back the entire housing estate and return it to a wildlife reserve for the little penguins.

A large upgrade to facilities went ahead in 1988 as more money was invested in protecting what was believed to be a declining population of penguins.

Philip Island Nature Parks continues to manage the penguin colony, which now has a cafe, live-streamed cameras and 500 m of boardwalks. Tourists flock to the reserve, and car parks and waste facilities have been upgraded to match this increased demand. In 2019, a $58 million revamp created a world-class tourism facility, adding fine dining and architecture to the tourism experience and further supporting the protection of the little penguins.

Penguin Beach, Summerlands, 1920s © Phillip Island and District Historical Society

Figure 5.6 A postcard from 1940 of Summerland Beach. It was an age of little environmental concern, when parking cars on the delicate penguin burrows was not given a second thought, and introduced foxes and pet dogs killed dozens of penguins.

Alamy Stock Photo/Joe Eldridge

Figure 5.7 Current facilities at the penguin parade

Kane Constructions Pty Ltd

Figure 5.8 Facilities at the penguin parade under development

QUESTIONS

1 How many visitors per year attend the penguin parade?
2 Describe the state government's role in the management of the facility.
3 The economic development of housing in the Summerland area eventually threatened the penguin colony. How did further economic development with the encouragement of tourism work to protect the penguins?
4 Is the penguin parade an example of a positive or negative impact of economic activity on an outdoor environment? Discuss with your class.

IMPACTS OF LARGE-SCALE ECONOMIC ACTIVITIES

Figure 5.9 Field of solar panels

Alamy Stock Photo/ Philip Game

Economic activities of a large-scale industrial nature can have positive and negative effects on the local and global environment. In Numurkah, private company Neoen has constructed the largest solar electricity farm in Victoria. The facility generates enough clean, renewable energy to supply 100% of Melbourne's tram network. During construction, 755 temporary and 33 ongoing jobs were created, all supporting growth in regional Victoria. This is an excellent example of a positive human impact via industrial activity, where electricity is produced by the sun rather than the use of **non-renewable** fossil fuels such as coal burning.

non-renewable
a resource that does not renew itself at a sufficient rate for sustainable economic extraction in meaningful human time frames

However, economic and/or industrial activities that create large amounts of jobs and income, thus boosting the Australian economy, can also have negative impacts on outdoor environments. The Adani Carmichael coal mine in Queensland's Galilee Basin is one example of an economic activity that could have a significant negative impact in the future by threatening the black-throated finch, an endangered species that lives in the region. On a larger scale, the mined coal will be burned for electricity production, releasing more greenhouse gases and further contributing to climate change. It is important to understand that we do need to utilise the natural resources that outdoor environments supply in order to sustain our species. However, if not regulated by laws and community action, then potentially significant negative consequences can occur.

Worksheet
5.1c Carmichael coal mine

LEARNING ACTIVITY 5.1C

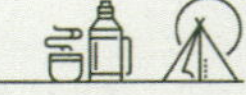

Carmichael coal mine

Research the Carmichael coal mine and construct a single-page fact sheet highlighting the impacts (both positive and negative) that the mine will have on its surrounding environment.

IMPACTS OF FORESTRY

Positive environmental impacts of forestry include:

- timber becoming a renewable resource through sustainable timber harvesting schemes where trees are continually replanted, resulting in less use of non-sustainable resources
- local timber production and harvesting resulting in reduction in pollution due to timber not being transported from elsewhere in Australia or from overseas.

Negative environmental impacts of forestry include:

- large-scale land clearing due to clear felling
- **habitat fragmentation** due to removal of live trees, dead/fallen timber and truck routes being built
- species endangerment and reduction in biodiversity

habitat fragmentation
habitat that is divided or broken down into smaller habitats (e.g. when a road is constructed in a swamp and the swamp is separated into two)

Figure 5.10 Topsoil disturbance and gully erosion are among the negative impacts of forestry.

CSIRO/Nick Pitsas

- poisons entering food webs and waterways due to pesticide and herbicide use
- salinity issues due to the rising water table
- topsoil disturbance and gully erosion
- soil compaction and loss of leaf litter
- introduction of weed species and diseases
- erosion issues in mountainous areas, where trees are slower to grow.

IMPACTS OF MINING

Positive environmental impacts of mining include:

- reclamation of natural environments due to mine rehabilitation programs.

Negative environmental impacts of mining include:

- total removal of vegetation and soil
- loss of biodiversity and species endangerment due to loss of habitats
- introduction of weed species and diseases due to deforestation and machinery use
- atmospheric pollution and increased greenhouse gas **emissions**
- soil and water pollution due to chemicals leaching into soils
- disturbance of marine life due to underwater testing and drilling
- extraction of fossil fuels from inaccessible and possibly inappropriate locations, such as the Great Barrier Reef
- damage caused to places of cultural significance.

iStockphoto/Adventure_Photo

Figure 5.11 Water pollution is among the negative impacts of mining.

emissions
gases that result from energy production and other industrial processes

IMPACTS OF FARMING AND AGRICULTURE

Positive environmental impacts of farming include:

- soil retention and prevention of erosion due to use of recycled wastewater on crops
- the development of suitable practices for different biomes, matching technique with climate and soil.

Negative environmental impacts of farming include:

- loss of vegetation and topsoil due to land clearing
- salinity issues due to the rising water table
- decline in biodiversity due to land clearing and loss of habitats
- water diversion from natural rivers and creeks to irrigation channels
- erosion of creek beds and loss of nutrients in soil due to movement of hard-hooved animals
- degradation of soil nutrient values
- decline in water quality due to fertiliser use
- introduction of weed species and diseases
- increased impacts on arid environments compared to higher-rainfall environments.

Alamy Stock Photo/Scott Bowman

Figure 5.12 Irrigation is vital to the fruit-growing industry of the Goulburn Valley, where water is diverted from creeks and rivers to human-made channels that have very low biodiversity.

Worksheet
5.1d Threats to koalas

Weblink
Australian Koala Foundation

LEARNING ACTIVITY 5.1D

Threats to koalas

1 Visit the Australian Koala Foundation website to learn about threats to koalas that are a direct consequence of human activity.
2 Create a short multimedia video that summarises this information and could become a short social media campaign to educate someone who doesn't know much about koalas.

Figure 5.13 Australian Koala Foundation logo

Australian Koala Foundation: https://www.savethekoala.com/

Recreational activities

Recreation is an activity for enjoyment, amusement or pleasure. Activities are undertaken during leisure or free time. They can facilitate the refreshment of one's mind and body after a period of work and often fulfil a basic human need. Recreation in outdoor environments includes a multitude of pursuits such as bushwalking, camping, rock climbing, cross-country skiing, four-wheel driving, fishing, hunting, photography and canoeing.

Recreational activities may be undertaken via two ways:
- nature-based – focusing on the natural environment
- adventure-based – focusing on the activity as opposed to the setting.

For example, a nature-based rock climber may gain great fulfilment at Summerday Valley in the Grampians (Gariwerd) and associate the activity with other positive factors, such as being with friends, camping, sitting around campfires or viewing the night sky. An adventure-based rock climber, however, may primarily focus on the particular benefits of the activity, such as physical fitness, so the setting is less relevant (whether at Summerday Valley or at a local indoor rock climbing gym).

When undertaking recreational activities, it is important to be aware that your presence may indirectly cause negative impacts on the particular environment you are visiting. However, if the activity facilitates a greater care and understanding of that environment, then a positive impact may follow whereby you contribute to, or advocate for, the future protection of that particular environment.

Recreational activities are often associated with negative impacts on outdoor environments, even though users often express a love of nature or of being outdoors. The increasing popularity of outdoor activity can lead to 'loving a place to death', a colloquial term that describes when visitation by humans to a popular location gets so high the location can no longer manage the impacts. However, general positive impacts of recreational activities can also include the benefits gained from experiential education. Recreational users who venture into outdoor environments will gain a better understanding for the environment they use and what effect they can have on it. Consequently, they may develop a greater respect for it, leading to development of practices that conserve and protect the areas they visit.

Recreational activities are regularly linked to tourism (and therefore economic activities). Often, the most sought-after places are those areas with high natural values. Yet, visitors often disrupt the environment by making noise and requiring services such as accommodation, food, transport and waste disposal. Tourism infrastructure can decrease natural values if facilities encroach on and destroy habitats, causing native animal feeding and breeding behaviours to be disturbed and pollution to occur.

Alamy Stock Photo/Craig Ingram

Figure 5.14 Rock climbing in the Grampians (Gariwerd)

Recreational impacts are also often unique to the specific activity and to the suitability of the location in regard to the level of its impact. An example of this is recreational four-wheel driving. When drivers use approved ramps to access beaches, they don't harm any sand dune ecosystems. They then enjoy driving endless beaches, such as 90 Mile Beach in Victoria, and their tyre tracks are washed away with the tide. However, illegal access through sand dunes damages fragile habitats.

Shutterstock.com/Jacek Chabraszewski

Figure 5.15 Your presence can impact the environments you visit, yet also lead to greater respect.

LEARNING ACTIVITY 5.1E

Impacts of activities on different environments

1 Research and compare the impacts of the same activity on contrasting outdoor environments from the list below.
   - mountain biking in a forest environment and an alpine environment
   - four-wheel driving in an arid environment and a coastal environment
   - bushwalking in a grassland environment and a heathland environment.
2 What might be some positive and negative impacts of doing the activity and of providing infrastructure (buildings, trails etc.) to support the activity?

Worksheet
5.1e Impacts of activities on different environments

Resource
Case study: Cultural heritage threatened by recreation

CASE STUDY

CULTURAL HERITAGE IMPACTED BY RECREATION

The Grampians (Gariwerd) and Mt Arapiles, in western Victoria, have long been popular destinations for recreational rock climbers, attracting visitors from all over the world. Over the years, management challenges have been faced relating to:

- sewage and rubbish
- reducing walking/access trails to cliffs
- erosion at the base of cliffs.

In 2019, attention was also given to the impact of these activities on Indigenous peoples' cultural heritage.

Alamy Stock Photo/Ingo Oeland

Figure 5.16 Mt Arapiles

QUESTION

1 Use the Parks Victoria website, news media articles and research by rock-climbing associations to analyse how the impact of recreational rock climbing has been managed in regard to cultural heritage. As a class, collect your research and share your findings.

Worksheet
5.1f Comparing recreational users

LEARNING ACTIVITY 5.1F

Comparing recreational users

1 Compare and contrast the different human impacts that nature-based and adventure-based recreational users might have within an outdoor environment of your choice. For example, you could compare the rock climbers of the Grampians user group with bushwalkers on the Grampians Peaks Trail.
2 Provide appropriate reasons to support your response.

NEGATIVE IMPACTS OF RECREATION

Negative environmental impacts from bushwalking, hiking and camping include:

- introduction of weed species
- fire scars, depletion of firewood and damage to vegetation due to campfires
- soil compaction and erosion due to repeated treading and tent pitching, and walking on fragile soil types compared to rocky ground
- pollution from inappropriate toileting and rubbish-disposal methods
- wildlife dependency on humans due to feeding fauna
- disruption of wildlife due to domestic animals and increased noise levels
- decreased biodiversity levels due to habitat loss
- landscape degradation due to construction of visitor facilities.

Negative environmental impacts from recreational vehicles (trail bikes and 4WDs) include:

- damage to trees due to vehicle extraction methods
- pollution of watercourses due to increased runoff from tracks and inappropriate toileting and rubbish-disposal methods
- soil compaction, erosion and track rutting due to repeated tracking
- track braiding due to drivers seeking new routes around boggy or unnegotiable areas.

Negative environmental impacts from alpine skiing (downhill and cross-country) include:

- landscape degradation due to the construction of roads and ski-resort facilities
- loss of biodiversity and species endangerment due to loss of habitats
- land clearing due to construction of ski runs
- trampling of exposed vegetation
- pollution from inappropriate toileting and rubbish-disposal methods (particularly during summer months when litter and faeces are exposed).

Negative environmental impacts from rock climbing include:

- impact on the natural environment due to permanent fixtures such as bolts, belay points and ladders
- removal of vegetation, such as moss, along climbing routes
- pollution from inappropriate toileting and rubbish-disposal methods
- soil compaction and erosion due to repeated trampling at the base of popular climbs
- defacing of rock surfaces due to chipping, smoothing, cementing and chalk residue
- damage to Indigenous cultural heritage as represented by sacred sites and rock art.

LEARNING ACTIVITY 5.1G

Indoor versus outdoor settings

1 Undertake an activity of your choice at an indoor venue and then in an outdoor setting. For example, you could rock climb at an indoor climbing centre and then at an outdoor location such as Werribee Gorge or the Organ Pipes.

2 Compare the impacts at each location – do the positive impacts outweigh the negative impacts at the indoor venue or within the natural setting? Why or why not?

Worksheet
5.1g Indoor versus outdoor settings

Worksheet
5.1h Minimal impact guidelines

Weblink
Parks Victoria Minimal Impact Guidelines

LEARNING ACTIVITY 5.1H

Minimal impact guidelines

Visit the Parks Victoria website and search for the Minimal Impact Guidelines for Schools document. Read the guidelines and then complete the following activities:

1 Think of an environment you have visited or studied this year, and then re-create the table below to identify specific conservation, economic and/or recreational activities that have occurred (or currently occur) there.

Activity	Impacts		Possible cause(s) of these impacts	Procedures and/or practices to reduce these impacts
	Positive	Negative		

2 While on an outdoor experience, record in your logbook the activities that you took part in each day and the observable positive and/or negative impacts that occurred as a result. At the end of the practical experience, explain why minimal impact strategies should be employed while experiencing the outdoors.

NOTES FOR THE EXAM

For the exam, you should be able to:

- evaluate a range of impacts on contrasting outdoor environments
- compare the impacts of conservation, economic and recreational activities on contrasting outdoor environments
- describe impacts you have observed or studied regarding specific outdoor environments.

5.1 KEY CONCEPTS

- Humans impact all environments they live in or visit and these impacts can be classified as positive or negative.
- Conservation activities are generally classified as having a positive impact on outdoor environments and involve a range of efforts focused on caring for, maintaining or improving an outdoor environment, and minimising the impacts humans have or are having on the location.
- Economic environmental activities are those that result in a profit or financial gain from using an outdoor environment and include tourism, industrial and primary industry activities. Economic activities can have a predominantly negative impact on outdoor environments, but also recognise that job creation and satisfying the needs of humans are both important.
- Recreational activities may be undertaken as nature- or adventure-based and have various positive and negative impacts on outdoor environments. In common with eco-tourism, recreational activities can have positive impacts of connecting people with outdoor environments, which may lead to caring and taking up conservation practices. However, they can also involve mass visitation and 'trashing' of outdoor environments and their cultural heritage.

5.1 CONCEPT QUESTIONS

Worksheet
5.1 Key Concepts

REMEMBERING

1 Describe one example of each of the conservation activities, economic activities and recreational activities you have observed or studied.
2 What is meant by contrasting environments?

UNDERSTANDING

3 Explain the relationship between economic impacts on environments and employment.
4 Compare the impact of bushwalking in the alpine environment with bushwalking in a coastal environment. Would they be similar?

APPLYING

5 Using an example not presented in the text, explain three negative impacts of tourism on outdoor environments.
6 Describe three impacts on an outdoor environment of a primary industry (agriculture) user.

EXTENSION CHALLENGE

7 In your opinion, what is the single largest impacting group on outdoor environments? Conduct research and provide data to support your answer. Remember, this could include both positive and/or negative impacts of an activity type.

5.2 COMMUNITY-BASED ENVIRONMENTAL ACTION

KEY KNOWLEDGE

- the role of community-based environmental action to promote positive impacts of humans on outdoor environments

KEY SKILLS

- analyse a range of impacts on contrasting outdoor environments

In recent years there has been greater concern for environmental issues in modern society and a drive to live more sustainably. This is evident in changing attitudes about the use of renewable energy sources and recycled materials, and the application of taxes or levies to make individual people and big businesses more accountable (e.g. a **carbon tax**). All of these are examples of community-based environmental action.

carbon tax
a tax charged to industries based on their level of greenhouse gas (primarily CO_2) production

Every living thing has an impact on the environment; therefore, it is inevitable that humans will play a role in impacting the environment. Human impacts can be widespread and large scale, such as deforestation and the production of greenhouse gases, or they can occur on a much smaller scale, such as litter found in a local park or walking off the track on your way to the beach. By recognising the positive and negative human impacts that certain activities can have, you should now understand more deeply the issue of conservation and the need to conserve and/or protect outdoor environments.

5.2.1 COMMUNITY-BASED ENVIRONMENTAL GROUPS

In the context of Outdoor and Environmental Studies, a community can be seen as a group of people who share a common interest in a local environment and work together to develop plans and goals to protect it. Community groups are typically smaller organisations and are based in localised areas. They may be incorporated as trusts or groups of volunteers who share a common goal of preserving a specific local environment, such as a national park or wildlife habitat. These groups are often passionate and knowledgeable about the unique biodiversity of their area, and effective in engaging volunteers to help manage and protect the land. Local community reserves are often treasured by local residents and serve as valuable learning opportunities for schools. Their activities could include tree-planting, litter removal, counting native species, refusing to use plastic shopping bags, and walking on designated tracks and boardwalks while visiting parks and reserves.

Alamy Stock Photo/Bill Bachman

Figure 5.17 Volunteers participating in the revegetation of a river bank

community-based environmental group
a group of people who share a common interest in a local environment and work together to develop plans and goals to protect it

Community-based environmental groups focus on the ecological health of the whole local environment, which can extend beyond physical community borders and also includes social, economic and environmental conditions. Examples of such groups include the Adopt-A-Roadside program, Landcare groups, supporters of Land for Wildlife or Planet Ark, and participants in Clean Up Australia Day.

Overall, communities play a crucial role in environmental protection by bringing together people with similar interests and goals to create positive change in their local environment.

Worksheet
5.2a Your local community

LEARNING ACTIVITY 5.2A

Your local community

1. Make a short list of the communities you are a part of.
2. Which of these groups are influenced by the condition of the outdoor environment in some way?
3. Who is responsible for supporting the ecological health and maintenance of the environments the community groups use?

regenerative farming
a system of farming principles and practices that seeks to rehabilitate and enhance the entire ecosystem of the farm by placing a heavy premium on soil health, with attention also paid to water management, fertiliser use, carbon sequestration and more

Community-based environmental actions can also include:

- conservation covenants
- following minimal impact guidelines and codes of conduct
- revegetation, rehabilitation and restoration programs
- implementing a whole-farm plan, such as **regenerative farming**
- disposing of waste appropriately, such as using recycling and green waste bins
- participating in National Tree Day (co-founded by Planet Ark and Olivia Newton-John)
- participating in various projects undertaken by Greening Australia.

Newspix/Matthew Bouwmeester

Figure 5.18 Clean Up Australia Day is an example of community-based environmental action.

Community versus individual action

While it is important to differentiate between individual or personal action and the action of a community, some overlap always exists. What one person might be able to do alone could also be conducted by numerous people living in the same place. For example, the act of an individual picking up a piece of litter would be considered a personal action, whereas the act of multiple families in the same neighbourhood meeting on a Saturday morning and picking up litter in their local park would be considered a community-based action. Likewise, some of the community-based actions previously listed could also be described as programs, organisations, projects or initiatives. The next section focuses on the positive human impacts that maintain, protect and enhance outdoor environments.

5.2.2 CONSERVATION ORGANISATIONS

Conservation-focused non-government organisations (NGOs) include independent land conservation bodies that operate as not-for-profit businesses, buying and protecting land of outstanding **conservation value**. Their role is critical as they complement the public reserves by increasing conservation of private land through the purchase of and/or protection of land where governments are unable to do so.

conservation value the significance of a natural environment, habitat or species, based on its contribution to biodiversity, ecosystem services, scientific research, cultural heritage or aesthetic value

Prominent conservation organisations include the following:

- Australian Wildlife Conservancy acquires land, and works with other landholders, to establish sanctuaries for the conservation of threatened wildlife and ecosystems.
- Bush Heritage Australia owns and manages reserves throughout Australia.
- Nature Conservancy supports other conservation NGOs to buy and manage high-priority land.
- Tasmanian Land Conservancy conserves, enhances and protects Tasmania's natural environment by purchasing and managing land.
- Trust for Nature encourages people to bequeath land or money for conservation and for the purchase of Victoria's bushlands.

LEARNING ACTIVITY 5.2B

Positive impacts Australia-wide

Visit the Landcare Australia and Greening Australia websites and watch the video clips, especially about the project at Haining Farm on the Greening Australia website (see What we do/Great Southern Landscapes/Haining Farm). Spend some time browsing these websites to understand more about these organisations.

1 Re-create the table below and use it to summarise what community-based actions and subsequent positive impacts Landcare Australia and Greening Australia are having on the environment (include at least four points for each).

Organisation	Community-based actions	Positive impacts
Landcare Australia		
Greening Australia		

Worksheet 5.2b Positive impacts Australia-wide

Weblinks Landcare Australia

Greening Australia

Auscape International Pty Ltd/Alamy Stock Photo

Figure 5.19 Greening Australia is working with Haining Farm in Victoria's Yarra Valley to protect and restore habitat for the critically endangered lowland Leadbeater's possum and helmeted honeyeater.

Worksheet 5.2c Research task

LEARNING ACTIVITY 5.2C

Research task

1 Select an environment local to you.
2 Research a community-based environmental group that promotes positive human impacts (e.g. at your school, local nature reserve or council area).
3 Include the following in your research:
   - name of the group
   - aims
   - geographical area covered (include a map)
   - work they do to promote nature conservation
   - any particular species this group is working to protect
   - the environmental pressures this group is trying to mitigate (e.g. invasive species)
   - how the group promotes positive impacts on this local outdoor environment
   - how local people can join this group (provide details).

Present your findings live to the class or via a pre-recorded podcast or video.

Environmental volunteering in Victoria

Victoria has a rich history of environmental volunteering, with volunteers playing a significant role in enhancing our environment, local communities and economy. A vast number of Victorians generously devote their time to various environmental causes and organisations, such as Landcare, Friends and Coastcare groups, and their contributions are often highly respected and appreciated. Currently, over 170 000 volunteers are involved in environmental volunteering initiatives, dedicating their time and energy to protect and conserve the environment.

Figure 5.20 Promotional poster by Environment Victoria encouraging environmental volunteering

NOTES FOR THE EXAM 

For the exam, you should be able to:

- define what a community of people is and their role in environmental protection
- describe impacts of community groups you have observed or studied regarding specific outdoor environments.
- evaluate a range of impacts on contrasting outdoor environments.

Weblinks
Conservation Volunteers
Planet Ark
Coastcare
Waterwatch Victoria

Worksheets
5.2d Further reading
5.2e What, why, who, when and how

LEARNING ACTIVITY 5.2D

Further reading

Visit the websites for the following organisations to learn more about community-based actions and their positive impacts:

- Conservation Volunteers
- Planet Ark
- Coastcare
- Waterwatch Victoria.

LEARNING ACTIVITY 5.2E

What, why, who, when and how

Select a well-known community-based environmental action, then answer the following:

1 What is the name of the environmental action?
2 Why is this environmental action important?
3 Who is involved in this environmental action?
4 When does this environmental action occur?
5 How successful is this environmental action?

Present this information in a series of five flash cards that can be laminated and kept for revision.

5.2 KEY CONCEPTS

- A community can be defined as a group of people who live in the same area, share common characteristics and have a sense of fellowship with each other.
- Communities play a crucial role in environmental protection by bringing together people with similar interests and goals to create positive change in their local environment.
- Community groups are typically smaller organisations and are often passionate and knowledgeable about the unique biodiversity of their area.
- Community-based environmental groups focus on the ecological health of the whole local environment.
- Environmental action to promote positive human impacts is essentially any action that people take to help the environment, such as following minimal impact guidelines and codes of conduct, maintaining appropriate environmental licences or undertaking revegetation, rehabilitation a nd restoration programs.
- Conservation-focused non-government organisations (NGOs) are independent land conservation bodies that operate as not for profit businesses; for example, Trust for Nature.

Worksheets
5.2 Key concepts

5.2 CONCEPT QUESTIONS

REMEMBERING

1 Describe the term 'community group'.
2 Identify three community groups that promote positive environmental impacts.
3 Choose one of the groups you identified and detail three strategies or actions this group implements to promote positive impacts on outdoor environments.

UNDERSTANDING

4 Analyse why some people don't seem to care for outdoor environments and how this can be improved.

APPLYING

5 Plan your own community-based environmental action. Research an environmental issue in your local community (perhaps one within or close to your school, if possible), then complete the following:
   a Working as a group, give each member of the class a role (such as project manager, community liaison officer, media correspondent, equipment supervisor, funding/fundraising officer or advertising consultant).
   b Plan to implement a brand-new and unique community-based environmental action.

EXTENSION CHALLENGE

6 Become involved in a local environmental action; for example, tree-planting, cleaning up around a waterway or joining a litter-removal program such as Adopt-A-Roadside or Clean Up Australia Day. (If you cannot physically be involved, speak to someone from your community who promotes a local environmental action and/or is a participant.)

5.3 IMPACTS OF TECHNOLOGY ON OUTDOOR ENVIRONMENTS

KEY KNOWLEDGE

- direct and indirect impacts of technologies that support human interactions with outdoor environments, including:
  - equipment manufacture
  - transport
  - snow making
  - recreational vehicles

KEY SKILLS

- compare direct and indirect impacts of technologies on outdoor environments

5.3.1 IMPACTS OF TECHNOLOGY ON OUTDOOR ENVIRONMENTS

Developments in technology have impacted significantly on the ways in which humans relate to and experience outdoor environments. In Chapter 2, we explored the influence of technologies on an individual's outdoor experience. In this section, we compare the impacts of those technologies on outdoor environments, exploring the impact of the manufacturing of this technology on outdoor environments through to the impact of the use of the technology in and on an outdoor environment.

Shutterstock.com/Taras Hipp

Figure 5.21 Snow vehicles are able to get to areas that were difficult to access in the past, and impact outdoor environments directly via compaction of the snow, noise and fumes.

technology
the application of scientific knowledge for practical purposes to extend our human abilities and to manipulate nature to satisfy our wants and needs

The term '**technology**' is a representation of the various ways people have modified the natural world to suit their own purposes. It is the application of scientific knowledge for practical purposes, to extend our human abilities, and to manipulate nature to satisfy our wants and needs. In relation to outdoor environments, technology refers to a multitude of advancements:

- machinery (e.g. associated with farming, agriculture, mining, forestry, water harvesting and other commercial enterprises)
- transportation (e.g. cars, buses, planes, boats, hovercraft, snow/sand transport vehicles and recreational vehicles)
- infrastructure (e.g. associated with recreation and industry)
- communication devices (e.g. radios, smartphones and EPIRBs)
- navigational devices (e.g. GPSs and PNDs [personal/portable navigation devices])
- specialised equipment (e.g. canoes, mountain bikes, climbing ropes, tents and PFDs [personal floatation devices])
- materials and clothing (e.g. Gore-Tex, Dri-FIT, Smartwool and other synthetic materials)
- filming devices (e.g. point of view [POV] filming via wearable cameras, smart devices and drone cameras).

Direct impacts

direct impacts
the impacts of technology on an environment caused by the action itself

Direct impacts from technology are those caused by the action itself. They occur at the same time and place and are usually observable. Examples include the use of recreational vehicles directly causing erosion off-road or snowmaking machinery producing human-made snow to cover a natural area.

iStockphoto.com/dm2912

Figure 5.22 Four-wheel drive vehicle camping on a beach

Indirect impacts

indirect impacts
the impacts of technology on an environment caused by the production or disposal of the technology itself; it can be immediate (e.g. pollution) or delayed (e.g. climate change from greenhouse gas emissions)

Indirect impacts from technology are those that are caused by the production or disposal of the technology and can be immediate (e.g. noise and pollution) or delayed (e.g. greenhouse gas emissions contributing to climate change). They happen either before or after the event of using the technology, but the impacts are still reasonably predictable.

Although impacts can be positive or negative, our focus here is on the impact of technology on the actual outdoor environment (not its effect on our experience).

Shutterstock.com/Gorodenkoff

Figure 5.23 Manufacturing on a large scale reduces the pollution per piece of technology

Worksheet 5.3a Technology audit

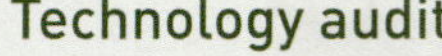

LEARNING ACTIVITY 5.3A

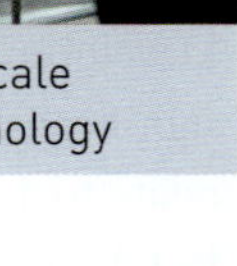

Technology audit

1 On an outdoor experience, choose a location that is visited by various user groups interacting with nature in different ways. Over a one- or two-hour period, note down:
   - the mode of transport for each person
   - any specialised clothing or equipment they are wearing or carrying
   - any direct impacts of their visit to the area you observe.

2 After your audit time elapses, collate your data.
3 Using the information you have collected and from your knowledge in the course, write a summary of the following:
- the direct and indirect impact of the activity the person is participating in
- the transport method and any specialised equipment they used.

EQUIPMENT MANUFACTURE

All equipment manufacture has an impact on the environment in terms of air quality, greenhouse gas emission, ozone depletion, water quality, use of natural resources and production of noise. Manufacturing equipment requires the use of non-renewable resources such as metals and fossil fuels. The extraction of these resources has a high energy demand, which results in atmospheric pollution. The recycling of any equipment is important as it decreases the one-way flow of these resources and reduces the amount of landfill material and associated soil and water pollution.

LEARNING ACTIVITY 5.3B

Worksheet
5.3b Equipment investigation

Equipment investigation

1 Select one of the following items of outdoor recreational equipment:
- canoe
- Gore-Tex clothing material
- mountain bike frame
- tent
- GPS device
- rock-climbing rope (or an alternative piece of equipment with teacher approval).

2 Investigate the following for your chosen piece of equipment:
- what the piece of equipment is made from
- where the parts/components come from (including how far they need to be transported)
- effects (positive and/or negative) on the environment from the manufacturing process
- what happens to the piece of equipment and/or its parts when it is no longer usable
- how this piece of equipment has changed the way humans interact with the environment (including increased or decreased participation rates)
- how this piece of equipment has changed the way humans impact on the environment (including positive, negative, direct and indirect impacts).

3 Link your chosen piece of equipment and its impacts to an environment you have visited recently by providing examples you observed while in the outdoors.

4 Create a presentation that can be viewed by your class using PowerPoint, Prezi or Glogster, or through a poster, brochure, video, song or simulation/role-play.

Outdoor equipment manufacturers such as Patagonia and One Planet are examples of companies that are very aware of the need to reduce their impact on the environment through their business practices. Each of these companies is committed to reducing the environmental impact of their products at every level of production. As discussed throughout this text, Patagonia donates 1% of its annual sales to environmental charities and organisations. Patagonia's commitment to environmental sustainability was highlighted in its switch from pesticide-laden industrially grown cotton to organic cotton in its products, which cost the company years' worth of profit. Patagonia made a promise that it will 'make great stuff, fix it when it breaks, and recycle it when you're done with it'. In return, its asks that the customer vows to 'buy only what you need, repair it when it breaks, and recycle it when you're through'.

Courtesy of Patagonia

Figure 5.24 Patagonia is committed to reducing its environmental impact.

Worksheet 5.3c The Footprint Chronicles

Weblink Patagonia, The Footprint Chronicles

LEARNING ACTIVITY 5.3C

The Footprint Chronicles

The Footprint Chronicles is an initiative of Patagonia that began in 2007. It allows you to explore the various impacts its manufacturing processes have on the environment.

Visit the website of the Footprint Chronicles. In pairs, select a product and record its manufacturing process, considering the following questions:

1 What is it made from?
2 Where do its component parts come from?
3 What impact is there on the natural environment as a result of this process?
4 How far has the product travelled between its place of manufacture and you?
5 What is its life cycle?

TRANSPORT

Transportation projects such as overpasses, ring roads and bridges can have lasting effects on the environment both through the development phase and into the future, with increased activity through the area. These projects are often attractive to people as they reduce travel time and improve accessibility to services. However, we often forget about the environment and what is at stake in relation to its health when these projects go ahead.

Direct environmental impacts of transportation include:

- habitat destruction and/or fragmentation
- increased stormwater runoff that pollutes waterways
- disruption to fauna
- increased noise pollution.

Indirect environmental impacts of transportation include:

- increased motor vehicle exhaust emissions, which can lead to further pollution of the environment over a period of time
- more land development and urbanisation, as more people are potentially attracted to the area due to increased access.

iStockphoto/mikulas1

Figure 5.25 Overpasses bring increased activity through an area and therefore have both a direct and indirect impact on the surrounding environment.

LEARNING ACTIVITY 5.3D

Vehicle impacts

Investigate the process of building, operating and maintaining a vehicle of your choice. Compile a range of facts specific to your vehicle (such as the specific amount of CO_2 emissions produced or fuel consumption).

Visit the following websites to begin your research:

- Green Vehicle Guide
- RACV.

Re-create the table below and complete it with your findings. Compare your results and impacts with those of your classmates.

Direct impacts	Indirect impacts

Worksheet
5.3d Vehicle impacts

Weblinks
Green Vehicle Guide

RACV

RECREATIONAL VEHICLES

The term 'recreational vehicle' covers a range of vehicles, including 4WDs, quad and trail bikes, specialised all-terrain vehicles, hovercraft, jet-skis and amphibious watercraft.

Since the early 2000s, there has been a dramatic increase in sales of 4WDs and SUVs (sports utility vehicles). The ability to handle different road types and access a greater range of environments, coupled with characteristics such as higher driving position and greater seating capacity, have attracted people to these recreational vehicles. They offer a means to escape the rigours of everyday life and access unique outdoor environments in comfort. Having vehicles that are capable of accessing areas that in the past have not been easily accessible means people are exploring previously untouched areas of the country.

There are positive impacts of increased access to unique environments, including visitors developing a great appreciation and therefore being more involved in the future protection and conservation of these areas.

Negative environmental impacts are more obvious and include:

- erosion and widening of tracks, causing a loss of native vegetation
- introduction of weeds and pests due to transmission via vehicles
- compaction of soil, reducing habitat potential
- disturbance to native fauna.

Vehicles impact on our water quality because different fluids (such as oil) and particles get deposited onto our roads. The runoff from roads is washed into stormwater drains, which feed into our creeks, rivers and larger waterways. Greenhouse gas emissions, such as carbon dioxide (CO_2), are also produced by vehicles, which contribute to global warming and climate change. While greenhouse gases occur naturally in the atmosphere, increases in these gases are occurring from the burning of carbon-based fuels, which is partly a result of vehicle use.

iStockphoto.com/samwaltenbergs

Figure 5.26 Four-wheel driving is a popular pursuit in Victoria.

SNOWMAKING

Artificial snow is produced by forcing water and pressurised air through snow cannons. These cannons require low temperatures, water pumps and air compressors to operate effectively. The machinery is large and expensive, and the process requires large amounts of energy and water.

You might initially be unaware of the negative impact that snowmaking machinery has on our environment because the water being utilised makes its way back to nature as the snow melts. However, moving large amounts of water through these machines can have harmful effects on plant and animal life, and on water reserves.

The period of intense water use for snowmaking occurs at a time when water levels are generally at their lowest. Our nation is already battling with the fact that water is our most precious resource. However, we continue to transport and store water for our ski seasons to ensure that tourism and our economy continue to boom. Putting in place snowmaking machinery such as tanks, pipelines and snow cannons means changing our landscape, altering ecosystems, increasing pollution and potentially destroying habitats. Trucks and generators required for water transportation require huge amounts of energy, further polluting our waterways and atmosphere.

Artificial snow is harder, denser and heavier than natural snow, resulting in a waterproofing effect on soil, subsequently making soil erosion easier. Artificial snow melts more slowly and therefore prolongs the snow season. The longer the artificial snow is maintained, the more impact there is on woody plants, snow-bed species and late-flowering species. This further impacts on bird diversity and populations due to the modification made to their natural habitat and food source.

In summary, direct environmental impacts of snowmaking include:

- less time for the environment to regenerate and replenish, as the snow season and the amount of time people are spending in alpine areas are extended
- increased soil erosion, as artificial snow is waterproof (unlike natural snow).

Indirect environmental impacts of snowmaking include:

- the requirement for large amounts of energy to produce artificial snow, with fossil fuels burnt, resulting in the pollution of air, water and ecosystems
- the requirement for large amounts of energy for the manufacturing process of snowmaking machinery and other equipment needed for the production of artificial snow
- increased power consumption, pollution and other associated impacts from a prolonged snow season, which results in more people populating alpine ski-resort areas for longer periods of time
- further pollution of the environment because snowmaking is usually required when the environmental water levels are at their lowest, which means that water has to be transported and stored before it is used to make snow, and large equipment such as trucks must be used to transport the water.

Resource
Case study: Can you tell the difference between artificial and natural snow?

CASE STUDY

Can you tell the difference between natural and artificial snow?

If you ask a bunch of weekend snow bunnies, chances are they will declare a preference for natural over man-made [sic] snow.

But is it better? And could they tell the difference on a typical run at any alpine resort in Australia, aside from the aesthetics of exposed grass beside their ski runs?

Probably not on both counts says Steve Lee, who has represented Australia at three Winter Olympics, and Geoff Lipshut, the chief executive of the Olympic Winter Institute of Australia, which is training our next crop of winter athletes.

Little beats the delicate beauty of snow flakes, which are about 35 per cent water, mixed with air. Artificial snow fired from guns at resorts is 50 to 60 per cent water.

The man-made snow, however, is 'way more durable', Mr Lee said. 'Once it's laid, it can last.'

Internationally, all major events are run on artificial snow because the courses often need to withstand 70 to 80 races, a load that would create 'massive holes' if left to nature, he said.

'In my day, if you were outside the top 20, it would be pretty much trashed,' said Mr Lee, who competed in the 1984, 1988 and 1992 Games.

Mr Lipshut said the dense artificial snow is also more resistant to rain, making for an ideal base. Except for fresh falls, most skiers won't be able to tell the difference, he said.

'You really have to be able to feel a lot through the edges of your skis to pick it,' he said.

That's especially the case on most ski runs, where resorts combine snow guns with other equipment to prepare the cover overnight.

'Once it's been cut up and groomed by the machines, no one can tell' what is artificial or natural, Mr Lipshut said.

Mr Lee said Australia's fickle weather means resort operators have become so expert in 'chasing pockets of cold air around the mountain' that snow-gun makers tend to test their newest technology here.

'If you can make snow in Australia, you can make it anywhere in the world,' he said.

...

Mr Lipshut said while climate change will likely leave resorts more reliant on snow-making to fill declines in natural snow, the use of snow guns is already an industry fixture.

'Except for the really high mountains ... most ski resorts anywhere would struggle without snowmaking in the first half of their season,' he said.

'Can you tell the difference between natural and man-made snow?'
Sydney Morning Herald, Peter Hannam, 10 July 2015

iStockphoto.com/Milkare

Figure 5.27 'If you can make snow in Australia, you can make it anywhere in the world': Snow guns replenish cover overnight.

QUESTIONS

1 How much higher in water content is artificial snow than natural snow?
2 Why is artificial snow better than natural snow for racing?
3 Name two direct and two indirect impacts of artificial snow.
4 What would be the future of snow-skiing in Australia without this technology?

UNCREWED AERIAL VEHICLES

Uncrewed aerial vehicles, better known as drones, are remote-controlled flying devices used in many applications, including:

- defence force (communications, reconnaissance and combat operations)
- film-making and photography
- recreational home-movie making
- agriculture (fence and property surveys)
- roof inspections
- hunting
- fun!

Domestic or recreational drones on average have a flying time of 30 minutes, allowing the pilot a flight range of up to 10 kilometres at speeds of up to 60–90 km/hour. The noise they emit by their tiny propellers is likened to a swarm of bees, equivalent to 80 decibels, which is louder than a vacuum cleaner and equal to a kitchen blender or smoothie maker. It's loud if you're close, but less so at a distance.

Direct environmental impacts of drones include:

- noise pollution to nearby users of an outdoor environment and its animals
- invasion of privacy of other users of outdoor environments
- disruption to fauna
- access to areas previously inaccessible, increasing the footprint of human impact through technology.

Indirect environmental impacts of drones include:

- embodied energy impact for the cost of the materials mined and manufactured to produce the product
- manufacturing impact through use of fossil fuels and climate change.

FUEL STOVES

Fuel stoves are lightweight, compact and versatile. Using a fuel stove for cooking in outdoor environments, rather than a campfire, is often mentioned in camping codes of conduct.

Direct environmental impacts of fuel stoves include:

- contributing to atmospheric pollution
- reducing the need to burn timber.

Indirect environmental impacts of fuel stoves include:

- requiring fuel that needs to be sourced and refined
- manufacturing processes that require a lot of energy, as well as creating a lot of waste.

iStockphoto.com/Multiart, iStockphoto.com/O Kemppainen©

Figure 5.28 Fuel stoves versus campfire cooking

LEARNING ACTIVITY 5.3E

Fuel stoves versus campfires

Using a fuel stove in fragile environments instead of a campfire is a standard minimal-impact practice. Fires are bad and stoves are good – right? But what are the actual impacts of a campfire on a local environment? And what are the impacts of a fuel stove?

Fuel stoves probably have a lower impact on the environment you visit, but what about deferred impacts? What about the mining, refining and processing to make the materials of the stove? What about the extraction of the fossil fuels? What about the transportation impacts of getting the stove to the shops for you to purchase? What about the impacts of the stove parts once it is old and thrown away?

1 Analyse the direct and indirect impacts of fuel stoves and campfires for cooking in outdoor environments. You have been provided with a few impacts from fuel stoves above, but you will need to undertake further research to put together a comprehensive analysis.
2 Once your analysis is complete, make a judgement about which cooking method actually has a lower impact on the environment overall.

Worksheet 5.3e Fuel stoves versus campfires

NOTES FOR THE EXAM

For the exam, you should be able to:

- compare direct and indirect impacts of technologies on outdoor environments, including:
  - equipment manufacture
  - transport
  - snow making
  - recreational vehicles
- describe impacts you have observed or studied regarding specific technologies in specific outdoor environments.

5.3 KEY CONCEPTS

- The term 'technology' is a representation of the various ways people have modified the natural world to suit their own purposes. It is the application of scientific knowledge for practical purposes, to extend our human abilities and to manipulate nature to satisfy our wants and needs.
- In relation to outdoor environments, technology can include machinery, transportation, infrastructure, communication devices, navigational devices, specialised equipment, materials and clothing, and filming devices.
- Direct impacts from technology are those caused by the action itself.
- Indirect impacts from technology are those that are caused by the production of the technology and can be immediate or delayed.
- Impacts examined here are equipment manufacture, transport, snow making and the use of recreational vehicles.

Worksheet
5.3 Key Concepts

5.3 CONCEPT QUESTIONS

REMEMBERING

1 Describe three direct impacts of technology.
2 Describe three indirect impacts of technology.
3 Using an example of each, distinguish between direct and indirect impacts of technology.

UNDERSTANDING

4 Compare how the indirect impact of equipment manufacture can affect outdoor environments.
5 Explain how technology can improve a human's experience in the outdoors. Give an example.

APPLYING

6 Using an example not presented in the text, explain an example of a direct impact of technology you have studied or observed.
7 Describe two technological improvements that have enhanced your experience of outdoor environments.

EXTENSION CHALLENGE

8 Four-wheel drive recreational vehicles are often condemned as having both a large indirect impact on outdoor environments (vehicle manufacture) and a large direct impact on outdoor environments (erosion). Explain how four-wheel driving can lead to positive impacts on outdoor environments and human relationships with them.

5.4 IMPACTS OF URBANISATION

KEY KNOWLEDGE

- the impact of urbanisation on outdoor environments

KEY SKILLS

- discuss and predict impacts of urbanisation on outdoor environments

5.4.1 URBANISATION IN AUSTRALIA

urbanisation
the development and physical growth of towns and cities, including residential areas, as people move to these locations

Urbanisation is the development and physical growth of towns and cities, including residential areas, as people move to these locations. This is influenced in Australia with an increasing proportion of Australians living in urban areas (cities and towns).

Australia is considered one of the most urbanised and coastal-dwelling populations in the world, as more than 80% of Australians live within 100 kilometres of the coast. Most of our population is concentrated in a few major metropolitan areas: Sydney, Melbourne, Brisbane, Perth and Adelaide. These cities have experienced significant growth in recent years due to factors such as migration, natural population growth and economic development.

Shutterstock.com/ymgerman

Figure 5.29 The trend towards larger homes on smaller blocks is having impacts on the urban environment.

Urbanisation in Australia has brought about both positive and negative impacts. Urbanisation has resulted in the development of modern infrastructure (e.g. public transport and road networks), improved access to services and amenities (e.g. hospitals and public gardens) and the creation of job opportunities. It has also led to issues such as urban sprawl, traffic congestion, air pollution and increased pressure on natural resources.

The process of urbanisation means that natural environments suffer as more space is required for the construction of houses and the development of industry. An example of this is the ever-growing **urban sprawl** of suburbs that surrounds Melbourne, which stretches approximately 150 kilometres from east to west.

Between 2011 and 2021, Victoria's population increased by 15%. The areas with the largest growth were:

- Wollert (up by 24 200 people) and Mickleham – Yuroke (22 200), both in Melbourne's outer north
- Cranbourne East–North (16 400) in Melbourne's outer south-east.

The areas with the highest **population density** in June 2021 were:

- Melbourne CBD – North (31 100 people per square kilometre)
- Southbank – East (20 600)
- Melbourne CBD – West (16 100)

Population density data is an important tool used to measure the amount of people living in a particular area. A larger number of people living in a smaller space places a greater demand on natural resources, and this puts the environment under more pressure.

FAST FACT

The Australian Bureau of Statistics (ABS) defines urban as centres with more than 200 people. Australia has over 1853 urban environments. However, 75% of people in Australia live in just 18 cities, each with more than 100 000 people.

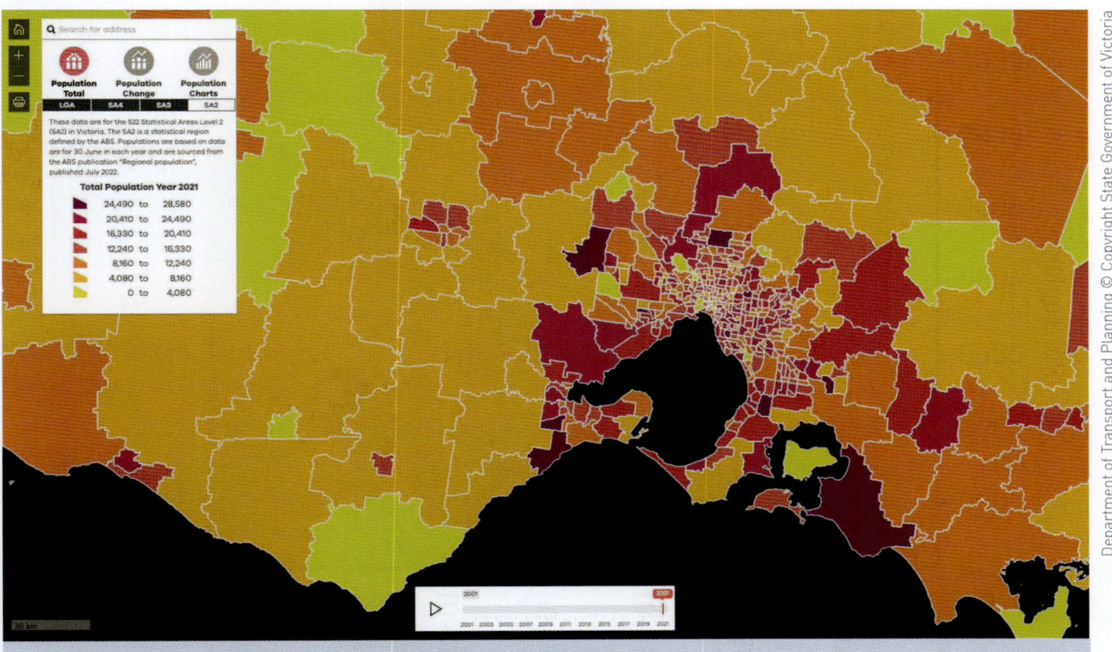

Figure 5.30 Victoria's population density, 2021

urban sprawl
the rapid expansion of the geographic extent of cities and towns, often characterised by low-density residential housing and increased reliance on private vehicles for transportation

population density
the number of people living in a given area

Characteristics of urbanisation

The process of urbanisation can result in the following impacts:

- Loss of habitat – the construction of buildings and infrastructure required for urbanisation often results in the destruction of natural habitats, such as forests and wetlands, which can have negative effects on biodiversity and ecosystem services.
- Increased pollution – urbanisation can lead to an increase in air, water and soil pollution due to the emissions from transportation, industrial activities and waste disposal.
- Increased surface temperature – urban areas often have higher temperatures than rural areas due to the large amount of heat-absorbing surfaces, such as concrete and asphalt, which can have negative effects on air quality.
- Reduced green spaces – urbanisation can lead to the reduction of green spaces such as parks and gardens, which can negatively affect biodiversity, recreation opportunities and mental health.
- Increased energy consumption – urbanisation can result in increased demand for energy for heating, cooling, lighting and transportation, which can contribute to climate change and other environmental issues.
- Water management – urbanisation often puts a strain on water resources, and can lead to issues such as water shortages, flooding and pollution.

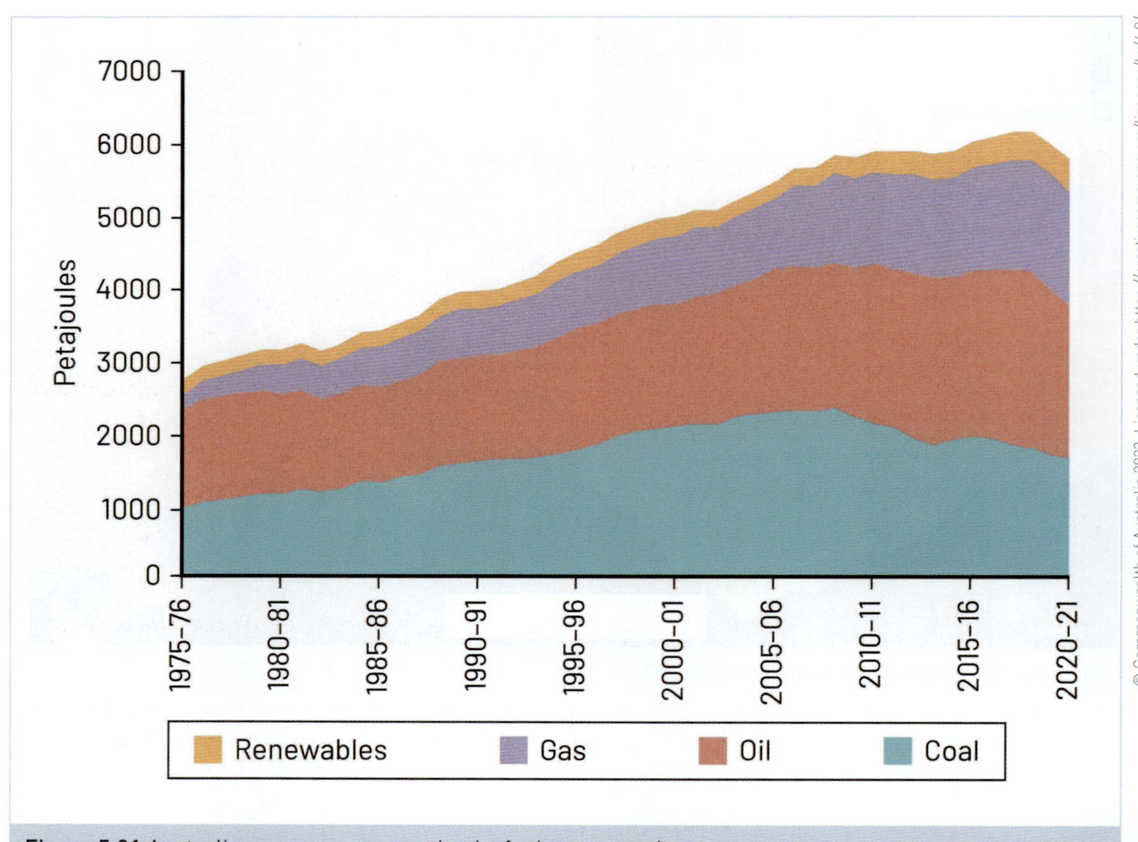

Figure 5.31 Australian energy consumption by fuel type over time

Shutterstock.com/Taras Vyshnya

Figure 5.32 Urbanisation requires infrastructure such as housing and roads, which impact on nature.

Impacts of urbanisation

WATERWAYS

Impacts from urbanisation on waterways can include:

- decline in riverine species
- disruption of ecosystems
- increased turbidity and siltation
- rising salt levels
- loss of streamside vegetation
- erosion of creeks and watercourses
- increased stormwater runoff
- alteration of flow patterns
- water pollution from litter, garden refuse, soap, petrol, oil products, animal faeces, sediment from rivers and streams, overflowing sewage and cigarette butts.

SPOTLIGHT

How urbanisation affects the water cycle

The City of Melbourne's Urban Water website discusses the impacts of urbanisation on the water cycle.

> The natural water cycle is impacted by buildings and sealed surfaces. As a result, natural water flows are altered and stormwater is created. Stormwater is rain that has collected on roofs, roads, footpaths and other sealed surfaces. It flows directly into our waterways via the stormwater drainage network. In urban areas, water cycle problems include:
>
> **Pollution**
>
> When water comes in contact with urban surfaces such as roofs, roads and footpaths, it becomes contaminated with oil, metals, litter and other pollutants. This is what we call stormwater. Stormwater drains do not usually have any treatment systems, so pollutants are carried directly into our waterways, bays and oceans.

Waterway flushing

When it rains, the volume of stormwater entering the waterways in urban areas increases. Water that would usually soak into the ground floods into the stormwater drainage network, where it is transported directly to our waterways. High volumes of stormwater impact our waterways by:

- damaging the habitat for aquatic animals, such as fish and invertebrates
- disturbing the breeding cycles of aquatic animals
- eroding stream banks
- increasing turbidity and pollution levels
- altering natural flood cycles.

Flooding

During heavy rainfall events, large volumes of stormwater collect on sealed surfaces and flow into the stormwater drainage network. Flooding can occur when the volume of stormwater exceeds the capacity of the stormwater drains. This can cause flooding in areas not necessarily close to waterways.

Decreased soil moisture

Most urban surfaces are sealed, or impermeable. They prevent rainwater from soaking into the soil as it does in the natural water cycle. Low soil moisture in urban areas can impede healthy growth of plants, so irrigation is required to keep trees, plants and grass healthy.

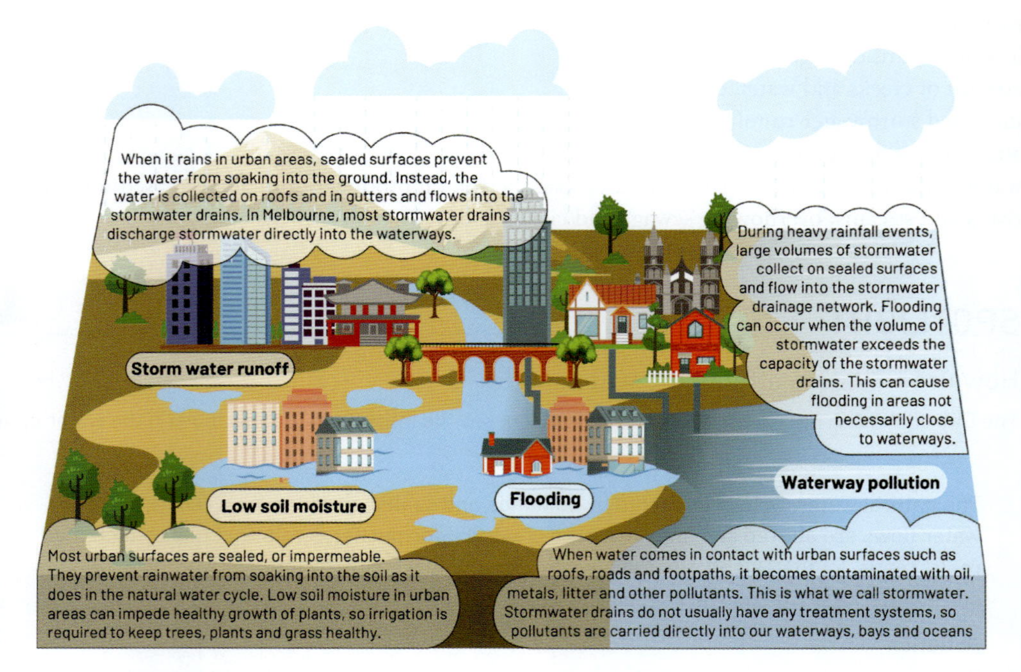

Figure 5.33 The impact of urbanisation on the water cycle

LAND

Impacts from urbanisation on land can include:

- habitat loss through timber harvesting, mining excavation and clearing of vegetation
- loss of habitat (hollow) trees
- lack of dead wood and forest litter
- introduction of weeds
- loss of topsoil
- decreased water absorption into the soil
- predation by introduced domestic animals
- decrease in biodiversity levels
- land disturbances due to deforestation, mining and farming.

While the impacts listed here have a predominantly negative effect on the environment, urbanisation can also bring about the development of parklands and areas where native **remnant vegetation** is protected and regenerated.

remnant vegetation small patches of native plants that remain after conversion of landscapes to agricultural or other use

Alamy/Davo Blair

Figure 5.34 An example of remnant vegetation left on productive farmland in Australia.

While it is pleasing to see landowners leave remnant vegetation, this example may have caused habitat fragmentation of some species as they would be unable to travel from this section to another section of remnant vegetation for survival.

Worksheet
5.4a Local investigation

LEARNING ACTIVITY 5.4A

Local investigation

1 Examine and explain the effects of urbanisation on the region where you live or attend school, taking into account the following aspects:
   - How has the local environment transformed over time? Use images where possible.
   - What were the main characteristics of the environment before European colonisation? Include environment type and native species details.
   - Are there any remaining habitats that exist?
   - What are the reasons for the occurrence of urbanisation?
   - How rapidly has your local area developed?
   - What are the main impacts of urbanisation in your local area, if any?

2 Present this information using Prezi or a similar interactive visual tool.

Management approaches to urban planning

The urban environment faces significant pressure from population growth, leading to increased resource consumption, waste generation, greater commuting distances and increased pollution. Adding to these pressures, climate change poses a significant challenge, causing rises in sea levels, higher temperatures and more extreme events like bushfires, droughts and floods. These pressures all adversely affect urban biodiversity and people.

To mitigate these challenges, many local and state governments are taking preventative measures. Urban planning policies are being recast to manage urban sprawl and preserve public green spaces. Governments are investing in integrated infrastructure by coordinating roads, public transport, cycle paths and walkways. There's also an increased focus on managing waste and reducing disaster risks.

The management of Australia's urban environments is divided among three levels of government:

- local
- state/territory
- federal.

Each level of government has its own set of policies, strategies and regulations to address urban planning and development issues.

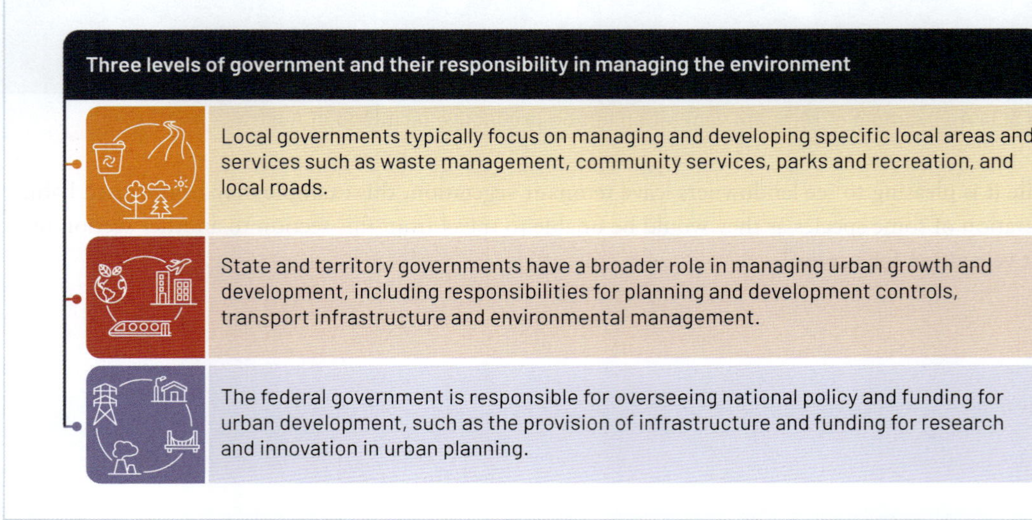

Figure 5.35 Local, state/territory and federal responsibilities for the management of urban environments

CASE STUDY

COOLING AND GREENING MELBOURNE

Victoria has the fastest-growing population of any state in Australia and is facing significant challenges as its population is projected to reach 11.2 million by 2056, with 9 million people living in Melbourne alone. The city's rapid growth, coupled with the effects of climate change, presents a complex set of challenges that must be addressed to ensure Melbourne's sustainability and liveability in the future. The Victorian government has made cooling and greening Melbourne a priority, and promoting sustainability and resilience through **green infrastructure** is a key component of this effort.

To achieve these goals, the state government has recognised the need to plan for green infrastructure, in the same way that grey infrastructure is planned. Grey infrastructure is the planning and implementation of gutters, pipes and drains in urban environments to remove stormwater. Green infrastructure will help to improve urban amenity and quality, enhance landscape connectivity and build resilience to climate change. To achieve more liveable outcomes, the government plans to protect existing green spaces, create new opportunities for urban greening, improve water-sensitive urban design, promote greening of buildings such as on roofs, facades and walls, and increase the use of permeable surfaces. By incorporating these measures into urban planning and development, Melbourne can become a more sustainable and liveable city for its growing population.

Shutterstock.com/trabantos

Figure 5.36 Green infrastructure will help build resilience to climate change.

Resource
Case study: Cooling and greening Melbourne

green infrastructure
natural or semi-natural systems (e.g. parks, wetlands, green roofs, trees or rain gardens) designed to provide ecosystem services and benefits to urban communities

Weblink
Cooling and greening Melbourne

NOTES FOR THE EXAM

For the exam, you should be able to:
- define the term 'urbanisation' and how it applies to the Australian population
- list the negative and positive impacts of urbanisation
- discuss and predict impacts of urbanisation on waterways
- discuss and predict impacts of urbanisation on outdoor environments.

5.4 KEY CONCEPTS

- Urbanisation refers to the development and physical growth of towns and cities, including residential areas, as people move to these locations.
- Australia is considered one of the most urbanised and coast-dwelling populations in the world, as more than 80% of Australians live within 100 kilometres of the coast.
- Urban sprawl is the rapid expansion of the geographic extent of cities and towns, often characterised by low-density residential housing.
- Characteristics of urbanisation include loss of habitats, increased pollution, increased surface temperature, reduced green spaces and increased energy consumption.
- Impacts from urbanisation on waterways can include disruption of ecosystems, increased turbidity and siltation, increased stormwater runoff, alteration of flow patterns, and water pollution from litter, garden and petrol products.
- Impacts from urbanisation on land can include loss of habitat (hollow) trees, lack of dead wood and forest litter, introduction of weeds, loss of topsoil and decreased water absorption into the soil.
- Green infrastructure refers to natural or semi-natural systems – such as parks, wetlands, green roofs, trees or rain gardens – designed to provide ecosystem services.

5.4 CONCEPT QUESTIONS

Worksheet
5.4 Key concepts

REMEMBERING

1 Describe 'urbanisation'.
2 Describe 'urban sprawl'.
3 Outline the connection between population growth and urban sprawl.

UNDERSTANDING

4 Explain three environmental impacts of urbanisation.
5 Discuss two techniques to reduce the impacts of urbanisation on an outdoor environment you have visited or studied.

APPLYING

6 Using the case study 'Cooling and greening Melbourne' in this section, explain two examples of green infrastructure.
7 Explain how green infrastructure could reduce the impacts of urbanisation.

EXTENSION CHALLENGE

8 Using an outdoor environment you have visited or studied:
   - research the population levels of this environment or nearest town (use the Australian Bureau of Statistics website to assist you)
   - identify potential threats to this environment due to population levels and urbanisation
   - suggest potential mitigation strategies to reduce the impact of urbanisation on this environment.

5.5 MANAGEMENT OF THREATENED SPECIES

KEY KNOWLEDGE

- identification and management of threatened species and/or ecological communities in an outdoor environment

KEY SKILLS

- evaluate the effectiveness of management strategies on an outdoor environment
- describe a local environmental policy that supports management of threatened species or an ecological community

5.5.1 MANAGEMENT OF THREATENED SPECIES

Australia has a diverse range of ecological communities, such as rainforests, wetlands, eucalypt forests, alpine areas and deserts. These communities are important for maintaining biodiversity and ecosystem services, such as clean water and air, nutrient cycling and climate regulation. However, many of these ecological communities are under threat due to human activities such as land clearing, agriculture, urbanisation and climate change.

Shutterstock.com/Steve Tritton

Figure 5.37 New housing estate subdivisions pose significant threats to ecological communities.

ecological community
naturally occurring and unique groups of plants and animals; their presence can be determined by factors such as soil type, position in the landscape, climate and water availability

To protect and restore **ecological communities**, various conservation and management strategies are employed, including the following:

- Protected areas – national parks, state parks and other protected areas provide legal protection to important ecological communities and their species. These areas are managed to reduce threats from activities such as logging, mining and hunting.
- Habitat restoration – ecological communities that have been degraded can be restored through activities such as revegetation, wetland rehabilitation and erosion control. Restoration activities aim to re-create the original ecological conditions and promote the recovery of native species.
- Threatened species management – many species of plants and animals are listed as threatened or endangered due to habitat loss and other threats. Management actions such as breeding programs, invasive species control and habitat restoration are used to protect and recover these species.
- Land-use planning – this seeks to balance the needs of development with conservation objectives. It involves identifying areas of high conservation value and reducing impacts on these areas through zoning.
- Community engagement – community engagement and education are important for raising awareness of the value of ecological communities and the threats they face. Engaging with local communities can also help to build support for conservation initiatives and foster a sense of stewardship for the natural environment.

ecosystem services
the benefits provided to humans through transforming environmental resources such as vegetation and waterways into essential goods such as food and clean water

Ecological communities are significant due to their blend of native biodiversity, unique landscape/seascape features and essential habitat characteristics that support a range of **ecosystem services**. These services include

- natural regulation of air, water and soil nutrients
- erosion and salinity control
- breeding and feeding habitats for various species, such as fish
- carbon storage.

How are threatened species and ecological communities protected in Australia?

All levels of government in Australia – federal, state/territory and local – have environmental policies to ensure the protection of native species and ecological communities. Each level of government also has its own classification process and related policy.

The Australian government's central environmental policy is the *Environment Protection and Biodiversity Conservation Act 1999* (EPBC Act). This provides a framework to protect and manage nationally and internationally important flora and fauna, ecological communities and heritage places.

As of 2023, the EPBC Act listed 18 protected ecological communities in Victoria. Of these 18 communities, 10 were critically endangered, seven were endangered and one was listed as vulnerable.

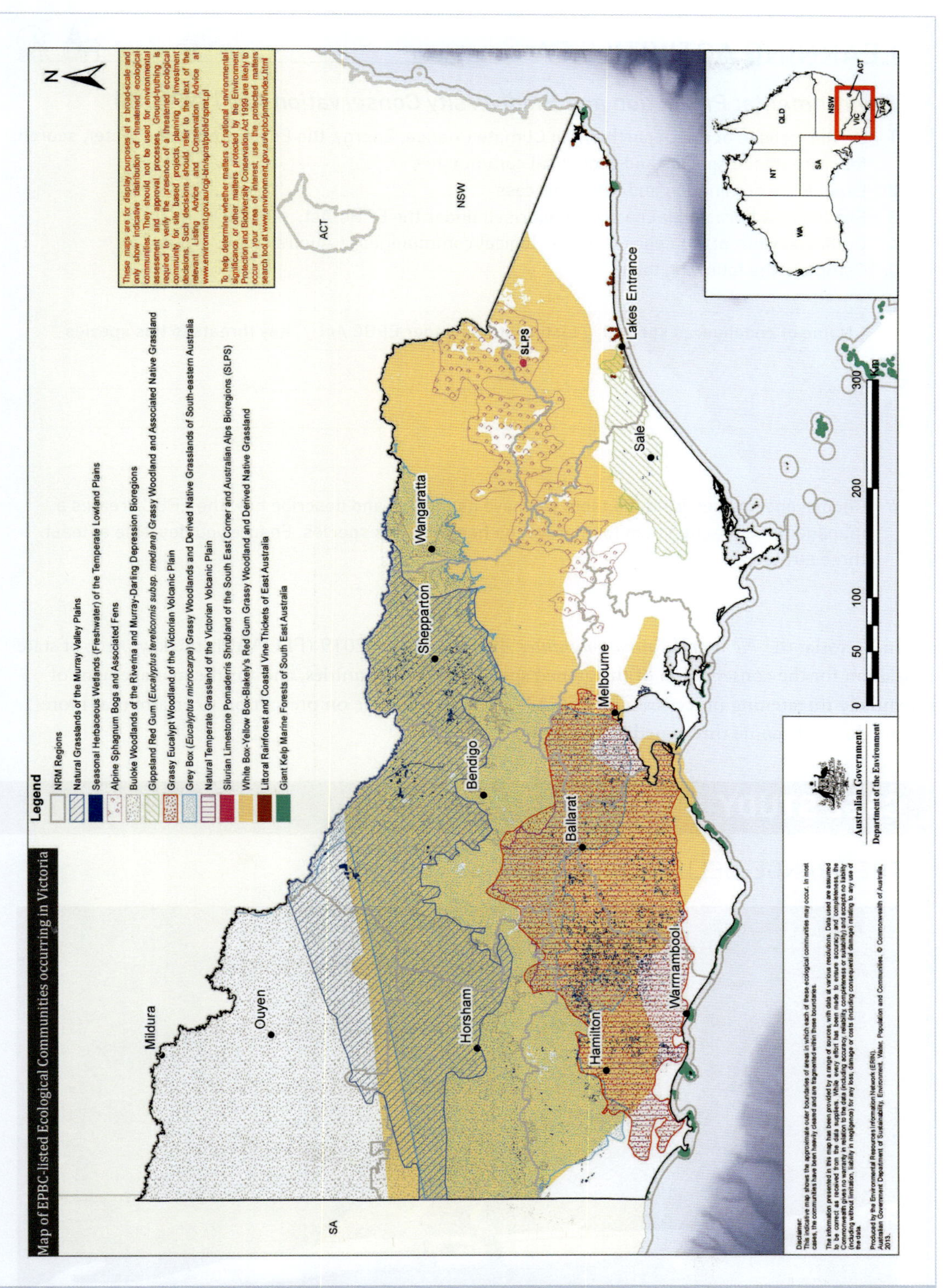

Figure 5.38 Ecological communities in Victoria listed under the Environment Protection and Biodiversity Conservation Act 1999 (Cth)

Worksheet
5.5a Environmental Protection and Biodiversity Conservation Act 1999

Weblink
Threatened species and ecological communities

LEARNING ACTIVITY 5.5A

Environmental Protection and Biodiversity Conservation Act 1999 (Cth)

1 On the website of the Department of Climate Change, Energy, the Environment and Water, search for 'Threatened species and ecological communities'.
2 Explain the nomination and listing process.
3 Explain how flora and fauna are categorised under the EPBC Act.
4 Outline key threats to species and ecological communities in Australia.
5 Complete the following table:

Name of endangered species	Listing status under EPBC Act	Key threats to this species

6 Choose one species from the table you have generated and describe how the EPBC creates a management or action plan to improve the health of this species. Ensure you describe at least three strategies.

In Victoria, the *Flora and Fauna Guarantee Amendment* Act 2019 (FFGAA) is the key piece of state legislation for the conservation of threatened species and communities, and for the management of potentially threatening processes. The FFGAA places importance on prevention to ensure that more species do not become threatened in the future.

Resource
Case study: The orange-bellied parrot

CASE STUDY

THE ORANGE-BELLIED PARROT

Shutterstock.com/Agami Photo Agency

Figure 5.39 The orange-bellied parrot

The orange-bellied parrot is listed as critically endangered and there are very few left in the wild. A migratory parrot that breeds on the west coast of Tasmania and flies to the south-east mainland of Australia during the winter months, the orange-bellied parrot is at risk of extinction.

Threats to the orange-bellied parrot include:

- alteration and destruction of its saltmarsh habitat for industrial and urban development
- vegetation clearance for agriculture
- inappropriate fire regimes
- drought on the mainland
- introduced predators such as foxes and cats
- lack of females in the wild
- beak feather disease.

What management strategies are being undertaken?

A National Recovery Team has been working to initiate wild and captive breeding programs. Captive breeding programs have been successful, with about 200 **fledglings** released into the wild. Wild breeding programs have been boosted with the supply of nesting boxes, supplementary food and management of predators in Tasmania.

fledglings a young bird that has just taken flight

Additional management strategies to improve orange-bellied parrot numbers include:

- habitat restoration
- public involvement, such as education displays
- excluding farming where orange-bellied parrots are known to exist
- providing funding for a vaccine to treat beak feather disease
- constructing a new captive breeding and research facility at Healesville Sanctuary.

There are two key skills applied to this key knowledge point. The first is asking you to describe a local environmental policy that supports management of threatened species or an ecological community and the second is asking you to evaluate the effectiveness of management strategies on an outdoor environment.

Below is an example of a local government policy that supports the management of a threatened native species.

CASE STUDY

SAVING THE SOUTHERN BROWN BANDICOOT

Resource Case study: Saving the southern brown bandicoot

Cardinia Shire Council is a municipality in south-east Melbourne. It has several environmental policies that work to protect and improve biodiversity within its boundaries. One example is the Biodiversity Conservation Strategy 2019–2029, which outlines strategies to protect native species, including the southern brown bandicoot.

Populations of the southern brown bandicoot exist in the Koo Wee Rup, Bayles, Cardinia, Lang Lang and Bunyip areas, although numbers continue to decline. Reasons for the bandicoot's declining population include land clearing for housing and farming, invasive species and habitat fragmentation.

The council has enacted a management plan from 2019 to ensure the long-term viability of this species, particularly in the former Koo Wee Rup Swamp area.

Management strategies to protect the bandicoot include the following:

- Bandicoot tunnels – seven trial bandicoot crossings have been installed under main roads to help the bandicoots to safely negotiate major roads. Cameras were installed to make sure the bandicoots were using these tunnels, and there were multiple sightings per week.

- Preferred habitat to be preserved – bandicoots like to live in dense understory vegetation such a shrubs and long grass. Weed eradication, such as the removal of blackberries, is vital to protect these habitats.
- Habitat restoration – linear strips of vegetation along drains, waterways, roads and railways, as well as on private property, assist to restore once fragmented habitat, and allow the bandicoot to travel from one area to another.

The council has additional information to encourage the community to preserve the southern brown bandicoot and its habitat:

- Practise **responsible pet ownership**, including desexing pets and keeping cats indoors.
- Gradually remove weeds from your property, and replace with indigenous plants.
- Protect and increase remnant vegetation on your property.
- Build a southern brown bandicoot hide (house).

responsible pet ownership
balancing of the needs of pets with the needs of the environment in which the pets exist (e.g. keeping cats from roaming, keeping dogs on a lead in nature reserves and bushland, and horse riding only in designated areas)

Shutterstock.com/Susan Flashman

Figure 5.40 The southern brown bandicoot

NOTES FOR THE EXAM

For the exam, you should be able to:

- define ecological communities and list Australia's diverse range of ecological communities
- describe an outdoor environment that you have visited or studied that is supported by an environmental policy
- describe a local environmental policy that supports management of threatened species or an ecological community.

LEARNING ACTIVITY 5.5B

Class presentation

1 Find your local council in the interactive map on the Vic Councils website.
2 Read about a species or ecological community within this local government area that is of interest to you.
3 Ensure you take notes on the following:
   - name of protected species
   - where the habitat of this species is within your local municipality
   - three threats to this species
   - three management strategies to protect and preserve this species.
4 Is this species protected under any additional state or federal environmental policies?
5 Do you think the current management plan in place is effective in managing this species? Explain why.

Worksheet 5.5b Class presentation

Weblink Vic Councils

5.5 KEY CONCEPTS

- Australia has a diverse range of ecological communities, such as rainforests, wetlands, eucalypt forests, alpine areas and deserts.
- Ecological communities are important for maintaining biodiversity and ecosystem services, such as clean water and air, nutrient cycling and climate regulation.
- To protect and restore ecological communities, a variety of strategies are employed, including protecting areas, habitat restoration, land use planning and community engagement.
- All levels of government in Australia – federal, state/territory and local – have environmental policies to ensure the protection of our native species and ecological communities.
- The Australian government's central environmental policy is the *Environmental Protection and Biodiversity Conservation Act 1999*, which works to protect and manage nationally and internationally important flora and fauna, ecological communities and heritage places.
- The Victorian legislation for the conservation of threatened species and communities is the *Flora and Fauna Guarantee Amendment Act 2019* (FFGAA).
- Ensure you can describe a local environmental policy that supports management of threatened species or an ecological community.

5.5 CONCEPT QUESTIONS

Worksheet 5.5 Key concepts

REMEMBERING

1 List the name of two environmental policies established to support the management of threatened species in Australia.
2 Explain the term 'ecological community'.
3 Identify three species and/or ecological communities that are protected in Australia, including their listing status.

UNDERSTANDING

4 Outline three common management strategies to prevent the further loss of native species and/or ecological communities in Australia.

5 Choose one of the strategies you identified above and analyse if you believe this strategy would be effective at conserving native species.

APPLYING

6 Referencing the southern brown bandicoot case study in this section:
   a explain two threats to this species
   b explain two management strategies to enhance the survival of this species.

EXTENSION CHALLENGE

Weblink Department of Energy, Environment and Climate Action

7 Visit the Victorian Department of Energy, Environment and Climate Action website and search for 'Action statements'. Choose a threatened species or ecological community of interest to you and research the following information:
   a name of the species
   b the habitat of this species
   c its listing status under the FFGAA act
   d two threats to this species
   e two management strategies to protect and preserve this species
   f the current management plan in managing this species, including its effectiveness.
   https://www.environment.vic.gov.au/

Worksheet 5.5c Falls Creek versus Mt Stirling

EXTENSION LEARNING ACTIVITY 5.5C

Class debate: Falls Creek versus Mt Stirling

1 Divide your class in two for a class debate. One side is Falls Creek and the other is Mt Stirling.

2 Using the articles below, the Falls Creek team is to promote their skiing experience, tantalising prospective visitors with the technology they've purchased. The Mt Stirling team should take the opposite, selling visitors on less technology, not more.

3 The winning sales pitch will be awarded by its descriptions of the direct and indirect impacts of the skiing experience at the different mountains.

Article 1: Mt Stirling

Backcountry skiing and snowboarding

Enjoy winter alpine escapes in this unspoiled alpine playground. Your adventure starts in the beautiful tall ash forests which lead on to scenic snow gum glades and open up the summit and the iconic Stanley Bowl area where your untracked turns await. Mt Stirling offers the unique balance of remote and peaceful backcountry style exploration, with the support of an experienced Mt Stirling Ski Patrol and a team of passionate locals and volunteers eager to share this special place with fellow alpine enthusiasts.

This is an uphill lover's paradise, so bring your skins or snowshoes and head out to unlock untracked, crowd-free touring. It's hard to believe we are just across the valley from the hustle and hum of Mt Buller. Mt Stirling moves at a different speed. We're all about quiet nights staring into a fire, sleeping under the stars, listening for lyrebirds as you break trail and enjoying the peace of this ancient Taungurung country.

No lifts, no hotels, no frills. No limit to your adventure.

'Backcountry Skiing & Snowboarding' © 2021 Mt Buller and Mt Stirling Resort Management Board

Article 2: Falls Creek

Falls Creek invests $1.88 million in new snowmaking upgrades!

The temperatures have dropped to minus 3.3 and our mountain team have started making snow. We are snow excited we may even catch snowflakes under the snow guns with our tongue ... maybe.

To celebrate, we thought we'd share some news, well, we were going to tell you last winter, but COVID kind of ruined the party. Over the past 2 years we have invested a whopping $1.88 million in snowmaking and snow grooming machines – it's the largest snowmaking investment in 10 years at Falls Creek!

The new kids on the block:

You'll find shiny new snow guns along Wombats Ramble, Main Street and at the base of Drovers. Our mountain team will sigh if we don't use the official terms, so, get ready to see these automated TechnoAlpin Snow Guns firing out tons of snow.

In addition to this, we have a PistenBully 400 Winch and a PistenBully 400 Park grooming machine joining the team. They're not really bullies, they're actually pretty cool Kats. For Falls Creek guests, this means even more flawless groomed runs to enjoy than ever before throughout the season.

This huge investment helps us get snow on the ground quicker across our perfect family friendly terrain, ensuring the best possible snow conditions all season long! So, it's kind of a big deal! Get up here and enjoy your snow holiday in Falls Creek this winter.

Our season opening is Saturday 12 June, and we cannot wait to get back to skiing and boarding. Hearing the whirling sound of the snow guns and watching white hit the ground sure has the village buzzing. Falls Creek Slope Operations Manager Danny Lucas said, 'It's really great to see the improvement this investment has made on the snowmaking system here in Falls Creek. With only 4 weeks to go before opening day, we will be making snow at every opportunity.'

'Falls Creek Invests $1.88 Million in NEW Snowmaking Upgrades!', FallsCreek, 2023 Falls Creek Ski Lift PTY LTD

Resource
Glossary – Chapter 5

Assessments
End of chapter exam

Glossary test

EXAM-STYLE QUESTIONS

1 Evaluate the impacts of recreational focused activities on outdoor environments. (4 marks)

2 Analyse a recreational activity you have participated in and how it impacts differently on two different environments. (6 marks)

3 Using a specific example, describe a direct impact of technology on your interaction with an outdoor environment on a practical experience this year. (2 marks)

4 Using a specific example, describe an indirect impact of technology on your interaction with an outdoor environment on a practical experience this year. (2 marks)

5 a Explain a community-based environmental action. (2 marks)

b Evaluate how this community-based environmental action promotes positive impacts of humans on outdoor environments. (4 marks)

6 Referencing an outdoor environment you have visited or studied, explain how a community-based environmental action has promoted positive impacts on this outdoor environment. (4 marks)

7 a Discuss the impacts of urbanisation on an outdoor environment you have visited or studied. (4 marks)

b Discuss and predict the impacts of urbanisation on outdoor environments. (4 marks)

8 How do we identify and manage threatened species and/or ecological communities in outdoor environments? (2 marks)

9 Describe a local environmental policy that supports the management of threatened species or an ecological community. (3 marks)

10 a What management strategies are effective in preserving and protecting outdoor environments? (2 marks)

b Evaluate the effectiveness of a management strategy on an outdoor environment you have visited or studied. (4 marks)

CHECK YOUR KNOWLEDGE AND SKILLS

Resources
Key knowledge and skills checklist

Checklist for students to mark and date, making judgements of their knowledge and understanding.
(D = developing C = consolidating ER = exam ready)

D	C	ER	Check your knowledge and skills
☐	☐	☐	Evaluate a range of impacts on contrasting outdoor environments
☐	☐	☐	Compare the impacts of conservation, economic and recreational activities on contrasting outdoor environments
☐	☐	☐	Describe impacts you have observed or studied regarding specific outdoor environments
☐	☐	☐	Define what a community of people is and their role in environmental protection
☐	☐	☐	Describe impacts of community groups you have observed or studied regarding specific outdoor environments
☐	☐	☐	Compare direct and indirect impacts of technologies on outdoor environments
☐	☐	☐	Define the term 'technology' and how this helps people modify the outdoor world or their experience
☐	☐	☐	Describe three types of direct and three types of indirect impacts of technology you have observed in an outdoor environment
☐	☐	☐	Discuss and predict impacts of urbanisation on land-based outdoor environments
☐	☐	☐	Define the term 'urbanisation' and how it applies to the Australian population
☐	☐	☐	List the negative and positive impacts of urbanisation
☐	☐	☐	Discuss and predict impacts of urbanisation on waterways
☐	☐	☐	Evaluate the effectiveness of management strategies on an outdoor environment
☐	☐	☐	Describe a local environmental policy that supports management of threatened species or an ecological community
☐	☐	☐	Define ecological communities and list Australia's diverse range of ecological communities
☐	☐	☐	Describe an outdoor environment that you have visited or studied that is supported by an environmental policy

CHAPTER 6

UNIT 2 – AREA OF STUDY 3

Independent participation in outdoor environments

KEY KNOWLEDGE

- how to conduct safe and sustainable peer-led outdoor activities, involving minimal impact strategies for groups, route planning, food and equipment planning, risk management planning and transport planning
- how to plan and adapt outdoor experiences due to weather, including weather patterns and extreme weather
- how to monitor observations of own and other groups' impacts on the outdoor environment during an outdoor experience.

KEY SKILLS

- plan and demonstrate safe and sustainable participation in outdoor environments
- describe the influence of weather and weather patterns on planning and conducting outdoor experiences
- analyse observations of impacts of groups on outdoor environments.

VCE Outdoor and Environmental Studies Study Design 2024–2028, p. 20.

How to conduct safe and sustainable peer-led outdoor activities

The aim of this chapter is to provide you with an overview of the knowledge and skills to participate in a range of outdoor experiences safely and sustainably in an independent manner.

How to plan and adapt outdoor experiences due to weather

We describe the influence of weather and weather patterns on planning and conducting outdoor experiences.

How to monitor observations of your own and other groups' impacts

We investigate different methods of monitoring and analysing impacts on the outdoor environment and investigate ways to reduce this and promote sustainable interactions.

KEY TERMS

Beaufort Wind Scale
cold front
contingency plan
El Niño
extreme weather
field sketch
high-pressure system
isobars
La Niña
low-pressure system
peer-led activities
remote camera
route card
scat
trough
warm front
weather forecast
weather map
weather pattern
wind chill
wind gust

Weblinks

- Apparent ('feels like') temperature **p. 283**
- Bureau of Meteorology **p. 287**
- Agriculture Victoria **p. 288**
- Discover Wildlife **p. 293**

Weblink – Videos

- Agriculture Victoria, 'Victoria's Climate 101 - A short animation of the latest science on our changing weather patterns' **p. 288**

Resources

- Planning and recording outdoor experiences **p. 268, 270, 271, 272, 274, 275, 276, 278, 282, 287, 288, 289, 293, 294**
- Key knowledge and skills checklist **p. 295**
- Glossary – Chapter 6 **p. 295**

Assessments

- Glossary test **p. 295**

Nelson MindTap

To access resources above, visit **cengage.com.au/nelsonmindtap**

SPOTLIGHT

Tourists told to 'clean up'

The summer influx of visitors to the Mornington Peninsula and the amount of rubbish being left on beaches and roadsides has prompted calls for a better tourism management planning and a fresh approach to litter control.

Community leaders have for years been demanding the council take proactive action on limiting the negative impacts of tourism, including rubbish, jet-ski misbehaviour and 'dangerous' parking near beaches.

Michelle Cheers, of Rye Community Group Alliance, said it was distressing to see the increasing amounts of littering on beaches this summer and the large number of jet skis 'crowding' peninsula beaches.

Cheers said there had been a strong community push for a tourism management plan on the peninsula that would provide strategies for better and easier rubbish disposal and management of visitor numbers.

'I went to the beach at Rye the other day and apart from rubbish everywhere there were 65 jet skis – that's unmanageable and dangerous.

'We've had petitions with 6000 signatures to ban jet skis from some of our beaches, but nothing gets done.'

Jet ski riders are banned from weaving, making sharp turns and doing doughnuts in zones where the speed limit is 5 knots, under rules introduced in 2019, but enforcement difficulties means many riders break the law.

But Cheers says the council's ongoing and overarching focus was the 'supposed' financial benefit of tourism, while ignoring the social and environmental costs.

Shutterstock.com/Maciej Bledowski

Figure 6.1 Rubbish left behind each day by lazy and thoughtless beachgoers.

'They go on about the money tourists bring in, but at what benefit to ratepayers and at what cost?' she said.

Environmental advocate and former citizen of the year, Josie Jones OAM, said the amount of litter left behind on beaches this summer was 'extremely concerning'.

Jones spearheaded a 'beach hand bin' pilot program for peninsula beaches last summer, with data showing the bins resulted in a 71.5% reduction in small litter.

Jones said simple solutions to litter prevention were needed to make it easier for people to discard of rubbish responsibly.

'I'd like to see greater emphasis on what is acceptable and communicate this via signs at the end of the freeway so everyone coming to the peninsula understands the awareness by community and the requirement as a visitor to be responsible with litter, and the local laws,' she said.

The mayor Steve Holland said the shire had 'very few' mechanisms to raise revenue from tourists.

'We are exploring some options because, right now, our residents and ratepayers are footing the bill for tourism infrastructure and services, which is quite unfair,' he said.

'Litter is always a problem at this time of year and unfortunately it seems some people were never taught to clean up after themselves.'

'Tourists told to "clean up"', Liz Bell, January 9, 2023, © 2023 Mornington Peninsula News Group.

INTRODUCTION

Shutterstock.com/Martin Helgemeir

Figure 6.2 What knowledge and skills are required for safe and sustainable participation in outdoor environments?

This chapter covers the key knowledge and skills for the final area of study for Unit 2. In this area of study you will analyse one specific outdoor environment explored during a practical experience. You will further develop your understanding of that environment and analyse how it has been impacted by different user groups. As shown in the opening article, impacts such as rubbish and noise pollution to popular beach-side environments are complex problems that require a range of solutions, including educating user groups and providing convenient waste management.

Before your practical experience, you are expected to contribute to the planning and take an active role in leading the outdoor experience with your peers. During the experience, you will analyse the impacts of your group and other users on the outdoor environment, and investigate ways to reduce these and to promote sustainable interactions. After the experience, you will suggest changes to the management of users in this outdoor environment and work to promote sustainable interactions into the future.

The aim of this chapter is to provide you with an overview of the knowledge and skills to participate in a range of outdoor experiences safely and sustainably in an independent manner.

The key knowledge covered in this chapter includes how to conduct safe and sustainable peer-led outdoor activities, how to plan and adapt outdoor experiences due to weather, and how to monitor observations of own and other groups' impacts on the outdoor environment during an outdoor experience.

In covering the requirements to successfully achieve Unit 2 Outcome 3, this chapter provides a review of some key knowledge introduced in Units 1 and 2, such as minimal impact strategies, route planning and risk management. It also introduces new knowledge and skills such as how to plan and adapt outdoor experiences due to the weather and how to monitor observations of impacts on the outdoor environment.

Logbook activities

The assessment task for this Area of Study is a practical demonstration of key skills with reference to outdoor experiences, in addition to ongoing logbook entries of outdoor practical experiences.

Throughout this chapter, you will find a series of logbook activities to help to consolidate your understanding of the key knowledge and skills, and to support you to participate in a range of outdoor experiences safely and sustainably in an independent manner. Below is a listing of the logbook activities in this chapter. Use this listing to check off activities as you complete them and to enable your teacher to keep track of your progress.

Resource Planning and recording outdoor experiences

Table 6.1 Logbook activities

Logbook activity	Complete	Date	Teacher Signature
1. Peer group roles and expectations	☐		
2. Minimal impact strategies presentation	☐		
3. Route cards	☐		
4. Outdoor experience menu	☐		
5. Equipment list	☐		
6. Risk assessment	☐		
7. Transport planning	☐		
8. Beaufort Wind Scale	☐		
9. Weather forecast	☐		
10. Weather graph	☐		
11. Weather contingency planning	☐		
12. Practice field sketch	☐		
13. Observations of own and other groups' impacts	☐		
14. Analysis of observed impacts on the outdoor environment	☐		
15. Ways to reduce impacts and promote sustainable interactions	☐		
16. Management of users in outdoor environments	☐		

6.1 HOW TO CONDUCT SAFE AND SUSTAINABLE PEER-LED OUTDOOR ACTIVITIES

KEY KNOWLEDGE

- how to conduct safe and sustainable peer-led outdoor activities, involving minimal impact strategies for groups, route planning, food and equipment planning, risk management planning and transport planning

KEY SKILLS

- plan and demonstrate safe and sustainable participation in outdoor environments

Some of the highlights of studying Outdoor and Environmental Studies are the opportunities to participate in a range of practical experiences with your peers in a variety of outdoor environments. Throughout these experiences, you will develop lifelong skills such as teamwork, responsibility, leadership and independence.

This first section of this chapter explores how to conduct safe and sustainable peer-led outdoor activities, including minimal impact strategies for groups, route planning, food and equipment planning, risk management planning and transport planning. It is vitally important that, in planning for an outdoor experience, we prioritise the safety of all participants. In addition, we need to ensure that 'proper prior planning' is done to 'prevent poor performance' (as discussed in Chapter 3), to minimise our impacts and to promote sustainable participation in outdoor environments. We start the chapter by discussing what is meant by and the benefits of peer-led activities.

Peer-led activities

Peer-led activities are those where students assume leadership roles and support each other in planning and facilitating an experience.

peer-led activities where students assume leadership roles and support each other in planning and facilitating an experience

Traditionally, many outdoor experiences are planned and facilitated by teachers and/or professional instructors. In this model, students are led by others, usually adults, and assume little responsibility for the activity. Through the careful guidance of your teachers (and, at times, activity professionals), you and your classmates will have opportunities to support each other and contribute to the planning and leading of practical experiences in outdoor environments. Working as an effective team is vital to the success of any group activity in an unfamiliar environment.

Travel organisations such as World Challenge and World Expeditions have been offering peer-led international trips to students for many years. On these trips, the students are responsible for many aspects of the experience, including booking accommodation, transport and restaurants; sticking to a budget; navigating through unfamiliar towns; and supporting each other physically and emotionally. The peer-led model has been shown to have significant benefits for developing teamwork, leadership, problem-solving, communication skills, self-confidence and independence. Many of these skills and attributes regularly appear on lists of desirable qualities that employers are looking for in prospective employees, and are also beneficial for people when facing life's challenges.

Adobe Stock/Avelino

Figure 6.3 Students leading students provides opportunities to develop lifelong skills such as teamwork, leadership, problem-solving, communication, self-confidence and independence.

Resource
Planning and recording outdoor experiences

LEARNING ACTIVITY 6.1A

Logbook activity 1 – Peer group roles and expectations

Throughout this chapter, a range of logbook activities will require you to work together as a team, and then support each other on the outdoor experience. The first task to promote success when working as a team is to establish clear roles and expectations.

1 Group introductions

Meet with the members of your group. Each group member is to introduce themselves to the group. Include the following information in your introduction:

- Your past experience in outdoor environments
- Your likes and dislikes of outdoor activities
- Your general strengths and areas for improvement
- What you are looking forward to in upcoming OES practical activities
- What do you think is most important when working in a group?

2 Group roles

Take a look at the listing of the logbook activities at the start of this chapter. Allocate one person to 'lead' the completion of each activity.

3 Group expectations

Create a listing of agreed-upon expectations for your group. Display these in a prominent place in your classroom or workbook. Group expectations may include the following:

- Respect all and the environment.
- Prioritise safety at all times.
- Take responsibility for your own roles within the group.
- Come prepared and on time.
- Maintain open and effective communication within the group.
- Be proactive in addressing challenges and finding solutions.
- Be an active listener and let everyone have their say.
- Critique ideas, not people.
- Support each other and ask for support when needed.

The following sections provide a review of the key knowledge for this area of study, including minimal impact strategies, route and equipment planning, and risk management. Refer to Chapter 3 for more detail on these approaches.

Minimal impact strategies for groups

Participating in activities such as bushwalking, surfing and mountain biking as part of this subject will provide many memorable experiences. However, it is important to recognise that the large size of school groups can result in significant negative impacts on the outdoor environments that we visit. It is the responsibility of all group members to implement minimal impact strategies to promote sustainable interactions and to protect the health of outdoor environments. Minimal impact strategies are practices that aim to have as little environmental impact as possible. In Chapter 3, we discussed minimal impact strategies, including codes of conduct for a range of outdoor activities. Refer to these for advice on how to minimise your impacts on an outdoor environment for specific outdoor activities.

Minimal impact strategies for safe and sustainable group interactions with outdoor environments include the following:

- Plan ahead and prepare.
- Educate the group on the characteristics of the specific outdoor environment, such as culturally sensitive locations, conservation and fauna breeding areas, plus management activities such as planned burns.
- Travel and camp in designated areas that are suitable for large groups.
- Use supplied toilet and washing-up facilities where possible.
- Dispose of waste properly – it is advisable to develop a group rubbish disposal system that includes separate bins for recyclables, glass, food waste and general waste.
- Be wary of moving around in groups near sensitive vegetation.
- Keep noise levels to a minimum, so as not to disturb other visitors and native animals.
- Only use designated fireplaces, keep fires small and contained, and ensure any fire is fully extinguished before leaving.
- Always remain on formed tracks (including when resting) to prevent the trampling of vegetation and the spread of invasive fungi, weeds and diseases.

Shutterstock.com/THP Creative

Figure 6.4 What can be done to encourage campers to follow minimal impact strategies?

LEARNING ACTIVITY 6.1B

Logbook activity 2 – Minimal impact strategies presentation

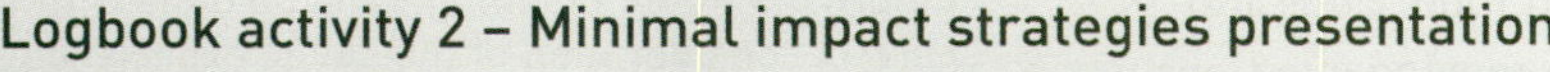

Work individually or in small groups to create a visual presentation of minimal impact strategies for one of the following aspects of a practical experience:

- transport to the outdoor environment
- camping
- campfire
- hygiene
- waste disposal
- hiking
- cultural sites
- native flora and fauna
- another recreational activity.

Resource Planning and recording outdoor experiences

Route planning

When moving through outdoor environments, groups tend to move at a slower rate and require more breaks. This occurs due to the need to accommodate the varied skills, experience and capabilities of all group members. When route planning for a group, there are many factors to consider to promote safe and sustainable interactions with the outdoor environment. These include the following:

- Capabilities of the group – the size, skills, fitness and experience of the group will determine the planned route, the activities you can include and the time you'll need to allow for each section of the route.
- Time of year – this will determine the sort of clothing, equipment and food required for the experience, and may be affected by the availability of water to supply the entire group.
- Possible weather conditions – some routes may be inaccessible, such as those that cross waterways during or after wet weather. Other tracks may be too difficult to attempt in extreme temperatures.
- Nature of the terrain – the sort of terrain you face will affect how easy (or otherwise) it is for a group to move through the environment.
- Location of campsites (if needed) – suitable places for campsites are a requirement not just for the comfort and convenience of the group, but also for the likely sustainability of your journey. It is easy for a large group setting up tents to trample, disturb and otherwise impact sensitive environments.
- Possible escape routes when facing emergencies – a planner should consider ways a group might escape from emergencies, such as extreme weather events, injuries, fire or flood, depending on their location while on a trip.

As a part of the planning for an outdoor experience involving travelling through an outdoor environment (e.g. a multi-day hike), it is important that all participants have a clear understanding of the intended route. One way of sharing the planned route is via the use of a **route card**. A route card is a document that is used by hikers as an aid to navigation by detailing information about a planned route. Description, direction, distance and duration are the 'four Ds' of a route card. Once shared with group members, route cards provide participants with a clear understanding of the planned route, allowing them to mentally prepare for upcoming aspects of the hike – such as difficult ascents – and providing them with an idea of times and distances to points of interest, lunch breaks and campsites. Shared route cards also reduce the chance of those annoying questions such as, 'Are we there yet?' and 'How far to go?'

route card
a document used as an aid in navigation by detailing information about a planned route

iStockphoto/Paul Feikema

Figure 6.5 Route planning must be done with careful consideration of the skills, experience and capabilities of all group members.

Resource
Planning and recording outdoor experiences

LEARNING ACTIVITY 6.1C

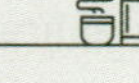

Logbook activity 3 – Route cards

This activity provides you with an opportunity to work together to create route cards for an upcoming outdoor experience.

1 Access a map of the outdoor environment for an upcoming outdoor experience.
2 Allocate a section of the experience to each student (or group of students), such as a hike, ski or paddle.
3 Working together, students are to complete route cards for their allocated section of the outdoor experience.

As there are no standard templates for route cards, the first task may be to decide on what information to include on the card. See below for an example route card.

Features of a route card

Practical experience: *e.g. Grampians bushwalk*				Day No.: *1*	Day: *Monday*	Date: *12/02/24*
Start location	**Direction**	**Distance**	**Duration**	**Description of stage**		**Finish location**

Food planning

Figure 6.6 Food for outdoor experiences should be nutritious and delicious.

Shutterstock.com/artem evdokimov

As discussed in Chapter 3, eating sufficient amounts of nutritious food is necessary to ensure your body is capable of producing the energy required to safely participate in planned outdoor experiences. Many students have made the mistake of not allowing for the greater energy demands of outdoor activities, compared to a normal school day. Running out of energy while mountain biking or surfing could have significant consequences.

High-performance sports research into the effectiveness of energy drinks has found that taste has a significant influence on the amount of drink consumed; and, of course, we all prefer to eat foods that taste good. So, it is important to select food and drink that not only provides the necessary hydration and nutrition, but also tastes great.

There is a range of things to consider when selecting appropriate food and hydration for an outdoor experience. These include the following:

- Ensure your planning considers all dietary and allergy requirements for all group members.
- Select food that is nutritious, light to carry, tasty, easy to prepare and easy to clean up afterwards.
- Aim for a variety of foods and a balance of flavours, carbohydrates, protein and fats.
- Include some fresh foods if possible, such as fruit, nuts, seeds and vegetables.
- Food that requires water to cook it (such as noodles, pasta and rice) can be very light and is useful to take on trips, but make sure you will have access to enough water, and note that some of these (particularly rice) can be tricky to cook on a trip.
- Pre-packaged dehydrated meals can be great in terms of weight and ease of preparation, but sometimes leave a bit to be desired when it comes to taste, and they can be very expensive.
- If your trip is likely to run at a time when a Total Fire Ban could be in force, make sure you have alternative meals that don't require the use of a portable stove or cooking fire – you don't want to be eating your pasta or noodles uncooked!
- Stay hydrated by drinking water slowly in the hours before participating in physical activity. You should consume about 1 litre of water for every hour of activity, and more if it's a hot day and/or you're participating in an intense activity. Think about having both water and sports drinks available to ensure an adequate intake of electrolytes.
- Hygiene is vital on an outdoor experience, so don't forget to allow time for the clean-up.

Resource
Planning and recording outdoor experiences

LEARNING ACTIVITY 6.1D

Logbook activity 4 – Outdoor experience menu

Working together with your group members, develop a menu for an upcoming outdoor experience. Ensure you include all meals and snacks, and provide a hydration plan.

Once you have a menu for each meal for each day of your trip, allocate individual menu items for each group member to bring on the trip.

Equipment planning

When heading into the great outdoors, it is important to be prepared for all that nature can serve up. This, of course, means ensuring you take all appropriate equipment to suit the location, activity, weather

Alamy/Peter Lewis

Figure 6.7 You don't need to spend a fortune on equipment. Borrow gear if possible and seek expert advice if you're unsure about the quality and type of equipment you need.

and time away, which will promote a safe, sustainable, enjoyable and comfortable experience. Also consider the following when deciding on what equipment to pack:

- Carry a first aid kit suited to the size and needs of the group, and the location and the nature of the activities undertaken.
- Have a range of charged communication devices such as mobile phones and personal locator beacons available in case of emergency.
- Protecting yourself against the elements – whether it be heat, wind or cold – is crucial to your safety when exploring outdoor environments.
- One of the best strategies to regulate your body temperature and protect yourself from weather extremes is to layer your clothing. Layers can be added or taken off as required.
- Your sleeping bag is one of the most important pieces of equipment to ensure a good night's sleep.
- Sleeping mats provide comfort and insulation from the cold hard ground.
- Familiarise yourself with your equipment before you need it in the outdoors. Ensure your first time setting up your tent is not on a cold, wet and windy night without a torch!
- Check out the huge range of online information and advice to help guide your equipment planning.

LEARNING ACTIVITY 6.1E

Logbook activity 5 – Equipment list

Working with your classmates/group members, develop an equipment list for an upcoming outdoor experience. Use the following headings to categorise your equipment:

- Essentials (e.g. medications)
- Emergencies (e.g. first aid kit)
- Health and hygiene (e.g. hand sanitiser)
- Navigation (e.g. map and compass)
- Clothing and footwear (e.g. boots and waterproofs)
- Sleeping (e.g. tent and sleeping bag)
- Food and water (e.g. meals and snacks)
- Cooking (e.g. stove and fuel)
- Activity specific (e.g. gloves and helmet).

Resource
Planning and recording outdoor experiences

Risk management planning

> Only those who risk going too far can possibly find out how far they can go.
>
> – T.S. Eliot

The safety of participants is the number one priority of any practical experience. Risk management is the process of identifying and evaluating risks and implementing control measures to minimise the likelihood of an injury/illness occurring.

There are a large number of possible risks inherent in outdoor experiences. We can break these risks into three main categories:

1 environmental risks (such as weather, terrain and native animals)
2 risks associated with people (such as skills, knowledge, experience, health and fitness)
3 risks associated with equipment (such as ropes, kayaks, surfboards and bikes).

Once risks have been identified, the next step is to implement control measures to minimise the likelihood of an injury/illness occurring. This process is known as a risk assessment.

A risk assessment is a systematic process of analysing the potential risks that may be involved in an activity. The purpose of a risk assessment in outdoor education is to ensure everyone involved in an activity is as safe as possible when participating in outdoor experiences. The removal of hazards and their associated risks is always the best option; however, this is not always possible. Safety procedures (known as controls) are implemented to reduce risks to an acceptable level.

Risk assessments should raise awareness about hazards and risks, identify who may be at risk and determine whether there are existing and adequate controls in place. It is important that a risk assessment is shared and discussed with all participants as a part of the planning for a practical experience.

iStockphoto.com/pacland

Figure 6.8 Risk assessments are essential for all outdoor experiences, including white water rafting

Resource Planning and recording outdoor experiences

LEARNING ACTIVITY 6.1F

Logbook activity 6 – Risk assessment

Using the following risk assessment template (with example provided) and risk-rating matrix, together with the instructions presented in Chapter 3, complete a risk assessment for an upcoming outdoor experience. You may need to seek advice from a teacher who has performed a reconnaissance trip to the specific outdoor environment to gain an understanding of the possible risks involved.

Risk assessment template

Identified hazard	Who may be harmed	Risk rating before controls	Controls	Risk rating after controls
For example: Dislodged rocks when climbing	All participants	High	• Review site for loose rock prior to climbing • Always wear helmets • Teach calls to warn of rock fall and appropriate protective response	Low

Risk-rating matrix

Likelihood	Consequences				
	Negligible	Minor	Medium	High	Catastrophic
Almost certain	Moderate	Moderate	High	Extreme	Extreme
Likely	Low	Moderate	High	High	Extreme
Possible	Low	Low	Moderate	High	High
Unlikely	Very Low	Low	Moderate	Moderate	High
Rare	Very Low	Very Low	Low	Moderate	Moderate

Transport planning

As discussed in Chapter 5, there are both direct and indirect impacts of transportation on outdoor environments. It is important that we keep this in mind when planning for practical experiences. Travel is a necessary part of all trips and, particularly if travelling long distances, it is most likely that it will have a greater impact on the overall environment than any other activity. Therefore, it is important to consider ways to reduce your travel impacts; for example:

Shutterstock.com/Jacek Chabraszewski

Figure 6.9 Riding to and from an outdoor environment (or some of the way) allows you to minimise your environmental impacts during an activity.

- Use as few vehicles as possible, or public transport if it's an option.
- Consider extending the trip to include more walking, cycling or canoeing to help you get to the main site of the activity.
- Avoid using eroded and damaged roads.
- Always use made roads and designated car parks where available.
- Think about ways to offset the impact of your travel, such as through tree-planting programs at school, at home or in your local community.

Resource
Planning and recording outdoor experiences

LEARNING ACTIVITY 6.1G

Logbook activity 7 – Transport planning

Research different transport options to travel to and from (and during, if required) your outdoor experience.

For each transport option consider the following:

- the time to complete the trip using this transport option
- the costs involved in this transport option
- the availability of this transport option
- the impact of this transport option on the environment (e.g. carbon emissions)
- the impact of this transport option on your activity
- a summary of the advantages and disadvantages of this transport option.

Taking all of this into account, is it a viable transport option?

6.2 HOW TO PLAN AND ADAPT OUTDOOR EXPERIENCES DUE TO WEATHER

KEY KNOWLEDGE

- how to plan and adapt outdoor experiences due to weather, including weather patterns and extreme weather

KEY SKILLS

- describe the influence of weather and weather patterns on planning and conducting outdoor experiences

6.2.1 THE INFLUENCE OF WEATHER AND WEATHER PATTERNS ON PLANNING AND CONDUCTING OUTDOOR ACTIVITIES

There is no such thing as bad weather, only poor clothing choices.

Alfred Wainwright

The saying above may bring a smile to your face, but it does highlight an important point – while it is never a bad day to go outside, you do need to take responsibility for knowing the expected weather and be prepared for it. The weather will not only significantly influence the activities of a practical experience, but also (and more importantly) the health and safety of participants. The weather can change quickly – particularly in the Melbourne region – and all participants need to be prepared for this. In fact, the rock band Crowded House has a song titled 'Four Seasons in One Day' in reference to Melbourne's changeable weather. Tim Finn explained in an interview: '"Four seasons in one day" was a common Melbourne phrase, because you go from a blazing-hot sunny day to raining, and then it'll be hailing that night'. This highlights the importance to be prepared for a range of weather conditions, not just what is forecast.

This section investigates the influence of weather and weather patterns on planning and conducting outdoor activities. The aim of this section is for you to develop your knowledge and understanding of weather forecasts and weather patterns so that you are able to plan and adapt outdoor experiences due to the weather, such as the need to change a walking route and seek shelter when hit by a storm.

iStockphoto.com/Paola Giannoni

Figure 6.10 Melbourne is famous for its great coffee and ever-changing weather conditions.

Weather forecasts

'Red sky at night, shepherd's delight. Red sky in the morning, shepherd's warning'. This old saying, which has long been used to forecast the next day's weather, has some truth to it, according to the British Met Office:

> A red sky at sunset means high pressure is moving in from the west, so therefore the next day will usually be dry and pleasant. A red sky in the morning means a high-pressure weather system has already moved east, meaning the good weather has passed, most likely making way for a wet and windy low-pressure system.
>
> Met Office, 'Red sky at night and other weather lore' © Crown Copyright

These days we have access to a large array of weather tracking technology to enable a more scientific forecast of upcoming weather conditions.

A **weather forecast** is a prediction of the conditions of the atmosphere for a given location and time. Weather forecasts are usually available one week in advance, but there are also long-term forecasts, such as those by the Bureau of Meteorology, that publish general patterns of weather for the coming season; for example, autumn is likely to be drier and warmer than usual. This is important information for those that rely on climatic conditions for their livelihoods, such as farmers or ski resort operators. That said, in terms of planning a practical experience, a detailed and accurate forecast is more valuable.

weather forecast
a prediction of the conditions of the atmosphere for a given location and time

istockphoto.com/JEN_n82

Figure 6.11 'Red sky at night, shepherd's delight. Red sky in the morning, shepherd's warning.'

Newspix/Eugene Hyland

Figure 6.12 Jane Bunn, meteorologist and 7NEWS Melbourne weather presenter

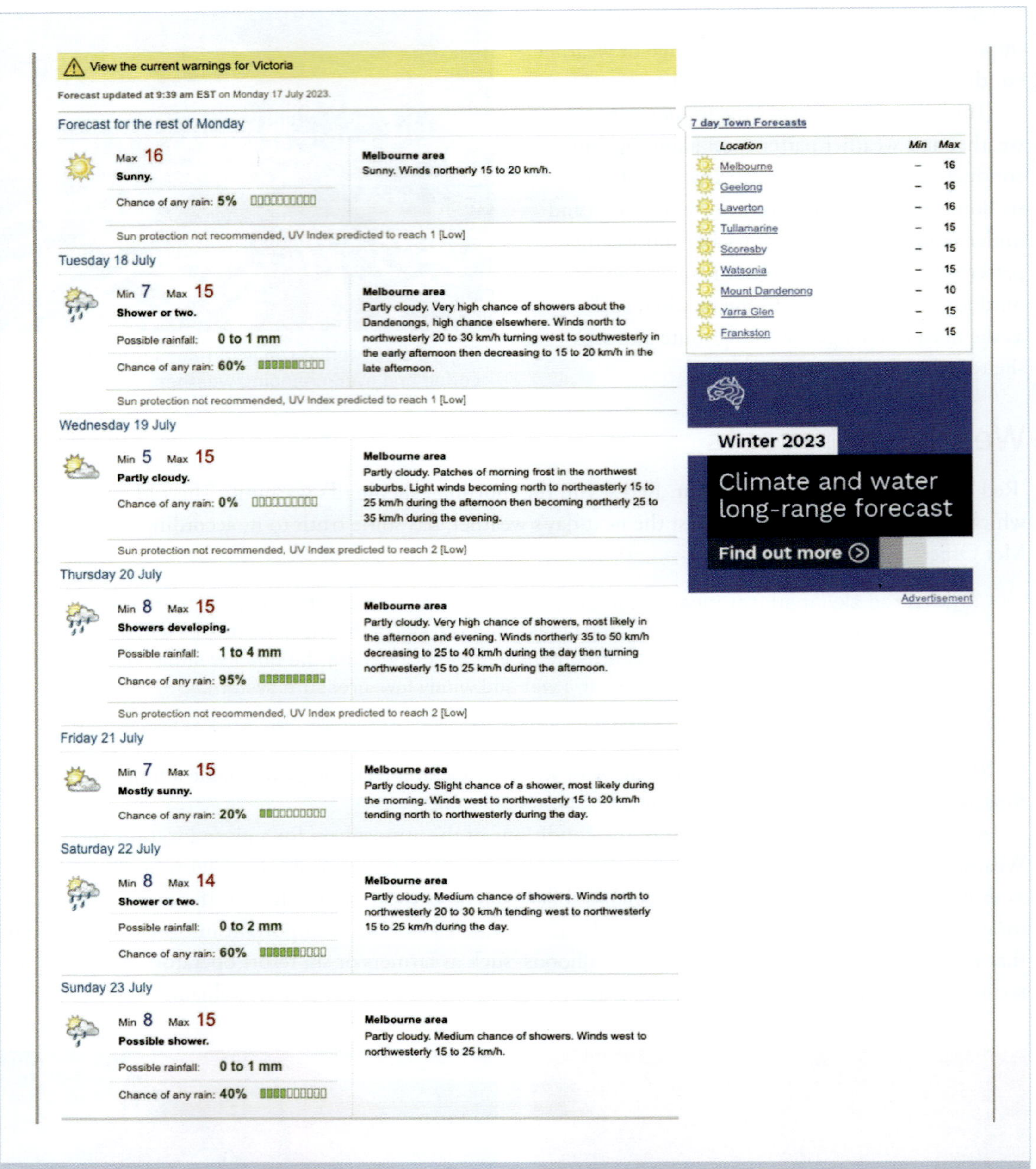

Figure 6.13 Example weather forecast from the Bureau of Meteorology

In terms of accuracy, a 10-day forecast is correct about 50% of the time, a seven-day forecast is correct about 80% of the time, and a five-day forecast is correct about 90% of the time. Therefore, as a part of the practical experience planning, it is important to keep checking the forecast in the days leading up to the trip, as the closer the trip, the more accurate the forecast will be.

Published weather forecasts present a large amount of information. The weather forecast image depicted in Figure 6.13 is from the Bureau of Meteorology website (http://www.bom.gov.au/). It provides information on temperature range, rain percentage, rainfall amount, wind speed, UV index and more.

An explanation of the information contained in the example weather forecast is presented in Table 6.2.

Table 6.2 Weather forecast features

Feature	Meaning
Maximum temperature	The highest point the temperature will reach during the day
Minimum temperature	The lowest point the temperature will reach before 9 am
Precis	A brief written description of the expected weather
Chance of any rain	The likelihood of any rain from midnight to midnight
Rainfall amount	The likely amount if it does rain
Text description	Information about sky cover, wet weather and wind
Sky cover	The presence of clear skies or clouds, and how much (e.g. partly cloudy)
Wet weather	The wet weather expected in the area and its likelihood
Wind	Wind direction and speed, and changes to this across the day

Wind can have a significant impact on any outdoor practical experience. Learning about how wind is measured and its likely impact on the land and sea can help you to make an informed decision on the weather suitability of an upcoming activity. The Bureau's forecasts of wind speed and direction are the average of **wind gusts** and lulls. Wind gusts are typically 40% higher than the average wind speed. The **Beaufort Wind Scale** measures wind speed according to the impact the wind has on the land and sea. Table 6.3 describes what can be expected for each level of the scale.

Shutterstock/Luciano Santandreu

Figure 6.14 High winds can create treacherous conditions.

wind gust a brief increase in the speed of the wind, usually for less than 20 seconds

Beaufort Wind Scale a measure that relates wind speed to observed conditions at sea or on land

Table 6.3 Beaufort Wind Scale

Beaufort Wind Scale	Descriptive term	Units in knots	Description on land	Description at sea
0	Calm	0	Smoke rises vertically	Sea like a mirror
1–3	Light winds	10 knots or less (19 km/h or less)	Wind felt on face; leaves rustle; ordinary vanes moved by wind	Small wavelets, ripples formed but do not break: a glassy appearance maintained
4	Moderate winds	11–16 knots (20–30 km/h)	Raises dust and loose paper; small branches are moved	Small waves – becoming longer; fairly frequent white horses
5	Fresh winds	17–21 knots (30–39 km/h)	Small trees in leaf begin to sway; crested wavelets form on inland waters	Moderate waves, taking a more pronounced long form; many white horses are formed – a chance of some spray

Beaufort Wind Scale	Descriptive term	Units in knots	Description on land	Description at sea
6	Strong winds	22–27 knots (40–50 km/h)	Large branches in motion; whistling heard in telephone wires; umbrellas used with difficulty	Large waves begin to form; the white foam crests are more extensive with probably some spray
7	Near gale	28–33 knots (50–62 km/h)	Whole trees in motion; inconvenience felt when walking against wind	Sea heaps up and white foam from breaking waves begins to be blown in streaks along direction of wind
8	Gale	34–40 knots (63–74 km/h)	Twigs break off trees; progress generally impeded	Moderately high waves of greater length; edges of crests begin to break into spindrift; foam is blown in well-marked streaks along the direction of the wind
9	Strong gale	41–47 knots (75–87 km/h)	Slight structural damage occurs –roofing dislodged; larger branches break off	High waves; dense streaks of foam; crests of waves begin to topple, tumble and roll over; spray may affect visibility
10	Storm	48–55 knots (88–102 km/h)	Seldom experienced inland; trees uprooted; considerable structural damage	Very high waves with long overhanging crests; the resulting foam in great patches is blown in dense white streaks; the surface of the sea takes on a white appearance; the tumbling of the sea becomes heavy with visibility affected
11	Violent storm	56–63 knots (103–117 km/h)	Very rarely experienced – widespread damage	Exceptionally high waves; small and medium sized ships occasionally lost from view behind waves; the sea is completely covered with long white patches of foam; the edges of wave crests are blown into froth
12+	Hurricane	64 knots or more (118km/h or more)	Very rarely experienced – widespread damage	The air is filled with foam and spray; sea completely white with driving spray; visibility very seriously affected

Resource
Planning and recording outdoor experiences

LEARNING ACTIVITY 6.2A

Logbook activity 8 – Beaufort Wind Scale

Using the information presented in Table 6.3, work with your classmates to create a classroom display that visually shows the impact on land and on the sea of each level of the Beaufort Wind Scale.

WIND CHILL

As wind increases, heat is carried away from the body, decreasing skin temperature and eventually the internal body temperature. This can lead to life-threatening medical conditions such as hypothermia. Warning signs of hypothermia include uncontrollable shivering, disorientation, slurred speech, drowsiness and exhaustion. If you suspect someone is showing signs of hypothermia, seek medical attention immediately.

The lowering of body temperature due to the passing flow of lower-temperature air is known as **wind chill**. The wind makes it feel much colder than the actual temperature. Sometimes weather forecasts report the 'feels like' temperature to provide the public with a better understanding of the expected outside conditions.

wind chill
the lowering of body temperature due to the passing flow of lower-temperature air

The wind chill temperature index (see Figure 6.15) combines air temperature with wind speed to indicate how cold it would actually feel on exposed skin. This can be used to provide advice on suitable clothing choices and the possible health effects of the cold. Learn more about the impact of windchill on apparent 'feels like' temperature at the Bureau of Meteorology's website (https://media.bom.gov.au/social/blog/1153/apparent-feels-like-temperature).

Weblink
Apparent ('feels like') temperature

WIND CHILL TEMPERATURE INDEX
Frostbite Times are for Exposed Facial Skin

Wind Speed (km/h)	Air Temperature (°C) 5	0	-5	-10	-15	-20	-25	-30	-35	-40	-45	-50
5	4	-2	-7	-13	-19	-24	-30	-36	-41	-47	-53	-58
10	3	-3	-9	-15	-21	-27	-33	-39	-45	-51	-57	-63
15	2	-4	-11	-17	-23	-29	-35	-41	-48	-54	-60	-66
20	1	-5	-12	-18	-24	-30	-37	-43	-49	-56	-62	-68
25	1	-6	-12	-19	-25	-32	-38	-44	-51	-57	-64	-70
30	0	-6	-13	-20	-26	-33	-39	-46	-52	-59	-65	-72
35	0	-7	-14	-20	-27	-33	-40	-47	-53	-60	-66	-73
40	-1	-7	-14	-21	-27	-34	-41	-48	-54	-61	-68	-74
45	-1	-8	-15	-21	-28	-35	-42	-48	-55	-62	-69	-75
50	-1	-8	-15	-22	-29	-35	-42	-49	-56	-63	-69	-76
55	-2	-8	-15	-22	-29	-36	-43	-50	-57	-63	-70	-77
60	-2	-9	-16	-23	-30	-36	-43	-50	-57	-64	-71	-78
65	-2	-9	-16	-23	-30	-37	-44	-51	-58	-65	-72	-79
70	-2	-9	-16	-23	-30	-37	-44	-51	-58	-65	-72	-80
75	-3	-10	-17	-24	-31	-38	-45	-52	-59	-66	-73	-80
80	-3	-10	-17	-24	-31	-38	-45	-52	-60	-67	-74	-80

Figure 6.15 The wind chill temperature is a measure of the combined cooling effect of wind and temperature.

CLOUDS

The size, shape, colour and altitude of clouds can be used as a means of predicting the weather. Cloud formation is influenced by water and moisture levels in the atmosphere, condensation of water vapour and air movement. The 10 main types of clouds and their common associated weather conditions are presented in Figure 6.16.

High-level clouds (above 6 kilometres)		
Cirrus clouds	**Cirrocumulus clouds**	**Cirrostratus clouds**

Cirrus: high-level, white tufts or filaments; made up of ice crystals. No precipitation.	Cirrocumulus: high-level, small rippled elements; ice crystals. No precipitation.	Cirrostratus: high-level, transparent sheet or veil, halo phenomena; ice crystals. No precipitation.

Middle-level clouds (2.5 to 6 kilometres)		
Altocumulus clouds	**Altostratus clouds**	**Nimbostratus clouds**

Altocumulus: middle-level layered cloud, rippled elements, generally white with some shading. Precipitation: may produce light showers.	Altostratus: middle-level grey sheet, thinner layer allows sun to appear as through ground glass. Precipitation: rain or snow.	Nimbostratus: thicker, darker and lower-based sheet. Precipitation: heavier-intensityrain or snow.

Low-level clouds (below 2.5 kilometres)	
Stratocumulus clouds	**Stratus clouds**

Stratocumulus: low-level layered cloud, series of rounded rolls, generally white. Precipitation: drizzle.	Stratus: low-level layer or mass, grey, uniform base; if ragged, referred to as 'fractostratus'. Precipitation: drizzle.
Cumulus clouds	**Cumulonimbus clouds**

Cumulus: low-level, individual cells, vertical rolls or towers, flat base. Precipitation: showers or snow.	Cumulonimbus: low-level, very large cauliflower-shaped towers to 16 km high, often 'anvil tops'. Phenomena: thunderstorms, lightning, squalls. Precipitation: showers or snow.

Figure 6.16 Main types of clouds and their associated weather conditions

As a general rule, high-altitude wispy clouds are an indication of good weather and towering, dense and dark clouds are an indication of rain, thunderstorms and high wind.

Weather maps

You may have noticed a common feature of a weather forecast is the weather map. Similar to weather forecasts, weather maps contain a great deal of information about upcoming weather. However, to the untrained eye, they can be difficult to understand.

The **weather map**, also known as a synoptic chart, is a visual representation of the locations and movement of weather patterns. Weather maps are useful as they can be used to forecast the predicted conditions, such as rainfall, wind and swell, and can be used in conjunction with a weather forecast to plan for an outdoor experience. Figure 6.17 presents an example weather map from the Bureau of Meteorology.

weather map a visual representation of the locations and movement of weather patterns

An explanation of the information contained in Figure 6.17 is presented in Table 6.4.

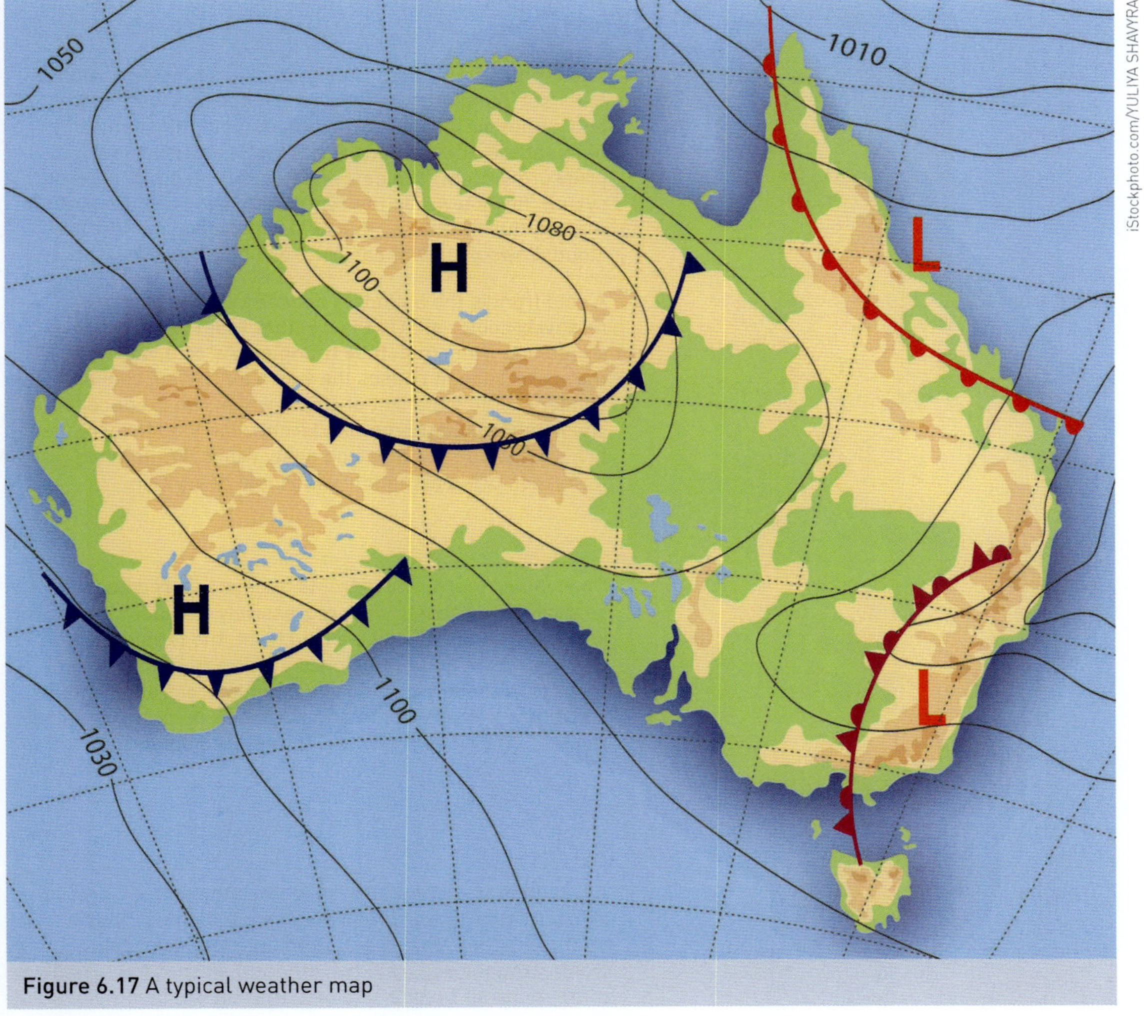

Figure 6.17 A typical weather map

Table 6.4 Typical weather symbols and explanations

Feature	Symbol	Meaning
Isobars	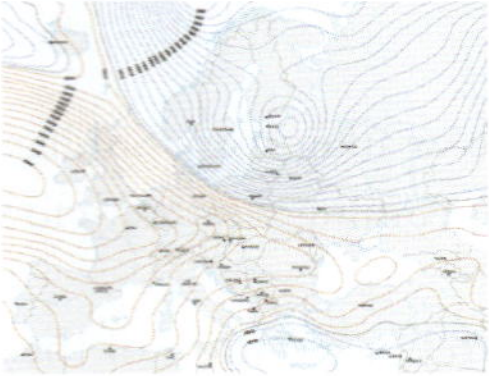	Isobars are the plain lines curving across the map. They connect points with the same mean sea level air pressure. Isobars indicate the flow of air around weather systems, meaning you can broadly interpret wind strength and direction from these maps. Generally, winds are strongest where the isobars are closest together and calmer where they're further apart.
High-pressure systems	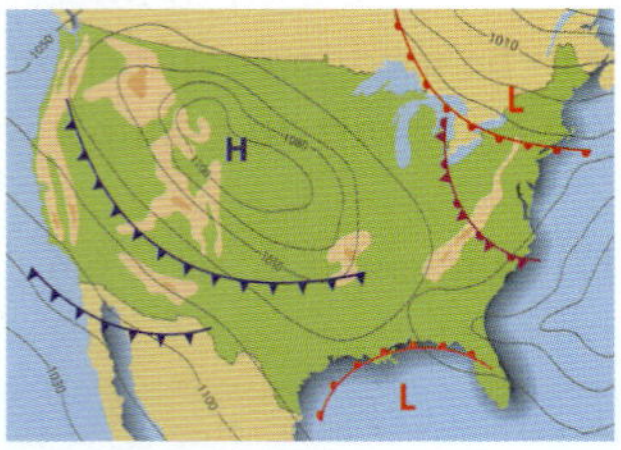	A high-pressure system is an area of high pressure relative to its surroundings. On the chart, it appears with an 'H' and a number indicating the pressure. In the southern hemisphere, the wind flows anticlockwise around and away from a high-pressure system. Air from the atmosphere above a high-pressure system sinks down and warms as it does so. The sinking air is very stable, so high-pressure systems mean dry, settled weather and light winds.
Low-pressure systems	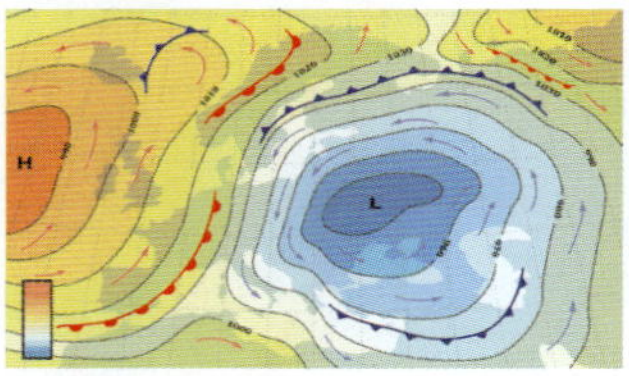	A low-pressure system is an area of low pressure relative to its surroundings. On the chart, it appears with an 'L' and a number indicating the pressure. In the southern hemisphere, the wind flows clockwise around a low-pressure system. The descending air from a high-pressure system flows towards the centre of a low-pressure system making the air rise. As the air rises it cools and forms cloud. Lows usually mean cold, wet and windy weather.
Cold front	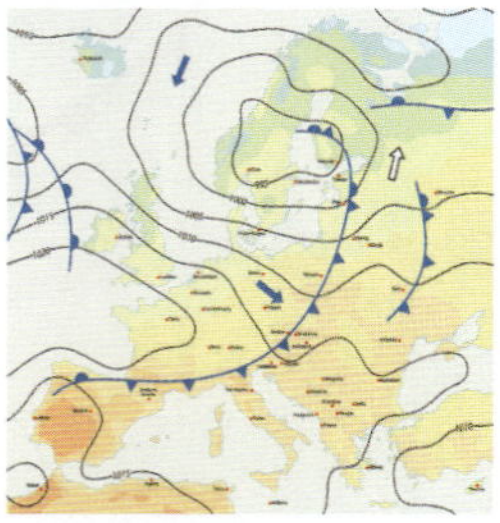	A cold front is the boundary between warm air and relatively cooler air. On the weather map it appears as a blue line with small blue triangles. As cold, dense air moves through, it causes the warm air to rise, cool and condense into a blanket of cloud, which can produce fairly consistent rain. Summer cold fronts can lead to increased fire danger due to producing gusty winds. Winter cold fronts can bring damaging winds and heavy rain.
Warm front	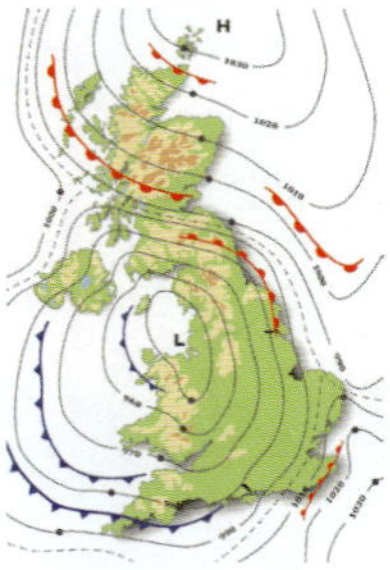	On weather maps, warm fronts appear as a red line with semicircles. Warm fronts progressively displace cool air with warmer air. Just like a cold front, the temperature change can be quite large once a warm front moves through, although it tends to happen more gradually than a cold front. Warm fronts bring warmer air, at times steady rainfall, grey skies and more humid conditions.
Troughs	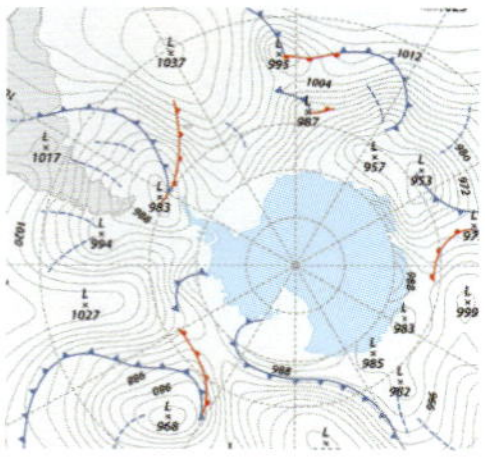	A trough is an elongated area where atmospheric pressure is low relative to its immediate surroundings. It appears on the weather map as a dashed blue line. Like cold fronts, troughs separate two different air masses (usually more moist air on one side and drier air on the other). As a trough moves towards moist air, it causes cloud or even showers and thunderstorms to develop.

iStockphoto/Rainer Lesniewski
iStockphoto/YULIYA SHAVYRA
iStockphoto/YULIYA SHAVYRA
iStockphoto/Rainer Lesniewski
iStockphoto/YULIYA SHAVYRA
iStockphoto/seamartini

Generally speaking, an approaching high-pressure system will result in dry, settled weather and light winds, whereas a low-pressure system usually means be prepared for cold, wet and windy weather. Learn more about information contained in a weather map at the Bureau of Meteorology website.

Resource Planning and recording outdoor experiences

Weblink Bureau of Meteorology

LEARNING ACTIVITY 6.2B

Logbook activity 9 – Weather forecast

Visit the Bureau of Meteorology website and access a weather map displaying weather patterns for an upcoming outdoor experience. Create a PowerPoint, Canva or Prezi presentation that displays the weather map accompanied by annotations explaining the meaning of each feature of the map and its probable impact on the weather for the trip.

Weather patterns

A common question on many people's minds each morning is, 'What is the weather going to be like today?' The answer may influence clothing choices, how to travel to and from school, or whether to pack a jacket or an umbrella for the day. We have all been caught out by a sudden change in the weather. After days of sunny weather, a sudden downpour can be considered a break or change in the weather pattern. A **weather pattern** is when the weather is consistent for a period of time.

weather pattern a period of time when the weather is consistent

Shutterstock.com/Ravi K Longia

Figure 6.18 A storm rolling in over the Twelve Apostles, Port Campbell National Park

Weather patterns are influenced by large global patterns in the atmosphere caused by changes in solar radiation, ocean currents, temperature, humidity and landmass topography. Generally, the four seasons are a way of understanding expected weather patterns. In a Victorian summer, we expect hot and dry conditions; in autumn, cooler, sunny days with light winds; in winter, cold and wet conditions; and in spring, unsettled weather that can quickly change from calm and sunny to cold and windy.

Resource
Planning and recording outdoor experiences

LEARNING ACTIVITY 6.2C

Understanding weather and climate

Visit the Agriculture Victoria website to watch a short animation of the latest science on our changing weather patterns, called 'Victoria's Climate 101 - A short animation of the latest science on our changing weather patterns'.

Weblink
Agriculture Victoria

Weblink
Video: Agriculture Victoria, 'Victoria's Climate 101 - A short animation of the latest science on our changing weather patterns'

Australia is the driest inhabited continent in the world, and our climate is highly variable – drier periods and wetter periods are natural cycles, and their extent varies from year to year. Therefore, we must learn to live and adapt to naturally occurring drought conditions. Bureau of Meteorology records since the 1860s show that a 'severe' drought has occurred in Australia, on average, once every 18 years.

The fluctuations in Australia's climate are influenced mostly by what is known as the El Niño–Southern Oscillation, which is the fluctuation between the natural climate cycles of La Niña and El Niño. As discussed further in Chapter 7, **La Niña** is the cooling of ocean surface temperatures in the central and east-central equatorial Pacific, and is associated with increased rainfall and cooler daytime temperatures. **El Niño** is the extensive warming of the central and eastern tropical Pacific, and is associated with an increased probability of drier conditions.

It is vital to have an understanding of the expected weather patterns for an upcoming practical experience to ensure all participants are appropriately prepared for changing weather conditions. Clothing and equipment lists may need to be changed, alternative activities may need to be considered or possibly the entire outdoor experience may need to be postponed or cancelled due to unsuitable weather.

La Niña
extensive cooling of the central and eastern tropical Pacific Ocean, associated with increased probability of wetter conditions in Australia

LEARNING ACTIVITY 6.2D

Logbook activity 10 – Weather graph

Construct a weather graph displaying the average annual rainfall and temperatures (minimums and maximums) for the location of your upcoming outdoor experience. Describe the possible influence of this expected weather (including La Niña and El Niño) on planning and conducting your outdoor experience.

Resource
Planning and recording outdoor experiences

El Niño
extensive warming of the central and eastern tropical Pacific, associated with an increased probability of drier conditions in Australia

Extreme weather

Tragically, each year lives are lost due to individuals underestimating the influence of **extreme weather** on their outdoor experience. Extreme weather is weather that is unexpected, unusual, severe or unseasonal based on conditions recorded in the past. Examples of extreme weather include heat waves, heavy rainfall, blizzards, drought, floods, bushfires and storm events, such as thunderstorms and tropical cyclones.

You may have heard the saying, 'Knowledge is power'. When it comes to planning an outdoor experience, it is vital to seek out detailed and up-to-date knowledge for all aspects of the experience.

Figure 6.19 Continue or turn back?

As can be seen in past flood events, rivers may flood many days after rain has fallen hundreds of kilometres away. The knowledge of this movement of water and its impact on connected waterways is an example of the detail required when planning experiences after extreme weather events.

extreme weather
weather that is unexpected, unusual, severe or unseasonal based on conditions recorded in the past

Climate change is increasing the intensity and frequency of many extreme weather events. According to the Australian Climate Council, examples of the influence of climate change on extreme weather events include the following:

- Heat – extreme heat is increasing across Australia. There will still be record cold events, but hot records are now happening three times more often than cold records.
- Bushfire weather – extreme fire weather has increased over the past 30 years in many parts of Australia, including southern NSW, Victoria, Tasmania and parts of South Australia.
- Rainfall – heavy rainfall has increased globally. Over the past three years, Australia's east coast has experienced several very heavy rainfall events, fuelled by record-high surface water temperatures in the adjacent seas.
- Drought – a long-term drying trend is affecting the south-west corner of Western Australia, which has experienced a 15% drop in rainfall since the mid-1970s.
- Sea-level rise – storm surges are higher than they were a century ago, increasing the risk of flooding along Australia's socially, economically and environmentally important coastlines.

Contingency planning

For all practical experiences, it is necessary to consider **contingency plans** to account for extreme (and unsuitable) weather conditions. A strong offshore wind on a day planned for a beginner's sea-kayaking class could turn treacherous very quickly. Plans to go caving in the days after heavy rainfall will need to be reconsidered.

contingency plan
a plan designed to take into account a possible future event or circumstance

The preparation of contingency plans to modify outdoor experiences due to extreme weather is important. Factors to consider will differ depending on the environment, group size and level of experience, recent weather events and their impacts, forecast weather and the nature of the planned activities.

Shutterstock.com/watcher fox

Figure 6.20 Visit the Bureau of Meteorology's website to learn more about the dangers of boating in high wind.

Contingency plans need to be well documented and shared with all participants. There have been situations where groups have been separated by flash flooding and have had to make emergency evacuation decisions to reach safety without contact with other group members. Planning for any outdoor experience requires all participants to know what to do if they are separated without the ability to communicate with the rest of the group.

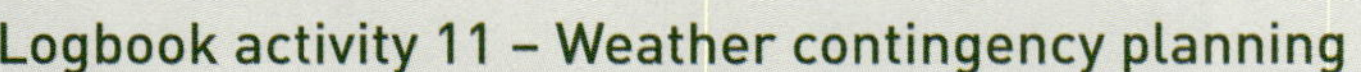

LEARNING ACTIVITY 6.2E

Logbook activity 11 – Weather contingency planning

Investigate past weather events that have affected the outdoor environment of an upcoming outdoor experience. You will be allocated one activity planned as a part of an upcoming practical experience. Create a detailed contingency plan for this activity to be followed if similar weather events occur, such as high winds, thunderstorms, heatwaves, whiteout conditions or flooding rains.

Resource
Planning and recording outdoor experiences

6.3 HOW TO MONITOR YOUR IMPACTS ON THE OUTDOOR ENVIRONMENT

KEY KNOWLEDGE

- how to monitor observations of own and other groups' impacts on the outdoor environment during an outdoor experience

KEY SKILLS

- analyse observations of impacts of groups on outdoor environments

As is evident in Figure 6.21, some people leave significant impacts when they venture into outdoor environments. For example, the seemingly harmless act of enjoying the serenity of a campfire not only adds to carbon emissions but also destroys habitat of native species, leaves a scar on the ground and removes nutrients bound for the soil. However, as we investigated in Chapter 5, not all impacts are negative. The rehabilitation and protection of sensitive vegetation through conservation activities are examples of positive impacts in many managed parks throughout Victoria.

This final section of the chapter covers information to prepare you on how to monitor and analyse observations of your own and other groups' impacts on the outdoor environment during an outdoor experience. We start by looking at the different ways of collecting evidence of impacts, including the use of a remote camera and advice on how to complete a field sketch. The section finishes with three logbook activities that ask you to suggest changes to the management of users in the investigated outdoor environment and recommend ways to promote sustainable interactions in the future.

Alamy/Washington Imaging

Figure 6.21 How many impacts can you spot on this outdoor environment?

6.3.1 HOW TO MONITOR OBSERVATIONS OF IMPACTS ON THE OUTDOOR ENVIRONMENT

Monitoring observations of one's own and other groups' impacts on outdoor environments is crucial for understanding the ecological consequences of activities and for making informed decisions about conservation and sustainability. Effective monitoring helps identify positive actions and areas needing improvement.

The following is a step-by-step guide on how to monitor impacts.

Table 6.5 Step-by-step guide on how to monitor observations of impacts on the outdoor environment

Step
1. **Define objectives** Clearly define what you want to monitor, such as habitat disturbance, water quality or changes in native vegetation
2. **Baseline assessment** Conduct an initial assessment of the outdoor environment before any activities occur
3. **Regular monitoring** Set a schedule for regular monitoring (e.g. daily, weekly, seasonally)
4. **Collect data** Collect data using consistent methods (observations, sketches, remote cameras, etc.)
5. **Record impacts** Record observed impacts, both positive and negative
6. **Analyse data** Analyse the collected data to understand patterns, trends and the potential causes of impacts
7. **Report findings** Share the findings with relevant stakeholders, including the group members, land managers, environmental agencies or local communities
8. **Educate and raise awareness** Educate your group and other users of the outdoor environment about the importance of monitoring impacts and how individual actions contribute to its overall health

It is possible to monitor observations of impacts by gathering a range of different data across various time periods and locations. Monitoring changes over time provides a clearer understanding of the impact of different interactions with the outdoor environment.

Some observations may provide inaccurate information in regard to the cause of impacts on the environment. For example, the erosion of a 4WD track could be inaccurately blamed on recreational vehicles, when in fact the damage could be due for the most part to recent severe flooding.

There is a range of different methods to collect observations of impacts while in outdoor environments, including taking photos and videos, completing sketches and noting descriptions of what is seen, heard and smelled.

Field sketches

Field sketches are simplistic drawings of a particular location and are useful to remember locations you have visited and for identifying, describing and explaining the characteristics of places. Field sketches can be used to record information in the field, including observations of impacts on the outdoor environment. Field sketches are not supposed to be works of art, but rather should highlight the key features of the place being observed.

field sketch
a simplistic drawing of a particular location

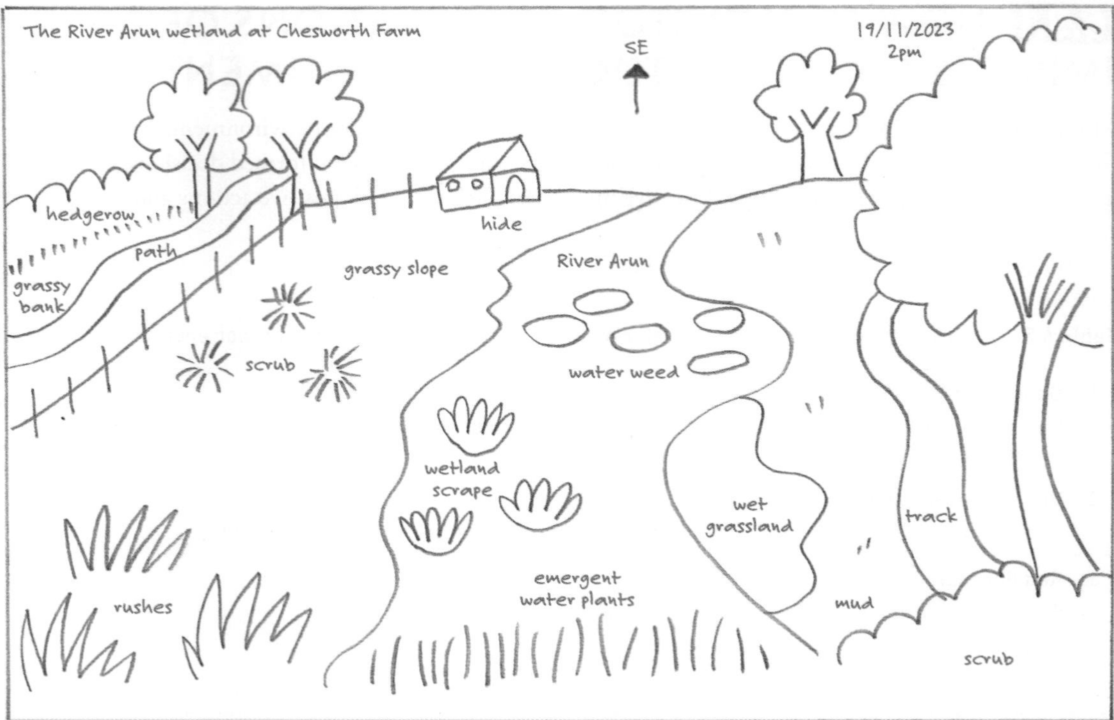

Figure 6.22 Field sketches are an effective way of developing an understanding of natural features and processes.

HOW TO COMPLETE A FIELD SKETCH

1 Choose a location to be sketched based on specific features such as examples of impacts.
2 Use a ruler to draw a border around your sketch.
3 Divide the scene into three parts:
   a. foreground
   b. middle distance
   c. background.
4 Sketch in the main features of the scene, such as the horizon, landforms and large objects.
5 Add other prominent features.
6 Add further detail (and shading) as required to highlight key features of the scene.
7 Add additional key features including a title, orientation, annotations, scale and time/date of sketch.

One way to remember the key features of a field sketch is using the acronym, TOASTIE. That is, field sketches should have the following features marked on them:

- **T** – Title (information about the location of your sketch)
- **O** – Orientation (the direction you are facing)
- **A** – Annotations (details and explanation of observations)
- **S** – Scale (estimation of the size of objects in the sketch)
- **T** – Time (the time of day and date your sketch was done)
- **I** – Information (detail added in your annotations to help with later analysis)
- **E** – Edge (use a ruler to draw a border around your sketch).

LEARNING ACTIVITY 6.2F

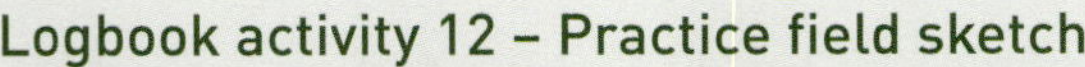

Logbook activity 12 – Practice field sketch

Find a location in the schoolyard that contains some natural habitat that has been impacted on. Develop your recording skills by creating a sketch of this area, ensuring that you add all TOASTIE features (title, orientation, annotations, scale, time, information and edge) to your sketch.

Resource
Planning and recording outdoor experiences

REMOTE CAMERAS

At times it is difficult to collect data on the number of native species in an area due to the animals being nocturnal and/or fearful of your presence. Therefore, you may need to collect covert observations, and this can be done by the use of a remote camera.

It is possible to collect multiple observations of an environment to identify the cause of impacts by setting up a remote camera. **Remote cameras**, also known as trail cameras or game cameras, are rugged and weatherproof cameras designed to take motion-activated images and videos without the need for a person to be present.

iStockphoto.com/Ralf Geithe

Figure 6.23 Motion-activated remote cameras are valuable in collecting evidence of threatened species.

remote camera
a rugged and weatherproof camera designed to take motion-activated images and videos without the need for a person to be present

Remote cameras can be used to monitor the positive impacts of conservation activities such as the success of a threatened species breeding program. By placing a remote camera in a known location frequented by members of the species, evidenced by the presence of **scat** (wild animal poo), it is possible to gather data on the number of species in a specific location.

scat
wild animal poo

LEARNING ACTIVITY 6.2G

There is a variety of wildlife livestreams that can be accessed from remote cameras around the world. Visit the BBC's Discover Wildlife website and search for 'The best wildlife livestreams to watch around the world' for a list of links.

Weblink
Discover Wildlife

LEARNING ACTIVITY 6.2H

Logbook activity 13 – Observations of own and other groups' impacts

During your outdoor experience, record observations of impacts on the outdoor environment through descriptions, sketches, water and soil samples, and photos.

Resource
Planning and recording outdoor experiences

9780170477123

Analysis of impacts and promotion of sustainable interactions

The following logbook activities provide an opportunity to analyse the impacts of other users and yourself on the outdoor environment and to investigate ways to reduce this and promote sustainable interactions. You will also reflect on your outdoor experience to suggest changes to the management of users in the outdoor environment and recommend ways to promote sustainable interactions in the future.

Resource Planning and recording outdoor experiences

LEARNING ACTIVITY 6.2I

Logbook activity 14 – Analysis of observed impacts on the outdoor environment

Create a presentation in PowerPoint, Canva or Prezi (or a similar program) that analyses your impacts and those of other users on the outdoor environment. In your presentation, include the following:

1 An introduction to your outdoor environment (including location, landscape features and common interactions)
2 A description of your environment's significant natural features, including threatened species and ecological communities
3 A collection of annotated photos (and/or sketches) of observed impacts on the outdoor environment (caused by you and by other users)
4 An explanation of the interactions causing impacts and the related consequences on the outdoor environment.

Resource Planning and recording outdoor experiences

LEARNING ACTIVITY 6.2J

Logbook activity 15 – Ways to reduce impacts and promote sustainable interactions

Create a pamphlet designed to inform users of ways to reduce their impacts and how to adopt sustainable interactions with your specific outdoor environment.

Resource Planning and recording outdoor experiences

LEARNING ACTIVITY 6.2K

Logbook activity 16 – Management of users in outdoor environments

Write a letter to the managers of your specific outdoor environment (e.g. Parks Victoria rangers) that details the information you collected on impacts on the environment. Also suggest ways that users could be managed to promote sustainable interactions into the future.

CHECK YOUR KNOWLEDGE AND SKILLS

Checklist for students to mark and date, making judgements of their knowledge and understanding.
(D = developing C = consolidating ER = exam ready)

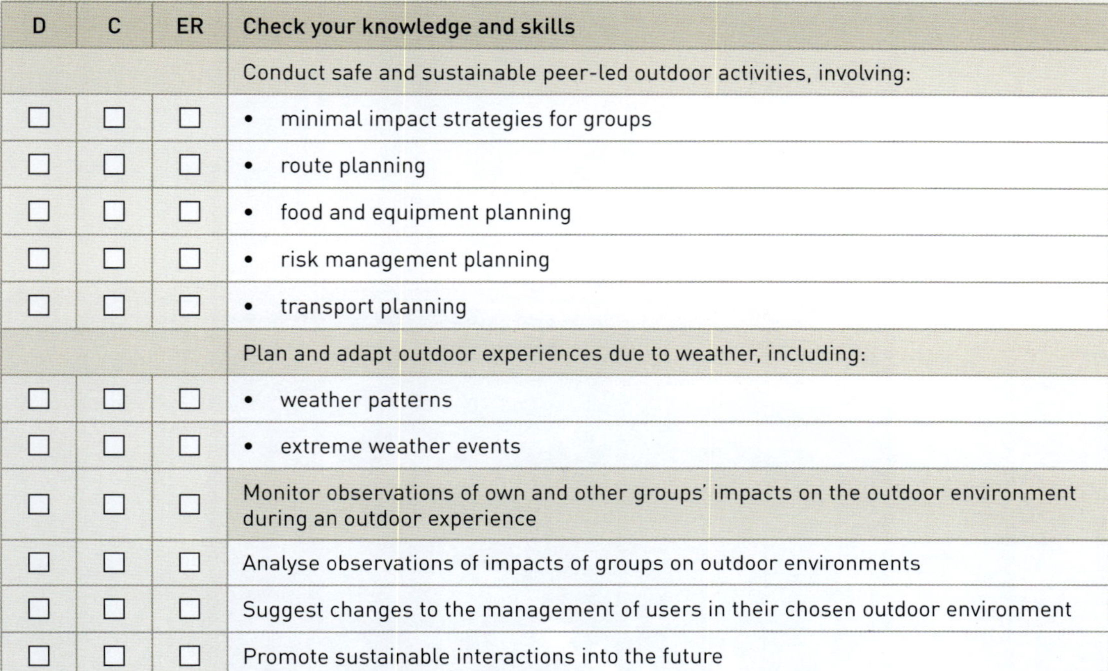

D	C	ER	Check your knowledge and skills
			Conduct safe and sustainable peer-led outdoor activities, involving:
☐	☐	☐	• minimal impact strategies for groups
☐	☐	☐	• route planning
☐	☐	☐	• food and equipment planning
☐	☐	☐	• risk management planning
☐	☐	☐	• transport planning
			Plan and adapt outdoor experiences due to weather, including:
☐	☐	☐	• weather patterns
☐	☐	☐	• extreme weather events
☐	☐	☐	Monitor observations of own and other groups' impacts on the outdoor environment during an outdoor experience
☐	☐	☐	Analyse observations of impacts of groups on outdoor environments
☐	☐	☐	Suggest changes to the management of users in their chosen outdoor environment
☐	☐	☐	Promote sustainable interactions into the future

Resources
Key knowledge and skills checklist

Glossary – Chapter 6

Assessment
Glossary test

Unit 3
Relationships with outdoor environments

AREA OF STUDY 1 **Chapter 7: Changing human relationships with outdoor environments**

AREA OF STUDY 2 **Chapter 8: Relationships with Australian environments in the past decade**

Unit 3 – Introduction

Unit 3 of the VCE Outdoor and Environmental Studies study design, titled 'Relationships with outdoor environments', focuses on the ecological, historical and social contexts of relationships between humans and outdoor environments in Australia. It examines the dynamic nature of these relationships between humans and their environment over time periods stretching back more than 60 000 years until the present.

Students are involved in multiple experiences in outdoor environments, including in areas where there is evidence of human interaction. Through these practical experiences, students make comparisons between, and reflect upon, outdoor environments, as well as develop theoretical knowledge and skills about specific outdoor environments.

Chapter 7 – Changing human relationships with outdoor environments explores the formation of the planet's land masses and the way in which humans have understood and interacted with Australian outdoor environments from various Indigenous peoples' cultural experiences. It also discusses the influence of several major historical environmental events and issues following European colonisation.

Chapter 8 – Relationships with Australian environments in the past decade examines conflicting values of human use and relationships with outdoor environments in the past decade and how these are represented in various forms of media. Social, cultural, economic and political factors that influence these relationships are explored along with how these conflicts are resolved.

Through their studies of Unit 3, students will develop theoretical knowledge and skills about specific outdoor environments and commence preparing for the assessment task of Unit 4, Area of Study 3. For this cross-unit task, students undertake an independent investigation into the changing relationships with, and sustainability of, the visited outdoor environments.

CHAPTER 7

UNIT 3 – AREA OF STUDY 1

Changing human relationships with outdoor environments

KEY KNOWLEDGE

- Australian outdoor environments before humans arrived, including characteristics of biological isolation, geological stability and climatic variations
- relationships with outdoor environments expressed by specific Indigenous peoples' communities before and after European colonisation
- relationships of non-Indigenous peoples with specific outdoor environments as influenced by and observed in local or visited outdoor environments, during historical time periods:
  - Early colonisation (1788–1859)
  - Pre-Federation (1860–1900)
  - Post-Federation (1901–1990)
- the beginnings of environmentalism and resulting influence on political party or policy, as observed in one of the following historical campaigns:
  - Lake Pedder
  - Franklin River
  - Little Desert.

KEY SKILLS

- explain characteristics of Australian outdoor environments before humans arrived
- analyse the changing relationships with Victorian outdoor environments expressed by specific Indigenous peoples' communities before and after European colonisation
- analyse the changing relationships of non-Indigenous peoples with Victorian outdoor environments observed during historical time periods
- describe the beginnings of environmentalism as observed in an historical campaign
- evaluate the influence of a historical campaign on the development of a government policy or political party.

VCE Outdoor and Environmental Studies Study Design 2024–2028

Australia before humans

The continent of Australia has been forming from its beginnings on the ocean floor and is home to rocks that are over 3000 million years old. The land mass was once connected to all land on earth and covered in ice, and has undergone separations and movements from as far as the South Pole. All of this movement has shaped our landscape and its inhabitants, including the unique flora and fauna that the world recognises us for.

Indigenous relationships with the environment

The Traditional Owners of what we now identify as Australia have a rich and unrivalled connection and relationship with this country. They were the first to walk upon this land and continue to do so as modern Indigenous peoples, celebrating their customs and beliefs, and sharing their Country with non-Indigenous peoples.

Non-Indigenous relationships with the environment

Colonisation of Australia began in 1788 and with it came new people with new perceptions of outdoor environments. The colonisers brought with them fears and ideals that influenced their interactions and impacts on outdoor environments, treating the bush as a resource to be plundered as if it were infinite; but these practices have changed in the ensuing 200 years.

Environmental movements in Australia

The new Australians - those who colonised, slashed, burned and built new lives on this continent - began to realise before Federation that these outdoor environments were fantastic to recreate in and were less threatening than initially thought. Conservation of natural environments began in these early times and less than 100 years later, environmentalism was born in Australia.

KEY TERMS

adversary
alienation
anthropocentric
contemporary
El Niño
endemic
environmental activism
fire-stick farming
hunting and gathering
hydro-electricity
impacts
industrialisation
interactions
Kinship
La Niña
metaphor
middens
mobile
nation building
neutral phase
perceptions
pyrolysis
reconciliation
relationship
sedentary
semi-sedentary
soil salinity
terra nullius
volatile

Worksheets

Weblinks

Weblink – Video

Resources and Templates

Assessments

To access resources above, visit **cengage.com.au/nelsonmindtap**

SPOTLIGHT

Charles Darwin and Australia

When the now-famous English scientist Charles Darwin visited the British colony of New South Wales in January 1836, he began to write the first observations of the similarities and confusions of Australian flora and fauna. Writing from a distinctly European perspective, he marvelled at the many natural wonders – floral, faunal and geological – of the great southern land. Darwin's revelations would eventually inform his well-known theory of evolution by natural selection. In the *Beagle Diary*, in which he recorded his observations and experiences during his trip around Australia, he wrote about an examination of a rat-kangaroo and a platypus. He noted that they occupied outdoor environments similar to those of the rabbit and water rat in the Northern Hemisphere. He wondered why a single creator or God would make such different animals for the same apparent purpose: 'Surely two distinct Creators must have been [at] work.'

Figure 7.1 Naturalist Charles Darwin

Shutterstock.com/Nicku

This became the key question of creation raised by Charles Darwin in what was to become the most important of Australia's contributions to his famous work, *On the Origin of Species*.

Figure 7.2 Darwin's voyage around Australia on the HMS Beagle in 1846

'General Chart of Australia', Discoveries in Australia, Volume 1, John Lort Stokes 1846

RELATIONSHIPS WITH OUTDOOR ENVIRONMENTS

When we separate and consider the origin of the word '**relationship**', it is read as 'relation' and 'ship'. The word relation, a noun (identifying word) speaks to identifying the connection between two things: 'How are they related?' is a question commonly asked. The 'ship' on the end is changing the concrete fact of the relation of a thing and creates a word that is measuring the quality or condition of the connection between the two things.

Andrew Mannion

Figure 7.3 What do your relationships with outdoor environments look like?

relationship the way in which two or more people or things are connected, or the state of being connected

In Unit 3, our study focuses largely on developing your understanding of the concept of relationships with outdoor environments and how they have changed since before our land mass became its own isolated island. We first analyse how flora and fauna have come to relate to Australian outdoor environments and then, in more depth, evaluate the changing nature of the way humans relate to outdoor environments. That is, what connections do humans have with outdoor environments and how do we describe these relations, their quality and their condition over time?

Describing relationships

Think about some of the relationships you have with other people – your friends, parents, siblings, classmates, teachers, and so on. How would you describe these? Relationships are hard to describe. They're fluid – constantly changing as we change, as others around us change and as the circumstances that connect us in these relationships change. But if you were forced to describe any of your own relationships with another person, you might include some (or all) of the following:

- feelings about the person
- experiences and history with the person
- things you like and don't like about the person
- things you do with the person.

DESCRIBING RELATIONSHIPS WITH OUTDOOR ENVIRONMENTS – A FIRST GO

You might have found describing relationships with other people a bit tricky, but if that was hard, doing the same thing for non-human objects (such as outdoor environments) may be even harder. One way to think about human relationships with outdoor environments (a way that we'll use in this book) is shown in Figure 7.4. The arrows connecting each of the three aspects in the diagram are an attempt to show how these aspects of relationships connect to and affect each other. For example, our perceptions help to determine

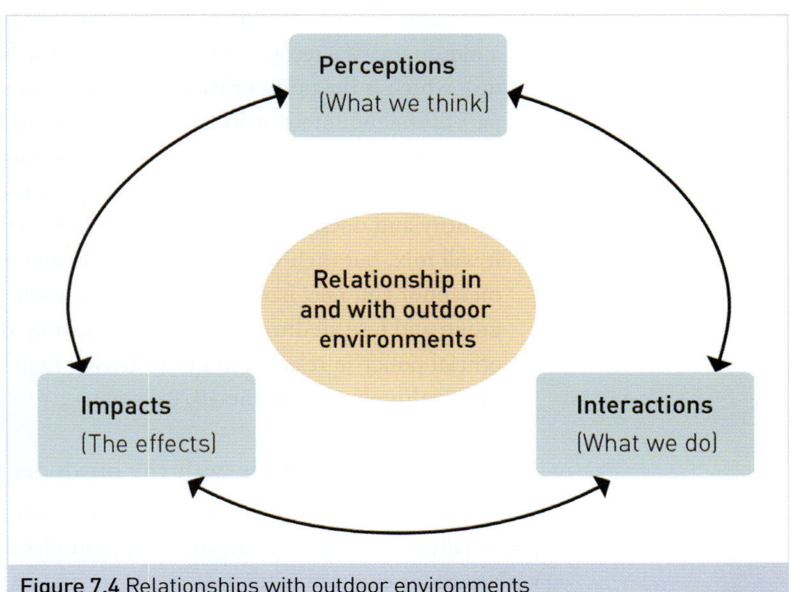

Figure 7.4 Relationships with outdoor environments

the interactions we have, and the interactions we have also help to influence our perceptions. In this view of relationships, we have three key aspects of the relationship:

1 **perceptions** – what we think about outdoor environments
2 **interactions** – what we do in, and with, the outdoor environments
3 **impacts** – what happens as a result of our relationships with outdoor environments. We will use this idea about relationships as we move through different parts of the course.

We will use this idea about relationships as we move through different parts of the course.

perceptions what we think about outdoor environments

interactions what we do in, and with, the outdoor environments

impacts what happens as a result of our relationships with outdoor environments

metaphor a figure of speech where a word or phrase is used to describe an object in question and the word or phrase has no literal connection to the object but aids in understanding the object

DESCRIBING RELATIONSHIPS USING METAPHORS

Having a model (like Figure 7.4) to understand relationships is one thing, but to write about them and describe them we need other tools – and this is where **metaphors** come in. As discussed in Chapter 4 of this text, a metaphor is when two unlike things are compared with each other because of something they have in common. They are used in lots of ways, but often when we are trying to understand something complex, we use a metaphor to compare it with something simpler. Here are some examples of metaphors:

- The foot of the mountain.
- Life is a journey.
- The USA is a salad bowl culture.
- Today was a roller-coaster ride.
- 'All the world's a stage, and all the men and women merely players, they have their exits and their entrances.' (Shakespeare from *As You Like It*).
- She ruffled his feathers.
- The world is my oyster.
- Your words cut deeper than a knife.
- The classroom turns into a zoo during recess.

A closer look at some metaphors

Let's look a little more closely at some metaphors.

1 When considering the metaphor, 'The foot of the mountain', we know mountains don't actually have feet – feet are something that humans have, and maybe we can attribute feet to some other animals too. But a mountain is not a human. So, what do we mean when we talk about the foot of a mountain? We recognise feet as being at the bottom of a human (assuming the human is standing up), which is the point of using this particular metaphor. The foot of a mountain is recognised as the part of the mountain at its base – it's the bottom of the mountain.
2 What about the metaphor 'salad bowl' when used to describe the multicultural aspects of towns, cities and countries? In this context this metaphor means that each vegetable in the salad maintains its identity and doesn't dissolve into a single item. You can toss around a salad, add new ingredients and it will mix, but each vegetable will stay distinct and separate. This metaphor is the opposite of another metaphor you may have heard of or read about: the 'melting pot', where all cultures melt into a single group and lose all distinct identity in forming one singular culture. In postwar Australia, this was official policy in relation to newly arrived migrants before the adoption and development of modern multiculturalism from the 1970s.

Describing relationships with the outdoors using metaphors

The beauty of a metaphor is that it can have many meanings – the one intended by the author, but also potentially many others. Here are some examples of metaphors used to describe human relationships with outdoor environments:

- **Outdoors as an adversary**: An adversary is an opponent in a contest or fight. This might be a useful metaphor when looking at a human–nature relationship that sees the environment as something we are fighting against; something to beat, control and dominate.
- **Outdoors as a museum**: In a museum we collect and store valuable things. We protect them and maintain them, and we use them to learn about the world. When describing human–nature relationships that involve conservation, protection or preservation, this can be a useful metaphor. When the environment is seen as a place to study or learn – whether it's the environment we're studying or ourselves, or something else altogether – this metaphor could also be useful.
- **Outdoors as a cathedral**: Cathedrals, temples, mosques, synagogues and so on are spaces where people worship and pray, and spend time to connect with their spirituality. The connection with spirituality and the outdoors is something we've discussed in earlier chapters and is what can make this a useful metaphor.
- **Outdoors as a gymnasium**: A gymnasium is a place where we challenge ourselves and look to physically develop our bodies. People go to gyms to work out and 'raise a sweat'. For many people, outdoor environments are places for them to do this (work out, sweat and physically develop), which makes this metaphor useful in this sense.
- **Outdoors as a storehouse**: A storehouse or a warehouse is a place where we find resources – things we need, and we need lots of them. This would be a useful metaphor for people who see the human–nature relationship as being made up of an environment that gives us access to many resources, such as food, air, water, plants, animals, minerals and more.
- **Outdoors as a mother**: A mother is the person who birthed us and who then goes on to protect and care for us. We expect a mother to love her children and to be cared for by her children when she ages. This metaphor, commonly cited by Indigenous communities (as we'll see shortly), is an important way to describe human–nature relationships that see the environment as the place from which we come, the place which provides for us, and the place we must help to protect.
- **Outdoors as a web**: A web (as in the web of a spider) is a connection of many strands that interweave and combine in many ways. This metaphor could be appropriate for someone who sees the environment as a place with many interconnecting aspects. When we describe food webs in science and biology, for example, we are connecting with this metaphor as a useful way of seeing aspects of the natural world.

adversary
an opponent in a contest or fight

Shutterstock.com/THP Creative

Figure 7.5 The spectacular Twelve Apostles on the Victorian coast

LEARNING ACTIVITY 7.1A

Metaphors

1. Are metaphors useful in helping to explain a relationship someone might have with an outdoor environment? Choose one or two metaphors discussed in this book and analyse how useful you think it would be in helping to describe and explain the relationship that an individual or group might have with outdoor environments.
2. Find examples of people whose relationships might be typical of those described in these metaphors.
3. Find examples of other metaphors (or come up with your own) that could be useful in helping to describe and understand the relationships that people can have with outdoor environments.

Worksheet
Additional Learning Activity: 7.1a Quotes

7.1b Metaphors

7.1 AUSTRALIA BEFORE HUMANS

KEY KNOWLEDGE

- Australian outdoor environments before humans arrived, including characteristics of biological isolation, geological stability and climatic variations

KEY SKILLS

- explain characteristics of Australian outdoor environments before humans arrived

To understand the relationships people have had, and are having, with the Australian environment, and the impacts associated with these relationships, we need a reference point from which to compare and make an assessment. Have humans had a negative effect? Have we had a positive effect? The only real way to know is to understand what Australia was like before there were any humans here at all.

Australia – a brief overview

Australia, with an area of just over 7,600,000 square kilometres, is the sixth-largest country in the world. Australia is the flattest continent, with the oldest and least fertile soils, and is also the driest inhabited continent.

The climate of Australia is significantly influenced by ocean currents, including the El Niño–Southern Oscillation (ENSO; discussed later in this section) and the seasonal tropical low-pressure systems that produce cyclones in the north. These factors mean rainfall patterns vary dramatically from year to year.

Although most of Australia is semi-arid or desert, it includes a wide range of habitats – from alpine heaths to tropical rainforests. Over a vast amount of time, the Australian outdoor environments have been exposed to biological isolation, geological stability and climatic variations, leading to the unique plants and animals found on this continent.

Prehistory of Australia before humans

The prehistory of Australia before humans was characterised by a dynamic and ever-changing environment, with different geological and climatic events shaping the continent over millions of years.

The geological time scale is a system used by geologists and palaeontologists to divide Earth's history into different periods. It begins 4.6 billion years ago, which is 460 million million years ago! In this section the focus is on the final six geological time periods that have shaped Australia:

- Triassic Period (252–201 million years ago)
- Jurassic Period (200–145 million years ago)
- Cretaceous Period (144–66 million years ago)
- Paleogene Period (65–23 million years ago)
- Neogene Period (22–2.6 million years ago)
- Quaternary Period (2.5 million years ago to present)

UNIT 4 AREA OF STUDY 3 LOGBOOK ENTRY REMINDER

Has your teacher communicated to you the four key knowledge points you are focusing on for Unit 4 AOS 3, which begins at the start of Unit 3? Use the planner similar to chapter 11 - 11.1A to begin this.

The Triassic Period (252–201 million years ago)

Land connections

The continents were joined in the supercontinent of Pangaea and what was to become Gondwana formed the southern part of the single continent, with the south pole sitting in what is now South Australia.

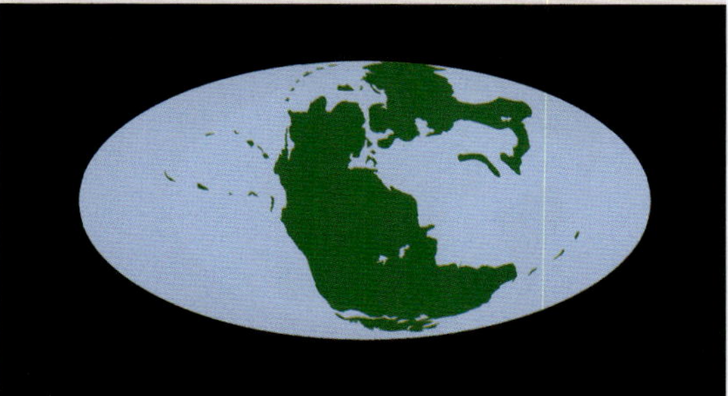

Climate

During this period, the area that would become Australia was dominated with monsoonal weather, while the rest of the world's climate was experiencing a hotter, drier period. Atmospheric carbon dioxide levels were roughly three times higher than today.

Flora and Fauna

The first dinosaurs could travel across the entire landmass dominated by ferns and conifers, while aquatic animals such as the early crocodiles, lizards and turtles could also be found, and primitive mammals began to appear.

The Jurassic Period (201–145 million years ago)

Land connections

200 million years ago, Pangaea began to separate, forming what became Laurasia to the north and Gondwana to the south. In a later period, Gondwana would eventually split up, becoming Australia, South America, Africa, India, Madagascar, Antarctica and New Zealand. When we reconnect these landmasses today on a map like a puzzle, the rocks and fossils are a match.

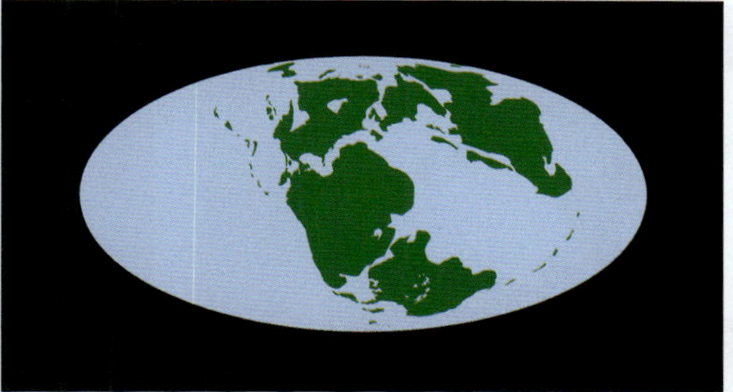

Climate

There were still no polar ice caps and the world's climate was wet and warm, with carbon dioxide levels about seven times higher than today.

Flora and Fauna

While dinosaurs were beginning to be found on all other continents in large numbers as the main group of fauna, Australia seems to have had very little dinosaur habitation as indicated in the fossil record for this period. Ferns and conifers remain the land based major species of flora.

Figure 7.6 Six geological time periods that have shaped Australia

The Cretaceous Period (144–66 million years ago)

Land connections

At the beginning of this period, Gondwana remained, with South America, Australia, New Zealand and Antarctica connected, with a large part of Australia close to the south pole. Then 80 million years ago, New Zealand went its own way, detaching from the remaining continents of Gondwana.

Climate

Due to how close the land that is now Australia is located to the south pole, it experienced an icy polar winter and, for the remainder of the year, a cool, wet climate.

Flora and Fauna

The dominance of conifer tree species reigned supreme but the first flowering plants began to appear. Dinosaurs became widespread in Australia during this period in the sky, land and water. The Platypus, our first mammal, also appeared.

The Tertiary Period (65–23 million years ago)

Land connections

Gondwana does a lot of moving and shaking in this period, but Australia, Antarctica and South America remained joined and a belt of forest stretches across Australia and South America until 50 million years ago, when Australia and Antarctica started to separate.

Climate

The area that was to become Australia had a largely wet and warm climate, despite being close to the south pole.

Flora and Fauna

Broad-leaved rainforests were becoming dominant over conifer forests, with a large volume of southern beech trees. Dinosaurs become extinct worldwide and in Gondwanan forests a variety of land animals, from frogs to the mammals we know today made up of monotremes, marsupials and placentals, began moving around. South America and Australia shared fauna through our shared forest belt. South America also had many marsupial groups in this period, including the ancestors of sabre-toothed marsupials.

Figure 7.6 Six geological time periods that have shaped Australia (*Continued*)

The Neogene Period (22–2.6 million years ago)

Land connections

Gondwana had completed breaking up by this period, separating Australia from Antarctica and South America. Our continent was slowly drifting northwards with the islands of New Guinea until eventually colliding with South-East Asia about 15 million years ago. By about 5 million years ago, the world's continents were close to their present positions.

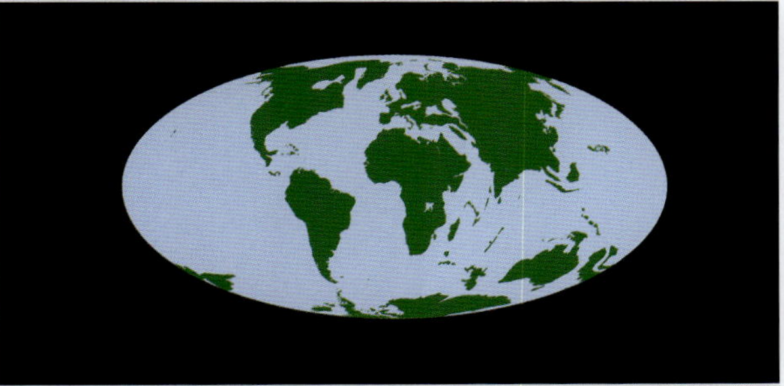

Climate

After it's collision with South-East Asia, a **rain shadow** effect occurred, impacting Australia's weather patterns, which dried the whole continent out. This, combined with global drying and cooling trends – involving ice rapidly accumulating at the poles, sea levels falling and rainfall decreasing – meant Australia was drying out.

Flora and Fauna

Many plant and animal groups died out and rainforests retreated. As Australia dried out, the dominance of acacia and eucalypt trees across much of the continent began. Grasslands were spreading and Australia's plants and animals began to gain their distinctive modern appearance.

The Quaternary Period (2.5 million years ago to present)

Land connections

Australia inched closer to its current position, with sea levels much lower at the beginning of this period creating temporary land bridges between mainland Australia and Tasmania in the south, and Australia and New Guinea in the north.

These land connections are indicated in the fossil record and modern archaeological techniques also indicate the likely arrival of the first humans in Australia at least 60 000 years ago (a topic covered further in the next key knowledge section).

Climate

Polar icecaps began to experience a waxing and waning, and this affected Australia's climate, cycling between icehouse phases (cold, dry conditions) and greenhouse phases (warmer, wetter conditions.)

During icehouse phases, sea levels fell, creating land bridges that would then be flooded during greenhouse phases.

The last ice age concluded approximately 10 000 years ago and the drying of our continent and erratic climate conditions continued, largely influenced by sea water temperatures that we identify as **La niña**, bringing wet and cool conditions to eastern Australia, and **El Niño**, bringing hot and dry conditions to eastern Australia.

Flora and Fauna

Australia's eucalypt and dryland specialist species thrived further in the quaternary period, but the pattern of forests, grasslands and deserts was constantly changing. Australia and the world had many examples of large animals that are collectively known as Megafauna.

Figure 7.6 Six geological time periods that have shaped Australia (*Continued*)

Figure 7.7 Zaglossus hacketti: a sheep-sized echidna

Figure 7.8 Procoptodon goliah: a 2–3 metre short-faced kangaroo

MEGAFAUNA

Many of the animals that lived in Australia before humans first arrived are known collectively as megafauna, which literally means 'large animals'. They roamed the earth during the Quaternary Period, from 2.5 million to 11 700 years ago. The term 'megafauna' is not only used in Australia; in the past, you could find megafauna in many parts of the world, such as dinosaurs, fish, reptiles and many mammals. Today, there are still many examples of megafauna around the world, including whales, elephants, hippos, rhinos, bears, crocodiles and others.

Diprotodon optatum, commonly known as the diprotodon, was a massive wombat 3 metres long and 2 metres tall. *Dromonis stirtoni* was a large flightless bird that grew up to 3 metres tall. *Varanus priscus* was a 7-metre long carnivorous goanna, and weighed in at almost 2 tonnes. You can research pictures of all of these. While all these examples of megafauna species are now extinct, others remain in Australia (although smaller in size when compared with their cousins). Kangaroos, wombats, emus, cassowaries, goannas and crocodiles are all examples of species that today are large enough to be considered megafauna.

Extinction theories

Evidence of Indigenous peoples of Australia having been on this continent for 60 000 years or more tells us that megafauna and humans lived side by side on this continent for tens of thousands of years. It is highly likely that these large animals played an important part in Indigenous peoples' social and spiritual cultures. It Is also very likely that the megafauna were hunted for food and other resources, such as tools, clothing and shelter.

Based on fossils that have been found to date, megafauna became scarce and then extinct in Australia by the end of the Pleistocene era, between 11 700 and 7000 years ago. This period marked the end of the last ice age and brought major changes to the Australian environment. The megafauna that had thrived in glacial cold-dry conditions struggled to adapt to the onset of warmer climates, which made surface water scarce and introduced mass changes to vegetation, including the growing dominance of the eucalypt species that we see in Australia today.

To this day, it cannot be determined how large a role Indigenous peoples played in the extinction of megafauna in Australia or the role played by the changing climate.

Getty Images/Paul Sinclair Photography

Figure 7.9 Exposed rocks often tell a story of the geology of a place.

PREHISTORY OF VICTORIA

The rocks and soils that make up Victoria's landscape have undergone the same sort of processes that occur all around the world – volcanism, uplift, sedimentation, folding, faulting, erosion and weathering. The result of these processes are features you'll come across throughout your experiences – from the coast, wetlands and Victorian Alps to rivers and valleys, forests, grasslands and deserts.

The rocks we see in outdoor environments (on the coast, in the mountains and elsewhere) tell a story of layering, folding, faulting and many other geological processes. There are some key features, events and issues in the prehistory of Victoria that are important to consider:

- For much of the history of Australia, Victoria and the whole eastern part of Australia didn't exist – they only formed about 500 million years ago. Western Australia has been around almost 10 times longer. In Victoria, some of the oldest rocks can be seen on the exposed coast at Waratah Bay, near Phillip Island.
- For much of its history, Victoria was a seabed. The evidence of this can be found in surprising places – fossilised sea life has been found high in the Victorian Alps.
- As Australia and Antarctica broke apart, the stretching and tearing of rock had some of its greatest effects in south-eastern Australia. Rift valleys formed between the two continents, including two in southern Victoria. Steep mountain ranges formed at the edges of these valleys. The Grampians and Mount Arapiles are the remnants of one. The Victorian Alps are the remnants of the other.
- The low-lying areas around Gippsland and Western Port Bay, formed from the separation of Australia and Antarctica, became swamps and marshes. These would later form the coal beds of the Latrobe Valley, and also include some of the best examples of dinosaur fossils found in Australia.

CHARACTERISTICS OF PRE-HUMAN AUSTRALIA

There are three key characteristics that were a feature of the Australian environment before humans arrived, and each has been shaped by our geological history:

1 biological isolation
2 geological stability
3 climatic variations.

All three of these characteristics have helped shape the land and its various forms, the variety and features of Australian ecosystems, and the specific behaviours and features of the plant and animal species found across the continent.

When humans arrived in Australia, these characteristics continued to shape the interactions people had with the environments they found. Indeed, these three characteristics still shape the interactions that we have today with the Australian environment.

Worksheet
7.1c Geology

LEARNING ACTIVITY 7.1B

Geology

Find out about the geology of a place you've visited or are about to visit, or of a place local to you.

1 How did it come to be like it is? Take some photos or make some sketches of the layout of the place, and find out about the time periods and processes that were involved in its formation.

2 How do we know what we know about this place?

GEOLOGICAL STABILITY

There are no major faults across the Australian mainland – earthquakes are rare and usually low in intensity. The boundaries of the Indo-Australian Plate, on which the continent sits, are nowhere near the mainland, and so Australia has been mostly geologically stable (at least since the separation of Gondwana).

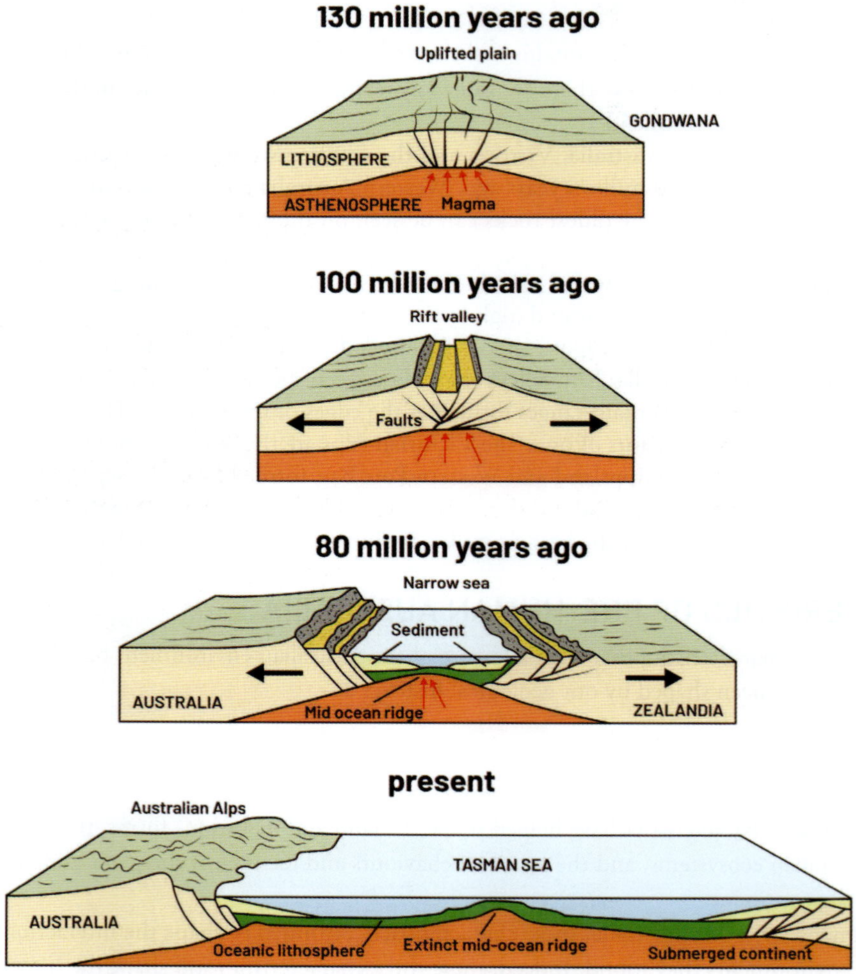

Figure 7.10 The development of the Australian Alps

One of the main consequences of this geological stability is that Australian soils are very nutrient-poor. Because of the relative lack of volcanic and other tectonic activities, which normally help to recycle rocks and soils, Australian soils have weathered and aged and become progressively poorer in the nutrients that are important for plants. Of course, Australia still has a lot of plants, but the relatively poor soils have led to the dominance of particularly tough species, especially eucalypts.

Resource
Additional Case Study: The geology of eastern Victoria

Geological stability has also meant that the dominant geological process in Australia has long been erosion due to water, ice and wind – large-scale examples of mountain building or other catastrophic processes are quite rare (although there are a number of volcanic sites across Victoria and in some other parts of the continent).

Recent research, however, suggests that the Australian Alps are not just flat old, eroded mountains but rather a dynamic new mountain range sitting atop fault lines that have re-awoken over the past six to 10 million years. Indeed, the very process of uplifting the Australian Alps that began as New Zealand began to shear off Gondwana 80 million years ago is still occurring and growing our Alps. Before the uplift caused by New Zealand's separation, the Australian continent experienced a 200-million-year period of geological stability while located in the middle of Pangaea and then Gondwana. The mountain ranges were slowly worn down to a low-lying plain and very few rocks of this period are preserved due to erosion, further demonstrating our continent's stability. Australian outdoor environments continue to be viewed as erosion-dominated largely flat places with old soil and little geological shaping characteristics.

BIOLOGICAL ISOLATION

The separation of Australia and Antarctica about 50 million years ago marked the time when Australian plant and animal species lost their ability to interact with other species. From that time, Australia has been an island. Apart from bird migrations, and perhaps the occasional swimming castaway from South-East Asia, the species of Australia haven't been able to naturally mix, compete and relate with any other species than those already here (or those introduced by humans).

One main and obvious implication of this characteristic has been the dominance of marsupial mammals in the Australian landscape. Marsupials are found in other parts of the world, but in Australia, with no significant competition from placental mammals, the marsupials (kangaroos, koalas, wombats, devils and others) had the whole place to themselves.

As Charles Darwin explained, where species modify over time in isolation, such as on an island, and favourable adaptations repeat, we will see a slow changing of species better adapted to a changing climate. Other consequences of biological isolation include:

- a high percentage of **endemic** species found in Australia (that is, species that are not found anywhere else). Species endemic to Australia include:
  - 95 per cent of our fungi
  - 85 per cent of our land mammals
  - 89 per cent of our reptiles
  - 94 per cent of our frogs
  - 91 per cent of our flowering plants
  - 57 per cent of our mangrove species
- a large diversity of different plant and animal groups
- wildlife of major evolutionary importance; for example, Australia has 12 of the 19 known families of primitive flowering plants, two of which occur nowhere else.

endemic
a feature or species that is unique to a defined geographic location

Figure 7.11 The *Austrobaileya* vine, a primitive flowering plant, is unique to Australia.

CLIMATIC VARIATIONS

Australia is generally regarded as the driest inhabited continent (Antarctica is the driest, but has no permanent human inhabitants). The large areas of desert and arid lands has been significant in the development of Australia's flora and fauna, but the variation in Australia's climate over time has perhaps been more important in shaping the ways plants and animals have evolved and adapted.

Australia (like the rest of the world) has endured many large-scale climate changes, including ice ages and periodic warming events, over many millions of years. However, for at least the past 10 000 years, a periodic change in ocean temperatures off the western coast of South America has led to an erratic climate felt on both sides of the Pacific Ocean, including the Australian continent.

El Niño–Southern Oscillation and Indian Ocean Dipole

The two most significant influences on climatic variation in Australia are the natural cycle of the El Niño–Southern Oscillation (ENSO), centred in the tropical Pacific Ocean, and the Indian Ocean Dipole (IOD) centred in the Indian Ocean.

El Niño–Southern Oscillation

neutral phase where warming winds towards the western Pacific keep the central Pacific Ocean relatively cool, associated with average rainfall and temperatures for eastern Australia

El Niño extensive warming of the central and eastern tropical Pacific, associated with an increased probability of drier conditions in Australia

The ENSO cycle loosely operates over timescales from one to eight years, shifting from the **neutral phase** to El Niño (think dry and hot) to La Niña (think wet and cool). Indirect evidence (such as coral samples) indicates that large century-to-century changes in ENSO activity are not unusual, and the International Panel on Climate Change report of 2021 notes 'low confidence that the recent high levels of ENSO activity represent a long-term trend beyond the limits of natural variability'.

The name **El Niño** comes from the Spanish term for 'Christ child', since in South America the effects of this pattern were initially more noticeable around Christmas. The ENSO results in drier-than-normal conditions in winter and spring in eastern Australia. Typically, this sees an increase in droughts and the risk of bushfires, which cause haze and a decrease in air quality. These events occur on average two or three times per decade. According to the Bureau of Meteorology:

> “Nine of the ten driest winter–spring periods on record for eastern Australia occurred during El Niño years. In the Murray–Darling Basin, winter–spring rainfall averaged over all El Niño events since 1900 was 28% lower than the long-term average, with the severe droughts of 1982, 1994, 2002, 2006 and 2015 all associated with El Niño.
>
> ”

See a diagram explaining El Niño Figure 4.10 in chapter 4 on page 169.

La Niña

La Niña extensive cooling of the central and eastern tropical Pacific Ocean, associated with an increased probability of wetter conditions in Australia

La Niña is the reverse process to El Niño. When it occurs, Australia has more significant periods of rainfall and cooler daytime temperatures. This climatic variation often leads to extensive flooding, as observed in Eastern Australia in the spring and summer of 2022. According to the Bureau of Meteorology:

> “The six wettest winter–spring periods on record for eastern Australia occurred during La Niña years. In the Murray–Darling Basin, winter–spring rainfall averaged over all 18 La Niña events (including multi-year events) since 1900 was 22% higher than the long-term average, with the severe floods of 1955, 1988, 1998 and 2010 [as well as 2022] all associated with La Niña.
>
> 

See a diagram explaining La Niña Figure 4.11 in chapter 4 on page 169.

Indian Ocean Dipole

The Indian Ocean is to Australia's west and, like the Pacific Ocean to the east, sea surface temperatures impact rainfall and temperature patterns over Australia. Warmer than average sea surface temperatures in the Indian Ocean can provide more moisture for frontal systems and lows crossing Australia.

The IOD has three phases: neutral, positive and negative. Events usually start around May or June, coinciding with the winter crop growing season, and peak between August and October. Positive phases of the IOD are associated with lower winter and spring rainfall in many parts of Australia, and negative phases with high rainfall. In 2019, which was Australia's driest year on record, one of the strongest known positive IOD events occurred; conversely, the strong negative IOD event of 2016 saw an exceptionally wet May–September over much of the country.

Our continent

Australia's variable climate and weather has led to a continent where:

- drought is common, leading to plants and animals developing strategies to conserve or reduce the requirement for water
- bushfires are common, and plants both fuel these and have adapted to them
- floods occur intermittently, leading to large-scale bird migration patterns to take advantage of occasional wetlands.

Archaeological studies suggest that fire has been a major part of the Australian environments since the final break-up of Gondwana. We can see the cultural importance of fire events in our recent history with the way that we name many of the biggest of these events:

- Black Thursday – 1851
- Red Tuesday – 1898
- Black Sunday – 1926
- Black Friday – 1939
- Ash Wednesday – 1983
- Black Saturday – 2009.

Australian plants are very heavily adapted to survive (and in many cases thrive from) large-scale fires. Some adaptations of Australian plants for fire conditions include the following:

- Eucalyptus species (gum trees) produce shoots from burned trunks.
- The ash and charcoal created helps to provide nourishment to young trees.
- Eucalypts also produce **volatile** leaf oils and highly flammable bark, and drop large amounts of leaves as litter. These features are thought to help promote and expand fires, which may help to kill off competitors for the soil nutrients, water and sunlight.
- Eucalypts are sometimes known as 'suicide trees'. *Eucalyptus regnans* (mountain ash trees) regenerate from seeds that germinate after fires. The sprouts require large amounts of light – more than would be available with a lot of tree cover, which of course a fire has likely removed.

iStockphoto/Photon-Photos

Figure 7.12 This area was affected by a bushfire and is showing signs of recovery

volatile
of a substance; easily evaporated to produce a flammable vapour

Resource
Additional Case Study: Platypus – Part bird, mammal and reptile?

Worksheets
Additional Worksheet: 7.1d Fire or other natural disasters in a specific environment

Additional Worksheet: 7.1e Melbourne Museum Forest Gallery

Table 7.1 Examples of adaptations of Australian plants and animals to this climate

Plants	Animals
• Hard, glossy leaves that resist water loss through transpiration • Evergreens that reduce energy usage • Long, narrow leaves that hang vertically, presenting a small surface area to the sun • Small leaves, or none • Dormancy	• Water storage • Metabolic adaptations to minimise water loss • Energy-efficient movement (e.g. hopping) • Energy-saving adaptations (e.g. koala's brain)

NOTES FOR THE EXAM

For the exam, you should:

- know some examples of adaptations of plant or animal species that suit Australian conditions
- be able to connect at least one of the three conditions (biological isolation, geological stability and climatic variation) to an environment you have visited.

Weblink
Melbourne Museum – 'Forest Secrets'

Worksheet
7.1f El Niño and La Niña

LEARNING ACTIVITY 7.1C

El Niño and La Niña

These ocean currents are well described by Australian weather services for their effects on the weather in eastern Australia. Using Figures 7.13 and 7.14 as a reference, explain to a person new to these concepts what is happening in the images and why.

AAP Photo/Mark Parden

Figure 7.13 Bushfires of 2019

Fairfax Photo/JUSTIN MCMANUS

Figure 7.14 Floods of 2022

7.1 KEY CONCEPTS

- Several significant events occurred in the prehistory of Australia:
  - 200 million years ago – Gondwana separates from Pangaea
  - 100 million years ago – Gondwana begins to split apart
  - 80 million years ago – New Zealand separates from Australia
  - 50 million years ago – Antarctica and Australia separate, creating an uplift and the Great Dividing Range
  - 50 000–10 000 years ago – Australia has a land connection with South-East Asia
  - 13 000 years ago – Tasmania disconnects from the mainland following the flooding of the Bass Strait
  - 8000 years ago – New Guinea disconnects from the mainland following the flooding of the Torres Strait.
- Megafauna roamed Australia in vast numbers up until the end of the last ice age (between 7000 and 11 700 years ago) when Australia became a warmer, drier continent.
- As an island country, Australia's flora and fauna have been isolated from species on other continents for over 60 000 years, meaning our species have evolved independently, with many endemic and unique species.
- Australia's location in the middle of a tectonic plate means the land hasn't experienced powerful landscapeshaping forces like volcanoes since it was part of Gondwana; equally, earthquakes are rare and of low intensity. This means our soils are not revitalised and renewed. The dominant geological process in Australia has long been erosion and weathering, which has created old and nutrient-poor soils, leading to dominant tough species like eucalypts.
- The climate of Australia has always changed, depending on our location in relation to the south pole and the rise and fall of global ice ages. In our present-day position (slowly shifting northward) our predominantly warm and dry climate is shaped by ocean temperatures and currents we identify as El Niño and La Niña.
- Evidence suggests fire has shaped Australian flora since Gondwana times, and is still seen today through heavily adapted species that dominate the landscape.

7.1 CONCEPT QUESTIONS

Worksheet
7.1 Key concepts

REMEMBERING

1 What is Charles Darwin's theory of evolution and how does it relate to Australia's biological isolation?

2 Name the ocean current that brings significant periods of rainfall and floods to Australia.

UNDERSTANDING

3 Explain how the El Niño ocean current influences Australia's climate.

4 Compare the characteristics of the Himalayan mountain range and the Australian Alps.

APPLYING

5 Using an example not presented in the text, explain how an Australian animal has adapted to its environment.

6 Describe the shaping characteristics that have led tough eucalypt plant species to dominate Australia.

EXTENSION CHALLENGE

7 Read the additional case study on the 'PLATYPUS – PART BIRD, MAMMAL AND REPTILE?' on Nelson MindTap. After referring to the platypus's genomic lineage in the case study, research another animal unique to Australia and construct a diagram that shows its lineage.

7.2 INDIGENOUS RELATIONSHIPS WITH THE ENVIRONMENT

KEY KNOWLEDGE

- relationships with outdoor environments expressed by specific Indigenous peoples' communities before and after European colonisation

KEY SKILLS

- analyse the changing relationships with Victorian outdoor environments expressed by specific Indigenous peoples' communities before and after European colonisation

The Traditional Owners of what we now identify as Australia have a rich and unrivalled connection and relationship with this continent. They were the first to walk upon this land and continue to do so as modern Indigenous peoples, celebrating their customs and beliefs, and sharing their 'Country' with non-Indigenous peoples. In this section, we will learn about:

- how specific Indigenous peoples express this relationship and how this relationship has changed since colonisation
- Indigenous peoples' relationships with the environment before non-Indigenous settlement
- the arrival of the first Australians, based on modern non-Indigenous understandings.

AAP Photo/ GUNDJEIHMI ABORIGINAL CORPORATION

Figure 7.15 The Madjedbebe rock shelter in Arnhem Land

In our modern context, researchers often seek to establish facts via the scientific method; that is, conducting research – often through observation, testing and experimentation – to objectively establish understandings in a particular field. This Western approach to knowledge allows a non-Indigenous person to navigate the world they live in and understand it on their own terms. For example, research and discovery to inform modern non-Indigenous understandings of the arrival in Australia of the First Peoples occupies a large amount of time and effort in many universities. Archaeological teams around Australia are spending hundreds of hours digging and identifying objects to provide evidence of the earliest human settlement of this continent.

The Madjedbebe (formerly called Malakunanja II) rock shelter in Arnhem Land has been dated to around 55 000 years old. It's located near a likely early entry point to the continent – given geological evidence of land bridges with South-East Asia – so it is currently our best indication of when humans first settled in Australia. Even if it underestimates the date, Madjedbebe still represents one of the longest continuous settlements of one place anywhere in the world.

However, despite the traditional use of the scientific method, researchers are starting to incorporate Indigenous knowledge drawn from oral histories into their work. For example, in their report to Parks Victoria about an archaeological study of a rock shelter at Mt Cope, on the Bogong High Plains, archaeologists highlighted the importance of balancing archaeological evidence with the oral traditions of Indigenous peoples. The report states:

Closer to current times, we have rock shelters on the Bogong High Plains that have been investigated and carbon dated as having been used as recently as 5000 years ago. This amongst hundreds of rock shelter (carbon dating middens) and art carbon dating projects across the nation confirm an arrival of the first Australians thousands and thousands of years ago.

The story, however, from the oral tradition of the Indigenous people of Australia tells us of a relationship with outdoor environments far older than any modern scientific measurement technique.

Mt Cope No. 6 Rockshelter, The Archaeological Investigation Shawrcross et al, 2006

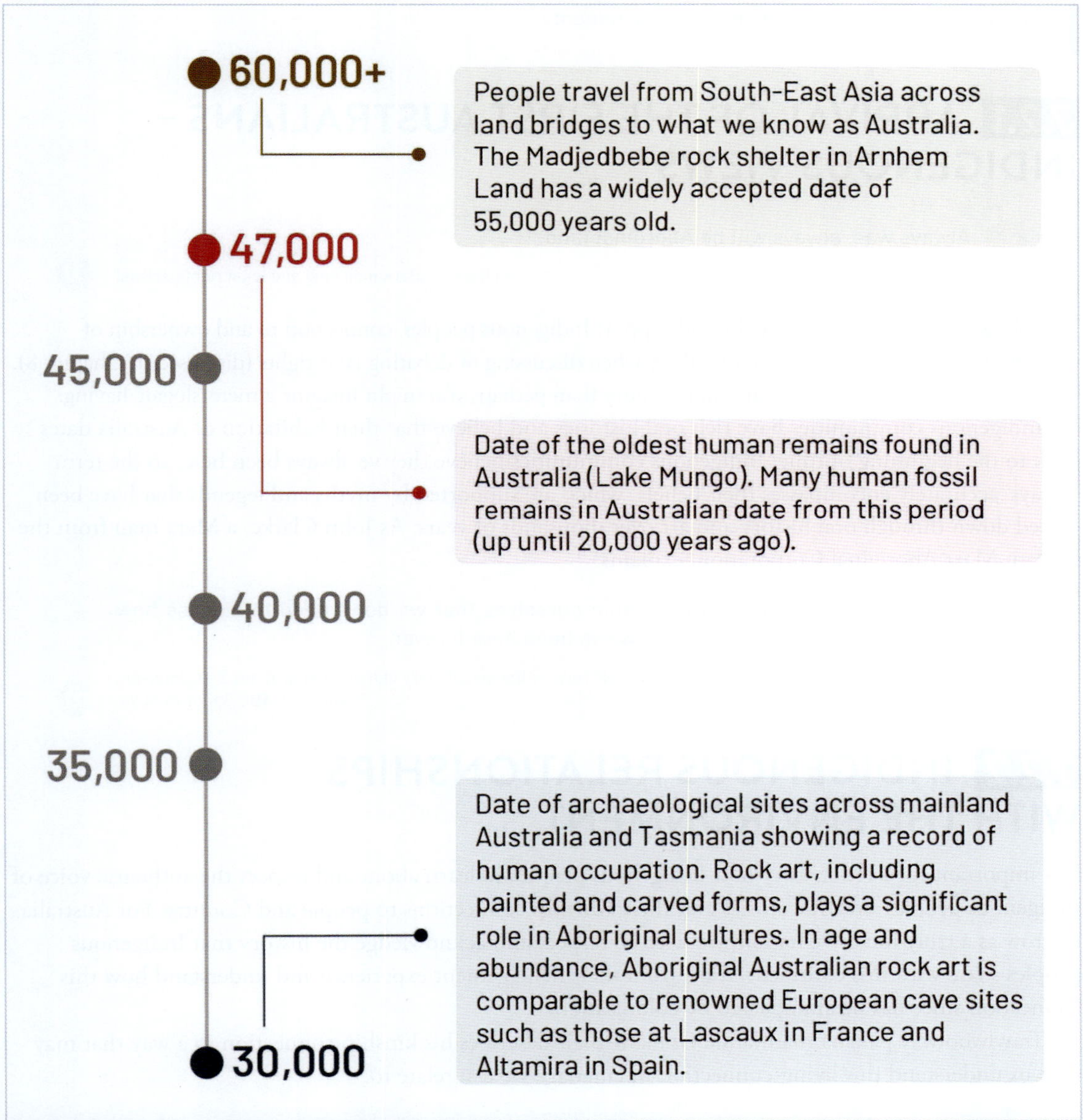

60,000+

People travel from South-East Asia across land bridges to what we know as Australia. The Madjedbebe rock shelter in Arnhem Land has a widely accepted date of 55,000 years old.

47,000

45,000

Date of the oldest human remains found in Australia (Lake Mungo). Many human fossil remains in Australian date from this period (up until 20,000 years ago).

40,000

35,000

Date of archaeological sites across mainland Australia and Tasmania showing a record of human occupation. Rock art, including painted and carved forms, plays a significant role in Aboriginal cultures. In age and abundance, Aboriginal Australian rock art is comparable to renowned European cave sites such as those at Lascaux in France and Altamira in Spain.

30,000

Figure 7.16 Timeline for human arrival on the Australian continent

Worksheet
7.2a Research

Weblink
CSIRO, 'The Moyjil site, south-west Victoria, Australia: prologue – of people, birds, shell and fire'

LEARNING ACTIVITY 7.2A

Research

Near Warrnambool, at a site named Moyjil, **middens** have been uncovered and are being dated via modern techniques. Early results indicate their age as between 115 000 and 130 000 years old.

Search on the CSIRO Publishing website for the article 'The Moyjil site, south-west Victoria, Australia: prologue – of people, birds, shell and fire' by John E. Sherwood. Read this paper, then do your own research to learn what other updates are being made with this project. While it is early days, this and other projects around the country are working to establish the rich timeline of Indigenous peoples' connections to this continent.

middens
a mound or deposit containing shells, animal bones and other refuse that indicates the site of a human settlement

7.2.1 ARRIVAL OF THE FIRST AUSTRALIANS – INDIGENOUS VIEWS

> Always was, always will be Aborginal land.
>
> Jim Bates, a Barkandji land and legal rights activist

This phrase is used widely to describe and support Indigenous peoples' connection to and ownership of Country in our modern context, particularly when discussing or debating land rights (discussed in Chapter 8). But this phrase has more power and more history than perhaps you might imagine a mere 'slogan' having.

Indigenous communities have rich oral histories and believe that their habitation of Australia dates back to the beginning of time. Indigenous communities believe they've always been here, so the term 'always' accurately encompasses their beliefs, which are supported by myths and legends that have been passed down through oral history and art over thousands of years. As John Clarke, a Mara man from the Eastern Maar Aboriginal Corporation, explains:

> There's an unwritten knowing within ourselves that we don't need to discuss how long we've been here – we know we've been here forever.
>
> 'Ancient Aboriginal site Moyjil could rewrite the global story of human migration', Sian Johnson, 5 Sep 2020, ABC South West Vic

7.2.2 INDIGENOUS RELATIONSHIPS WITH THE ENVIRONMENT

Kinship
an Indigenous person's relationship and responsibilities to other people, to their Country and to natural resources

It is important for 21st-century non-Indigenous peoples to learn about and respect the authentic voice of Indigenous peoples' creation stories and their **Kinship** connections to people and Country. For Australia to grow as a truly inclusive nation, we need to respect and acknowledge the history that Indigenous peoples share with this continent, learn from their management experience and understand how this connection since has been impacted by colonisation.

Trawlwoolway pakana man Jamie Graham-Blair describes his kinship connection in a way that may help us understand this living-connection in language we can relate to.

> Every molecule in our body, every atom, every protein, every acid comes from somewhere, right? Our physical being is made up of hundreds and thousands, millions and billions of little tiny components that we get from the air we breathe, the water we drink and the food we eat. Every plant you see here, every bird you hear singing ... this stump that we're sitting on. It's very possible that we're sitting on atoms that were within my Old People, my actual ancestors.
>
> 'Indigenous people often talk about the importance of Country. But what does it mean?' Anna Salleh, ABC Science, 21 Sep 2021

DREAMING

'Dreaming' or 'Dreamtime' are English words that describe a rich Aboriginal and Torres Strait Islander concept. In reality, it is impossible to find words that adequately capture this core element of who we are but it's something you feel when you sit with us on our Country and hear our stories with an open mind and heart.

Dreaming is more than a mythical past; it prescribes our connection as Aboriginal people with the spiritual essence of everything around us and beyond us. Dreaming stories are not in the past, they are outside of time – always present and giving meaning to all aspects of life.

Across Australia, we have different words for this concept, including Tjukurpa in central Australia, Bugari in Broome, and Wongar in north-east Arnhem Land. Many Aboriginal and Torres Strait Islander people prefer the use of the word Dreaming over Dreamtime. The Dreaming is passed from generation to generation through stories, song, dance and art. This knowledge gives us special responsibility and is seen as a great honour. For example, when we are taught a story or song about the travels of ancient creation spirits and ancestors from water-hole to water-hole, we become the holder of this essential knowledge.

LEARNING ACTIVITY 7.2B

A Dreaming story

Read the Ngurunderi Dreaming creation story of the Murray River on the Murray River website.

Write your own story of the Dreaming to explain the creation of an environment, or a particular feature in an environment, that you've visited or studied.

Worksheet
7.2b A Dreaming story

Weblink
Ngurunderi Dreaming creation story of the Murray River

INDIGENOUS RELATIONSHIPS – PERCEPTIONS

Indigenous peoples' measurement of the time that they have lived here is informed by the Dreaming. With this wealth of time available, their perceptions are founded in the Dreaming and they have grown and developed intricate knowledge about the Australian outdoor environments that they lived in.

There are over 200 distinct First Nations cultural groups across Australia, encompassing a host of different languages. Narratives about the origin of the universe differ between groups, but there are also similarities. For example, many of these origin stories feature giant serpents, but each group has a unique name and narrative. Many Australians will have heard of the Rainbow Serpent, which is discussed in Chapter 1 of this text, and also discussed later in this chapter. A common feature of Victorian creation stories is Bunjil, the wedge-tailed eagle, as discussed in the Learning activity 7.2C.

To see the AIATSIS Map of Indigenous Australia and explore it in detail visit the Australian Institute of Aboriginal and Torres Strait Islander Studies (AIATSIS) website.

Weblink
Map of Indigenous Australia

Worksheet
7.2c Bunjil, the 'creator'

LEARNING ACTIVITY 7.2C

Bunjil, the 'creator'

Read the creation story about Bunjil at the beginning of Chapter 1. Many Aboriginal groups in Victoria have creation stories that focus on their own special relationship with Bunjil and are described differently from group to group.

For this activity, research and identify a specific Aboriginal peoples' group whose land you live on or have visited for an outdoor experience and discover and share a creation story specific to that group.

Indigenous Australians have established effective ways to use and sustain resources, informed by their belief that their ancestors are a part of each feature of the landscape as well as living within the landscape for thousands of years. The following quote from the Gunaikurnai Land and Waters Aboriginal Corporation explains how central Country is to their culture:

> We have one of the oldest cultures in the world and that culture has been passed on through many generations. Our culture is embedded in our Country, which is vital to our identity. Our stories and songlines link us to our ancestors, who travelled across Country practising the customs that make us Gunaikurnai.
>
> © 2023 Gunaikurnai Land and Waters Aboriginal Corporation

- A key feature of Indigenous communities across the world is a strong focus on oral history – the passing of knowledge and beliefs through Elders and the use of song and traditional dance.

- Indigenous peoples of Australia complemented their oral culture with the use of cave paintings and other forms of rock art as a means of recording thoughts, ideas, events and histories.

- Rather than expressing ownership of a particular area, Indigenous peoples operate in a system of belonging within the land. Dreaming beliefs describe the right of certain people to control the use of resources in a particular area.

- Totemism is a deep cultural and spiritual value that plays an important part in Indigenous peoples' perceptions of outdoor environments. It connects human beings to other animals, plants and aspects of nature. Groups and individuals are assigned a particular animal that they are related to and have to care for. This gives them a profound sense of connection to and responsibility for the natural world.

- The pre-colonial life of Indigenous communities helped to fuel this connection with the land. They lived completely in an environment. There are no permanent walls separating people from the land, and no disconnection between land and people. As a result, the land is not separated from other aspects of life.

- This deep connection with the land helped to build a perception in Indigenous communities that the land was simultaneously both their protector and something to be protected.

Figure 7.17 Key points of Indigenous cultures

Note: These general themes, which may not be relevant for every Indigenous nation, shift and change with the effects of colonisation and contemporary values.

LEARNING ACTIVITY 7.2D

Worksheet
7.2d Little Things

Little Things

Listen to or watch Ziggy Romo's song 'Little Things' , which is a reinterpretation of the famous Paul Kelly and Kev Carmody song 'From Little Things Big Things Grow'.

How does Ramo describe the arrival of the First Fleet and colonisation of Australia?

Weblink
Video: Ziggy Ramo 'Little Things'

INDIGENOUS RELATIONSHIPS – INTERACTIONS

Studies of Indigenous societies show that they spend only a small number of hours each day actually hunting for and gathering food, so there's a lot of downtime. We can imagine that the sort of things that take up some of our leisure time – conversations, arguments and jokes with friends and others, singing along to favourite tunes, and so on – likely took up some of the leisure time of people in early Indigenous communities as well.

For this course, we are particularly interested in the practices of Indigenous communities that were connected to their interactions with, and uses of, the environment around them. These practices, of course, varied from place to place. As a simple example, for the communities living in and around the Victorian Alps, it made sense to use the furs from animals they hunted to clothe themselves. In the hot north and west of the country, this would have made no sense.

For many Indigenous communities, there are six key practices we can examine that shed some light on their relationships with their environment.

Food security

Hunting and gathering in more modern communities is often divided on a gender basis, with the men spending time hunting and women and children spending their time gathering. This seems to be the way most Indigenous communities allocated these tasks as well.

hunting and gathering of a community or society that hunts animals for meat and other useful materials, and gathers wild fruit, vegetables, roots, nuts, grasses and other edible plants

Exactly what communities would hunt for and gather depended on where they were and what types of plant, fish and animal species were present in these areas, but we see evidence for hunting of kangaroos, wallabies and other mammals; emus, crocodiles, lizards and snakes; and fish. Gathering included a huge range of edible plants, many of which we now recognise by the term 'bush food'. However, many of these (e.g. the Yam daisy) have all but disappeared from the landscape due to grazing by introduced animals.

Hunting for and gathering food develops a very close connection with a place. When you can't just drop into the local supermarket to buy your next meal, but must instead find it, kill it and prepare it, this changes the way you look at the environment around you. Where are the best places to hunt for a particular animal? Where might be the best places to look for a tasty plant? And at what time of year? Hunting and gathering requires that people know these things intimately. If they don't, they die.

LEARNING ACTIVITY 7.2E

Worksheet
7.2e Bush food

Bush food

Find out about some plants from your local area that are considered edible and might be part of what we would call bush food.

Develop a menu that could sustain an active person for a day using only bush food that could be gathered locally.

Mobile, semi-sedentary and sedentary communities

mobile
communities that move across large distances and to many different locations within their Country, mostly in arid environments

The majority of Aboriginal people who did not live in Australia's arid regions were not nomadic or **mobile**. As we know and now understand from extensive Indigenous sharing of their knowledge, Traditional Owners lived on Country that was their tribal lands – their property or estate.

This meant that communities would move around in clans made up of extended family groups and utilise its resources for food and shelter as required season by season, living lightly on the land, and meeting and trading with other clans of their tribe, with clear knowledge of their geographical boundaries, sacred sites and territorial rules. For example, if a particular animal species that was being hunted was moving into a different area, the community would move to follow it. Or they might find that hunting and gathering in one place has depleted the resources available, and so try to find better supplies somewhere else. Or maybe the environment changed, perhaps with a change in season (getting hotter or colder), and a move to another place might improve a community's level of comfort. The early records of European settlers and explorers show that many Indigenous communities across Australia moved from place to place following resources or as the seasons changed, erecting shelters and structures as they travelled. The exception was where permanent camps were constructed in places with favourable resources and climate (e.g. at Lake Condah in western Victoria by the Gunditjmara people). Such **sedentary** communities also allowed infirm or younger members of a tribe to remain year-round.

sedentary
of communities or community members that remain in a single permanent location (referred to as tribal base camp or village) with fresh water and plentiful year-round food resources

Alpine Indigenous communities and the Wurundjeri around Melbourne are good examples of **semi-sedentary** groups. The following quote from the Taungurung Land and Waters Council website provides an insight into their traditional migration across their Country.

semi-sedentary
communities that move from one location to another and back again in regular cycles within their Country

> The Taungurung people occupy much of central Victoria. Our country encompasses the area between the upper reaches of the Goulburn River and its tributaries north of the Dividing Range. From the Campaspe River to Kilmore in the West, eastwards to Mount Beauty, from Benalla in the north down to the top of the Great Dividing Range, our boundaries with other Aboriginal tribes are respected in accordance with traditional laws. Traditionally, our people lived a hunter/gatherer existence. The various clan groups migrated through their territory dependent upon the seasonal variations of weather and the availability of food.
>
> Taungurung Clans, 2021 Taungurung Land & Waters Council

To categorise Indigenous peoples' migration patterns as simply drifting aimlessly across the landscape, as was the case with early European settlers, diminishes their connection to and ownership of Country. It also reflects a lack of understanding of this complex connection and a lack of respect towards land rights.

Agriculture

In recent years, a number of historical and archaeological findings have given rise to the ongoing development of a new perspective on the practices of many pre-European Indigenous communities, including those spread throughout Victoria.

Historian Bill Gammage, in his book *The Biggest Estate on Earth*, and Indigenous writer Bruce Pascoe, in his book *Dark Emu*, both describe and discuss the growing evidence that many Indigenous communities practised large-scale and long-term forms of agriculture. Pascoe, citing much of his evidence from the eyewitness accounts of early European settlers and explorers, describes communities that were practising plant domestication, sowing, irrigation, harvesting and storing of foodstuffs.

For Gammage, this is evidence of a sophisticated form of land management developed and refined over many thousands of years of engagement with Australian environments. Pascoe goes further to suggest that we need to reconsider the use of the term 'hunter–gatherer' for Indigenous practices.

His view is that by calling the pre-European Indigenous communities hunter–gatherers, we are minimising, and in some ways trivialising, their engagement as land managers of Australian environments (also see the 'Mobile, semi-sedentary and sedentary communities' section earlier).

In his 2022 book, *Gunyah Goonide + Wurley: The Aboriginal Architecture of Australia*, Paul Memmott richly describes the diversity of permanent and semi-permanent structures constructed by Indigenous peoples all over Australia, including Victoria, as further evidence of a more static inhabitation of Australia than the established and narrow view of a 'nomadic' way of life. Memmot asserts:

> "The mobile hunter-gatherer lifestyle often resulted in relatively impermanent architecture. But for those in localities where there were plentiful food resources, there was a tendency to establish more permanent villages, which were often seasonally occupied."

In south-western Victoria, evidence of thousands of years of aquaculture can be seen across the landscape, where the Gunditjmara people used volcanic rock created by the Budj Bim lava flow to create a complex of fish traps, weirs, dams and channels. This organised practice ensured year-round food and encouraged the Gunditjmara people to settle, build stone houses and develop more sophisticated fish traps. These eel traps are now part of a UNESCO World Heritage site and represent the practice of keeping more permanent shelters by some Indigenous groups; in this case, stone houses with timber and thatch roofs strong enough for a human to stand on.

Samuel Calvert, ***Natives Fishing*** (c.1873) wood engraving National Library of Australia

Figure 7.18 A wood etching from 1873 by colonial artist, Samuel Calvert provides insights into Indigenous fishing techniques

9780170477123

Figure 7.19 Gunditjmarra Aboriginal site Budj Bim is considered one of the oldest aquaculture systems in the world. The Budj Bim Cultural Landscape is a UNESCO World Heritage site.

Figure 7.20 Replica stone shelters built and used by the Gunditj Mirring people.

Hardware and clothing store

Outdoor environments provided all the raw resources that Indigenous peoples needed, and they applied their own techniques to forming these materials into products that made life easier, tastier or warmer.

- **Clothing** – Many parts of Australia get cold, especially alpine areas in Tasmania and Victoria and other southern parts of mainland Australia. Clothing prevents heat loss from the body. First Nations Australians have long made cloaks from animal furs to control heat loss. In cool climates, clothing was made from the furs of a variety of animals, including wallabies, kangaroos, possums, platypuses and quolls. In the past, this clothing kept people warm and dry, but today these garments are used to signify status and are worn during ceremonies and special occasions. The possum skin cloak is often adorned with designs, sometimes displaying totemic identity and stories.

- **Separating techniques** – Indigenous communities selected, processed and used natural material for many purposes, including cooking and medicine. These techniques include hand-picking, winnowing, yandying, sieving, filtering, straining, cold-pressing and steam distillation.
- **Use of chemical and physical changes** – Indigenous peoples developed many sophisticated chemical processes, including fermentation and **pyrolysis**, which allowed them to produce required substances that were either scarce or unavailable, such as resins, pigments and salts.

AAP Photo/Julian Smith

Figure 7.21 Possum-skin cloak

pyrolysis the thermal decomposition of organic matter in a low-oxygen environment

Fire-stick farming

Indigenous peoples have shaped the landscape in Australia for thousands of years, but they did this in a way that was intrinsically sustainable, ensuring that it thrived and remained healthy. Indigenous peoples developed a continent-wide land management system using fire, often called '**fire-stick farming**', a practice that evolved over millennia.

fire-stick farming the consistent and repeated use of fire to clear vegetation and create open forests to ensure food for both people and wildlife

Indigenous peoples have used fire intentionally and frequently for tens of thousands of years at a relatively low intensity, and did not burn large areas. Lighting small areas of grass and scrubland in the cooler shoulder seasons to summer meant the fires burned lower to the ground in the understorey, clearing out the undergrowth to make way for fresh plants to prosper. This made it a natural part of the landscape and aided their life on the land. Fire-stick farming was used to:

- reduce fire loads and therefore reduce the risk of large, intense and dangerous wildfires, and to increase productivity of preferred foods
- create grasslands for kangaroos and for yam and grain harvest, interspersed with forested areas for possum habitats and for fruit harvest
- improve access through thick and prickly vegetation
- maintain a pattern of vegetation to encourage new growth and attract game for hunting
- encourage the development of useful food plants, for cooking, warmth, signalling and spiritual reasons
- hunt using 'the fire drive'. Fire could be a useful hunting tool because fire is dangerous. We recognise this and other animals do too. When faced with a fire, animals will try to escape, and if the fire is constructed in a suitable pattern, then the escaping animals may be herded directly into the path of a group of hunters.

Figure 7.22 Joseph Lycett, *Aborigines Using Fire to Hunt Kangaroos* (c. 1817)

This type of constant burning produced what we call 'cool burns', since the small amount of undergrowth tended to limit fires so they quickly burned out. No more land was burned than necessary. Burning was also more than just sound land management; it was also evidence that the land was healthy and being fully utilised. There was also a religious significance to burning based on the belief that ancestral spirits of the Dreaming still inhabited the land, and the burnings provided these spiritual inhabitants with lands on which they could hunt.

Sacred sites

All societies have sacred places. In our modern society, these sacred places might be a cathedral, church, mosque or synagogue (or perhaps the local Apple store!). Indigenous communities also had sacred places – what we call sacred sites. These places had many purposes for Indigenous Australians. Some were for the burial of their Elders, while others were for initiation ceremonies for adolescent boys and girls, marking their transition into adulthood.

Another purpose for these places has been discovered in recent years. Scientists and archaeologists have found interesting connections between some Indigenous sacred sites and the breeding grounds of particular species. At least some of these sacred sites were used as conservation zones or protected areas. Indigenous communities banned access to certain places that were breeding grounds for important food species, both plant and animal. This makes sense, since if you hunt in breeding grounds, you will eventually hunt a species to extinction. Alternatively, if you make the breeding grounds off limits for any sort of activity, then you allow your food animals to grow and mature, and move out into the rest of the environment – where you can hunt them. This is a smart and sustainable use of resources.

Sacred sites can be found throughout Australia, such as in the Kakadu National Park in the Northern Territory (see Figure 7.23). In Victoria, Gariwerd (Grampians National Park) is particularly significant with sacred sites spread throughout caves and rock formations, including shelters and rock art.

Gariwerd contains 90% of the rock art sites in Victoria, and is central to the Dreaming of the Djab Wurrung and the Jardwadjali peoples, the Traditional Owners of this area. Their descendants are still at Gariwerd and are involved in maintaining the culture and the stories of the land.

As students of Outdoor and Environmental Studies, and citizens of 21st-century Australia, we all need to continue to learn about the relationships of the early Australians to this ancient land, and use their example as models for our own engagement with environments.

AAP Photo/ DEAN LEWINS

Figure 7.23 Aboriginal rock art at Gariwerd (the Grampians).

LEARNING ACTIVITY 7.2F

A day in the life

Write a day in the life of an Indigenous person in an environment you've visited or studied. Include practices and perceptions relevant to the communities that lived in and around this environment. Consider researching how they made resins, used pyrolysis or created clothing from the pelts of animals.

Worksheet
7.2f A day in the life

Resource
Additional Case Study: Bogong moth harvesting

7.2.3 INDIGENOUS RELATIONSHIPS – IMPACTS

Hunter–gatherer societies are typically what we would call today 'low impact' or 'minimal impact', as is evident from the practices discussed above. These practices developed to be sustainable, which allowed them to be passed down from generation to generation. Certainly, beyond the occasional scarred tree, rock-art site, midden, stone hut walls, stone eel and fish traps or other small-scale feature that still exists, we see little obvious evidence of the incredibly long Indigenous associations across this country. Or do we?

Fire

We have looked at fire-stick farming as a practice. As mentioned, one of the key features that early colonisers noticed about the Australian environment was its park-like appearance, with trees and shrubs separated by large swaths of grass. Our bush today appears to be much more dense. One of the key differences between then and now is that the practice of controlled or cultural burning is highly restricted today, and occurs with far less frequency than it did for thousands of years under Indigenous management. The result is that we now experience more frequent and higher-intensity bushfires, fires that burn all the way into the crown of the forest trees and threaten homes and communities.

Could this process of cultural burning (sometimes known as the 'fire regime') have changed the Australian environment, perhaps even permanently? Certainly, there are many species of plants that are adapted to fires, needing the heat of a fire for seed germination and the ash afterwards for nutrients. Did Indigenous communities create this situation? The most likely answer is no, since the evolution of these fire-dependent features probably occurred many millions of years before humans arrived here. The constant burning of the early Indigenous communities may well have accelerated or enhanced features that were common in Australia, however, and the use of fire could have had large-scale impacts on the land.

Dingos

Over the course of many thousands of years, the Indigenous communities of northern Australia had ongoing trading relationships with communities in the countries that are now known as Indonesia, Timor and Papua New Guinea. South-East Asian fishermen are known to have travelled to Australia to trade fish, turtles and other food and goods.

DNA studies estimate that the dingo arrived on the Australian continent between 4700 and 18 000 years ago, representing perhaps the earliest example of human-assisted oceanic migration, and are closely related to native dogs in Asia. Dingo pups were taken from the wild when very young, and were highly valued as a ritual food source. Other pups were adopted into human society, growing up in the company of women and children. They were used as hunting aids and to provide warmth as a living blanket. They also guarded against intruders. More recent research also points to the importance of domesticated dingoes as companions to households of older people who no longer had children at home, where a deep emotional attachment existed between human and animal. The dingo also plays an important role in many Indigenous peoples' Creation stories.

As dingos are not native to Australia, they quickly became a threat to many other native species. It seems that the thylacine (Tasmanian tiger) probably became extinct on the mainland close to the time when dingos were introduced – it is likely that the competition for food and habitat from these dogs may have contributed to their disappearance, and perhaps to that of other species too.

istockphoto/SolStock

Figure 7.24 Dingo in the wild

LEARNING ACTIVITY 7.2G

Worksheet
7.2g Indigenous impact on an environment

Indigenous impact on an environment

1 Walk through a local environment and try to imagine what it might have been like 10 years ago. What about 100 years ago? And 1000 years ago? How about 10000 years ago? What signs are there of Indigenous practices in this area?
2 Using the same environment, or a more urban/built environment, try to imagine into the future 10 years, 100 years, 1000 years or even 10 000 years. What signs might there still be of the existence of modern Australian society?

Summarising Indigenous relationships before Europeans

While Table 7.2 is a useful starting point to thinking about Indigenous relationships with Australian environments, all relationships, Indigenous or non-Indigenous, are more complex than a few simple lists. At the very least, you should be able to take some of these points and think about them, analyse them, and consider what they mean both for the nature of the people who thought and acted in these ways, as well as their implications for people, including you and me, in Australia today.

Table 7.2 Summary of Indigenous relationships with outdoor environments

Perceptions	Practices and interactions	Impacts
• Spiritual connection with land • Land as mother • The creators of the land are their ancestors and they live in the features of the land • People are related to and part of the land	• Hunting and gathering • Mobile, semi-sedentary and sedentary • Fire-stick farming • Sacred sites and conservation zones • Agricultural practices	• Largely a low impact on the environment • Hunting may have contributed to the extinction of some of the megafauna • Introduction of the dingo may have contributed to impacts on some marsupials • Fire-stick farming and other agricultural practices may have changed forest environments

LEARNING ACTIVITY 7.2H

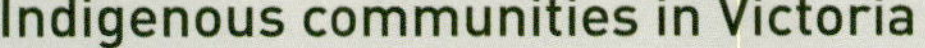

Worksheet
7.2h Indigenous communities in Victoria

Indigenous communities in Victoria

Find out about the relationships that a local Victorian Indigenous community had, and continue to have, with their environments on their Country. Find out about specific perceptions (including stories of the Dreaming), practices specific to this community, and some of the impacts on the environments local to this community.

After the arrival of Europeans

While much of the focus of this part of the course is to examine the relationships Indigenous communities had with outdoor environments before Europeans and others arrived, it's not as if these communities have just disappeared. Indigenous people had lives after colonisation and settlement by farmers and convicts, and they continue to have lives in Australia today. According to some historians, there were between 300 000 to 950 000 Indigenous people living in Australia when the British arrived in 1788. The most recent Census, in 2021, estimated the Indigenous population to be 881 600.

COLONISATION OF INDIGENOUS PEOPLES' LANDS

By the early 1800s, the growing colonies were restless for more land to expand their settlements and to use for grazing and to grow food. Over time, they began to explore to the north, west and south of the original British settlement in Botany Bay in search of space and pasture, and they found it – but at the expense of Indigenous peoples.

One of the most well-known expeditions is that of the Hume and Hovell, who travelled from greater Sydney (Appin) all the way to what is now known as Geelong.

The printed map depicted in Figure 7.26, published by the Victorian Department of Lands in 1924, shows the track taken by the expedition and their camp sites. This expedition travelled across the mouths of the north-east Victorian Kiewa and Ovens valleys, where William Hovell noted in his diary about navigating through rough, mountainous country, swollen rivers and impenetrable bush, and dealing with plagues of mosquitoes and sandflies, and the irregular supply of fresh game, such as kangaroos and waterbirds. He also recorded encounters with the Indigenous peoples of the region, noting land management techniques such as grass burning and damming of rivers to catch fish.

Figure 7.25 George Edwards Peacock, *The South Australian Alps as first seen by Messrs. Hovell and Hume on the 8th November 1824* (undated)

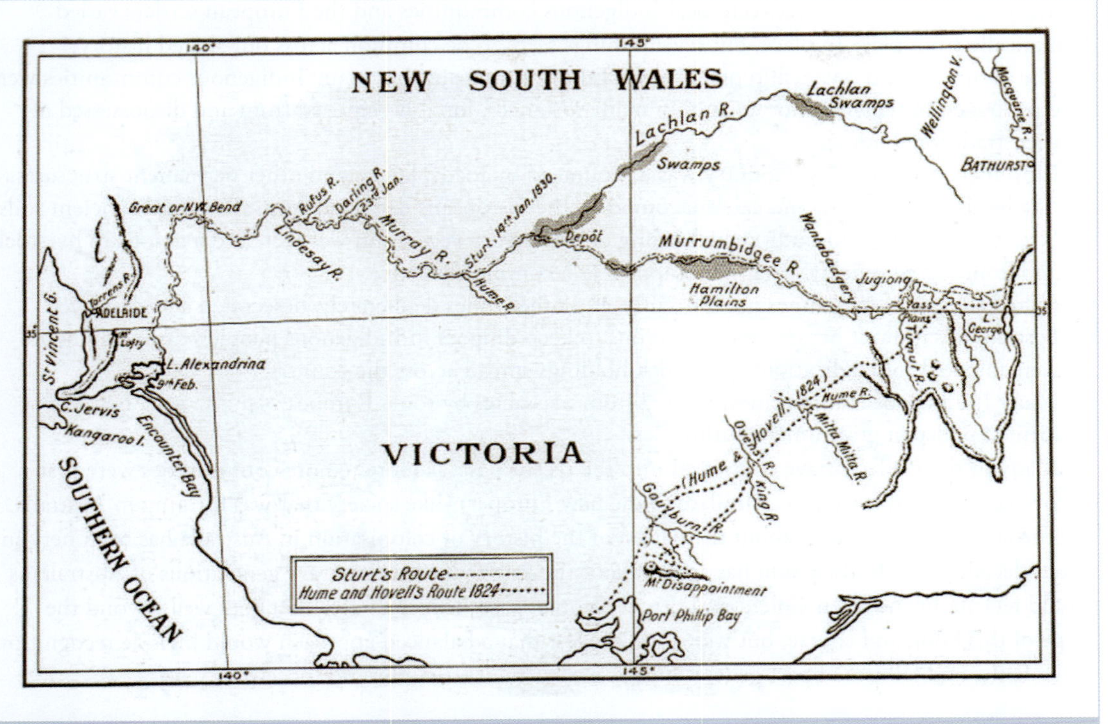

Figure 7.26 Hume and Hovell's route across Victoria in 1824

The Explorers of Australia and their Life-work Ernest Favenc, 1908

Many graziers established their properties following the route of Hume and Hovell. They seized and cleared land and built fences to demarcate property ownership. They added tracks, roads and fences. Fighting broke out between the settlers and Aboriginal peoples who once moved freely across the land. Tens of thousands of Aboriginal people were massacred in more than a century of frontier warfare.

WHO ARE THE TAUNGURUNG CLANS?

When Europeans first settled the region in the early 1800s, the area was already occupied by Taungurung people. From that time, life for Taungurung people in central Victoria changed dramatically and was severely disrupted by the early establishment and expansion of European settlement. Traditional society broke down with the first settler's arrival and soon after, Aboriginal mortality rates soared as a result of conflict, introduced diseases, denial of access to traditional foods and medicines.

Who are the Taungurung Clans? © Copyright 2021 - Taungurung Land & Waters Council

What came with this extending settlement of colonisers across Australia was an approach to land management that was in direct conflict with the long-established practices of the Indigenous peoples. As Bruce Pascoe recounts in *Dark Emu*, little visible evidence remained of Aboriginal civilisation by the 1860s. The members of Indigenous clans who previously occupied the land had perished in large numbers due to 'warfare, murder and disease', their agricultural lands wiped out by sheep and cattle, and their villages and storage facilities destroyed by nature or, in many cases, arson.

The conflicts that arose between local Indigenous communities and the European settlers varied dramatically from place to place, but several things seem to be common across our shared history:

- The nature of land ownership practised by Europeans meant that many Indigenous communities were encouraged initially to relocate and, in many instances, forcibly removed from and dispossessed of their traditional homes.
- Movement around tribal Country was discouraged by roads, fences and other permanent structures.
- The hard-hooved sheep and cattle favoured by the Europeans degraded Australia's fragile ancient soils.
- Fire-stick farming and traditional burning of understory vegetation was seen as dangerous to livestock and homes, and quickly began to disappear as a common practice.
- Sacred sites were sometimes inadvertently, and other times deliberately, desecrated and destroyed.
- Post-colonisation, it became too dangerous to keep dingoes in Indigenous peoples' camps. Dingoes were targeted for eradication as livestock holdings spread across the country.
- Many Indigenous communities suffered from massacres by some European landowners, newly formed police or government authorities.

Many of these effects have continued into the recent past, as more remote communities were first discovered and then forced to assimilate to the new European-like society that was forming in Australia.

contemporary events or actions that have occurred within the past 15 years

The language and debate about the nature of the history of colonisation in Australia has been fiery in recent decades. Much discussion has arisen about the way that **contemporary** generations of Australians should feel about the often violent and brutal colonial past. This is an area that goes well beyond the scope of this book and course, but we would suggest that a balanced approach would include recognition of the faults and failures of past generations, as well as their strengths and successes.

7.2.4 CONTEMPORARY INDIGENOUS RELATIONSHIPS WITH THE AUSTRALIAN ENVIRONMENT

The second half of the 20th century saw improvements (although agonisingly slow) in the ways Indigenous peoples and communities were treated in Australia. The assimilation policy that led to the practices of removing Indigenous children from their parents (known as the Stolen Generations) changed to the notion of 'integration' and led to the successful 1967 referendum to enable the Commonwealth to count Indigenous people in the national census, and make laws for Indigenous peoples.

reconciliation restoration of friendly relations, especially between Indigenous and non-Indigenous individuals and communities in Australia

More recently, the notion of **reconciliation** has characterised many efforts between Indigenous and non-Indigenous individuals and communities to forge improved relationships.

Figure 7.27 The beginning of the Bullawah cultural trail, Wangaratta. In many places across Australia, Indigenous Australian connections with the land are being recognised, moving forward from the challenges of the past.

Health, education and social issues

Colonisation has resulted in conditions where indigenous communities around the world suffer from poor physical and mental health, low education and income levels, lower life expectancy, alcohol and drug abuse, increased incarceration rates, domestic violence, and other related social issues.

Reduced access to Country and sacred sites, relocation of whole communities and the prevention of traditional practices are some of the factors that created further trauma. Conversely, the reinstatement of access to Country and the development of a new appreciation for Indigenous relationships of the past can help to revitalise and repair the relationships that Indigenous people have, not only with Country, but also with other Australians.

TRUTH AND JUSTICE IN VICTORIA

Victoria's truth and justice process will recognise and address historic and ongoing injustices and form a key part of the treaty process.

The Yoorrook Justice Commission is investigating historical and ongoing injustices committed against Aboriginal Victorians since colonisation, across all areas of social, political and economic life.

Aboriginal Victorians have been clear and consistent in the call for truth-telling as an essential part of the treaty process. In June 2020, the First Peoples' Assembly of Victoria (Assembly) passed a resolution seeking commitment from the State to establish a truth and justice process. In response, in July 2020 the Victorian Government committed to working with the Assembly to formally establish a truth and justice process

…

What is Truth-Telling?

It is widely acknowledged among First Peoples in Australia that we cannot talk about our shared future until we acknowledge our shared past. Through decades of activism, First Peoples have fought for truth-telling, to recognise the impacts of colonisation and address historic and ongoing injustices.

Truth commissions offer a formal and legitimate process for this to happen. Establishing a formal truth-telling process will assist reconciliation and healing for people harmed and their communities.

On 12 May 2021, the Governor of Victoria signed the letters patent, as required under the Inquiries Act 2014, to legally establish the Yoorrook Justice Commission as a Royal Commission and set its Terms of Reference.

…

Yoorrook is the Wemba/Wamba word for 'truth'.

NOTES FOR THE EXAM

For the exam, you should:

- be able to name an Aboriginal group or community that lived in an area you visited or are familiar with
- know about Indigenous relationships relevant to places you've visited or are familiar with, both before and after European colonisation.

Resource
Case Study: An invitation to understand Country

CASE STUDY

AN INVITATION TO UNDERSTAND COUNTRY

INDIGENOUS PEOPLE OFTEN TALK ABOUT THE IMPORTANCE OF COUNTRY. BUT WHAT DOES IT MEAN?

Under an hour from Melbourne, on Wurundjeri Country, is Organ Pipes National Park. It has basalt columns resembling organ pipes, which were formed a million years ago by one of the world's largest ancient lava flows.

This is the eastern edge of where the lava reached.

Wurundjeri elder, Aunty Di Kerr, talks about how she feels about her Country.

Family features strongly.

'My Country is my mother, so I belong to her.

'The land is her. The waterways are her veins.'

This sense of reverence and connection extends beyond her own Country and wherever she goes in Australia, she finds out whose Country she's on.

'The main thing is to always acknowledge that you're on Country, because acknowledging that and acknowledging their ancestors, it actually helps you feel safer on Country.'

At Aunty Di's feet sits Indi Clarke.

Indi is head of the Victorian Koorie Youth Council, whose ancestors are Mutti Mutti, Wemba Wemba and Boon Wurrung from across Victoria and New South Wales, and Lardil from north Queensland.

But he feels a strong connection to this place, deepened by spending a lot of time with Aunty Di, who says she has known Indi 'since he was in nappies'.

'I discovered this place about two years ago and ever since, I've been here probably once a month,' Indi says.

For Indi, it's the perfect salve for a chaotic busy life.

'There's so many times where we just need to press pause and really replenish our spirit.

'Aunty Di and other elders ... have taught me that no matter where we go, our ancestors are always watching over us.

'You can come here, you can listen to the birds, the fairy wrens and watch them fly around and for me, it takes me to my family.'

On Ngarigo Country, in the foothills of Mount Kosciuszko, the Snowy River twists through valleys and foothills a couple of thousand metres beneath its birthplace.

Wiradjuri man Richard Swain, a river guide, echoes Aunty Di's respect for Country.

He invites people to listen to the voice of the land, as Indigenous people have done for so long.

'You know if you're ever walking through the bush and you're really relaxed, birds are chirping, and you might get off track and you don't feel as good and the bush goes quiet?

'That's Country to speaking to you. And when you get back on track, things feel good again.

'The Country's crying out for people to listen to it.

Richard says he remembers nearly every day he's ever been on the Snowy River.

'Every river's the river of life,' he says.

'Water is sacred. And it is alive. It's magical.'

And before he takes a trip down the river, Richard likes to say hello to Country.

As he stands on the bank, he takes a handful of sand and mixes it with the sweat in his armpits.

'I sprinkle it into the river to let the river know that I'm here, and to introduce myself to the land.

'You'll have a better day if you introduce yourself to Country.'

'Indigenous people often talk about the importance of Country. But what does it mean?', Anna Salleh, ABC Science

Figure 7.28 Organ Pipes National Park, Melbourne

QUESTIONS

1 How does Aunty Di refer to Country?
2 What does it mean to acknowledge Country?
3 Have you ever had a time in the bush where you've felt spooked? How did you unspook things?
4 Richard Swain suggests saying hello to Country before taking a trip down the river. How can you 'say hello to Country' when you visit outdoor environments as an individual or a class?

7.2 KEY CONCEPTS

- There is strong evidence of Indigenous relationships with Australian outdoor environments dating from more than 60 000 years ago.
- Indigenous perceptions of outdoor environments include:
  - a spiritual connection with land
  - viewing the land as mother
  - that creators of the land are their ancestors and they live in the features of the land
  - that people are related to and part of the land.
- Interactions pre-colonisation involved:
  - hunting and gathering
  - nomadism and semi-nomadism
  - fire-stick farming
  - sacred sites
  - various agricultural practices like grain and eel/fish traps.

- Impacts on outdoor environments pre-colonisation include:
  - largely low impacts
  - that hunting may have contributed to the extinction of some of the megafauna
  - introduction of the dingo, and possible impacts on some marsupials
  - fire-stick farming and other agricultural practices that may have changed forest environments.
- Interaction and impacts post-colonisation involved:
  - cessation of fire stick farming and traditional understorey burning
  - change of movement of groups due to fencing and settlement
  - soil compacted by sheep and cattle
  - disease and conflict (massacres).

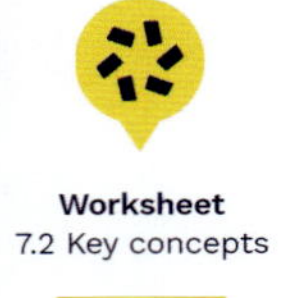

Worksheet
7.2 Key concepts

7.2 CONCEPT QUESTIONS

REMEMBERING

1 How do Indigenous Australians describe their relationship to Country?
2 List five examples of Indigenous peoples' interactions with outdoor environments pre-colonisation.

UNDERSTANDING

3 Describe two examples of how Indigenous peoples' traditional practices were changed by colonisation.
4 Explain the relationship between Indigenous peoples and the bogong moth.

APPLYING

5 Using an example not presented in the text, describe a Traditional Owner group and the extent of their Country.
6 Describe an agricultural practice utilised by Indigenous peoples, noting the location of the example and the evidence.

EXTENSION CHALLENGE

7 Massacres against Indigenous peoples were a horrible reality of early European colonisation. Research the Faithful Massacres and describe the perceptions of the opposing cultures and why conflict occurred.

7.3 NON-INDIGENOUS RELATIONSHIPS WITH THE ENVIRONMENT

KEY KNOWLEDGE

- relationships of non-Indigenous peoples with specific outdoor environments as influenced by and observed in local or visited outdoor environments, during historical time periods:
  - Early colonisation (1788–1859)
  - Pre-Federation (1860–1900)
  - Post-Federation (1901–1990)

KEY SKILLS

- analyse the changing relationships of non-Indigenous peoples with Victorian outdoor environments observed during historical time periods

7.3.1 EARLY COLONISATION (1788–1859)

The first non-Indigenous settlers to Australia were Europeans – specifically British and Irish farmers, explorers, representatives of the British government (including administrators and the military) and convicts. Australia's history shows several different instances of 'discovery' of the continent by the Portuguese, Chinese and Dutch, among others, but the First Fleet's arrival in Botany Bay in 1788 marked the beginning of a new phase of permanent settlement by the British.

Colonisers' perceptions of outdoor environments

FEAR

The British and Irish colonists were used to January being cold and dark and wet – Australia was hot and bright and dry. The colonists were used to bright green plants and small furry animals as the backdrop to their land – Australian plants were dull greenish grey, and while there were some furry animals, there were also snakes and spiders. The colonists came from a place that was 'civilised', where people wore clothes and lived in houses and farmed in 'appropriate' ways – the Indigenous people that they first encountered and their understanding of how they lived didn't match any of these preconceived ideas.

In many countries today, when groups of new immigrants from a particular country or society arrive, they tend to stay close together in small 'island' communities. This is a result of that same fear of the unknown that we can see in the first European settlers arriving in Australia.

OPPORTUNITY

What mindset allows someone to come to a new place where people already live and take over their land to use for themselves? As we've already seen, Indigenous communities were often on the move and so there were times when early settlers might have assumed that no one lived in a particular place. But Indigenous peoples could be found across the country, so this assumption was flawed and based on a Eurocentric mindset that settlements required roads and permanent structures.

The early settlers were unable to find in the many Indigenous communities anything resembling the governmental authority that they were familiar with in Europe. Of course, the harsh nature of the Australian environment combined with the semi-nomadic lifestyle of Indigenous communities didn't support the formation of anything like a central government. But the lack of such an organisation, coupled with the apparent lack of any sort of organised use of the land (such as European-style farming), meant that it was easy to simply see the land as without owners. As mentioned earlier, this came to be known as ***terra nullius*** – land belonging to no one.

While many of the first settlers (i.e. convicts) had no choice in their journey to Australia, there were many others who chose to come here as a chance to start a new life. And so another common perception among the first settlers was the wonder and opportunity that this new place represented.

terra nullius
land that is legally deemed to be unoccupied or uninhabited

To be saved

Related to the previous perception of opportunity was the view among many of the European people at this time that humans were given divine dominion over the environment – a sense that it was right to develop and 'improve' lands that were 'wild' and 'untamed'. So, not only did the European settlers see the land as belonging to no one, but since it seemed to be largely wild and untamed, many of them also saw it as their divine purpose to improve it, farm it, build on it and tame it.

Alamy/World History Archive

Figure 7.29 A sketch of an early settler's hut from 1849

HOW THE COLONIALISTS INTERACTED WITH THEIR NEW ENVIRONMENT

The perception of fear contributed to a consequence that we still live with today – a society that is heavily concentrated around the coast. The coast was close to the ships that had brought the settlers and represented the last connection with their homes. There were a number of expeditions inland, and as confidence with the Australian environment grew, many farmers and others moved inland. But the comfort that settlers found near the coast meant that it was here that the first towns, and later the large cities, formed.

The perception of opportunity, of starting a new and perhaps more successful life, led some to push inland and reach the wealth of land ownership, a much prized status. Explorers were followed by graziers, farmers and gold prospectors, all looking for something that would make the long and harsh journey south worthwhile.

The combination of the development of towns around agriculture and, in the 1850s, gold mining, along with the introduction of new species and the opening up of the land to settlers, led to the sort of practices that we are familiar with today:

- land clearing – for houses, towns and farms
- infrastructure development – such as roads and other transport features, and public buildings and services.

Worksheet
7.3a How was land cleared?

LEARNING ACTIVITY 7.3A

How was land cleared?

Early colonists had no sophisticated machinery for felling large trees and removing them from properties. How did they complete this task? Identify a specific outdoor environment where this interaction occurred in Victoria.

Figure 7.30 A sketch of early Melbourne from 1841

Detail from Panoramic Sketch of Melbourne Port Phillip from the walls of Scots Church on the Eastern Hill July 30th 1841, Samuel Jackson 1807 - 1876

COLONISERS' IMPACTS

The impacts on Australia of the early colonisers' practices were probably small, since the numbers of settlers was relatively small. However, these small impacts increased as the population grew, both through births of new Australians and ongoing waves of immigration.

The following factors have had major impacts on the Australian environment as the numbers of people increased:

- **Cessation of cultural fires** – fear of the casual way Indigenous communities lit fires to manage the landscape led to a cessation of this practice and a subsequent increase in the levels of undergrowth. Whether the large fires that periodically rage through many parts of Australia (particularly the south-eastern corner) are a result of a lack of these cultural or controlled burns is being actively debated.
- **Forest clearing** – clearing of forests began to meet the need for timber and land for growing crops and providing pasture for sheep and cattle. There must have seemed an endless abundance of trees and, therefore, a presumption arose that it could not possibly have any impact. Before European settlement in the early 1800s, around 88% of the 23.7 million hectare colony of what was to become the State of Victoria in 1851 was tree covered. But this clearing continues today and has severely reduced the native vegetation cover found across Australia.

iStockphoto.com/PatrikStedrak

Figure 7.31 Remnants of early forest clearing can still be found in the form of large tree stumps. Those stumps that couldn't be removed by hand or by using strong bullock teams were left behind.

- **Introduction of species** – introduced species have created ongoing problems for land managers across the country. Problems include the hard hooves of sheep, cattle and horses; the incredible expansion of rabbit and fox numbers; the predation of cats and dogs; and the unchecked growth of many exotic plants.

Table 7.3 Summary of early colonisation (1788–1859) relationships with outdoor environments

Perceptions	Practices	Impacts
• Fear of this harsh, new, distant and different place – seasons, weather, local inhabitants, quality of light • Opportunity – to start fresh, make a fortune, build a family, get away from struggle in other places • No one owned the land – *terra nullius* • Humans as improvers and tamers of the land	• Colonisation usually meant bringing most things with you – settlers struggled to live off the land as the Indigenous communities did • Development of towns and first cities – mostly near the coast • Exploration of land for farms and for treasure (gold) and adventure • Introduction of many species – animals for farming, hunting, companionship; plants for farming and homesickness	• Initially small, due to numbers, but began an attitude that would create larger effects as the population grew • Reduced use of fires meant heavy undergrowth • Clearing of forests for farming and construction • Impacts of introduced species • Beginning of gold discovery and mining

Figure 7.32 Map of Victoria from 1922 showing settlements

Pre-Federation (1860–1900)

After the early period of colonisation and leading up to Federation, convicts and free settlers became more resourceful and exploratory, moving further from the eastern coasts of mainland Australia. In search of what this land could provide, they discovered more bountiful stands of trees to harvest, land to clear for farming and shiny gold in creeks, rivers and underground. It was a time of rapid growth, with money to be earnt, and whole towns springing up around this growth.

Different parts of the country saw rapid increases in population, often as a result of this resource boom. In Victoria, the largest increase in population occurred in the 1860s as a result of the discovery of gold in a number of different places around the state. Melbourne and several regional centres, including Ballarat, Bendigo and Geelong, saw enormous increases in population in a very short period. Fortune-hunters were followed by merchants and others looking to make money from the gold seekers.

State Library of NSW/Edward Roper; a9298798. Edward Roper, Gold Diggings, Ararat (c.1858)

Figure 7.33 Edward Roper, *Gold Diggings, Ararat* (c. 1858)

Population estimates show there was a steady increase of population in Australia from 1800 (although this ignores the Indigenous population at that time) through to 2010. If you look carefully at the figures for 1850 and 1860, you'll notice an almost threefold increase – the settler population of Australia in 1860 was almost three times bigger than it was 10 years earlier. In modern terms, that would be like the population of Australia in 2020 increasing from 25 million to around 60 million people. This massive relative increase is not seen in any of the later figures, which is why we focus particularly on what was happening around this time in Australia.

In the years following this boom, pre-Federation Australia underwent constant development – existing towns and cities grew and new ones were founded, agricultural enterprises expanded, land clearing and development accelerated, and new industries began. As with other events, the timing of this phase of Australia's non-Indigenous settlement and development varied across the continent.

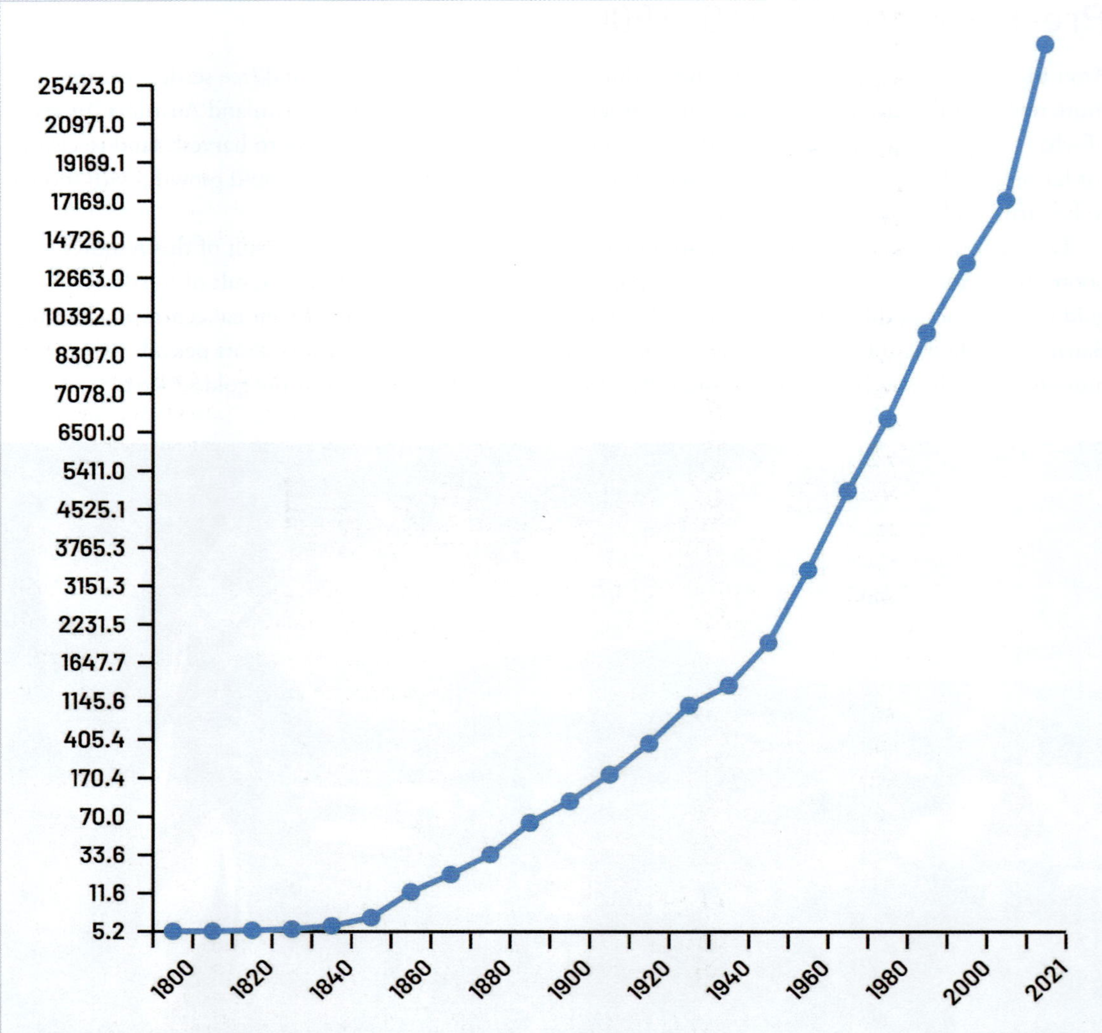

Figure 7.34 Population of Australia (estimate from 1800 to 2021)

Note: In 1967, a referendum passed to count Indigenous Australians in the national census for the first time. Prior to this, states had only intermittently measured or included statistics relating to Indigenous communities in their census taking. The figures that are included in the graph are only estimates – the figures before 1960 should be read as a count of non-Indigenous Australians only.

PERCEPTIONS DURING PRE-FEDERATION (1860–1900)

Many of the same sorts of perceptions towards the Australian environment continued during the stage of pre-Federation. Fear of the bush remained common, particularly among new arrivals, most of whom were either from Europe, North America or China.

The environment, however, was becoming increasingly seen as a resource – one that could make your fortune. This perception of the Australian environment remains today and continues to drive the search for new veins of mineral deposits. During the initial stage of increasing population, seeing the environment as a resource encouraged many of the expeditions that set out to explore the Australian environment.

SPOTLIGHT

Environment as a resource

Despite the rapidly increasing population in Victoria and across the country, there were still huge areas of land that were uninhabited. State governments saw these places as important in the economic development of the state and land (for urban development and agriculture). Minerals, forests and waterways were all seen as resources that could be extracted and used to profit all of the population.

Environment in danger

While there had been some concern about environmental degradation in the early years following the European settlers, the size of the land that the new colonies had to move into meant that any damaging impacts could essentially be hidden from the sight of small towns and farms. But with increasing population came increasing pressures on environments and a new perception arose (particularly coming from a similar trend in the United States) of the need for preservation and protection of environments.

PRACTICES DURING PRE-FEDERATION (1860–1900)

The sort of practices that occurred during this stage of Australia's non-Indigenous settlement were like those from any other stage. Farming and clearing of land continued to grow, and development and settlement expanded. The mass boom of gold-rush-related population growth had stabilised, which led to non-Indigenous people to really expand their influence on the land and increase the scale of change.

Getty Images/Zu_09

Figure 7.35 An engraving from 1897 depicting an early sheep farm in Victoria

It is the scale of practices that occurred during this time that we are particularly interested in, rather than the practices themselves. During this stage, with a dramatically increasing population, small frontier towns went through a boom. Shops, libraries, schools, theatres, police stations and many other buildings that the miners and their families needed had to be built. Communities expanded dramatically as a result, and the requirement for food prompted farmers to expand their land-clearing efforts for crops and grazing animals. The construction of mines and mine equipment led to dramatic deforestation of the lands around these towns as the worldwide technology boom we now know as 'industrialisation' made it to Australia.

Worksheet
7.3b Gold mining

LEARNING ACTIVITY 7.3B

Gold mining

The discovery of gold in Victoria in the 1850s sent a boom of prospecting around the state during that decade and beyond.

Ballarat was the famous boom town of Victoria, but gold was also discovered and mined elsewhere. Research a lesser-known location of gold discovery in Victoria and establish:

1 its location, environment and date of discovery
2 how long the 'boom' lasted for
3 the methods used to find or extract gold
4 the impacts of this interaction that can still be observed or 'felt' today.

THREE KEY EVENTS OF PRE-FEDERATION PERIOD

Machinery

industrialisation the development of industry on an extensive scale

Industrialisation normally refers to the expansion and development of a state's or country's manufacturing system; that is, rather than simply extracting and selling raw materials, other things are made out of the raw materials and are then sold. Victoria and Australia were in the midst of a manufacturing boom at this time, but we'll extend the notion of industrialisation for this first key change to consider the machinery and subsequent processes that were being introduced and used. Mining for gold and other precious minerals, for harvesting timber and for agricultural production were all being boosted and expanded by the introduction of machinery. The necessary jobs could be done faster, longer and more efficiently than human labour, but, of course, there were more expansive impacts as well.

Arguably, some of the new industrial machine-driven processes (e.g. dredging rivers for gold) had more long-term impacts on the Australian environment than the previous small mine and panning operations.

In Victoria, the first irrigation projects were started at this time. In 1886, the Canadian Chaffey brothers were invited by the Victorian Government to begin an irrigation project at Mildura. This would be the start of many such projects that fundamentally changed the nature of agricultural practices across the state and throughout Australia.

Alamy/The Print Collector

Figure 7.36 A river dredge working the Ovens River (c. 1890s)

State Library of Victoria/V.A.N. Hood collection

Figure 7.37 The development of much of north-western Victoria in the era of industrialisation was based on irrigation schemes, as seen here near Mildura.

Working conditions

The second change that occurred at around this time was the improvement in the conditions of workers. The rise of labour movements and trade unions throughout the 19th century (particularly in its last two decades) brought about significant improvements in both the working conditions and pay of the many people that were finding employment in the era of industrialisation. The successful campaign for the eight-hour working day (which we celebrate in Victoria on Labour Day in March) led to higher standards of living and workers with both more time on their hands and more disposable income to spend. For many workers, this was a chance to spend time with their families exploring the parks and bushlands on the fringes of the cities and towns, and this led to the foundation of many naturalist and recreational organisations. Bushwalking, skiing and cycling all became popular pastimes during this period.

National parks

The third and most positive change that occurred around this time was the founding of the nation's first national parks. Yellowstone in the United States was founded in 1872 as the world's first national park, and was followed by the Royal National Park in Sydney in 1879. In Victoria, the first national park was declared at Tower Hill in the state's west in 1892, followed by Mount Buffalo and Wilsons Promontory in 1898. These began a process of protection, preservation and management that has seen more and more of the world's natural environments set aside in conservation areas.

Alamy/Piter Lank

Figure 7.38 Tower Hill Reserve volcano crater lake in lush vegetation in Victoria

IMPACTS DURING PRE-FEDERATION (1860–1900)

With the boom of towns and cities came the associated impacts of increasing numbers of people on a land not used to supporting this many. Land clearing scaled up and produced dramatic and (in some cases) devastating erosion and runoff effects. The introduction of new species continued with impacts that still reverberate today. The impacts that had begun to appear during the early settler stage increased significantly now, as the numbers of people influencing these impacts increased.

The impacts of the increased use of machinery during this time also extended – ongoing erosion and loss of topsoil chief among them. From the use of irrigation and other agricultural practices, new impacts could be seen with the beginnings of the dryland **soil salinity** problem and the decrease in the health of inland waterways across the country.

The rise in recreation saw both positive and negative impacts on the environment. More people had a chance to visit outdoor environments and could begin to see why it was so important to protect these places. But, equally, more people could visit them – and with more people came greater burdens on these places.

soil salinity
the salt content in the soil; the process of increasing the salt content is known as salination

Table 7.4 Summary of non-Indigenous relationships with outdoor environments during pre-Federation (1860–1900)

Perceptions	Practices and interactions	Impacts
• Many saw the land as an opportunity to make a fortune • Fear of 'bush' remained among many – particularly in the cities and towns • Connection to the land was growing with the rise of nationalism – this helped with early formation of preservation societies • Environment seen as a resource by some to use for profit • Environment seen as a place to protect by some	• Large scale increase in mining, forestry, farming, grazing, river dredging • Expansion of cities and development of regional towns and centres • Development of transport infrastructure – roads and railways • First national parks created • Corporatisation of primary industries – farming, mining and forestry • Irrigation used to foster more introduced crops including cotton and rice • Recreation developed – including bushwalking and skiing	• Land clearing and deforestation impacts – rising salinity levels, erosion and runoff • Increasing urbanisation and industrialisation – pollution, unrestrained use of resources • More erosion and other negative effects of land use • Decreasing environmental health of rivers with excessive use of water and change in flows from irrigation • Positive effects of first national parks

Post-Federation (1901–1990)

Federation of the individual states in 1901 saw the foundation of the Commonwealth of Australia and, with it, pressure to build a self-sustained and independent nation. Australia continued to grow on the back of the industrial age, which brought new technology and machines to further increase harvesting of the land's resources. This process of **nation building** continued as a major force in the development and events of Australia, and included further planned immigration of people to Australia and the establishment of massive scale government-funded infrastructure and industrial projects.

nation building
the process of constructing a national identity including the development of national myths as well as major infrastructure development

PERCEPTIONS DURING POST-FEDERATION (1901–1990)

In 1900, the state forests were still commonly regarded by the general public, and by most of their parliamentary representatives, as inexhaustible and ready for disposal via **alienation** into freehold property for the purposes of agricultural settlement.

Perceptions of the Australian environment as a fearful place largely disappeared during this time, at least as anything other than a nostalgic look back to the early settlers and their views. The perception of the environment as a place to protect continued from the beginnings of industrialisation, but during

alienation
transfer of the ownership of property rights

Worksheet Additional Learning Activity: 7.3c Big projects you've visited

the first half of the 20th century they tended to be drowned out by the noise and drive for a strong and prosperous nation. This came from, and resulted in, perceptions of the Australian environment as a place on which to shape the national identity.

From our perspective, many decades later, this is sometimes seen as a creative process – making something new (the Australian nation) from separate pieces (the states). Nation building is sometimes described in metaphorical terms, with the landscape as a blank canvas on which to paint the new Australian nation, and to foster patriotism and national pride.

PRACTICES DURING POST-FEDERATION (1901–1990)

Farming, other agricultural practices, mining and logging, as well as the ongoing expansion of cities and towns, all continued throughout this period of nation building – growing as the population continued to grow.

There were a couple of new developments that were important both for the expansion of Australia as a nation and for the impacts that they were to have on the environment and the way people saw the environment: national transportation and electrical networks.

Transportation network

Following on directly from the beginning of industrialisation came the creation of a national transportation network. The individual states had already developed extensive railway systems, but these were all slightly different – particularly in the gauge (distance between tracks). One of the first jobs of the federal government was to begin the process of trying to standardise this system to allow easier interstate transportation of people and goods.

The nation-building period also saw the continuing development and upgrading of the road network throughout the country. The post-World War II period of prosperity saw a dramatic increase in the private ownership of cars, leading to a massive development of highways and other road systems from the 1950s.

In Victoria, however, one of the major developments in the transportation network came after the Great War (World War I). With the return of many Australian soldiers came the need to find them suitable employment. In Victoria, many of these soldiers formed the basis of work crews that helped to build the Great Ocean Road. Constructed between 1919 and 1932, the road, which links Torquay and Warrnambool and travels along some of the most spectacular coastal cliffs and scenery anywhere in the world, was conceived as both a project for returned soldiers and as a memorial to their fallen comrades. It now stands as the largest war memorial anywhere in the world.

Figure 7.39 *The Diggers*, a sculpture commemorating the 3000 Australian returned soldiers and sailors of World War I who built the Great Ocean Road

Electrical power

Apart from transportation, the other key requirement in the development of 20th-century Australia as a nation was electrical power. Expanding industries and growing cities had a huge appetite for electricity, and, just as they had after the Great War, the returning soldiers from World War II were seen as a rich resource to tap.

Combined with the immigration of refugees from war-ravaged Europe, particularly Italy and Greece, federal and state governments saw a way to continue the nation's postwar development by constructing massive dams and hydro-electric power stations. In the 1940s and 1950s, the Snowy Mountains Scheme (in New South Wales and Victoria) and the smaller Rocky Valley and Pretty Valley **hydro-electricity** schemes (in Victoria) started a period of dam construction that had major impacts both on Australian environments and on the people of Australia for many years to come.

Alamy/Taras Vyshnya

Figure 7.40 Jindabyne dam on Jindabyne Lake and the Snowy River is part of the Snowy Hyrdro power generation scheme

hydro-electricity the generation of electricity using water power

INTRODUCED SPECIES

This period also saw some major introductions of non-native species. In particular, the cane toad, a native of South America, was introduced into Australia in 1935 in an effort to control the beetles that were impacting lucrative sugar cane crops. Over 100 young toads were released; they now number more than 200 million. (Even worse, the toads failed to have any impact on the cane beetles.)

FIRE LOOKOUTS

With the forestry industry and land clearing for agriculture continuing at full pace and encouraged by the machination of the process, the population of Victorians was spread throughout the state. However, uncontrolled bushfires had become a common summer experience and fear of them was increasing. In the 1930s, fire lookouts and towers began to be built atop hills and staffed as early-warning systems. By the time of the Black Friday fires of 1939, three towers had been erected. However, with fires lit by landholders and being generated by lightning strikes, in combination with more than a century of understorey growth (due to cultural burning being declared illegal), fire had become omnipresent.

Alamy/Zeytun Travel Images

Figure 7.41 A fire tower at the summit of Mt Buller in Victoria, Australia

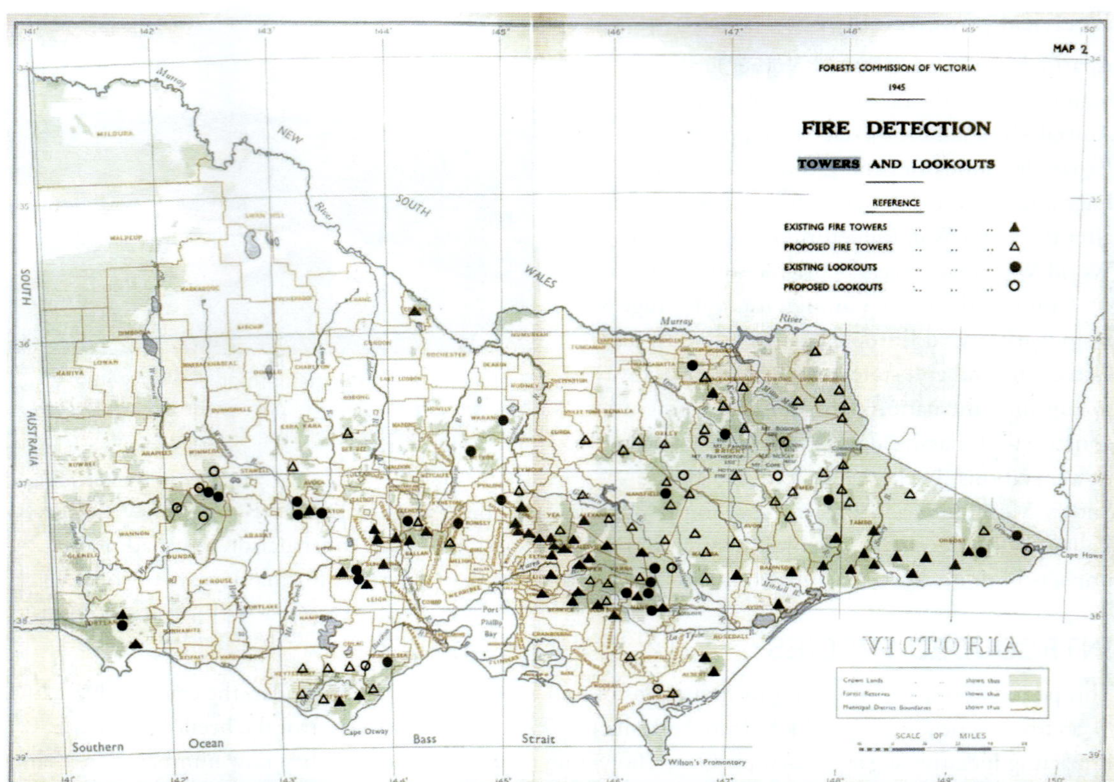

Figure 7.42 The Forests Commission established a network of fire lookouts (hill tops) and fire towers (built structures) throughout Victoria from the 1930s and accelerated after the Black Friday bushfires in 1939.

IMPACTS DURING POST-FEDERATION (1901–1990)

Salinity and land degradation

The environmental impacts of ongoing development practices continued to increase during this period, particularly dryland soil salinity. By the 1920s, salinity was beginning to be recognised as a major environmental problem. Coming up with a solution to it was another thing altogether, however, and the pressure to develop and build in nature meant that protecting the environment was only a secondary concern. This is still a debate that we have today.

To recognise their efforts during the war, many returned soldiers were given access to cheap land so that they might develop farms and Australia might reap the agricultural rewards. Often these farms were on marginal land, and coupled with continuing practices of irrigation and large-scale clearing, led to widespread land degradation – topsoil erosion, declining water quality and dust bowls.

Road and rail

The expansion and nationalisation of road and railway networks had both negative and positive impacts. Negative impacts included the development of land for the infrastructure itself, and the larger scale of impacts from the associated industrial, commercial and urban practices. The positive impact included one we covered earlier – increasing people's access to the natural environment. It was becoming easier and easier to see the beautiful outdoor environments, and this helped with the ongoing development of environmental and conservation perspectives.

Water

The construction of dams in the great hydro schemes of the 1940s and 1950s had a number of impacts, both positive and negative:

- The reduction and change in flow of many alpine rivers had devastating effects on sensitive ecosystems. Some of these are only now being repaired, while others remain threatened.
- The construction of the dams sparked a series of similar developments across the country, which eventually led to major environmental conflicts and important changes in the nature of many Australians' relationships to the outdoor environment. (We examine this more in the next section.)
- The construction of the many hydro dams represented the first and still the largest renewable energy schemes in Australia.

State Electricity Commission of Victoria, Frank Hurley, Australia, National Library of Australia, an23503829-v7

Figure 7.43 Rocky Valley construction camp and dam site, 1947

Introduced species

The impact of introduced species increased throughout this period. Rabbit plagues led to serious erosion problems and reduced native plant cover. The introduction of the cane toad reduced food sources for native insect eaters, and also poisoned native animals, as well as pets and humans. The crown-of-thorns starfish had a devastating impact on corals across Australian reefs.

iStockphoto/tracielouise

Figure 7.44 An introduced wild rabbit

Positive impacts

While it's easy for us to focus on the negative and environmentally damaging impacts, we should remember that every time period has negative and positive impacts. Perhaps most positive during the nation-building period was the increasing level of concern for environments and the growing awareness of the need for environmental protection.

Native fauna advisory committees were established in many states after dramatic programs of culling koalas, Tasmanian tigers and Tasmanian devils, leading to early legislation to protect some native animal species. Field naturalists, recreationalists and others began to lobby for further environmental protections, including more national parks. Many of these groups, and the people who worked within them, would help to ignite the larger-scale environmental movements that were to come.

Table 7.5 Summary of relationships with outdoor environments during post-Federation (1901–1990)

Perceptions	Practices and interactions	Impacts
• Environment seen as the canvas on which to paint a nation • Belief that humans can control nature	• Nation building took place after Federation, featuring large-scale developments • Conservation through national parks continued • Postwar projects and infrastructure projects included the building of Great Ocean Road (WWI) and Snowy Mountains Scheme (WWII) • Major dams were built • Cities continued to grow with most of the population residing in major centres	• Excessive farming created dust bowls in some parts of southern Australia • Dryland soil salinity became a major problem • Water, soil and air pollution continued to grow, especially around cities • The conservation movement grew and protected areas increased • Introduced species began to threaten biodiversity and agriculture

Worksheet
7.3d This land is mine

LEARNING ACTIVITY 7.3C

'This land is mine'

The song 'This land is mine' by Paul Kelly and Kev Carmody is well worth listening to, as the lyrics offer a good comparison between the two groups – Indigenous and non-Indigenous – based on land ownership. Make a list of the key differences that the song highlights and share with a classmate.

Summary of non-Indigenous relationships

> The problem with comprehension is that it often comes too late.
>
> Rasmenia Massoud

As we wrap up this section, it's worth considering a common oversimplification about non-Indigenous relationships – that Indigenous relationships were good for the environment and non-Indigenous relationships were bad. While the effects of these relationships are varied, we need to recognise several important points:

- Indigenous communities were relatively small and lived mostly through a time in human history when the overall human population on Earth was small. European and subsequent settlers to

Australia, on the other hand, arrived here at a time when the world population was rapidly rising. More people almost always means more negative impacts on environments, compared to fewer people.

- The Indigenous communities that the early settlers came across were the product of more than 60 000 years of learning to live in this place – probably the longest uninterrupted habitation by a group of humans anywhere on the planet. We shouldn't be surprised that they had developed skills and practices through this lengthy experience that would allow them to survive sustainably.
- Non-Indigenous settlers had agricultural, social and cultural practices that had held them in good stead, allowed the development of a rich and capable society, and had worked well for them in the places these practices developed – Europe. It was only natural for them to take these practices and continue to use them when they arrived in Australia. From our vantage point over 200 years later, we recognise the error. Hindsight is always a wonderful thing.
- Finally, and perhaps most importantly, keep in mind the driving force behind the early colonisers – to live, to survive and to thrive. No one goes into an environment with the sole purpose of destroying or damaging it. Damage and destruction are by-products of the things that some people do, and have done, in outdoor environments, and we should understand and use that understanding to help us to not repeat such mistakes.

Resource
Additional Case Study: One hundred years of the Great Ocean Road

NOTES FOR THE EXAM 

For the exam, you should:

- know about a variety of non-Indigenous relationships relevant to places you've visited or are familiar with
- be able to identify connections with
  - early colonisation (1788–1859)
  - pre-Federation (1860–1900)
  - post-Federation (1901–1990).

7.3 KEY CONCEPTS

- Relationships with outdoor environments during early colonisation (1788–1859)
  - Perceptions – fear; opportunity; *terra nullius*
  - Interactions – make Australia look like Europe; development; exploration; introduction of species
  - Impacts – cessation of cultural burning; mass clearing; mining impacts commenced
- Relationships with outdoor environments pre-Federation (1860–1900)
  - Perceptions – land of opportunity; less fear; rise of connection with nature; outdoor environments as resource by some and to protect by some
  - Interactions – large-scale mining, forestry and agriculture; expansion of cities and towns and supporting businesses/populations; recreation in outdoor environments developed
  - Impacts – mass population growth means larger impacts on outdoor environments and waterways; land clearing, salinity and urbanisation; first national parks created
- Relationships with outdoor environments post-Federation (1901–1990)
  - Perceptions – outdoor environments as canvas to paint a nation on; humans can dominate nature; fire and fear of fire continues
  - Interactions – mass infrastructure projects like the Great Ocean Road and Snowy Mountains Scheme; growth of cities and city populations away from outdoor environments; fire towers constructed
  - Impacts – intensive farming 'playing out' some land; salinity continues; pollution of air, water and soil; conservation movement begins and continues; introduced species threatening biodiversity (mass extinctions commenced of native species)

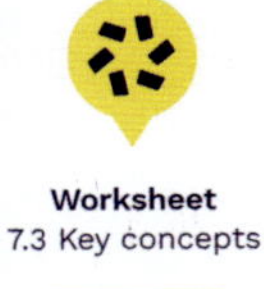

Worksheet
7.3 Key concepts

7.3 CONCEPT QUESTIONS

REMEMBERING

1 Using an example of each, distinguish between the perceptions of groups of people across the three key time periods.
2 Why was fear of Australian outdoor environments a common perception that continues to this day?

UNDERSTANDING

3 Explain the potential relationship between the cessation of Indigenous peoples' fire practices and the increase of uncontrolled bushfires in Victoria.
4 Compare the scale of the impacts in each of the three key time periods.
5 Describe the benefits of improved working conditions for workers in pre-Federation Australia and its potential impact on attitudes towards outdoor environments.

APPLYING

6 Using an example not presented in the text, explain the benefits of a large infrastructure project that was commissioned post-Federation.
7 Describe three impacts of colonisers and how you can observe them to this day in outdoor environments.

EXTENSION CHALLENGE

8 Biodiversity and forest cover in Victorian outdoor environments has plummeted since colonisation. What are the statistics relating to number of species made extinct, the decrease of forest cover and the increase in uninhabitable land in Victoria?

7.4 ENVIRONMENTAL MOVEMENTS IN AUSTRALIA

KEY KNOWLEDGE

- the beginnings of environmentalism and resulting influence on political party or policy, as observed in one of the following historical campaigns:
  - Lake Pedder
  - Franklin River
  - Little Desert

KEY SKILLS

- describe the beginnings of environmentalism as observed in a historical campaign
- evaluate the influence of a historical campaign on the development of a government policy or political party

7.4.1 A BRIEF ENVIRONMENTAL HISTORY OF AUSTRALIA

Non-Indigenous Australians – those who colonised, cleared trees and built new lives on this continent – began to realise in the late part of the 19th century and into the 20th century that these outdoor environments were not only here to provide raw materials for the growth of society but could also be fantastic to recreate in. The perception of non-Indigenous Australians of outdoor environments, as we studied in the previous section, had changed from fear of those natural spaces towards a dual recognition that outdoor environments were there both as a resource to plunder and as a setting for recreation and appreciation. Environmental movements were born in Australia in the 1860s and became part of the Australian consciousness for two somewhat opposing motivations.

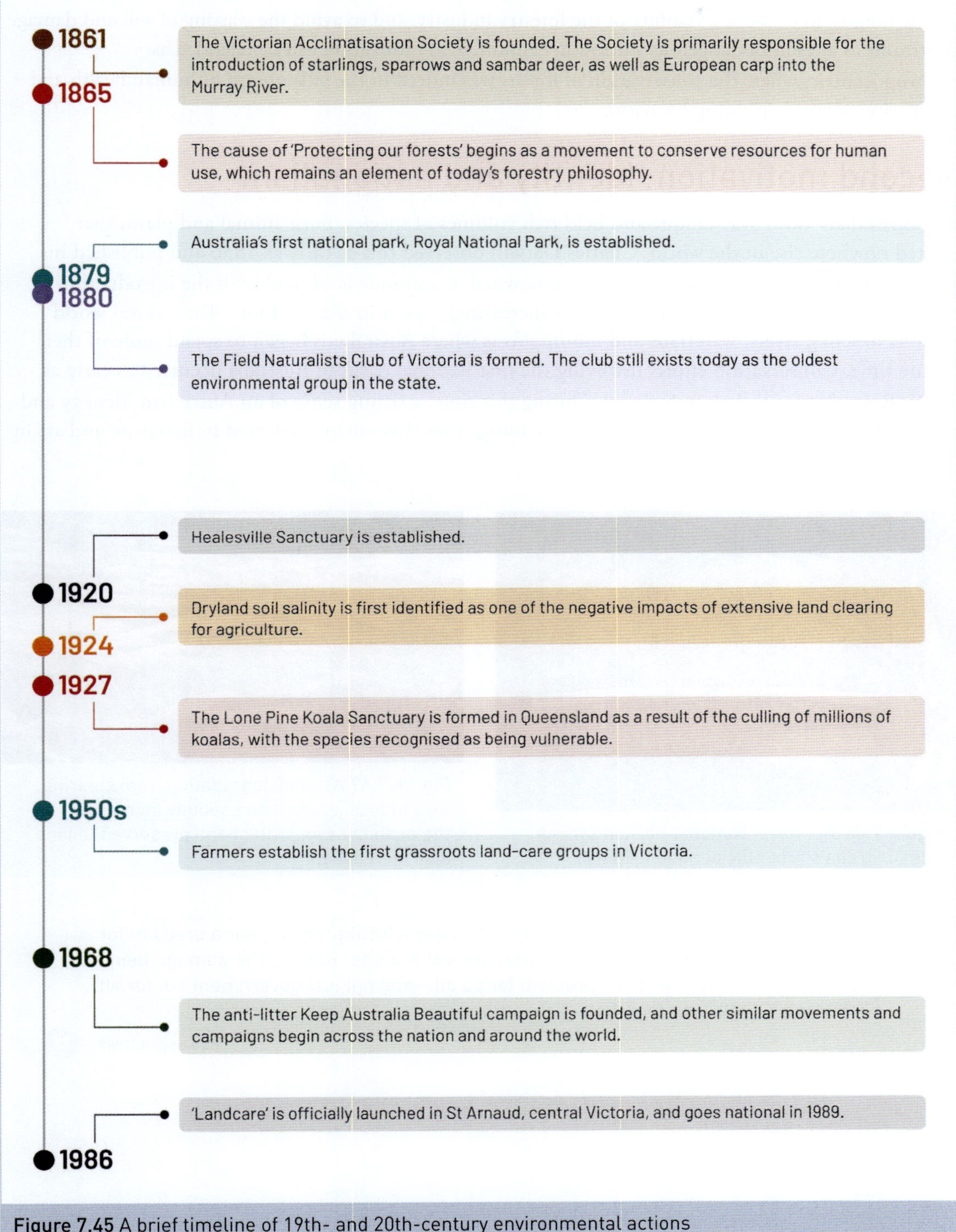

Figure 7.45 A brief timeline of 19th- and 20th-century environmental actions

First motivation: economic

The first motivation was driven by economic factors. Colonisers realised that if they felled timber indiscriminately in pursuit of land clearing for mining, agriculture and settlement, what seemed to be never-ending forests would indeed end. Some conservation was required to save the resource in order to ensure long-term economic viability of the forestry industry, and to avoid the wasting of soil and damage to streams as a result of dangerous stormwater runoff. This motivation can be regarded as **anthropocentric**, where the conservation of forests was indeed in the self-interest of human health and prosperity, rather than caring for nature.

anthropocentric regarding humankind as the central or most important element of existence

Second motivation: identity and nationalism

The Australian 'bush' was unique and held rich volumes of species, both animal and plant, that existed nowhere else in the world. Charles Darwin observed this as early as 1836 and published his thoughts in 1859. Australia became world renowned as a unique land, and with the introduction of the five-day working week, 'time-off' was increasingly spent in the outdoors. The natural world with its beaches, rivers, waterfalls and mountains is where Australians began to spend more of their leisure time. Conservation efforts involving the first national parks in Australia occurred as early as 1879 (Royal National Park in Sydney). During this time, a strong sense of an Australian identity and nationalism also began to emerge among non-Indigenous Australians, reflected in literature and art in the lead-up to Federation.

Alamy/Karen Black

Figure 7.46 The Royal National Park in Sydney was Australia's first official national park.

CSIRO/Willem van Aken

Figure 7.47 As land degradation from clearing, topsoil loss, erosion and salinity increased, the push to save, protect and preserve the land also increased.

> The degradation of our environment is not simply a local problem, nor a problem for one state or another, nor for the Commonwealth alone. Rather, the damage being done to our environment is a problem for us all – and not just government but for all of us individually and together.
>
> Bob Hawke, 1989

Environmentalism worldwide

It was a combination of a number of issues – both in Australia and in other parts of the world, particularly the United States – that saw the beginnings of the modern environmental movement in the 1960s:

- A number of books were published critiquing the use of manufactured chemicals and over-population issues, including Rachel Carson's *Silent Spring* in 1962 and Paul Ehrlich's *The Population Bomb* in 1968.
- The post–World War II baby boom and economic progressions saw a huge rise in the number of affluent teenagers and young people.
- The Cold War stand-off between the United States and its allies (including Australia) and the Soviet Union and its allies flared up in a number of regional conflicts, particularly Vietnam.
- Tensions began to develop between the youth and the generation of their parents over the war in Vietnam, race relations, women's rights and human sexuality.
- Experimentation with psychoactive drugs increased and new trends in music, art and culture were prominent.

Courtesy of Keep Australia Beautiful National Association

Figure 7.48 Keep Australia Beautiful is a not-for-profit organisation that aims to inspire and educate communtities to take action on sustainabiity and keep their environment clean and beautiful.

Environmentalism in Australia

International events and pressures led to tensions over the uses and abuses of natural environments in Australia. Following on from earlier forays into environmental care locally and statewide, these tensions boiled over from the late 1960s through the early 1980s in a number of important environmental protests and campaigns that would dramatically impact the way Australians viewed natural places. This period signalled the beginning of **environmental activism** in our country.

environmental activism the actions of individuals or groups that aim to protect or support the environment

- Victoria – the struggle over the use of Crown land at the Little Desert (finishing in 1969)
- Tasmania – the battles over dams at Lake Pedder (lost in 1972) and the Franklin River (won in 1983)
- New South Wales – the disputes over logging at Terania Creek (in 1979) and the South East Forests (in 1989)
- Queensland – the battle to stop sand mining on Fraser Island (stopped in 1974) and the ongoing effort to protect the Daintree Rainforest (throughout the 1980s)
- South Australia – the attempt to prevent uranium mining at Roxby Downs (in 1983)
- Northern Territory – the efforts to prevent uranium mining at the Ranger mine next to Kakadu National Park (opened in 1981)
- Western Australia – campaigns to stop or reduce logging in old growth jarrah forests of the South-West (through the 1970s and 1980s).

Let's dive deeper into the Lake Pedder, Little Desert and Franklin River campaigns and learn about the beginnings of each movement, and the influence of each movement on the development of a government policy or political party.

The Little Desert dispute

Figure 7.49 Dry, rocky and scrubby, the Little Desert doesn't seem initially to be worthy of protection, but this arid environment is home to a huge variety of plant and animal species. It also helps reminds us that environments have intrinsic value, not just value based on what humans can take from them.

The Little Desert is a part of the dry, nutrient-poor, far west of Victoria. The Wotjobaluk people are the traditional Indigenous occupants of the region. It's sandy and arid but it is not bare, and supports a dense vegetation of heath and wildflowers. The following outlines the key events in the battle to save this region from development:

- Across Australia, large infrastructure building projects occurred at the state level as public lands were released for development, particularly agricultural.
- In neighbouring regions across the border, South Australian governments, combined with new scientific and agricultural techniques, expanded farming into marginal arid lands.
- A number of developments were proposed in the Little Desert area during the 1950s.
- Conservationists urged the development of reserves to protect species before development occurred.
- A planned development in the mid-1960s was abandoned due to low wheat and wool prices, and the difficulty in establishing roads into the area.
- In 1968, the Victorian government decided to go ahead with the development, seeing it as a part of their priority to continue the state's economic growth.
- Economists, scientists and public officials argued the development was not economically viable.
- Conservationists, including the recently formed VNPA (Victorian National Parks Association), wanted to preserve the area.
- The government scaled back their plans from 44 new wheat farms to only 12 sheep farms, and a massive extension of the existing national park from 945 hectares to over 35000 hectares.
- Conservationists were unhappy, claiming the proposed increases in the national park were not representative of the region, and they demanded scientific surveys before any development.
- A push from conservationists in Melbourne continued pressure on the government, with a number of public meetings attracting over 1000 people.
- The media joined the campaign, with all three major newspapers opposing the development.
- Opposition political parties used the dispute to wedge the government by setting up a parliamentary inquiry into the development.
- In the 1969 state election, the government was re-elected but the Lands Minister lost his seat.
- The new Lands Minister was restyled as the Minister for Conservation and helped to develop the Land Conservation Council (LCC) to act as an independent body charged with guiding government decisions. LCC recommended that the development not go ahead. The LCC later became the ECC (Environment Conservation Council) and more recently the VEAC (Victorian Environment Assessment Council).

The Little Desert was saved from development, and the introduction of a new body (the LCC) to help guide government decision-making on the use of environments marked a major change in the way land was considered. Before this, the environment had often been viewed as a frontier to tame, develop and control, so money could be made and the nation (or, in this case, the state) improved. After this, the environment came to be seen as fragile and needing our protection and management.

Little Desert development protests' influence on politics

In 1971, the Liberal government of Henry Bolte formed the Land Control Council (LCC), with the purpose of providing independent advice on the use of public land. To emphasise the importance of their work, the Minister for Lands and Conservation at the time, Bill Borthwick, told the newly appointed LCC councillors to make their recommendations about land use 'as if for a thousand years'.

The LCC and later incarnations – the ECC (Environment Control Council) and VEAC (Victorian Environmental Assessment Council) – pioneered a new way of making decisions about the use of public lands across the state. Instead of ad hoc decisions made to suit small interest groups, the LCC instigated systematic reviews based on comprehensive information and including public input and feedback. Flora and fauna surveys were conducted by scientists for the LCC as a part of their information gathering, and helped broaden the understanding of environments and ecosystems across the state.

The Victorian approach was less confrontational and more bureaucratic than the efforts going on in Tasmania (discussed next), but the Victorian government is generally recognised today as world-leading in its planning for the use of public lands.

Summary of policy or party influence

The influence of the Little Desert environmental movement led to the formation of the Land Control Council (LCC) in 1971, with the following consequences:

- It introduced an independent body with the responsibility of advising the government on the use of public land.
- Systematic reviews were now carried out with independent advice collected and delivered to government to better inform decisions.
- The LCC system provided a process to follow for future disputes on the use of public land. This process is now administered by the Victorian Environmental Assessment Council (VEAC).

The flooding of Lake Pedder

> If we can revise some of our attitudes to the land under our feet; if we can accept a role of a steward, and depart from the role of the conqueror; if we can accept the view that man and nature are inseparable parts of the unified whole – then Tasmania that is truly beautiful can be a shining beacon in the dull, uniform and largely artificial world.
>
> Olegas Truchanas, wilderness photographer and conservationist 

David Nielson, Snowgum Press

Figure 7.50 Flooding had started in the Serpentine River but had not yet reached the lake.

Lake Pedder was a small natural glacial lake ringed by mountains in the south-west wilderness of Tasmania. A prominent feature of the lake – and what drew many bushwalkers, photographers and artists to the place – was the pale pink quartz sand beach that each summer emerged from the waters to one end of the lake as the water level dropped.

Figure 7.51 Bushwalkers and their blue tent at Lake Pedder, 1968

Bushwalkers and their blue tent at Lake Pedder, Tasmania in 1968 [photo by Olegas Truchanas]. Source: National Library of Australia with permission Melva Truchanas vn-3885846-s13

Lake Pedder was to become a key battleground in what would eventually lead to new ways of thinking about and interacting with the Australian environment. The following outlines the key events in the battle to save this lake from the proposed dam:

- The lake was the centrepiece of the Lake Pedder National Park, which the Tasmanian Government declared in 1955.
- Decades before the electrical connection to Victoria across Bass Strait (Basslink) was established, Tasmania was keen on development (like the rest of the nation), and access to electrical power was a key part of this. Tasmania was rich in mountains and water, and so dams and hydro-electric schemes were a natural option.
- The state-run Hydro-Electric Commission (HEC) planned, built and managed the dams and the hydro power stations – similar to concurrent developments in Victoria (Rocky and Pretty valleys) and New South Wales (Snowy Mountains Scheme).
- In 1967, Tasmanian Premier Eric Reece announced that the Middle Gordon Power Scheme, then under development, would require 'some modification to the Lake Pedder National Park'.
- Concerned bushwalkers and artists soon discovered that the 'modification' was in fact the complete flooding of the lake under a much larger Middle Gordon scheme.
- The Southwest Tasmania Action Committee and the Lake Pedder Action Committee (LPAC) were both formed to protest the damming.
- The protests failed, and the incredibly beautiful lake was submerged under the rising dam waters in 1973. The dam was called Lake Pedder, much to the disgust of the protesters.

Lake Pedder protests' influence on politics

The attempts to stop the damming of Lake Pedder helped to give rise to modern Australian environmental movements. In 1972, the United Tasmania Group (UTG) was formed and joined by the Lake Pedder Action Committee in March – one of the first 'green' political parties in the world at the time. When Kevin Lyons (an independent member and holder of the balance of power in the Tasmanian Parliament) resigned, the state government collapsed and a general election was called for April. The Lake Pedder Action Committee called a public meeting in late March, forming the UTG and fielding 12 candidates across Tasmania's electorates. They were unsuccessful but continued to pressure members of the Tasmanian Parliament to act on conservation and environmental issues. Dr

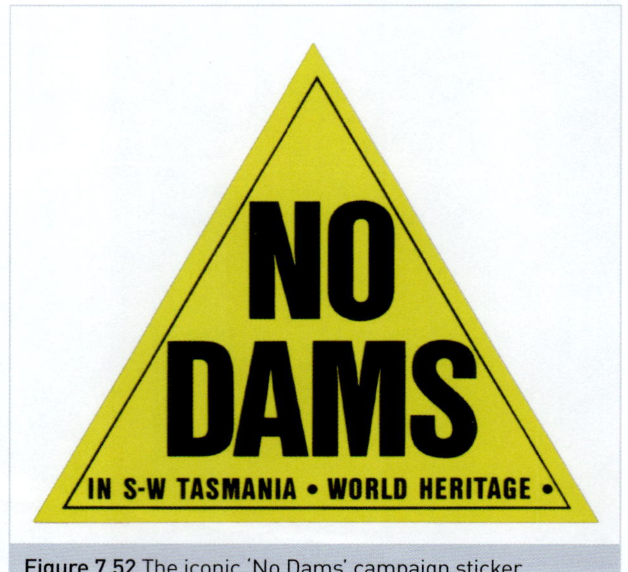

Figure 7.52 The iconic 'No Dams' campaign sticker

Bob Brown joined the UTG briefly in 1975, and the party continued to field candidates unsuccessfully until the early 1990s. In 1976, the Southwest Tasmania Action Committee changed their name to the Tasmanian Wilderness Society (later becoming just the Wilderness Society) after the flooding of the lake.

Summary of policy or party influence

The influence of the Lake Pedder environmental movement led to the formation of UTG in 1972, which heralded the formation of the Wilderness Society. The UTG left a twofold legacy:

- It demonstrated that development should proceed only after considering the long-term benefits of conserving environmental and aesthetic resources.
- It was the catalyst for the creation of new environmental groups, such as the Wilderness Society in 1976, in Australia and elsewhere and paved the way for later green political parties around the globe.

Stopping the damming of the Franklin River

The Franklin Dam (or the Gordon-below-Franklin Dam) was a proposed part of the Tasmanian Government's hydro schemes, following on from the damming of Lake Pedder.

AAP/Peter Dombrovskis

Figure 7.53 One of the most iconic wilderness photographs, Morning Mist, Rock Island Bend was captured by Australian landscape photographer Peter Dombrovskis, and was instrumental in raising awareness of the beauty of the Franklin River across the world.

The Franklin River is one of the major tributaries of the Gordon River, rising in the south-west of the island and eventually flowing out to sea at the west coast town of Strahan. Initially, it was relatively unknown, very remote and largely inaccessible. The campaign to stop the dam made it an iconic environment around the world. The following outlines the key events in the battle to save this river from the proposed dam.

1978

- The Tasmanian Hydro-Electric Commission announced its plans to dam the river as a part of a new hydro scheme in 1978.
- Tasmanian society was immediately polarised, with 70% of the population in favour of the construction, and a very vocal minority of 30% against.
- Immediately after the plans had been announced, the Tasmanian Wilderness Society and Australian Conservation Foundation began a publicity program to raise awareness of the environment. This began initially in Tasmania, but quickly moved to the mainland and eventually became an international campaign.

1980

- In June 1980, 10 000 people marched through the streets of Hobart demanding an end to the project (Tasmania's largest ever rally).
- The Labor Premier, Doug Lowe, proposed a modified Gordon-below-Franklin scheme, which would save most of the Franklin River from the dam. The environmental groups rejected the modification.

1981–82

- In December 1981, the state government held a referendum – the power referendum – asking the Tasmanian people to decide. They had only two options: the original dam or the modified proposal; 47% voted for the original plan, 8% for the modified scheme and 45% voted informally, supporting neither proposal.
- In the Tasmanian Parliament, the modified proposal was defeated anyway and a period of chaos began. In May 1982, a pro-dam Liberal government, under Premier Robin Gray, was elected.
- The new premier ordered the dam construction to proceed.
- Meanwhile, in the Australian Federal Parliament, a senate inquiry began into the natural values of south-west Tasmania, and what responsibility the federal government had to preserve the area.
- On the mainland, rallies were starting to be held, and people voting in elections in Sydney and the ACT wrote anti-dam slogans on their ballot papers.
- In November 1982, Bob Brown, head of the Wilderness Society, announced a blockade of the dam construction by peaceful protesters; 2500 people showed up, many from interstate and overseas, and over 1200 were arrested, including Brown, who spent 19 days in jail.
- At about the same time, 2500 people rallied in Hobart in support of the dam. The issue continued to divide the population.
- Folk rock singer Shane Howard, a member of the band Goanna, wrote the song 'Let the Franklin Flow', which became an anthem for the campaign.
- Liberal Prime Minister Malcolm Fraser struggled to come to terms with division among members of the federal and state Liberal parties on this issue.

1983

- In March 1983, Fraser and his party lost the federal election to Labor, led by Bob Hawke.
- Hawke had vowed to stop the dam, and one of his first acts as the new prime minister was to introduce legislation to create a new World Heritage Area blanketing the disputed site.
- The Tasmanian Government ignored the legislation. Robin Gray at one point threatened that Tasmania would secede from Australia if they were forced to stop the dam.
- The issue was brought to the High Court. In July 1983, it ruled (with a vote of four to three) that the federal government was within its constitutional right to protect the Franklin environment.
- The dam construction ended. The conservation protests had won a significant victory, which became a template for future campaigns around the world. It also marked an end to most large-scale dam construction across Australia.

Fairfax Photos/The Age

Figure 7.54 The Gordon River was used to transport materials and machinery to the site of the dam construction. A blockade organised by opponents of the dam involved protesters in rafts and canoes spread across the Gordon River to stop the supply boats.

Franklin River dam protests' influence on politics

In 1983, Dr Brown was elected to the Tasmanian Parliament mid-term after the outgoing member resigned to run for the Australian Senate. Dr Brown was elected as an independent, in large part due to the prominence he gained as a result of his role in the Franklin Dam protests and blockade (he completed his prison sentence from the Franklin blockade the day he was elected).

Newspix/Greg Newington

Figure 7.55 In 1983, Bob Brown was elected to the Tasmanian Parliament as an independent and three years later helped to form the Green Independents.

When Bob Hawke became prime minister in 1983, he recognised the growing electoral power of environmental movements and the people that joined them. It is probably a stretch to claim that the environmental movement helped elect him to power in the wake of the Franklin River events, but certainly he recognised the importance to the Labor Party and government of supporting conservation efforts. Bob Hawke's campaign to win this election involved him travelling to Tasmania and addressing the protest crowd directly, expressing his support and gaining further national media attention to the conflict.

Fairfax Syndication/Joe Sabljak

Figure 7.56 Prime Minister Bob Hawke (far left) plants the first of the Labor Party's 'One Billion Trees', with ACT Chief Minister Rosemary Follett, Victorian Premier John Cain, NSW Premier Nick Greiner and SA Premier John Bannon in July 1989.

Summary of policy or party influence

The influence of the Franklin River protest and campaign to 'stop the dam' created the largest ripple of environmentalism out of the three campaigns you've studied and was felt across the nation.

In 1983, Dr Bob Brown was elected to Tasmanian Parliament on the strength of his campaigning against 'the dam', but the Tasmanian government was in support of the dam. Also in 1983, the federal Labor Government led by Bob Hawke was elected, and its first act was to introduce legislation to create a new World Heritage area encompassing the Franklin River and effectively stopping construction (as you can't destroy World Heritage areas).

Worksheet 7.4a Timeline of an environmental movement

LEARNING ACTIVITY 7.4A

Timeline of an environmental movement

Research the conflicts and events associated with Lake Pedder, the Little Desert or the Franklin River. Create an annotated timeline that details the conflicts and events. Compare the work of other students on the other environments. What similarities and differences can you see in the conflicts and events associated with these three environments?

LEARNING ACTIVITY 7.4B

Environmental movement role-play

Role-play some of the key events or debates that occurred between groups involved in one of the environmental movements – Lake Pedder, the Little Desert or the Franklin River. Have groups of students in the class take on the role and positions of the different groups involved in the movement.

Worksheet 7.4b Environmental movement role-play

7.4.2 ONGOING INFLUENCE OF ENVIRONMENTAL ISSUES ON POLITICS

Additional Worksheet Additional Worksheet: 7.4c Election Candidate

Resource Additional Case Study: Celebrating 40 years of world heritage in Lutruwita/ Tasmania

After Dr Bob Brown's election to Tasmanian parliament as an independent in 1983, he was joined in 1986 by four other independent members and they became known collectively as the Green Independents. This small group would become the first politicians of the Tasmanian Greens party, which was particularly concerned with protecting the environment. Throughout the 1980s, 'green' political parties formed in other Australian states.

Following the 1983 election win, Bob Hawke and his environment ministers promoted policies that:

- phased out ozone-depleting chlorofluorocarbons
- set up the national Landcare program
- recognised climate change as a pressing problem
- developed programs focusing on reducing pollution and deforestation
- improved efforts at conserving endangered flora and fauna.

In opinion polls after the 1990 federal election, concern for the environment was rated as the second-most important reason for choosing a particular candidate. The 1990 election result (where the Hawke government was returned for a fourth term) was very much influenced by the environmental movement. The activism and support for threatened but remote places in the 1960s and 1970s had created a movement that now influenced the most powerful people in the country. Concern for the environment, fuelled by environmental movements, had helped give rise to political movements that continued to push this agenda.

In 1992, the New South Wales, Tasmanian and Queensland Greens parties agreed to form a national Greens Party, led by Bob Brown. The Greens are now the third force in national politics after Labor and Liberal–National Coalition.

The three campaigns discussed above are historically important, but they're not the only ones that have occurred in the past 40 years. Across Australia, using the lessons learned from these three campaigns, people have protested and fought to protect and preserve pieces of the landscape, or to reduce what many have seen as abuses of the environment. From protests to stop logging or uranium mining, to movements to prevent freeway construction or a new fast-food restaurant in their local neighbourhood, many environmental campaigns have been fought.

We continue to see the ongoing influence of these historical campaigns and the impact of environmentalism on political policies and parties. In further chapters you will learn specifically about current environmental issues and their related policy in federal political parties (Chapter 8) and also consider international agreements to combat climate change (Chapter 9).

9780170477123

NOTES FOR THE EXAM 

For the exam, you should:

- know about the history of a particular environmental movement – perhaps including conflicts and associated events, and the development of organisations and interest groups. This should include one of the following:
  - Lake Pedder
  - The Little Desert
  - The Franklin River
- be able to evaluate the influence of a historical campaign on the development of a government policy or political party.

7.4.3 NON-PROFIT LAND CONSERVATION ORGANISATIONS

While this section focuses heavily on the influence of the 'big three' environmental campaigns on political policy and parties, it is important also to note that a significant component of these campaigns were the result of the creation and organisation of not-for profit conservation groups. These groups remain to this day a vital component of influencing politics.

Non-profit land conservation organisations such as the Wilderness Society and Australian Conservation Foundation grew out of environmental campaigns and became the formalised arm of the protests. What began as a few like-minded individuals banding together informally grew to be structured groups of people working and coordinating state- and nation-wide campaigns to conserve natural areas.

Australian Conservation Foundation and the Wilderness Society

The Australian Conservation Foundation (ACF) was launched in 1965 as Australia's first conservation organisation, encouraged by the World Wide Fund for Nature and the Duke of Edinburgh, Prince Philip, with the purpose of campaigning on a wide range of environmental issues. ACF was involved in both the Lake Pedder and Franklin River campaigns, in the latter working alongside the Wilderness Society, as it does to this current day. On its website, the ACF identifies its main focus today:

> Right now, a pollution and extinction crisis threaten our natural world. Climate pollution and habitat destruction are our biggest challenges.
>
> We're living with the consequences of bad decisions, discredited ideas and short-term thinking. The big polluters. The rigged rules. The politicians who forget they represent the people.
>
> But we don't accept the story that we must sacrifice nature for a quick buck. People made this crisis and together we can solve it.
>
> Australian Conservation Foundation

Beginning in the home of Dr Bob Brown in 1976, the Wilderness Society emerged to save the Franklin River from being dammed. The Wilderness Society has grown since then and is a leading environmental advocacy organisation in Australia. It works to spotlight environmental issues and bring them to the attention of the public and politicians to create change to care for or conserve the natural world.

7.4 KEY CONCEPTS

- Conservation began in Australia in the 1860s with two motivations – to conserve resources for future use and to conserve the natural beauty and significance of outdoor environments.
- Environmentalism began worldwide in the 1960s, with the first major campaign in Australia being the Little Desert dispute in 1968.
- The Little Desert dispute involved the planned development of wheat and sheep farms at the location and was resolved through the bureaucratic creation of the Land Control Council by the Victorian government to help guide decision-making. In this case the LCC recommended the development not go ahead. This body is now known as the Victorian Environment Assessment Council (VEAC).
- The Lake Pedder campaign began in 1967 to stop the expansion of the Middle Gordon Power Scheme that would flood the existing natural lake. Campaign efforts failed and the damming went ahead in 1973.
- Lake Pedder protests influenced the creation of the Tasmanian political party United Tasmania Group in 1972, and what is now known as the Wilderness Society in 1976.
- The Franklin River campaign involved Tasmanian Hydro-Electric Commission planning to dam the Franklin River. The Wilderness Society and Australian Conservation Foundation quickly mobilised and began a publicity campaign from Tasmania to the mainland.
- In 1982, Bob Brown led a blockade of construction, and national protest rallies occurred. The dam was stopped by federal government intervention after the election of the Hawke Labor government.
- The Franklin River protest's influence on politics was to gain national attention with its campaign and elevate the issue to a 1983 federal election level, leading to Bob Hawke 'stopping the dam' by passing legislation to create a new World Heritage site.
- The campaigns collectively contributed to the creation of state 'greens' parties, along with non-profit land conservation organisations, the Wilderness Society and Australian Conservation Foundation, and later the Australian Greens party, today the third force in Australian politics, after Labor and the Liberal–National Coalition.

7.4 CONCEPT QUESTIONS

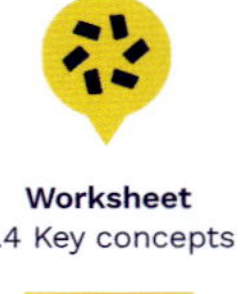

Worksheet
7.4 Key concepts

REMEMBERING

1 Distinguish between the two motivations for conservation in 19th-century Australia.
2 What does it mean to be a non-profit land conservation organisation?
3 List the key dates and people/organisations involved in the conclusion of each of the Little Desert, Lake Pedder and Franklin River campaigns.

UNDERSTANDING

4 Explain the relationship between protesting and politics.
5 Compare the campaigns of Lake Pedder and the Franklin River. Why was one successful and one not?
6 Describe the benefits of a national spotlight coming upon a remote conservation campaign.

APPLYING

7 Using an example not presented in the text, explain a current conservation campaign and their similarities with any of the 'big three' campaigns discussed in this chapter.
8 Describe three strategies that you could utilise to influence political policy or parties.

EXTENSION CHALLENGE

9 Many of the Franklin River protesters were arrested and taken to prison. Describe their hopes in being incarcerated and the potential impact this could have on their permanent criminal record. Why did they participate regardless?

Resource
Glossary - Chapter 7

Assessments
End of chapter exam questions

Glossary test

EXAM-STYLE QUESTIONS

1 Explain how Australian flora and fauna evolved in a unique way prior to humans living in Australia. (6 marks)

  a Name a specific example of an Australian flora species and identify two adaptations this species has developed to cope with Australia's fire regime. (3 marks)

  b Select one characteristic of the Australian environment and explain how that characteristic influenced fauna species prior to human settlement. (2 marks)

2 Indigenous Australians have been living on the Australian continent for more than 60 000 years, and comprise many different and distinct groups, each with their own culture, customs, language and laws.

  a Name a Victorian Indigenous group that you have studied this year. (1 mark)

  b Analyse their relationship with their traditional land prior to colonisation. (4 marks)

  c Describe the impact of the arrival of the first non-Indigenous settlers on your group and their relationship with their traditional lands. (4 marks)

3 Discuss the perceptions the first non-Indigenous settlers may have had of Australian outdoor environments. (3 marks)

4 Analyse how the interactions of the non-Indigenous people during the early colonisation period with outdoor environments changed between the early colonisation and pre-Federation periods (4 marks)

5 During the post-Federation period there was a focus on big projects, and many people viewed the environment only as a resource. Describe a relationship with a specific outdoor environment during the post-Federation time period where the focus was not on viewing the environment as a resource. (3 marks)

6 The beginning of environmentalism in Australia was marked by three historical campaigns. Choose one.

  - Lake Pedder
  - Franklin River
  - Little Desert
  - Name a specific example of an

  a Explain how your chosen environmental movement was formed. (4 marks)

  b Evaluate the influence of your chosen campaign on the development of a government policy or political party. (6 marks)

CHECK YOUR KNOWLEDGE AND SKILLS

Resources
Key knowledge and skills checklist

Checklist for students to mark and date, making judgements of their knowledge and understanding.
(D = developing C = consolidating ER = exam ready)

D	C	ER	Check your knowledge and skills
☐	☐	☐	Explain characteristics of Australian outdoor environments before humans arrived
☐	☐	☐	Understand the geological timeline of Australia that explains why Australia is the way it is in each of the three characteristics: biological isolation, geological stability and climatic variations including El Niño and La Niña
☐	☐	☐	Identify dominant flora and fauna species of specific outdoor environments as a result of the three characteristics
☐	☐	☐	Analyse the changing relationships with Victorian outdoor environments expressed by specific Indigenous peoples' communities before and after European colonisation
☐	☐	☐	Understand modern scientific measurement of Indigenous peoples' settlement in Australia
☐	☐	☐	Appreciate Indigenous peoples' perceptions, interactions and impacts on outdoor environments
☐	☐	☐	Understand the impact of colonisation on Indigenous peoples' relationships with outdoor environments
☐	☐	☐	Describe specific Indigenous peoples' communities relationships with outdoor environments you've studied
☐	☐	☐	Analyse the changing relationships of non-Indigenous peoples with Victorian outdoor environments observed during historical time periods
☐	☐	☐	Describe non-Indigenous relationships with outdoor environments during early colonisation (1788–1859), including examples related to a specific outdoor environment you have studied
☐	☐	☐	Describe non-Indigenous relationships with outdoor environments pre-Federation (1860–1900), including examples related to a specific outdoor environment you have studied
☐	☐	☐	Describe non-Indigenous relationships with outdoor environments post-Federation (1901–1990), including examples related to a specific outdoor environment you have studied
☐	☐	☐	Describe the beginnings of environmentalism as observed in a historical campaign
☐	☐	☐	Evaluate the influence of a historical campaign on the development of a government policy or political party
☐	☐	☐	Understand the motivations behind the beginnings of the conservation movement environmentalism
☐	☐	☐	Describe the Little Desert dispute and how it was resolved
☐	☐	☐	Describe the Lake Pedder dam campaign and result
☐	☐	☐	Describe Franklin River dam campaign and result
☐	☐	☐	Evaluate the influence of these campaigns on politics

CHAPTER 8

UNIT 3 – AREA OF STUDY 2

Relationships with Australian environments in the past decade

KEY KNOWLEDGE

- Indigenous peoples' custodianship of outdoor environments including the formation of land and water councils and Registered Aboriginal Parties (RAP)
- conservation, recreation and economic relationships with outdoor environments
- methods used by individuals and groups to influence decisions about two conflicts over the use of outdoor environments, and the processes followed by land managers to resolve said conflicts, including at least one from the following list:
  - feral species in the Alpine National Park
  - commercial logging in Victoria
  - establishment of new mountain bike parks
  - Southern Ocean whale hunting
  - Murray–Darling Basin water allocations
- an environmental issue in Australia and related policy from two federal political parties or representatives, including at least one of:
  - Labor party
  - Liberal-National Coalition
  - The Greens
- the influence of social debates on relationships with outdoor environments, including one of:
  - climate change
  - renewable energy
  - water management.

KEY SKILLS

- compare different human relationships with outdoor environments, including Indigenous and non-Indigenous peoples' relationships
- describe two conflicts and evaluate the methods used by conflicting parties to influence decisions in their favour, and the processes followed to resolve or potentially resolve said conflicts
- analyse differing environmental politics in Australia
- analyse the influence of social debates about environment issues on relationships with outdoor environments.

VCE Outdoor and Environmental Studies Study Design 2024–2028, pp. 23-24.

Indigenous peoples' formal custodianship

Indigenous peoples have nurtured and managed the natural resources of Australia for tens of thousands of years and have strong cultural connections to Country that value respect for all creatures, as practised through totemic spiritual beliefs. In contemporary Australia,we are part of a system and culture that is beginning to acknowledge the rights of Indigenous peoples to Country be formally recognised as custodians of their land and holders of native title.

Three types of relationships

We can describe human relationships with outdoor environments over the past decade as referring to how people perceive, interact and impact these outdoor environments. To assist us to examine the variety of the relationships that people have with outdoor environments, we look at three categories:

1. conservation
2. recreation
3. economic.

Conflicts over the use of outdoor environments

Many individuals or groups of people have different opinions about how natural environments should be used, and this results in conflicts. These conflicts reflect different perceptions of how to interact with an outdoor environment and often need resolving by a land manager. In this chapter, we learn about five specific conflicts and how the interested parties work to influence a land manager to allow them to use the land or waters for their own purposes.

Environmental politics in Australia

Australians care about the uniqueness and beauty of our outdoor environments, and political parties have learned to listen to our wishes. In this chapter, we compare the policies of each party or representative across three major environmental issues facing our country, while also building our knowledge of these issues.

Social debates about environmental issues

Australia has been facing various issues relating to the environment over the past decade, most of which are complex and multifaceted. Whenever a complex issue includes many potential solutions, we debate the merits of the different ways to approach a solution. When we do this, we influence the outcome of not only the issue but also of relationships with these specific outdoor environments.

KEY TERMS

Aboriginal Water Program
Barmah Choke
biofuel
capitalism
carbon capture and storage
carbon tax
clean coal
Climate 200
climate change
climate sceptic
coalition
commodity
COP
conflict
conservation
Country
cultural burning
custodianship
economic
Emissions Reduction Fund
emissions trading
extinguished
fossil fuel
greening
human condition
interest groups
Kinship
management plan
mass fish death
mitigate
net sink
political spectrum
positive feedback loop
recreation
renewable energy
snow line
socialism
sovereignty
statutory authorities
synergy
terra nullius
urbanisation

Worksheets

Weblinks

Videos

Resources and templates

Assessments

To access resources above, visit **cengage.com.au/nelsonmindtap**

UNIT 4 – AREA OF STUDY 3

Unit 4 – Area of Study 3 requires you to investigate four key knowledge points selected from Units 3 and 4, while visiting at least two outdoor environments.

You must retain a logbook for ongoing assessment throughout the year, containing observations of outdoor environments and experiences you have investigated.

It would be ideal for you to begin planning your logbook at this stage of the course. See Chapter 11 for information and activities related to planning and completing logbooks.

RELATIONSHIPS WITH OUTDOOR ENVIRONMENTS IN THE PRESENT

The previous area of study explored relationships with outdoor environments in Australia in a historical setting. We have investigated how these relationships have changed, some key events and issues that influence the relationships, and the effects of these relationships on people and on the Australian environment.

In this area of study, we will continue to examine relationships, but we will concentrate on a range of more recent relationships; that is, relationships that have existed in the past decade (from 2014 onwards) through the interconnectedness of:

- perceptions – what we think about the outdoor environments
- interactions – what we do in, and with, outdoor environments
- impacts – what happens as a result of our relationships with outdoor environments.

The focus here is on the relationships that occur within the wider Indigenous and non-Indigenous society, rather than a reflection on your own personal experiences. We also build an understanding of how conflicts can occur over the uses of outdoor environments and how they are resolved, and then the influence this all has on politics and the big nation-shaping policies created to manage environmental issues.

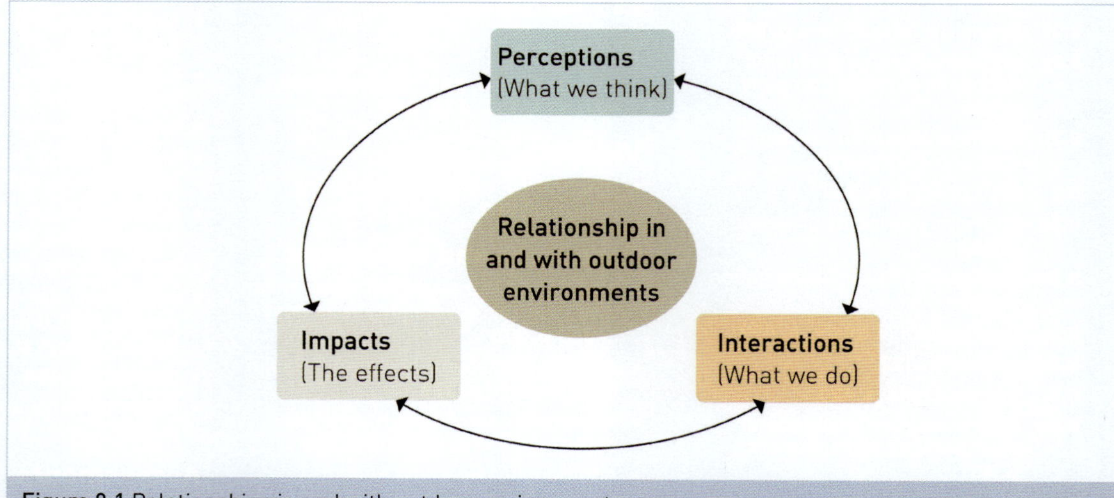

Figure 8.1 Relationships in and with outdoor environments

To shift our thinking and learning into a near-present mindset, we need to move our focus from land-making forces and Indigenous peoples' relationships with Country over millennia and fast-forward to the present day.

Until recently, Australia as a nation had not respected the ancient beliefs, customs and practices of Indigenous peoples, and had ignored and overlooked their rich knowledge about living in a sustainable relationship with Country. The near-present has brought a growing understanding and recognition of Indigenous peoples and their profound connection to and knowledge of Country. This has begun to be reflected in small but significant ways in both government policy and practice, underpinned by a growth in opportunities for Indigenous and non-Indigenous Australians to work together in a collaborative way in caring for our country. The following example of Indigenous people working together with a government agency to care for an outdoor environment is a great story for our near-present and future.

MITCHELL RIVER NATIONAL PARK, BRABRALUNG COUNTRY

The Gunaikurnai have a deep spiritual connection with the Mitchell River landscape through ceremony, songs and dreaming. For Gunaikurnai, many spirits still live in the landscape, their signs in the rocky outcrops and other features. Rock art, rock shelters, canoe trees, surface scatters, men's and women's sites, campsites, massacre sites, burials and many sacred places occur within the park.

The area in and around the Mitchell River Gorge was a major stopping off point for Gunaikurnai travelling between the high country and the lowlands. The rocky terrain, steep dropoffs and lookout points provided excellent vantage points for safety and defence. There are important places throughout this park – Angusvale was a good source of food, medicine and materials, Billy Goat Bend had reliable water. Deadcock Den is an important women's place. It was, at one time, proposed as the site of a mission but it was found to be too cold in winter so Lake Tyers was chosen instead.

Mitchell River National Park is one of the 10 jointly managed parks and reserves within Gippsland. The Joint Management Agreement recognises the fact that the Gunaikurnai people hold Native Title and maintain a strong connection to Country. As custodians of the land, they are the rightful people who speak for their Country. These parks and reserves are cultural landscapes that continue to be part of Gunaikurnai living culture.

iStockphoto/Michael Garner

Figure 8.2 Mitchell River National Park, Victoria

8.1 INDIGENOUS PEOPLES' FORMAL CUSTODIANSHIP

custodianship
the responsibility for taking care of or protecting something

Country
the term often used by Indigenous peoples to describe the lands, waterways and seas to which they are connected; the term contains complex ideas about law, place, custom, language, spiritual belief, cultural practice, material sustenance, family and identity (Source: AIATSIS)

Kinship
an Indigenous person's relationship and responsibilities to other people, to their Country and to natural resources

KEY KNOWLEDGE

- Indigenous peoples' custodianship of outdoor environments, including the formation of land and water councils

KEY SKILLS

- compare different human relationships with outdoor environments, including Indigenous and non-Indigenous peoples' relationships

8.1.1 FORMAL RECOGNITION OF INDIGENOUS PEOPLES' CUSTODIANSHIP

What is custodianship?

Custodianship is a word that has multiple meanings, depending on the context in which it is being used. For example, Indigenous peoples of Australia have practised custodianship of their land over millennia. People can also be custodians of buildings and objects, and in the USA a custodian in a school is the cleaner!

In the context of Outdoor and Environmental Studies, the use of the word 'custodianship' applies to Indigenous peoples' care of outdoor environments. Indigenous peoples have nurtured and managed the natural resources of Australia for tens of thousands of years, and have strong cultural connections to **Country** that include a deep respect for all creatures, as practised through totemic spiritual beliefs. As discussed in the previous chapter, Indigenous people's **Kinship** with country reflects their belief that their ancestors make up every atom of the land, resulting in a deep care and responsibility for the land.

Worksheet
8.1a What land do you live on

LEARNING ACTIVITY 8.1A

What land do you live on?

1 Research the Traditional Owners of the country you live on. What formal status does that community hold within the Victorian or national system?
2 Using your knowledge from Chapter 7, describe how custodianship of Country has become more problematic for Indigenous peoples of Australia since colonisation.

Sovereignty

sovereignty
a state or a governing body [that] has the full right and power to govern itself without any interference from outside sources or bodies

When two nations meet on the field of battle where one nation wants to take the land of the other nation, the end point (after the horror of conflict) is often a treaty. A treaty describes the terms of surrender and defeat, and involves the conditions on which the defeated nation will then relinquish its **sovereignty** to the conqueror.

In Australia after 1788, no formal war was fought between colonists and Indigenous peoples. Therefore, no nation was defeated formally, so sovereignty over the land now known as Australia was never relinquished. Indigenous peoples' beliefs and practices over thousands of years before colonisation demonstrated their self-governing abilities and processes, and thus their sovereignty of their country.

In contemporary Australia, we are all part of a system and culture that is waking up to the rights of all peoples of our nation; in particular the rights of Indigenous peoples to formally be recognised as custodians of their land and holders of native title.

As this is a dynamic space that is subject to rapid change, this module will focus on the current and most up-to-date systems of formal custodianship in Victoria. We look forward to future updates involving stronger Indigenous voices in decisions affecting our country and perhaps even treaty between Indigenous peoples and the Australian federal government.

Resource
Additional Case Study: What we mean when we say 'sovereignty' was never ceded

The rough road to recognition of land rights and sovereignty

The campaign to win formal recognition of Indigenous peoples' sovereignty and their rights as Traditional Owners of their country has been long and has been shaped by a horrifying history of dispossession and disruption since the first arrival of Europeans in Botany Bay in 1788. Characterised by massacres, disease, displacement and forced removal of children, this experience has played a large role in changing Indigenous peoples' relationships with outdoor environments and weakening their connection to, and active custodianship of, the land.

The Australian Government, in response to a growing understanding of land rights and Traditional Owner sovereignty, began to recognise and support the organisation of Indigenous peoples' groups in the 1970s under the Whitlam Labor government. The first land councils were created in the Northern Territory under the *Aboriginal Land Rights Act 1976*, with the states later creating their own legislation and system of land councils. Also known as Aboriginal Land Councils, or Land and Sea Councils, they are organised by region and are commonly formed to represent the Indigenous peoples who occupied their particular region before the arrival of European settlers. Each state now has a different system relating to Indigenous-owned land and waters, with the representative bodies given varying names.

The journey to native title ownership in Australia began with the Mabo decision in the High Court of Australia in 1992. The Court's decision recognised that the Meriam people had proven their unbroken connection and ownership of their traditional lands in Mer (Murray Island) in the Torres Strait Islands. In acknowledging the traditional rights of the Meriam people to their land, the court also held that native title existed for all Indigenous peoples. This landmark decision dismissed the claim of ***terra nullius*** that was the basis of Britain's claim to the possession of the continent and launched a new era of application and awarding to Indigenous peoples the native title ownership of their Country.

terra nullius
land that is legally deemed to be unoccupied or uninhabited

Meriam elder Eddie Mabo led the campaign to bring the case to the High Court and his name is now synonymous with the ground-breaking case. What became known as the 'Mabo ruling' or simply 'Mabo' led to the *Native Title Act 1993*, which provides a process through which Indigenous peoples can lodge an application to seek a determination of native title.

Fairfax Photo/Fairfax Media

Figure 8.3 Eddie Mabo (looking at camera) and other Aboriginal and Torres Strait Islander representatives speaking about native title at the Collingwood Town Hall in 1986.

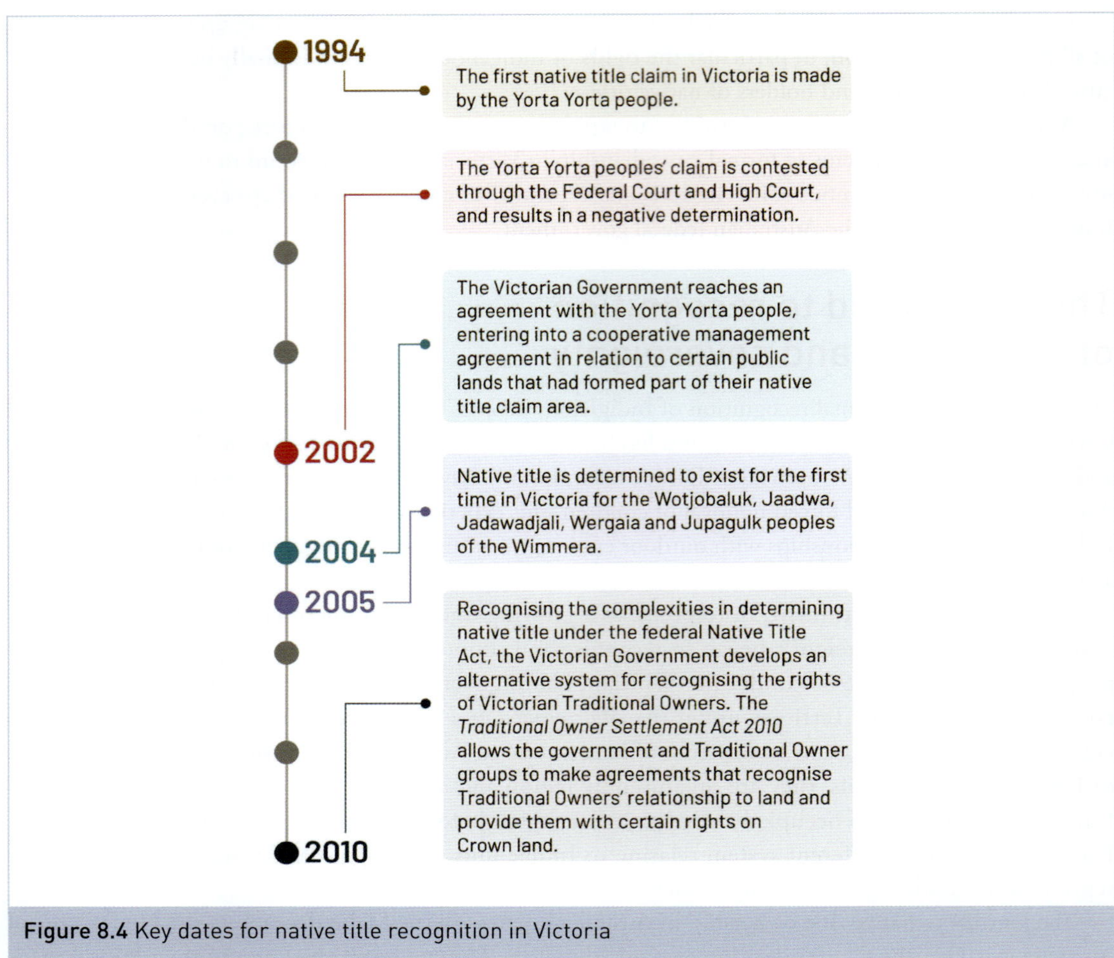

Figure 8.4 Key dates for native title recognition in Victoria

Formal recognition of custodianship

Figure 8.5 details how the Victorian Government formally recognises Traditional Owners.

Right people	Right Country	Decision making	Capability	Incorporation
Groups need to demonstrate that they are inclusive and representative of all Traditional Owners with interests and rights to Country.	Groups need to demonstrate a connection to a defined area of Country. Agreeing on boundaries with neighbours is encouraged.	Groups need to have agreed decision-making structures and processes in place.	Groups need to have the organisational capability to sustainably manage the legal responsibilities that flow from formal recognition.	Groups need to appoint a corporation to act for them under the *Corporations (Aboriginal and Torres Strait Islander) Act 2006*.

Figure 8.5 The requirements and considerations for formal recognition of Traditional Owner groups in Victoria – the first step to gaining recognition of formal custodianship.

Once the requirements of the first four steps have been met (Right people, Right Country, Decision making and Capability), Indigenous peoples organise together to form an incorporated group, known as a Land Council, Land and Water Council or Land and Sea Council.

The next step is to seek formal recognition, which in Victoria can be done in one of three ways (see Figure 8.6).

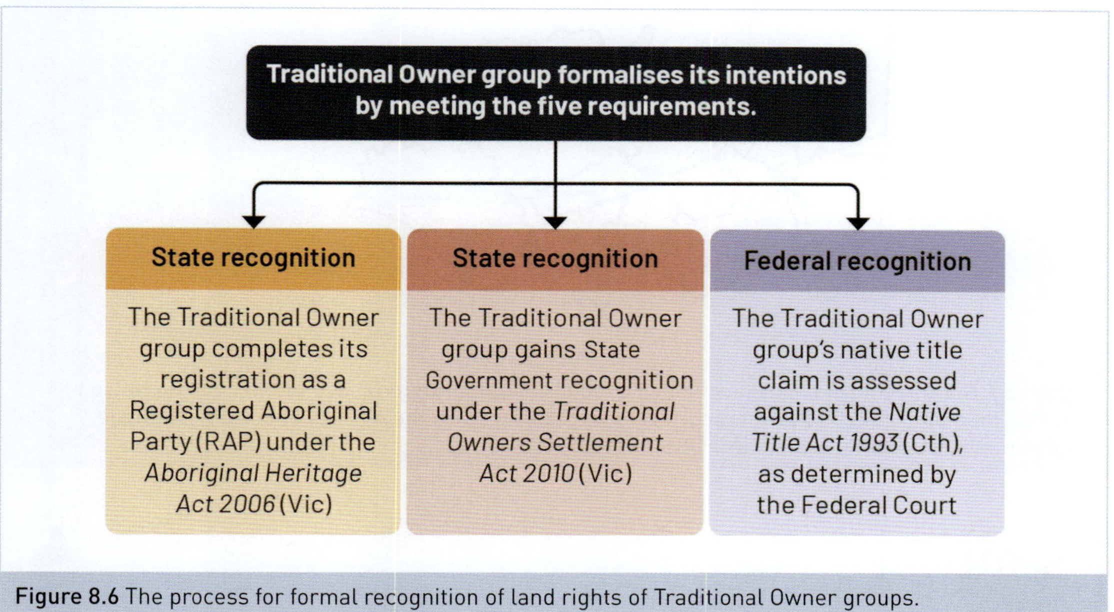

Figure 8.6 The process for formal recognition of land rights of Traditional Owner groups.

REGISTERED ABORIGINAL PARTIES

Registered Aboriginal Parties (RAPs) are responsible for managing all land, water and Aboriginal cultural heritage within their appointed areas. They are appointed by the Victorian Aboriginal Heritage Council, established under the *Aboriginal Heritage Act 2006*, which is made up of Victorian Traditional Owners.

RAPs are the primary source of advice and knowledge on all matters relating to Aboriginal places or Aboriginal objects in their region. Their core functions include:

- evaluating Cultural Heritage Management Plans
- assessing Cultural Heritage Permit applications
- making decisions about Cultural Heritage Agreements
- providing advice on applications for interim or ongoing Protection Declarations
- entering into Aboriginal Cultural Heritage Land Management Agreements with public land managers
- nominating Aboriginal intangible heritage to the Victorian Aboriginal Heritage Register and managing intangible heritage agreements.

As of January 2023, there were 11 RAPs in Victoria and that number is growing as other Traditional Owner groups complete the steps to formalising their native title claims. With the traditional ownership of 25 per cent of Victoria still to be negotiated, this is an evolving space.

Figure 8.7 The current traditional ownership boundaries within Victoria. See the weblink on Nelson MindTap for a direct link to the interactive version of this map on the Aboriginal Cultural Heritage Register and Information System (ACHRIS) website.

SPOTLIGHT

Example of a recent Registered Aboriginal Party in Victoria

In February 2020, the Eastern Maar Aboriginal Corporation group was approved as a Registered Aboriginal Party after nine years of negotiations around decisions and amendments.

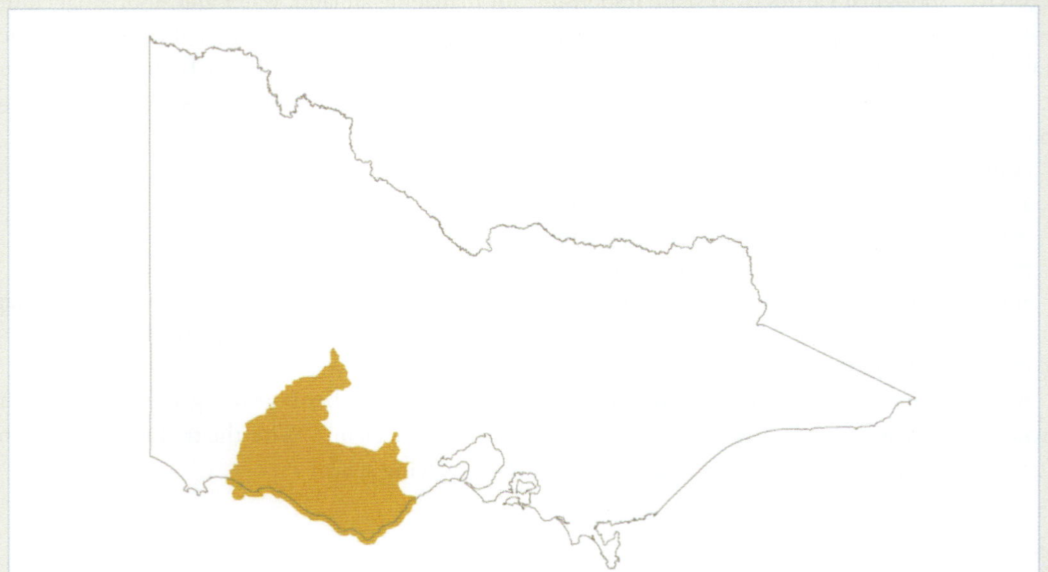

Figure 8.8 Registered Aboriginal Party boundary of the Eastern Maar Aboriginal Corporation

LEARNING ACTIVITY 8.1B

Registered Aboriginal Party applications

1 Research the approval process of RAPs and explain the reasons why some are denied or take longer to process regarding land boundaries.
2 Complete up-to-date research on any Traditional Owner Groups seeking RAP status now.
3 Choose a far-off destination in Australia that you have visited. Had it been determined as part of a native title claim?

Resource
Additional Case Study: Barengi Gadjin Land Council

Worksheet
8.1b Registered Aboriginal Party applications'

NATIVE TITLE DETERMINATION

OUTCOMES OF FORMAL RECOGNITION

A positive native title determination involves recognition by the Federal or High Court that a group's rights continue from before European colonisation to the present day.

It also lists the native title rights determined; for example, to camp, hunt, fish, gather food, and teach law and custom on Country.

Native title holders and registered native title claimants have rights under the Future Acts regime (such as the right to comment on or negotiate agreements) in relation to activities on Country that affect native title rights and interests.

The *Native Title Act 1993* also allows for claims for compensation from the Commonwealth or state governments for acts that have 'extinguished' or 'impaired' native title rights and interests.

Source: First People's State Relations, Outcomes of Formal Recognition © State Government of Victoria

Native title applies to public land and waters, except where it has been '**extinguished**' (meaning a group's rights to native title have been withdrawn). Native title includes property rights held by Aboriginal and Torres Strait Islander peoples under their traditional laws and customs, which pre-dates colonisation and is recognised by Australian law.

extinguished
in relation to native title, that native title holders are no longer able to fully exercise their traditional rights in an area as the result of governments granting freehold land or leases, or constructing public works

Native title may be claimed in the following areas:

- vacant (or unallocated) Crown land
- parks and public reserves
- beaches
- some leases (such as non-exclusive pastoral leases)
- land held by government agencies
- some land held for Aboriginal and Torres Strait Islander communities
- oceans, seas, reefs, lakes, rivers, creeks and other waters that are not privately owned.

To initiate this process, an application is made to the Federal Court by Traditional Owners.

Alamy Stock Photo/ Cephas Picture Library

Figure 8.9 Rainforest Tarra Bulga national park Victoria Australia, Gunaikurnai Country

Resource
Additional Case Study: Native Title determination in Victoria

9780170477123

Weblink
Victorian government, 'Outcomes of formal recognition'

Resource
Additional Case Study: Recent recognition and settlement agreement in Victoria

RECOGNITION AND SETTLEMENT AGREEMENTS

The *Traditional Owner Settlement Act 2010* (Vic) provides a framework for the recognition of Traditional Owner rights and settlement of native title claims in Victoria.

Recognition and Settlement Agreements apply to public land and waters, with some exceptions, and are negotiated by Traditional Owners with the Victorian Government. They provide recognition and a financial and land management package.

For more information on settlement packages, see the 'Outcomes of formal recognition' on the First Peoples–State Relations page of the Victoria government's website.

Fairfax Photo/Justin McManus

Figure 8.10 Aboriginal Narjong Ceremony at Long Plain in the Kosciuszko National Park

FORMAL CUSTODIANSHIP IMPLEMENTATION

After formal recognition is determined, Indigenous peoples' groups in the form of RAPs or Land and Water Councils are able to provide a strong voice for their people and for the care of their land and waters.

Involving and actively collaborating with Indigenous peoples in the care and management of outdoor environments:

- demonstrates respect for the sovereignty of Traditional Owner groups in Victoria (ownership of the land was never ceded)
- connects and reconnect Indigenous peoples with their Country and their Dreaming.

> Nevertheless, more can be done to empower the ownership and control of Traditional Owners over their water and waterways, and more can be done to ensure that Traditional Owners receive equitable benefits over the use and management of their resources. In fact, in some circumstances, waterways are left at active risk of harm through a failure to recognise their cultural and environmental significance.
>
> Aboriginal Heritage Council
> Waters are our spirit, Victoria Aboriginal Heritage Council © Copyright State Government of Victoria

Initiatives to promote Indigenous people's custodianship at work in Victoria

As noted in the timeline in Figure 8.4, the first formal custodianship agreement of land in Victoria was granted in 2004 with the Yorta Yorta people. Today, 11 RAPs have been formed and an increasing amount of care and management for Country is involving Indigenous peoples through their direct and collaborative management of land and waters. Some important examples of initiatives to promote and develop opportunities for Indigenous peoples' custodianship, include:

- Indigenous representatives consulting directly with landowners, including mining activity, employment and development, including in national parks
- designated Aboriginal positions as part of the government's recruitment policy
- bush food and medicine identification and promotion
- water management
- sacred site identification and management
- work within other organisations to maintain and enhance Aboriginal culture, identity and heritage.
- traditional or cultural burning practices and education.

CASE STUDY

CUSTODIANSHIP AT WORK

The Taungurung people are recognised as the traditional owners of approximately 913 000 hectares of public land, including the complete or parts of many national and state parks and reserves in Victoria through their Recognition and Settlement Agreement. Due to this large amount of land, an agreement was made between the state government and the Taungurung Land and Water council called the Traditional Owner Land Management Agreement (TOLMA), resulting in a designated Aboriginal position as a ranger within Parks Victoria based in the Ovens Valley.

The following is an extract from a position description for a Taungurung park ranger from August 2020.

> The objectives of joint management are to establish a partnership that:
>
> - ensures Taungurung involvement in the management of the nine designated areas subject to Aboriginal Title
> - benefits the Taungurung people by recognising, valuing, promoting and incorporating their culture and knowledge in all decision-making processes
> - identifies employment opportunities for Taungurung people in the day-to-day operations of the relevant parks and reserves
> - enhances the experience of all Victorians and visitors through the provision of Aboriginal cultural education, services and information
> - conserves, protects and enhances natural and cultural values
> - ensures the wellbeing of Country and the wellbeing of people.
>
>

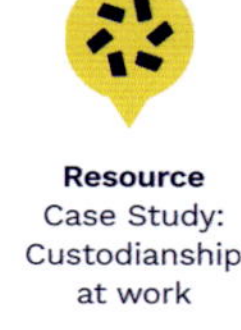

Resource
Case Study: Custodianship at work

Resource
Case Study: Water management

Weblinks
Victorian government, 'Water for Traditional Owners a Victorian First

Gunaikurnai Land and Waters Aboriginal Corporation, 'Water'

CASE STUDY

WATER MANAGEMENT

The Gunaikurnai Land and Water Aboriginal Corporation (GLaWAC) has a Water Team that works to reconnect Traditional Owners and Aboriginal community members to waters in the Latrobe Valley and South Gippsland. In 2020, GLaWAC was able to achieve recognition of the importance of gaining rights to water, directly receiving the benefit of its formal custodianship recognition.

Water for traditional owners a victorian first

Water ownership in a river system will be made available to Traditional Owners for the first time in the state's history thanks to the Victorian Government.

In a significant step recognising Traditional Owners' deep connection to water, the Gunaikurnai Land and Waters Aboriginal Corporation (GLaWAC) will receive two gigalitres of unallocated water in the Mitchell River.

Minister for Water Lisa Neville confirmed Southern Rural Water will make a further four gigalitres of unallocated water available on the open market – half in the coming months and the remainder next year.

Traditional Owners' connection to water is a key part of Water for Victoria – the government's long-term strategy for managing the state's water resources.

Water for Traditional Owners a Victorian First, 12 November 2020 © Copyright State Government of Victoria

Also see the short video 'Water' on the Gunaikurnai Land and Waters Aboriginal Corporation website.

Resource
Case Study: Cultural burning initiative

Weblink
Video: Djandak Wi – Traditional burning returns

CASE STUDY

CULTURAL BURNING INITIATIVE

Cultural or traditional burning is a highly valued practice by Traditional Owners and a skill that has been built over thousands of years. The formalisation of custodianship provides greater opportunity for Indigenous peoples' groups to share their knowledge and skills.

The Dja Dja Wurrung people, combined with other elders and Traditional Owner groups, have aspirations to increase collaboration with land and fire management agencies to facilitate the re-introduction of **cultural burning** in Victoria's natural landscapes.

Partly as a result of their input into the Victorian Traditional Owner Cultural Fire Strategy of 2019, they have secured funding (as the Dja Dja Wurrung Clans Aboriginal Corporation) to employ Forestry and Fire Team Leaders who will lead a team of fire crew to support fire and forestry works on Djandak through the delivery of culturally informed fire activities. See the short video, 'Djandak Wi – Traditional burning returns'.

cultural burning a traditional practice developed by Indigenous peoples to enhance the health of the land and its people; it includes burning (or prevention of burning) for the health of particular plants and animals

Fairfax Photo/LOUISE KENNERLEY

Figure 8.11 Illaroo Bunda program cultural burning of bush in Shoalhaven Victoria, to manage land using Aboriginal burning techniques

NOTES FOR THE EXAM 

For the exam, you should:

- know the process of formal recognition of Indigenous peoples' custodianship of land in Victoria
- know some examples of Land and Water Councils and how they are providing care for their lands.

8.1 KEY CONCEPTS

- Custodianship is the care and management of land and is related to Kinship.
- Custodianship is the process of the formal recognition of Indigenous peoples' custodianship of land or land and water.
- The five requirements for Traditional Owners to gain formal recognition are Right People, Right Country, Decision Making, Capability and Incorporation
- There are three processes for Traditional Owners to pursue formal recognition:
  - by becoming a Registered Aboriginal Party (RAP) as approved by the Aboriginal Heritage Council
  - through a native title determination by the federal government (*Native Title Act 1993*)
  - through a Victorian Recognition and Settlement Agreement under the *Traditional Owner Settlement Act 2010* of Victoria, created as an alternative and more direct Victorian pathway to native title.

- Examples of formalised Traditional Owner groups caring for outdoor environments include:
  - consulting with landowners including mining activity, employment and development, including in national parks
  - designated Aboriginal positions
  - water management
  - traditional burning practice and education
  - bush food and medicine identification and promotion
  - sacred site identification and management
  - working within other organisations to maintain and enhance Aboriginal culture, identity and heritage.

Worksheet
8.1 Key concepts

8.1 CONCEPT QUESTIONS

REMEMBERING

1 Describe the difference between custodianship and Kinship.
2 What does it mean to cede sovereignty?
3 List the five requirements for an Indigenous peoples' group to be considered for formal recognition.

UNDERSTANDING

4 Explain the process of a group becoming a Registered Aboriginal Party.
5 Explain the two pathways for a Registered Aboriginal Party to achieve a Native Title determination in Victoria.

APPLYING

6 Explain why the Mabo decision of 1992 was a catalyst for future land title claims.
7 Explain some practices of a formalised Indigenous peoples' group and how they benefit outdoor environments and people, including a specific example and group.

EXTENSION CHALLENGE

8 Sovereignty has been described as a dynamic space for Indigenous relationships in Australia, with possible future changes, including stronger Indigenous voices in decisions affecting our country and a treaty between Indigenous peoples and the Australian federal government. Research and describe the current state of these two areas in Australia.

8.2 THREE TYPES OF RELATIONSHIPS

KEY KNOWLEDGE

- conservation, recreation and economic relationships with outdoor environments

KEY SKILLS

- compare different human relationships with outdoor environments, including Indigenous and non-Indigenous peoples' relationships

8.2.1 THREE TYPES OF RELATIONSHIPS

We can describe human relationships with outdoor environments over the past decade as referring to how people perceive, interact with and impact these outdoor environments. To assist us to examine the wide variety of the relationships that people have with outdoor environments, the study design identifies three categories:

- conservation
- recreation
- economic.

As outdoor enthusiasts, lovers of outdoor environments and people earning an income from outdoor environments, it is important to be aware that many of our experiences in the outdoors cross over these three broad relationships simultaneously, and identifying and comparing these relationships individually can be problematic. For example, as a student of Outdoor and Environmental Studies you may ride your bike (recreation) to participate in an activity of tree planting (conservation) that you pay a fee to attend (economic). All three categories are included in the same broad experience, and each has an impact on the other. Your relationship with outdoor environments changes depending on your focus or intent.

iStockphoto.com/Robert Daly

Figure 8.12 Tree planting

In order to compare human relationships with outdoor environments, we must first consider these relationships in isolation, regardless of their ability to occur simultaneously by the same person in the same environment. Let's dig deeper into this.

LEARNING ACTIVITY 8.2A

Which is better and why?

Think of three people you know (one can be you) that each share different interactions with outdoor environments.

1. Write a short description of their interaction and what motivates them to participate in a relationship with outdoor environments in this way.
2. Provide an overall judgement of which in your opinion is the 'best' relationship with outdoor environments, and explain what criteria you used to describe what 'best' is.

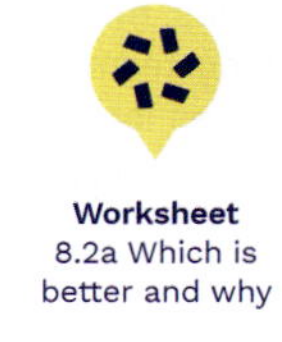

Worksheet 8.2a Which is better and why

Conservation

conservation the preservation, protection, management or restoration of the natural environment, inclusive of ecosystems, vegetation, wildlife and natural resources, such as soil and water

A person who identifies primarily as having a **conservation** relationship with outdoor environments will likely be taking part in activities focused on maintaining or improving the health of outdoor environments. They are motivated by a desire to care and feel responsible for said environment. They can be acting either in a voluntary capacity or as a paid service, and generally gain a feeling of reward intrinsically as part of their interaction.

The perception of a person in a conservation relationship with an outdoor environment is based on the belief that nature needs their help; that they need to intervene, often as a result of the damage caused by other humans or introduced species.

Interacting with an outdoor environment for conservation involves human efforts that are aimed at protecting environments from potential damage. It can also involve restoring an environment or elements of an environment; for example, physical actions performed by private land managers such as revegetation of an area that has previously been cleared, or planting trees along waterways through grazing properties.

iStockphoto.com/jax10289

Figure 8.13 Paved walking track leading to Port Campbell foreshore, Victoria.

management plan a document that contains guidelines on how an area of public land is managed; it articulates the vision, goals, outcomes, measures and long-term strategies for parks within planning areas

synergy the interaction of two or more elements that produces an effect that's greater than the sum of its parts

Increasingly, with the formalisation of Indigenous peoples' custodianship relationships with outdoor environments, we are also seeing traditional practices being integrated into conservation relationships. In the previous section, we learnt about Indigenous people working as rangers for Parks Victoria in the management of Crown land. This relationship is producing a great **synergy** that should only grow as respect for Indigenous culture and management methods increases.

It is not only the 'hands-on' efforts that protect and restore our outdoor environments. Strategies and policies are developed and implemented by the government and other land managers that focus on conservation. For example, **management plans** are drafted by Parks Victoria for all the parks and reserves it administers. These plans outline long-term strategies that guide the way public lands are managed.

CASE STUDY

PROJECT FIREFIGHTER, PARKS VICTORIA

Mingma Sherpa is employed as a project firefighter by Parks Victoria in the Bogong Unit based out of Mt Beauty, Victoria.

'My background is growing up with nature around me all of the time. I grew up in a vertical village in the Himalayas near Mount Kanchenjunga, the third highest mountain in the world. My family were self-sufficient farmers. We grew all of our food except rice and salt. When I was a child, I had to live in the jungle to look after the animals. When I was only 12 years old I would stay in a hut by myself for the summer and care for them. I would always look and learn about the many species of plants, and gather plants to eat as well as for medicine. I needed to go into the jungle to get food for the buffalo too. Even as a child I learned to protect small plants from the buffalo by cutting bamboo and making a guard. Nature makes me feel happy.

'My job at Parks Victoria is project firefighter. This is controlling bushfires in the national parks, creating firebreaks, clearing tracks and caring for the park plants and infrastructure. To get this job I first started volunteering with Landcare. At that time, I gained knowledge of Australian native plants. After that, I started to study horticulture and became interested in conservation. I volunteered with Parks Victoria. I needed to have knowledge of environmental science or related fields. It was also helpful to have volunteered or had experience in working in parks or Landcare.

'I have a Certificate III in Horticulture. I am also studying a Bachelor of Environmental Science and Management at university. English is my second language, so study is hard. I have not qualified as a ranger yet, but I hope to. I have volunteered and worked with the rangers in the national park. This led to my position as a project firefighter. When I complete my studies, I hope to work as a ranger.

'I enjoy caring for nature because it is part of our life. When I work, there is an abundance of plant species to see. There are also many birds and other wildlife. I also enjoy working physically, and that benefits my body and mind. There is a metaphor of healthy nature is a healthy life, and I feel this describes what I like.

'My favourite conservation project is protecting an endangered Alpine lizard. I volunteered with an ecological researcher constructing lizard habitats on the high plains, and I enjoyed learning how to protect them. I enjoyed knowing that I was helping to rehabilitate their natural environment.

'I have volunteered for Landcare for about four years. I volunteer every Saturday for a few hours. This protects native species and controls pests. I also connect with local people, to make friends and understand the local community culture.

Louise Hayes

Figure 8.14 Mingma Sherpa working hard for conservation

Resource
Case Study: Project firefighter, Parks Victoria

QUESTIONS

1 Why does Mingma care for outdoor environments?
2 How does Mingma describe his relationship with the natural world? How has that changed over his life?
3 Mingma writes of being involved in Landcare projects before he became a project firefighter. Do you think this is solely a conservation activity for him or is it also recreational? Explain your reasoning.

Conservation-focused relationships with outdoor environments

Perceptions	Interactions	Impacts
• Outdoor environments need protection and care • 'Outdoors as a museum' or 'outdoors as a mother' metaphors	• Revegetation • Erosion control • Weed and pest control • Habitat restoration • Track development and maintenance • Clean-up programs • Wildlife sanctuaries • Ranger activities • Junior ranger programs • Activists	• Encourage human visitation by managing a park • Grow more trees and make more oxygen • Stabilise soil with revegetation • Improve biodiversity through pest management • Develop vehicle maintenance tracks

Figure 8.15 Conservation-focused relationships with outdoor environments

Worksheet 8.2b Conservation interactions

LEARNING ACTIVITY 8.2B

Conservation interactions

For three of the conservation-focused interactions outlined in Figure 8.15, list a specific outdoor venue that would be appropriate for that interaction.

Figure 8.16 Revegetation of non-productive grazing land in Gippsland

Figure 8.17 Tracks and lookouts in popular parks and reserves are designed to direct visitors to places where impact can be managed.

Recreation

Recreation can be defined as pastimes that are a diversion from day-to-day routines. Recreation may include active or passive activities that provide the participant with fun, relaxation, enjoyment or fitness. For many of us, our interest in the environment begins with such experiences. It is these experiences in the outdoors that can lead to an appreciation and understanding of outdoor environments. Examples of popular recreation activities include bushwalking, surfing, rock climbing, skiing and snowboarding, canoeing, mountain bike riding and fishing.

recreation pastimes that are a diversion from day-to-day routines, including active or passive activities that provide the participant with fun, relaxation, enjoyment or fitness

In Chapter 7, we discussed the use of metaphors to assist us in our descriptions of human relationships with outdoor environments. Metaphors are also a handy way to describe how humans perceive the outdoors.

When participating in recreational activities, the outdoors can be perceived as a playground, or venue, that provides the scope for activity. It might be a route through a national park that someone takes during a hike, or a child climbing a tree in the backyard. The outdoors can also be perceived as a gymnasium, or a range of apparatuses that provide a series of problems and challenges to be completed during physical activity. A cliff face can become a collection of holds and bollards that are used to climb. The climber attaches ropes and finds handholds and footholds in order to achieve the objective of reaching the top. Here, the outdoors is also perceived as an object that provides risks for those that seek the thrill of participation. In these examples, people are separating themselves from the outdoor environment. This does not always need to be the case. Recreation interactions can also provide a more intimate human–nature relationship. Bird-watching has at its core an ecological focus on the study of another species.

Shutterstock.com/adirekjob

Figure 8.18 Impacts from climbers at 'The Gallery' in Gariwerd (Grampians National Park)

Recreation activities will inevitably have an impact on the outdoor environments in which they take place. Quite often, these impacts are negative as venues are altered or under continual pressure from regular use. The rock-climbing site may require modification to cater for visitor numbers. Paths and climbing sites can be constructed, resulting in clearing of vegetation and habitat loss. Anchors can be provided for climbers and magnesium chalk is often left behind, defacing the rock face

Resource Additional Case Study: Warburton mountain bikers remove in excess of five tonnes of rubbish from the bush

However, not all impacts need to be negative. Those who interact with the outdoor environment for recreation will often develop a bond with the specific environments they use. This bond can foster an allegiance between people and those environments that is based on sustaining their interactions to enable future participation for themselves and others. By promoting and adopting minimal-impact behaviours, such as leaving a venue in better condition than when you arrived, environments can sustain regular human use more resiliently. Have you performed an 'emu parade' before leaving a campground while on a school trip? If so, you have probably picked up a lot of rubbish left behind by previous campers.

Allies for a particular environment don't always stop there. Recreation-based **interest groups** are often among the first to take up the fight against threats to the environments they love, so cross over readily to the previous group: conservation.

interest groups groups of individuals with similar values who aim to promote their views about an issue (note that government or government agencies are not considered to be interest groups)

Resource
Case Study: Sharon Lewis– Open-water swimmer, Point Roadknight

CASE STUDY

Sharon Lewis: Open-water swimmer, Point Roadknight

I have always loved the outdoors and learning about, and caring for, the environment. I studied environmental science after leaving school and then worked for many years in jobs where I was looking after the environment – mainly working with farmers to protect remnant vegetation and wetlands on their farms. I have transitioned into a different type of career now, but I think ocean swimming has helped to keep me close to nature.

Sharon Lewis

Figure 8.19 Sharon Lewis pictured prior to the 2023 Lorne Pier to Pub open-water swim

Swimming is an addiction for me now. It's absolutely essential for my mental and physical health. When I swim, most of the time I'm really focused on everything that is going on with my stroke. Where is my hand entering the water? How hard am I pulling through the water? How fast am I kicking? At other times, I get completely lost in my thoughts while I process something that has been on my mind. Before I know it, I've swum 100 metres and I can't remember anything about it. So for me it's very meditative. But it's also really physical as well. Swimming tones my muscles and I love feeling strong in the water as well as the endorphin rush that I get at the end of a swim.

The Pier to Pub is a 1.2 km open-water swimming event held annually in Lorne, which is on the Great Ocean Road in Victoria. It is one of the largest open-water swims in the world and attracts up to 4000 competitors. The event starts at the pier in the town of Lorne, and swimmers follow buoys to cover 1.2 km, finishing at the Lorne Surf Life Saving Club on Lorne main beach (across the road to the Lorne Hotel – the pub). I have swum the Pier to Pub about five times now, and this year, my husband and daughters all swam it with me too.

I swim in both the pool and the ocean. I live about 30 minutes away from the sea, so it's not always convenient to swim in the ocean. There are also times when the weather is too cold, or I don't have anyone to swim with (because you shouldn't ocean swim alone), so the pool still gives me a swimming experience. It's also a different type of swimming in the pool. The fitness I can achieve in the pool helps me to perform well when I swim in the ocean.

What I love most about swimming in the ocean is that I can be immersed in the natural world. There might be hundreds of people on the beach, but I can escape the crowds by swimming out from the beach and exercising with just the sounds of the waves. You appreciate everything so much more when you are out in the ocean: the sunshine, the clouds, the cliffs, the animals you might see. All your senses are heightened and your adrenaline is pumping when you are out in the ocean.

This year I will be doing a swimming tour in Croatia. This will involve swimming about 3 km each day. I will also be swimming off Mozambique in Africa, where we might encounter some whale sharks. I am very excited about the adventures I will have with swimming this year.

I attend conservation-focused activities when I can. But often I'm not available when these events are on. My way of contributing to protecting our oceans is to be really careful about what I put down the sink when I'm at home, and I pick up any rubbish I see when I'm at the beach or when I'm walking anywhere. When rubbish gets to the beach or gets into the sea, it really spoils these beautiful places. It's such an easy thing to do, and quite empowering to think that you can make a small difference every day. I encourage everyone to do it.

Sharon Lewis

QUESTIONS

1 Why does Sharon swim in the ocean?
2 Would you classify Sharon as a recreation user of the ocean or a conservation user? Why?
3 How does Sharon describe her relationship with the natural world? How has that changed over her life?
4 Sharon notes being unavailable to attend most conservation focused activities. Is this due to her not wanting to attend, or the timing of the events? Suggest how conservation activities can be better designed to allow busy people to participate in them too.

LEARNING ACTIVITY 8.2C

Your favourite place

Choose an outdoor environment you love in which you have participated in a recreational activity, then complete to the following:

1 Make a list of the human impacts on this environment that you can recognise at this venue.
2 What has been done by managers of this venue to try to minimise human recreational interactions?
3 Have the efforts of managers been effective in minimising human impacts? Explain.
4 List three things you could do to minimise your impact on this outdoor environment when participating.

Worksheet 8.2c Your favourite place

Recreation-focused relationships with outdoor environments

Perceptions	Interactions	Impacts
• Outdoor environments are for having fun in • Outdoor environments are for exercise • 'Outdoors as a gymnasium' metaphor	Active recreation activities: • bushwalking • canoeing • fishing • mountain-bike riding • rock climbing • snowsports • surfing • swimming Passive recreation activities: • bird watching • sightseeing • strolling	• Erosion and rubbish through high visitation • The need for amenities to be provided (toilets, roads/trails) • Encourages visitation to outdoor environments, which improves health and wellbeing and can lead to conservation interactions

Figure 8.20 Recreation-focused relationships with outdoor environments

Worksheet
8.2d Recreation interactions

LEARNING ACTIVITY 8.2D

Recreation interactions

For three of the recreation-focused interactions outlined in Figure 8.20, list a specific outdoor venue that would be appropriate for that interaction.

Economic

economic
relating to or based on the production, distribution and consumption of goods and services

Economic interactions describe any instance where a person's motivation to be in the outdoor environment is linked directly to an economic benefit to themselves. This can be through:

- growing something to be sold for profit
- digging something from the Earth to be sold for profit
- harvesting native forests to be sold for profit
- providing tourism services to access recreation, beautiful views or pristine nature for profit.

This view of outdoor environments is based on the belief that nature is here to provide something to be sold. We return to a metaphor again – nature as resource – where nature provides the product to be sold, be it raw materials (like coal for creating electricity) or a beautiful waterfall for tourists to visit. It is very much a perception that nature is here for us to use for our own benefit. However, that's not to say that this is all bad.

A farmer's perception of their grazing land is indeed as a resource, but many farmers share the desire to care for their land so it can be used into the future. A farm is not a mine where you dig something out of it once and then it is done. Rather, it is a cycle of investment and return through earnings.

Tourism also provides the opportunity for commercially based recreation as adventure companies take advantage of the visitor who is seeking the outdoor experience during their stay. Besides passive or physical activities, the outdoor environment itself can be the focus of tourism.

Alamy Stock Photo/Roger Fletcher

Figure 8.21 Tourists at summit of Mount Buffalo

The buying and selling of the outdoor experience by commercial companies reflects the perception of the environments they operate in as a resource to be exploited, with the company as the main financial beneficiary. They might be selling the experience or the 'specialness' of an environmental feature, and in doing so they are treating it as a **commodity**. The negative impacts of these perceptions can be profound when tourism is rationalised. That is, if profits can be increased by methods such as increasing participant numbers or revisiting the same place regularly for efficiency of operation, the environment could be neglected, leading to negative impacts.

commodity
something that can be used for commercial advantage (i.e. it can be bought and sold)

Granted, the interactions of an economic-focused relationship with outdoor environments can be harmful to other aspects of the environment and have lasting impacts. For hundreds of years, non-Indigenous peoples have been exploiting 'nature as a resource' to increase survival, comfort and growth at the expense of the health of outdoor environments, and continue to do so. This can be evident in the way we as a society continue to impact environments and landscapes that provide for us, despite our awareness of the extent of the damage that we are doing. Consider the economics of the impact of burning fossil fuels on **climate change,** for example. Victoria's main source of emissions is burning fossil fuels – like coal, oil and gas – for energy and transport. In 2019, the energy sector accounted for 70% of Victoria's emissions, while transport accounted for 25%. Today, we still rely on coal for 60% of our electricity generation, despite the recent closure of power stations.

climate change
a change in global or regional climate patterns, in particular a change apparent from the mid to late 20th century onwards and attributed largely to the increased levels of atmospheric carbon dioxide produced by the use of fossil fuels

Alamy Stock Photo/ krutopimages

Figure 8.22 The Loy Yang coal mine in the Latrobe Valley, Victoria. The Latrobe Valley contains 25% of the world's known brown coal reserves.

Although we are aware of the impacts this industry has on atmospheric carbon levels and associated climate changes, as well as on the environments around the open-cut mines, the industry is set to expand in the future to meet society's increasing demand for energy and to reap the economic profits of its continuation.

Many of the environmental impacts of economic relationships are profound and long-lasting; for example:

- the alteration of landscapes through mining practices
- the removal of habitat for the harvesting of native timbers
- the modification of habitat and reduction of biodiversity that results from clearing for grazing and other agricultural practices
- pressures on specific species from excessive harvesting such as commercial fishing
- pollution of the atmosphere and other environments through energy production and other industries.

Many of these impacts are continuing and increasing in scale today as our demand for resources increases throughout our expanding society. Although we are making some efforts to make these industries more sustainable and to maintain the environments on which they depend, our level of consumption continues to deplete our resources at a rate never seen before on this planet.

Shutterstock.com/Bjoern Wylezich

Figure 8.23 Tantalum is used in capacitors, and as acoustic wave filters in mobile phones to achieve high audio quality. A mobile phone has, on average, 40 milligrams of tantalum inside.

Worksheet
Additional Learning Activity: 8.2e Alcoa Anglesea power station and open cut mine closure

Tourism interactions refer to people travelling to visit outdoor environments that are away from their usual surroundings. A tourism operator's perception of an environment needs to have a high value of care and reverence as a sustainable operation. If the tourism operator is relying on the pristine nature of the waterfall being visited, then their economic relationship would fail if they destroyed the access trails and allowed litter to accumulate in the swimming hole. Likewise, those in the timber industry of Victoria noted in the late 1800s the need to conserve the forest resource so it could be used into the future. We can't paint all economic relationships with outdoor environments as simply 'bad'.

Economic-focused relationships with outdoor environments

Perceptions	Interactions	Impacts
• Outdoor environments provide the material for making money • Outdoor environments are there to be accessed and used for the survival and comfort of humans • 'Outdoors as a resource or storehouse' metaphor	• Agriculture • Commercial fishing • Cultural tours • Dams • Desalination • Ecotourism • Energy production • Forestry • Grazing • Mining • Outdoor adventure companies • Sightseeing • Tour companies • Tourist attractions	• Long-term damage to outdoor environments including erosion, loss of topsoil, salinity and loss of biodiversity • Strong bond with outdoor environments leading to greater care and conservation attitudes (e.g. farmers working to reduce climate change with their practices) • Recreation also occurs on farms and promotes care for outdoor environments • Landcare and other programs are successful in rehabilitating agricultural or mining lands • Tourism often promotes wonder and care for outdoor environments, which can lead to conservation attitudes

Figure 8.24 Economic-focused relationships with outdoor environments

Worksheet
8.2f Economic interactions

LEARNING ACTIVITY 8.2E

Economic interactions

For three of the economic-focused interactions outlined in Figure 8.24, list a specific outdoor venue that would be appropriate for that interaction.

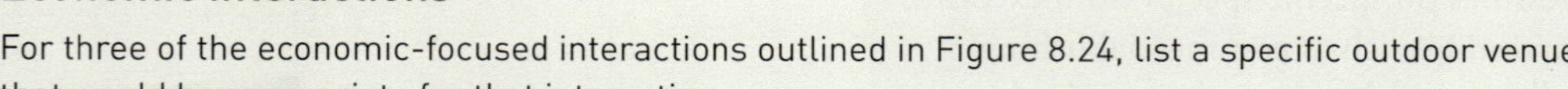

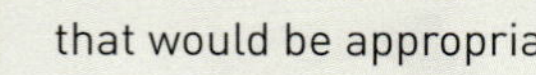

Resource
Case Study: Economics-focused relationship

CASE STUDY

ECONOMICS-FOCUSED RELATIONSHIP

Michael Bonhote, farmer

I grew up in Sydney, and after studying at university had a career in accounting and funds management. When I was 25, and living and working in the UK, I had a chance to work over the harvest period for a farmer in the north of England. I loved it and I think it sowed the seeds for my later move.

After nearly 50 years living in big cities both here and overseas, I made the decision to move to the country and try my hand at farming.

Farming is unlike many jobs you might do and is subject to a different set of rules, if you like. You are largely dictated to by the environment, weather, seasons, international markets and many other things outside of your control. As well, it can be hard physical work in difficult weather. When things have to get done, they need to be done, no matter when they happen (such as tending to sick animals). I enjoy many aspects of farming: producing a commodity that feeds people, being your own boss (but really working within the rules dictated by the natural world, which is challenging), living in a wonderful part of the world, and being in charge and responsible for a little corner of the country.

While they're two totally different occupations (farming and finance), I do use a lot of the skills learnt in my previous life in the city in the management of the farm business. I was ready for a more lifestyle-orientated second career and farming is giving me this.

The farm business is the right size for us in our stage of life, and with two teenagers at home. I am always looking to improve the sustainability of the farm, so am focused on soil health, grazing management, water and shade and shelter. These activities not only improve sustainability but also increase efficiency and the value of the enterprise. We are members of Landcare, which is a great movement and has opened my eyes up to many improvements that we have implemented on the farm.

The farm is where we live and work. Other than the odd bonfire with friends or having a day enjoying our own section of the river, we keep the farm for farming. Anything inconsistent with or detrimental to that doesn't really get a look in.

Interview by Jarrod Paine

Michael Bonhote

Figure 8.25 Michael Bonhote, cattle farmer in Tawonga, Victoria

QUESTIONS

1 Why did Michael move from finance to farming as a career?
2 What suggests Michael's relationship with his land has an economic focus? What part of his story describes this the best?
3 How does Michael describe his relationship with the land? Is it purely financial, purely care for the environment, or somewhere in between?
4 Michael writes that he enjoys many aspects of farming. What would you say he enjoys the most? Why do you think that?

Worksheet Additional Learning Activity: 8.2g Patagonia clothing company – Earth is our only shareholder

Resource Additional Case Study: Relationships with the Victorian Alps

NOTES FOR THE EXAM

For the exam, you should:

- know about a variety of examples of each type of interaction with environments you have visited or are familiar with; that is, examples of:
  - conservation interactions
  - recreation interactions
  - economic interactions
- be able to compare different human relationships with outdoor environments, including:
  - how the specific outdoor environment is being perceived
  - possible impacts on the outdoor environment
  - similarities and differences between various relationships.

Worksheet
8.2h
Relationships with outdoor environments

LEARNING ACTIVITY 8.2F

Relationships with outdoor environments

1. What is meant by a 'conservation' interaction with the outdoor environment? Give an example of this interaction with a specific venue.
2. What is meant by a 'recreation' interaction with the outdoor environment? Give an example of this interaction with a specific venue.
3. What is meant by an 'economics' interaction with the outdoor environment? Give an example of this interaction with a specific venue.
4. Choose two of the interactions above and compare the relationships with the chosen outdoor environment reflected in them. (In other words, how are the perceptions of the environment, interactions with the environment and impacts on the environment similar and different between the two examples chosen? You may find it helpful to start with a Venn diagram.)

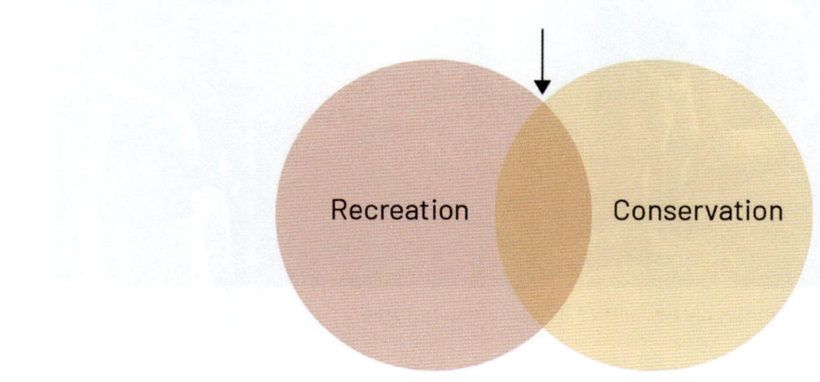

Figure 8.26 What perceptions, interactions and impacts do these two users have in common?

8.2 KEY CONCEPTS

- There are three types of relationships with outdoor environments. Each needs to be considered individually but can occur simultaneously by an individual.
- Conservation includes the perception that nature needs our help.
  - interaction – revegetation, weed and pest control, protests/activism and clean-ups, etc.
  - impacts – generally positive for the environment but can be negative for other users or users with economic goals
  - can simultaneously overlap with an economic relationship in an eco-tourism activity.
- Recreation includes the perception that nature is our gymnasium:
  - interaction – outdoor activities
  - impacts – damage to outdoor environments from activity and high user numbers, and can affect other users' experiences
  - can lead to a person becoming involved in a conservation relationship with an outdoor environment as they begin to appreciate its natural qualities or their activity is at risk of being 'cancelled' or 'locked out' due to damage or poor practice
  - can simultaneously overlap with an economic relationship in an eco-tourism activity or paying an activity provider.
- Economic includes the perception that nature is a resource to benefit us with money, sometimes despite the impacts caused:
  - interaction – grow, dig, harvest or provide tourism
  - impacts – can be localised or 'climate change' scale in damage
  - can be well intentioned and can also be positive (e.g. Patagonia).

8.2 CONCEPT QUESTIONS

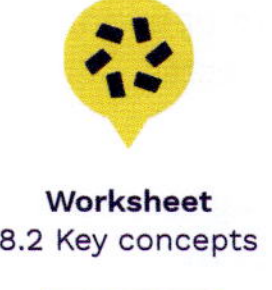

Worksheet
8.2 Key concepts

REMEMBERING

1 Using an example of each, distinguish between the perceptions of conservation, recreation and economic relationships with outdoor environments.

2 What does it mean to have two types of interactions occurring simultaneously? Provide an example.

3 List five examples of economic relationships with outdoor environments.

UNDERSTANDING

4 Explain the relationship between economic interactions, their impacts and our needs as humans.

5 Compare the three types of interactions and provide your opinion on which has the largest impact on humans.

6 Describe the benefits of a person who loves to mountain bike becoming involved in a group that cares for outdoor environments. Refer to the Warburton clean-up day.

APPLYING

7 Research somewhere that a recreation group has worked together on a conservation-focused task (use an example not presented in the text).

8 Describe three changes in interactions that could reduce the impact of three economic-focused interactions with outdoor environments.

EXTENSION CHALLENGE

9 Construct an infographic that portrays the three types of relationships occurring in the Victorian Alps.

8.3 CONFLICTS OVER THE USE OF OUTDOOR ENVIRONMENTS

KEY KNOWLEDGE

- methods used by individuals and groups to influence decisions about two conflicts over the use of outdoor environments, and the processes followed by land managers to resolve said conflicts, including at least one from the following list:
  - feral species in the Alpine National Park
  - commercial logging in Victoria
  - establishment of new mountain bike parks
  - Southern Ocean whale hunting
  - Murray–Darling Basin water allocations

KEY SKILLS

- describe two conflicts and evaluate the methods used by conflicting parties to influence decisions in their favour, and the processes followed to resolve or potentially resolve said conflicts

8.3.1 WHAT IS A CONFLICT?

conflict
a serious disagreement or argument, typically a protracted one; a serious incompatibility between two or more opinions or interests

human condition
all of the characteristics and key events of human life, including birth, learning, emotion, aspiration, morality, conflict and death

Humans seek **conflict**. Well, do we actually seek conflict or are we just obsessed with it? You only have to watch a nightly news service, which is researched to reflect what news 'we' want to view, and the stories are full of conflict. Neighbour to neighbour conflict, conflict with the government and within the government, international conflict and horrific conflict resulting in loss of human lives.

We are not saying conflict is something to be celebrated, but that conflict is part of the **human condition** and part of what sets our species apart – we are always trying to improve a situation to our particular liking or opinion.

Many individuals or groups of people have different opinions about how natural environments should be used. Conflicts occur because different individuals or groups have different interests in a specific outdoor environment and different beliefs about how to use, or interact, with it. For example, the perception that the outdoor environment is a resource that humans use primarily to provide for their own needs would be at odds with the notion that humans should be trying to live in harmony with the environment. Usually, competing views regarding the use of the outdoor environment involve quite specific aspects about a particular environment and can occur between groups who have very different types of interactions with the outdoors.

Figure 8.27 A sticker outlining the Otway Ranges Environment Network's campaign against logging old-growth forest for the production of woodchips. These stickers were placed on toilets in the region to provide community awareness about the brand of Kleenex toilet paper that utilised woodchips from Otway Ranges operations and to promote the use of alternative brands.

Otway Ranges Environmental Network

The use of an outdoor environment for primary industry focuses first and foremost on what this environment is able to provide in the way of resources for human needs. While the impacts of such interactions might be considered and even minimised as much as possible, they tend to be tolerated in favour of production. However, if this specific environment is also valued by others in society as a place worth protecting and preserving with conservation the focus, then conflict may arise.

Conflicts can also occur between groups who have similar uses or types of interactions with the outdoor environment. In this instance, the same venue could be used by people with very similar perceptions of how the particular environment should be used – such as surfers and fishers – but still conflict over access and impacts.

Shutterstock.com/Tom Grundy

Figure 8.28 Entanglement in discarded fishing gear has been identified as a threat to sea birds and smaller cetaceans. For large ocean-dwelling mammals, swallowing such refuse is also emerging as a serious cause of disability.

Getty Images/ Ian Waldie / Stringer

Figure 8.29 Discarded litter disrupts the experience for people in natural environments.

Individuals, groups and government agencies

When conflict over the use of the environment arises, individuals with similar values may form an interest group with the aim of promoting their views about the issue. The value of forming an interest group is the ability to draw on a greater degree of resources than might be available to an individual.

The largest environmentally focused interest group in Australia is the Australian Conservation Foundation (ACF), which was formed in 1964 by a group of scientists, businesspeople and public servants. It has addressed a range of national environmental issues such as protected areas, sustainable agriculture and land management, climate change and energy, nuclear issues and water management.

Not all interest groups are formed with conservation as their primary focus. The Australian Forest Products Association declared, after its creation in 2011, that it aimed to be 'a single voice … to present the forest products industry to governments, the media and the people of Australia in a united fashion'.

Government agencies are also considered as groups involved in conflicts over the uses of outdoor environments, despite their own interests and their influence during the decision-making about specific environmental issues. Government agencies are often part of governments, but can also exist as an entity across multiple governments. The perceptions and views of these stakeholders in regard to specific outdoor environments can often be in line with a government's policies, which in turn are intended to reflect the views of their supporters. The conflict case studies outlined later in this chapter include descriptions of the role of these types of government stakeholders as well as specific interest groups.

Courtesy of the Australian Conservation Foundation: www.acf.org

Figure 8.30 The Australian Conservation Foundation (ACF) is the largest environmentally focused interest group in Australia.

Methods used to influence decision makers

> First they ignore you, then they ridicule you, then they fight you, and then you win.
>
> Mahatma Gandhi 

Most Australians over the age of 18 are eligible to vote to elect governments in Australia at three levels: local, state and federal. This system allows members of the community to apply pressure to elected representatives within the relevant level(s) of government and, therefore, potentially influence decision-making processes. Interest groups and individuals adopt a range of methods to influence decision-making processes about the use of the natural environment. Governments and their agencies are also in the business of gaining support for their environmental policies. While they generally do not engage in the same sort of publicity as environmental groups, they can employ their own methods to influence public opinion on their intentions for a particular outdoor environment.

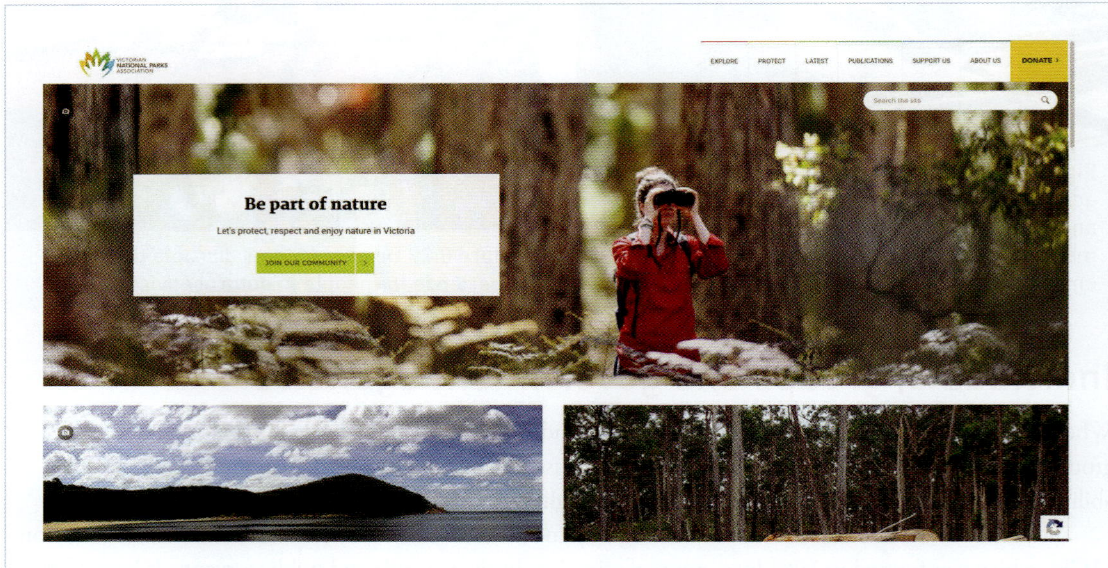

Victoria National Park Association, www.vnpa.org.au

Figure 8.31 The Victorian National Parks Association website

The methods that can be used effectively to influence decision-making processes about outdoor environments are many and varied. The effectiveness of these methods often depends on what decision-making process is being applied to the particular conflict and the parties involved. Various methods are used by local associations, social movements, neighbours, recreational groups and, less commonly, political parties or trade unions that have been driven to action. These groups work to raise community awareness and support, and to apply direct pressure to decision-makers. Which methods are used and the extent of their use will often depend on the scale of the conflict and the resources at the disposal of the interest group. It is not uncommon, for example, for campaigns to wane in their intensity if the ability to fight is hindered by lack of finances or costly legal battles. This is especially so in the case of smaller interest groups involved in localised issues against highly resourced corporations or governments.

Research into more than 2700 environmental conflicts across the world found that methods to persuading decision makers to take action to protect the environment ranged from non-violent (preferable and more successful) to potentially violent actions. Of the methods (see Table 8.1), the top 10 most popular are depicted in Figure 8.32.

Table 8.1 Summary of methods of influencing decision-making

Method	Advantages	Disadvantages
Non-violent protest and persuasion: • formal petitions • public campaigns including meetings and information evenings • street protests • development of collective action networks • involvement of NGOs • media-based activism • rights of nature argument • appeals to economic valuation • artistic actions	• Petitions are easy to conduct and gain a lot of support • Online petitions can reach a high number of people and gain awareness of an issue • Members of wider community are able to ask questions and become better informed • Shows the level of support on the issue • Clear message and information can be sent to a large number of people through media	• People sign petitions with little knowledge about the issue, so petitions have little impact unless numbers of signatures are high • Meetings are often poorly attended • Interviews are hard to organise • Advertising is very expensive
Non-cooperation: • strikes • boycotts of official processes • refusal of compensation • boycotts of companies and products	• Gains free media attention • Brings the issue to the public eye • Shows level of support • Can disrupt earnings of a company, encouraging a compromise	• Small groups may have little impact • May alienate groups in the community • Can require significant preparation and organisation • Time-consuming and expensive • Opposing groups/companies may gather opposing evidence negating the work
Non-violent interventions: • lawsuits • alternative knowledge creation • blockades • Environmental Impact Association objections • alternative proposals • participative research • occupation of public spaces • land occupation • referenda • hunger strikes/self-immolation • financial activism	• Conducted directly by individual/interest group • Ensures decision-makers have a clear understanding of your position • Scientific reports provide credibility to group's position • Arguments for position can be clearly understood through reports	• Can be difficult to access decision-makers • Can be time-consuming
Potentially violent actions: • property damage • sabotage • threats to use arms.	• Gains media attention	• May gain the wrong media attention • Illegal and dangerous • May risk the decision being made against its view due to poor behaviour

9780170477123

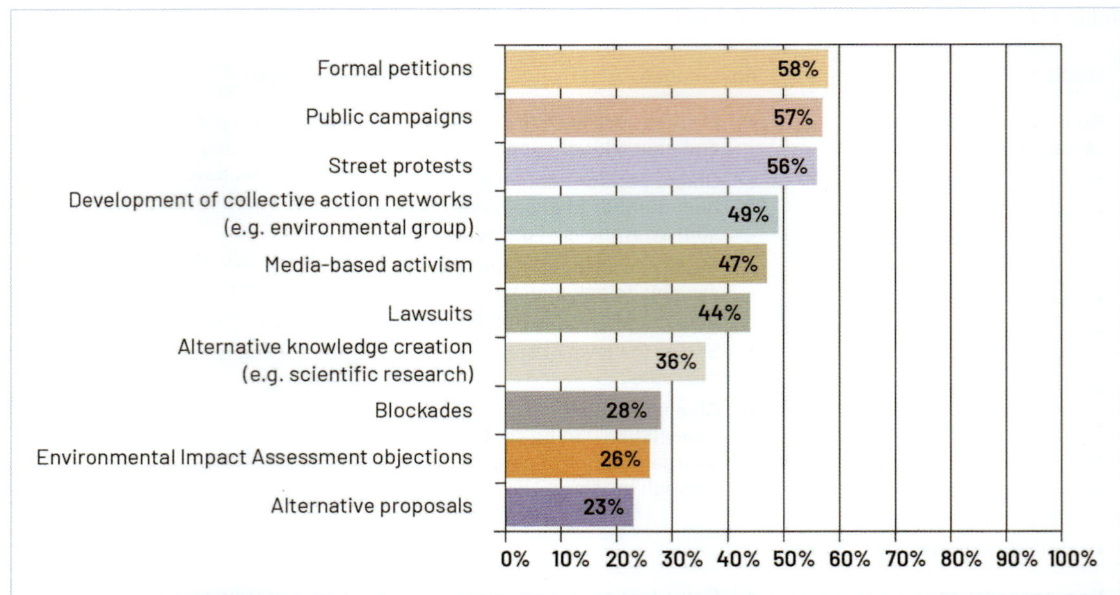

Figure 8.32 The top 10 most popular methods used by environmental activists to influence decision makers

Figure 8.33 Tree climbing, to draw attention to and disrupt logging operations, is a form of direct action.

Worksheet 8.3a Promote your point of view

LEARNING ACTIVITY 8.3A

Promote your point of view

Design your own campaign poster to promote one point of view for a conflict of your choice. The poster will need to contain some key points of information to enable someone with no prior knowledge of the conflict to understand your point of view. For your poster:

- create a catchy campaign slogan that could capture the attention of the public
- use images that relate to the environment involved and that support your argument
- include information about where someone can go to find out more about the issue.

Land manager processes for resolution of conflicts

When a conflict occurs over the use of an outdoor environment, a decision has to be made to resolve it in some way. When the conflict is over the use of Crown land (e.g. grazing in the Alpine National Park), the decision is the responsibility of the organisation or authority in charge of the management of that land. For example, Parks Victoria and the Victorian Environmental Assessment Council (VEAC) are **statutory authorities** that were established by the Victorian government to manage public land. Statutory authorities are not interest groups and they report to the government.

statutory authorities a government organisation established to exercise specific powers; for example, to manage outdoor environments such as a state parks or national parks

When conflict occurs over the use of outdoor environments, a process must occur through which some form of resolution can be achieved. The ideal resolution to most conflicts is a compromise between opposing interest groups that ensures all groups have been heard and can live with the decisions that are made. This is not always the case. More often than not, a compromise cannot be achieved, and interest groups may find themselves having to accept a decision made through another process (see Table 8.2).

Table 8.2 Different forms of decision-making processes

Process	Explanation	Advantages	Disadvantages
Community consultation For example, VEAC report of recommendations on the protection and sustainable use of Victoria's system of marine parks	• A consultative group can be formed or enlisted to consult interest groups, public and experts with specialised knowledge of the issue	• It provides the opportunity for all views to be heard • Accurate information is gathered • It promotes the possibility for compromise between groups, creating a win–win situation	• It may be time-consuming and expensive, and may result in no agreement being found • It requires skilled and respected mediator (VEAC)
Use of court systems For example, Waratah Coal v Youth was a court case heard in the Land Court of Queensland where it was recommended by the court that applications for a proposed coal mine be rejected on grounds that included human rights and climate change.	• The court system is used to clarify existing laws regarding the legal and appropriate use of an environment • It is often used when laws exist that relate to the conflict and need some clarification or interpretation	• It results in a clear decision • An independent decision is made by courts	• Court processes can be expensive and can take a long time • After a decision is made, government can change laws, creating more uncertainty • It creates a win–lose situation
Legislation (creating laws) For example, the federal government's creation of new laws in 1983 putting all World Heritage sites under the control of the federal government, allowing it to halt damming of the Franklin River	• Laws may be created to allow something to take place or to prevent something from occurring	• A clear decision is made • A definitive decision avoids a drawn-out dispute	• One interest group gets what they want while the other remains opposed to the decision made • It may strengthen the fight against the law, promoting further conflict • It creates a win–lose situation

Process	Explanation	Advantages	Disadvantages
Management plans For example, the Greater Alpine National Parks Management Plan guiding the management of Alpine, Baw Baw, Errinundra, Mount Buffalo and Snowy River national parks, Avon Wilderness Park, Tara Range Park, and Walhalla, Howqua Hills, Mount Wills and Mount Murphy Historic Areas (900000 hectares in total)	• The *National Parks Act 1975 (Cth)* requires that Department of Energy, Environment and Climate Action prepares a plan of management for each national and state park; park management plans articulate the vision, goals, outcomes, measures and long-term strategies for parks within planning areas • Essentially, it's a document containing guidelines on how an area of public land is managed • Zones may be created with specific focus for different areas of the environment	• Plans adopt a landscape-wide approach, taking into account elements bordering the park that influence how a park operates • They confine impacts to certain areas, enabling efficient management • They can address different needs through zones with different levels of restriction • Wide range of values can be protected, including Indigenous, environmental and historic	• It can create conflict between user groups, especially if one is excluded from their activities • Management is left to one organisation (e.g. Parks Victoria), rather than sharing the workload • Management has a narrow focus that doesn't allow other uses (e.g. conservation versus primary industry)
Use of the political system For example, the federal government's Murray–Darling Basin Plan	• The government formulates ideas and policies on how the environment is to be used • Political parties use parliamentary processes to influence decisions made by government	• It can be used to represent views of the majority of the public • It can be debated by different sides of an argument in a conflict	• Decisions can be held up by disagreeing political parties • Policies and decisions can be good for some and not others • Majority governments might make decisions despite opposing views
Referendum For example, 1981 referendum for the Tasmanian public to decide on whether the Franklin Dam should be built	• A vote by all registered voters decides whether to accept or reject a proposed change to the Constitution • It's generally used for larger-scale decision-making • Conflicts often become part of a larger process (e.g. a political party may use a conflict as a part of their political campaign to help gain the votes of a particular group of voters)	• It's a democratic process where everyone has their say • Clear decisions can be made	• It's a very expensive process where outcomes can be manipulated through media groups with lots of money • The decision-making process is often at the mercy of the political process and compromises, and backdowns can occur • The process is not generally suitable for small, local conflicts

LEARNING ACTIVITY 8.3B

Management plan investigation

Visit Parkweb, the main information website of Parks Victoria. Navigate to a national park of your own choice and read the information page. In the related publications section, access the management plan for this national park and respond to the following:

1 Outline the range of interactions available to visitors of this venue.
2 Describe any pressures on this outdoor environment.
3 Outline the different sections of the management plan.
4 Outline any management-based zoning of the national park.
5 List strategies for the management of visitors to the national park.
6 Referring to the management plan for a park or reserve that you have visited, outline the key management objectives of the park or reserve.
7 Provide practical examples of actions that are in conflict with the management objectives.

Worksheet
8.3b Management plan investigation

Weblink
Parkweb

EXAMPLES OF ENVIRONMENTAL CONFLICTS

The following case studies summarise some current high-profile environmental conflicts and some that have occurred in the past decade.

CASE STUDY

Resource
Case Study: Commercial logging in Victoria

COMMERCIAL LOGGING IN VICTORIA

Proposed Great Forest National Park

The proposal for the Great Forest National Park (GFNP) was influenced by the 2013 findings and recommendations of the Leadbeater's Possum Advisory Group. The proposal involved adding 355000 hectares of protected forest to existing 170000 hectares of parks and protected areas in the Central Highlands of Victoria, within Wurundjeri Country.

Fairfax Syndication/JOE ARMAO

Figure 8.34 The critically endangered Leadbeater's possum

While political parties and other stakeholders generally agreed with many of the recommendations, which were aimed at the support and recovery of Leadbeater's possum while maintaining a sustainable timber industry, the GFNP proposal created significant conflict. There was wide public support in 2014 (89% of Victorians, according to one opinion poll), including the support of many interest groups and prominent advocates for the environment (e.g. David Attenborough, Tim Flannery and Bob Brown). This conflict remains unresolved. As of 2023, the forestry industry continues to have access to native timber harvesting. Supplying jobs and income for commercial logging towns is deemed to be of higher importance to governments than the protection of these forests and the vulnerable species they contain.

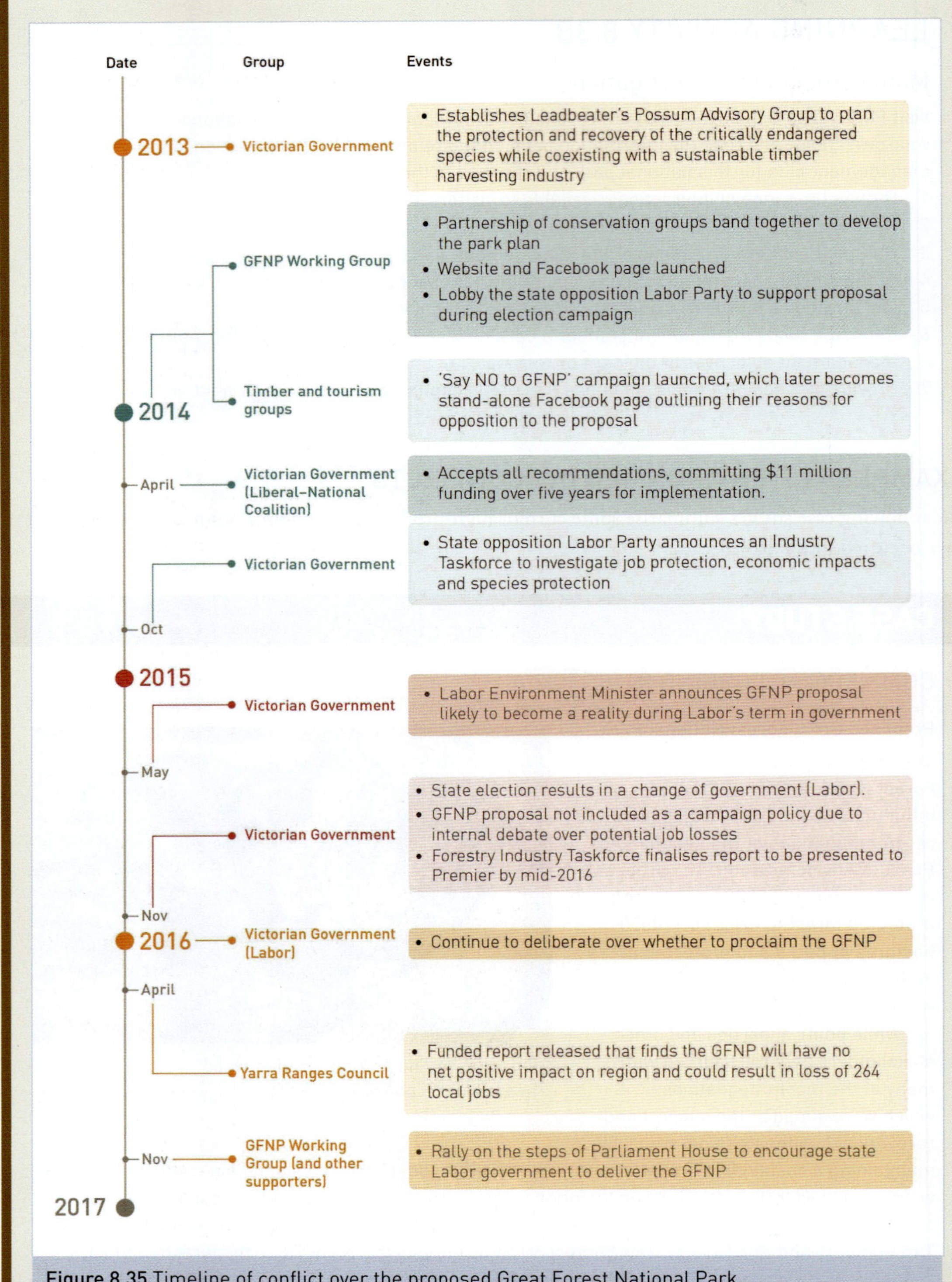

Figure 8.35 Timeline of conflict over the proposed Great Forest National Park

Table 8.3 Summary of conflict over the proposed Great Forest National Park

Environment	Central Highlands rainforests – roughly 60 kilometres east of Melbourne in an area that ranges from Kinglake in the east to Mt Baw Baw in the west, Eildon in the north and Warragul in the south	
Issue	Proposal to declare a national park of over 5000 square kilometres that incorporates existing national parks, reserves and state forests	
Parties to the conflict	• Great Forest National Park Working Group • Victorian National Parks Association (VNPA)	• Forestry industry • Some recreational interest groups
Interaction	Campaigning for the creation of the GFNP that focuses on a balance of conservation management and broad recreational possibilities, and the discontinuation of exploitive interactions	Campaigning for timber harvesting and exploitive recreational activities (e.g. deer hunting)
Position on issue	The creation of the Great Forest National Park would not only safeguard the habitats of endangered species and protect these critical forests, but also protect carbon stocks and water supplies, and lower the risk of bushfires. Adding a proposed 355000 hectares to the existing protected forests will combine a number of individual national parks to form the Great Forest National Park	State forests should remain to enable continued primary industry, and the hunting of deer and other introduced species, as these provide better environmental management and the greatest benefit for the Victorian public
Methods used to influence decision makers	• Formed this partnership to develop a joint statement of more than 30 environmental groups in favour of the proposal • Website and multimedia campaign • Gathered evidence to produce the GFNP plan • Published GFNP Summary Report • Organised tours of the environment to assist people to experience the environment • Direct action protests and blockade of logging sites (e.g. Toolangi State Forest) • Organised opinion polls to express support for GFNP • Organised petitions in support of GFNP • Lobbied state government to deliver the GFNP • Enlisted prominent people to support the GFNP (e.g. David Attenborough, Tim Flannery, Bob Brown, Peter Garrett, Dave Hughes, Jane Goodall and Beau Miles)	• Launched Facebook page and campaign against GFNP • Gathered evidence into economic impact of abolishing local timber industry, potential threats (including fire) to Leadbeater's possum, control of invasive species through hunting, and value-added tourism of recreational activities • Convened community meetings to provide information on potential local employment and economic impacts of GFNP • Lobbied political parties to consider the opposing position on the GFNP • Held rallies to voice opposition to GFNP
Processes followed by decision makers	• Community consultation – an advisory group was established to consult with community and other groups to investigate the potential social, economic and environmental impacts and benefits of an issue or proposal. Recommendations to be provided to government.	
Evaluation, including discussion surrounding the positives and negatives, and an overall judgement call	The resolution of this conflict is not the fairy-tale ending for conservationists and nature lovers. While it has not yet ended (technically, the proposal is still active), it has failed to gain support from either major Victorian political party. The methods used by the forestry industry to influence the government's position focused on wealth creation – local jobs, timber products and the flow-on effects to communities – and this was a stronger motivation for the government than polls, petitions, letter writing or prominent people writing books in support of the creation of the park.	

The Great Forest National Park would be a boon for the region. Investment in nature tourism is the next big thing for growing centres such as Healesville and Warburton, and will invigorate smaller towns such as Toolangi, Noojee and Rawson.

Figure 8.36 Location of the proposed Great Forest National Park

Figure 8.37 The Knitting Nannas of Toolangi, shown here protesting against logging, are in favour of the proposed Great Forest National Park.

Resource
Case Study: Feral horses in the Alpine National Park

CASE STUDY

FERAL HORSES IN THE ALPINE NATIONAL PARK

Horses were introduced in the area of the current Alpine National Park as early as the mid-19th century and represented the first hard-hoofed animals to live there. This land has been shared by multiple Indigenous peoples' groups and can be referred to as Dhudhuroa, Jaitmathang or Taungurung Country.

The original release of the horses was either deliberate for cheap summer grazing, or as a result of farm escapes. Over the decades, numbers grew and shrank with the seasons, external pressures and the beginnings of control programs by land managers. The current population includes horses escaped from, or abandoned by, nearby farms.

While the community accepts the control of most feral animals in the national park, not all members of the public support some methods of feral horse control, and this has continued as a conflict.

The current conflict began in June 2017, when Parks Victoria published its *Feral Horse Strategic Action Plan 2018–2021* for the Alpine National Park. It set a target to remove 1200 horses over three years from the eastern section of the park, and completely remove the smaller Bogong High Plains/Cobungra population. This plan was met with strong opposition from the Australian Brumby Alliance, and it took Parks Victoria to court in 2018 over their belief that it violated the protection the horses were due.

As indicated by the timeline for this conflict (see Figure 8.40), this conflict remains dynamic. On 5 June 2020, a motion to call on the state government to cease the planned shooting of feral horses in the alpine areas was passed, receiving the backing of the Victorian Upper House. It was led by the Member for Northern Victoria, Wendy Lovell, and sent a clear message to both the Labor Government and Parks Victoria that further consultation was required to come to a solution that still respects the Federal Court's ruling of the removal of the horses. Two years later, in June 2022, the Victorian Liberal opposition vowed to stop horses being killed across the state if it were successful in winning the state election. It did not win. This issue can be described as a 'political football', with each party attempting to score points off the other and win votes, and meet judicial and legislation requirements to care for the alpine environment. The issue remains ongoing.

Fairfax Photo/NICK MOIR

Figure 8.38 Brumbies on Bogong High Plains

The Australian Brumby Alliance

Figure 8.39 Australian Brumby Alliance campaign poster

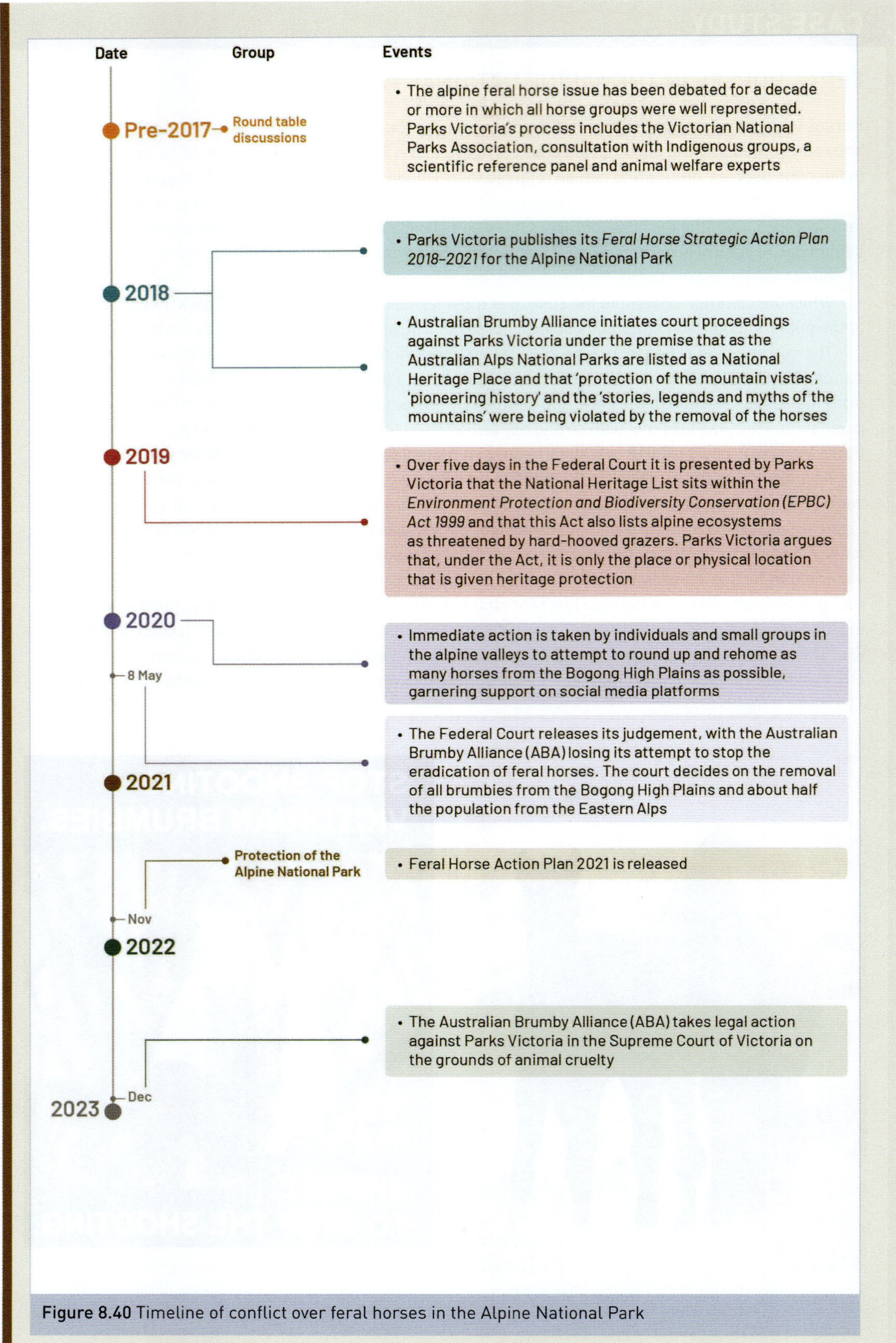

Figure 8.40 Timeline of conflict over feral horses in the Alpine National Park

Table 8.4 Summary of conflict over feral horses in the Alpine National Park

Environment	Sub-alpine and alpine environments of the Alpine National Park.
Issue	Introduced horses are growing in numbers and scientifically observed to be impacting threatened flora and fauna communities. The process of how to control the animals while respecting historic grazing culture and humane methods is sparking strong conflict

Parties to the conflict	• Australian Brumby Alliance (ABA) • Victorian Brumby Association (VBA)	• Victorian National Parks Association
Interaction	That 'brumbies' have grown as mountain creatures as part of an evolutionary process and should remain there to celebrate first cattlemen settler heritage	That the alpine environment should be protected so it can be enjoyed for recreation and conserved for future generations
Position on issue	The number of horses in the park should be managed. Animals should be rehomed and not exterminated	Feral horses should be removed from the Alpine National Park through the most effective means
Methods used to influence decision makers	• Created a website and multimedia campaign, including posting videos of the shooting of the horses in previous management • Published newsletters • Lobbied government • Participated in round table discussion	• Created a website and multimedia campaign • Wrote media releases • Conducted scientific research • Participated in round table discussion • Promoted community feedback forums

Processes followed by decision makers	• Round table discussions • Strategic Action Plan produced by land manager, Parks Victoria • Federal court and Victorian Supreme courts used to contest decision to remove horses on cultural heritage and animal cruelty platforms, but both failed, • Legislation – *The Environment Protection and Biodiversity Conservation* (EPBC) Act 1999 led to a decision by the Federal Court for the removal of horses • Further management plans produced: *Protection of the Alpine National Park: Feral Horse Action Plan 2021 released*
Evaluation, including discussion surrounding the positives and negatives, and an overall judgement call	• The Australian Brumby Alliance effectively uses dialogue (written and verbal) with Parks Victoria and government ministers as a positive method to influencing the ban on removal and shooting of horses in the Alpine National Park. • The Victorian National Parks Association, in a supporting role to land manager Parks Victoria, is using the method of law (through courts and legislation) to ensure care is being taken of threatened species in the Alpine environment. • This issue is still evolving but is perhaps entering an end point since the Federal Court ruled on the removal of all horses from the Bogong High Plains; what remains to be seen is the method used and whether shooting will return as an active control technique. • The Victorian National Parks Association, in a supporting role to land manager Parks Victoria, is using the method of law (through courts and legislation) to ensure care is being taken to protect threatened species in the Alpine environment, resulting in High court ruling to remove horses as per the management plan. • This issue still creates conflict in the community. Since the Victorian Supreme court ruling in February 2023 to continue with the plan to remove and reduce numbers of horses from the Alpine National Park, horse numbers have been drastically reduced, using approved methods including ground shooting.

Resource
Case Study: Establishment of Warburton Mountain Bike Destination

CASE STUDY

ESTABLISHMENT OF WARBURTON MOUNTAIN BIKE DESTINATION

The Warburton Mountain Bike Destination, just 80 kilometres east of Melbourne on Wurundjeri Woi Wurrung Country, is a project that began in 2016. The Yarra Valley Council proposed the construction of 186 kilometres of mountain bike trail in the Yarra Ranges National Park and surrounds, built across three hillsides. Its aim is to provide a world-class mountain biking destination, following proven methods of other towns' trail construction across Australia (generally old primary industry focused towns with highly impacted environments from past practices – in this case, logging), and to bring income and vibrancy to a 'sleepy' town. The council's primary reason for building the park is economic, with hopes thousands of tourists will flock to the trails as a solution to the boom-bust cycle the town has experienced as a result of its reliance on logging.

Since the proposal, the project has faced unprecedented opposition from local individuals and conservation groups. This has led the project to come under the spotlight from land managers, which in this case is the Victorian government through the department that cares for Crown land, the Department of Energy, Environment and Climate Action (DEECA).

Alamy/Andrew Bain

Figure 8.41 Cycling on the Lilydale to Warburton Rail Trail

While the project followed all the correct planning approval processes, pressure applied to government ministers from opposition groups meant the project kept meeting new obstacles. To this day, it is the only mountain bike trail project in Australia that has been required to undergo an Environmental Effects Statement. These are usually reserved for major infrastructure projects like mines, roads and pipelines.

With various interested groups working to influence the decision of the government, the process to come to a decision included the Environmental Effects Statement, four weeks of public hearings and more than 2700 submissions made by individuals or organisations. This information was presented to an advisory committee appointed by the Victorian Planning Minister, and this committee then presented a final report to the Minister.

After receiving all this information, a decision from the government was released on 28 October 2022:

> The Minister's assessment concludes that the majority of the project has acceptable environmental effects, subject to some specific project modifications and an amended environmental management regime.
>
> Warburton mountain bike destination © Copyright State Government of Victoria

This meant that the issue was resolved, with the vast majority of the network approved as proposed, minus some of the trails within the Yarra Ranges National Park. The amendments to the network were to shorten the project by four trails and to avoid trail construction where it would have impacted on the Leadbeater's possum and the Mount Donna Buang wingless stonefly.

Environment Effects Statement (EES)

Yarra Ranges Council has commenced a comprehensive and transparent planning, consultation and assessment process for the Warburton Mountain Bike Destination through an Environment Effects Statement (EES).

The EES for the Warburton Mountain Bike Destination will be exhibited publicly in early to mid 2021 and marks a significant project milestone.

What is an EES?

An EES is a well-established process, under the *Environment Effects Act 1978*, that provides a comprehensive framework for assessing the environmental effects of major projects in Victoria.

The Ministerial guidelines for assessment of environmental effects defines the environment as the physical, biological, heritage, cultural, social, health, safety and economic aspects of human surroundings. All of these are considered in the EES.

The guidelines state the objective of the assessment process is to provide for the transparent, integrated and timely assessment of projects capable of having a significant effect on the environment.

The EES process is managed by the Victorian Department of Environment, Land, Water and Planning (DELWP) on behalf of the Minister for Planning. Information about the process is available on the DELWP website www.delwp.vic.gov.au.

THE WARBURTON MOUNTAIN BIKE DESTINATION EES

Yarra Ranges Council's aim is to deliver a range of positive opportunities for the region through the Warburton Mountain Bike Destination. Protecting the area's significant natural and cultural environment and minimising potential impacts continue to be central to the project's development.

In May 2020, the Minister for Planning determined an EES would be required for the project. In June 2020, the Commonwealth Government determined that approval would be required under the *Environment Protection and Biodiversity Conservation Act 1999* (EPBC Act) and accredited the EES process for the related assessment.

In September 2020, DELWP issued draft Scoping Requirements for the project's EES for public comment. Following consideration of feedback, the final Scoping Requirements will be issued and will identify the matters to be investigated and documented in the EES.

The Minister for Planning will provide a final assessment of the effects of the project following public exhibition of the EES and a subsequent inquiry.

Warburton Mountain Bike Destination
A WORLD-CLASS MOUNTAIN BIKE DESTINATION IN THE HEART OF THE YARRA RANGES
rideyarraranges.com.au

Figure 8.42 Warburton Mountain Bike Destination Environmental Effects Statement

Warburton Mountain Bike Destination, Yarra Ranges Shire Council, November 2020

Table 8.5 Summary of conflict over Warburton Mountain Bike Destination

Environment		
Issue	The construction of 186 km of world-class mountain bike trail in the Yarra Ranges National Park and surrounds, built across three hillsides, attached to and encompassing some of the town of Warburton. Staunch opposition from many in the community, who fear the new mountain bike park will ruin the environment, increase bushfire risk and radically change the Warburton community.	
Parties to the conflict	• Ride Yarra Ranges • Yarra Valley Council	• Warburton Environment conservation group • Other interested individuals
Interaction	Planning and advocating for the creation of the Warburton Mountain Bike Destination that focuses on a balance of development of tourism infrastructure with environmental concerns	Conservation of Yarra Ranges environment so it can be enjoyed for recreation and conserved for future generations, as well as working to slow development of the town of Warburton
Position on issue	The Council proposed the project, and plan for it to have minimal impact on the natural environment and vulnerable species. They suggest that: • the trails could be some of the world's best • it offers solution to the town's boom-bust cycle • it will boost visitor numbers to the region by 130000 each year	They believe the project should not be constructed as the negatives are too large, such as: • serious health and safety concerns in fire-prone mountains – the trails envelop many houses and there is only one road out of Warburton in an emergency like a bushfire • fears for native wildlife – trails will enter the Yarra Ranges National Park, which is home to the critically endangered Leadbeater's possum and the Mount Donna Buang wingless stonefly, a tiny insect endemic to the region • it will put increasing pressure on a small town's resources
Methods used to influence decision makers	• Social media • Meetings and information evenings • Gathering evidence and presenting reports • Submission to EES by over 2700 individuals and groups • Direct action – the mountain bike club cleaned up more than five tonnes of rubbish from the area	Meetings and information evenings • Submission to EES by over 2700 individuals and groups • Petitions • Letter writing • Lobbying decision-makers • Forming partnerships with like-minded conservation groups, such as Victorian National Parks Association
Processes followed by decision makers	• Government spotlight and oversight, which led to an Environmental Effects statement being prepared • Extensive community consultation – an advisory group was established to consult with community and other groups to investigate the potential social, economic and environmental impacts and benefits of an issue or proposal. Recommendations were then provided to government to decide • Government planning minister decision	
Evaluation, including discussion surrounding the positives and negatives, and an overall judgement call	• This issue divided the community of Warburton and the greater Yarra Valley region. On the one hand, it is set to bring income to the town, in the form of visitors and tourism dollars, while showcasing the stunning outdoor environments of the Yarra Valley Ranges. But on the other hand, its impact on threatened and vulnerable species could be detrimental, as could the effect of the local human population with the influx of mass tourism. • The Planning Minister's decision – after an exhaustive planning approvals process never before seen on a similar project in Australia – delivered a compromise on the project by approving the larger trail network but removing trails with impact on those species within the Yarra Ranges National Park. Hopefully, this was seen as an amicable process and reasonable decision for all involved parties.	

Project timeline

- $60 thousand in funding secured for project feasibility
- Stakeholder and community consultation to inform feasibility

- Stakeholder and community consultation to inform Preliminary Master Plan
- Environmental and cultural heritage investigations

- Community consultation
- Decisions made on the scope of stage one with community feedback

- $11.3 million in funding secured
- World Trail appointed to complete trail network design & construct the trails

- Project referred for assessment under the *Environment Effects Act 1978* (Vic) and *Environment Protection and Biodiversity Act* (Cth)

- Minister for Planning decides the project needs to be assesed under the *Environment Effects Act 1978 (Vic)* and an (EES) needs to be prepared
- Community consultation to inform EES
- EES impact assessments commence:
  - Biodiversity and habitats
  - Surface water, groundwater & geotechnical hazards
  - Cultural heritage
  - Socio-Economic
  - Transport
  - Land use and planning

- EES submitted to Planning Minister

- EES exhibited for public comment

- EES Panel Inquiry assesses EES and prepares report for the Minister for Planning

- Receive Minister's decision on planning approvals
- Submit final approvals:
  - Planning Scheme Amendment
  - Cultural Heritage Management Plan
  - Emergency Management Plan
  - Construction & Operation Environment Management Plan

- Approvals secured
- Construction scheduled

Figure 8.43 Timeline of conflict over Warburton Mountain Bike Destination

Resource
Case Study: Ocean whale hunting

CASE STUDY

SOUTHERN OCEAN WHALE HUNTING

Whale hunting is an activity that has rewarded early explorers and industrious fishing villages for thousands of years. It is recorded that Norwegians were among the first to hunt whales, as early as 4000 years ago. In early times, this was a rich resource for humans: the meat, skin, blubber and organs were eaten, the baleen was woven into baskets and used as fishing lines, and the bones used for toolmaking.

In 1791, Southern Ocean whale hunting began in earnest after colonisation of Australia brought stronger seagoing craft than Indigenous peoples built and used. Commercial whaling existed in Australia until the last boat returned to port at Albany, Western Australia in 1978, marking the end of this practice for our nation. Whaling steadily became more unpopular, as its gruesome and cruel practices were publicised across the world, although both Japan and Norway continued to hunt.

The conflict of Southern Ocean whaling was based on Japan continuing to hunt and kill whales in the Southern Ocean near Antarctica, while its main conservation opponent, Sea Shepherd, rallied to influence decision makers to either stop Japan through international law, by force or by supporting the Sea Shepherd's work. Japan, through its Institute of Cetacean Research, maintained its purpose for hunting and killing was purely scientific. However, Sea Shepherd and its supporters asserted the number of whales being killed far exceeded what is needed for science.

In 2014, the United Nations International Court of Justice agreed that the numbers being slaughtered 'for science' by the Japanese whalers showed they had 'not adequately considered non-lethal methods of research or the feasibility of a smaller lethal harvest to achieve its research objectives', and had produced 'limited scientific output to date'.

The Sea Shepard is a direct-action international ocean conservation movement and its Australian chapter worked relentlessly to put the spotlight on Japan's slaughter of whales in our backyard and work to eventually stop Japan from operating in the Southern Ocean, a goal they were successful in achieving in 2018.

Figure 8.44 The Institute Of Cetacean Research whale hunting in the Southern ocean

Table 8.6 Summary of conflict over Southern Ocean whale hunting

Environment	Southern Ocean	
Issue	Despite a global moratorium on commercial whaling since 1986, Japan continued to hunt whales in the Southern Ocean until stopped by a court of law. Today, Japan continues to hunt whales in its own waters.	
Parties to the conflict	Sea Shepherd	Institute of Cetacean Research
Interaction	Sea Shepherd, an international direct-action ocean conservation movement founded in 1977 to fight against illegal fishing in international waters	The Japanese Whale Research Program (JARPA), initiated in the Antarctic Ocean mainly to estimate the biological parameters of minke whales
Position on issue	International and national laws protecting the world's oceans exist but are often not enforced. Sea Shepherd uses direct action to fill this enforcement void on the high seas	Advocates for the sustainable use of whales through research and analysis
Methods used to influence decision makers	• Website and multimedia campaigns • Merchandise sales • Direct action on the high seas being filmed and promoted on social media and news outlets • Forming partnerships with national governments • Use of prominent people to promote their aims • Gathering evidence on whale kill numbers and reporting these to the International Whaling Commission	• Meetings with decision makers, including the International Whaling Commission • Manipulating the media via whaling for science
Processes followed by decision makers	• Use of court systems – in 2014, in a case brought by the Australian and New Zealand governments, the International Court of Justice deemed the hunting of whales by the Government of Japan was illegal. This process began the end to Japan's whaling in the Southern Ocean. Japan kept whaling in the Southern Ocean for four more years and it was the constant direct action work by the Sea Shepherd that eventually led Japan to stop this practice.	
Evaluation, including discussion surrounding the positives and negatives, and an overall judgement call	• This issue is unusual because it involved the Government of Japan openly flouting an international moratorium on whaling and continuing this practice in international waters under the subterfuge of scientific research. No government intervened, as the cost would have been prohibitive, but the Sea Shepherd fundraised and then took to the seas to apply direct pressure throughout the whaling season, making it difficult for Japanese boats to continue to hunt and kill. Overall, this issue is interesting for us to study as it involves a private non-government organisation emerging as the most powerful force against a practice not condoned by the United Nations. It was largely through the efforts of this not-for-profit organisation, rather than other countries, that whale hunting in the Southern Ocean ended.	

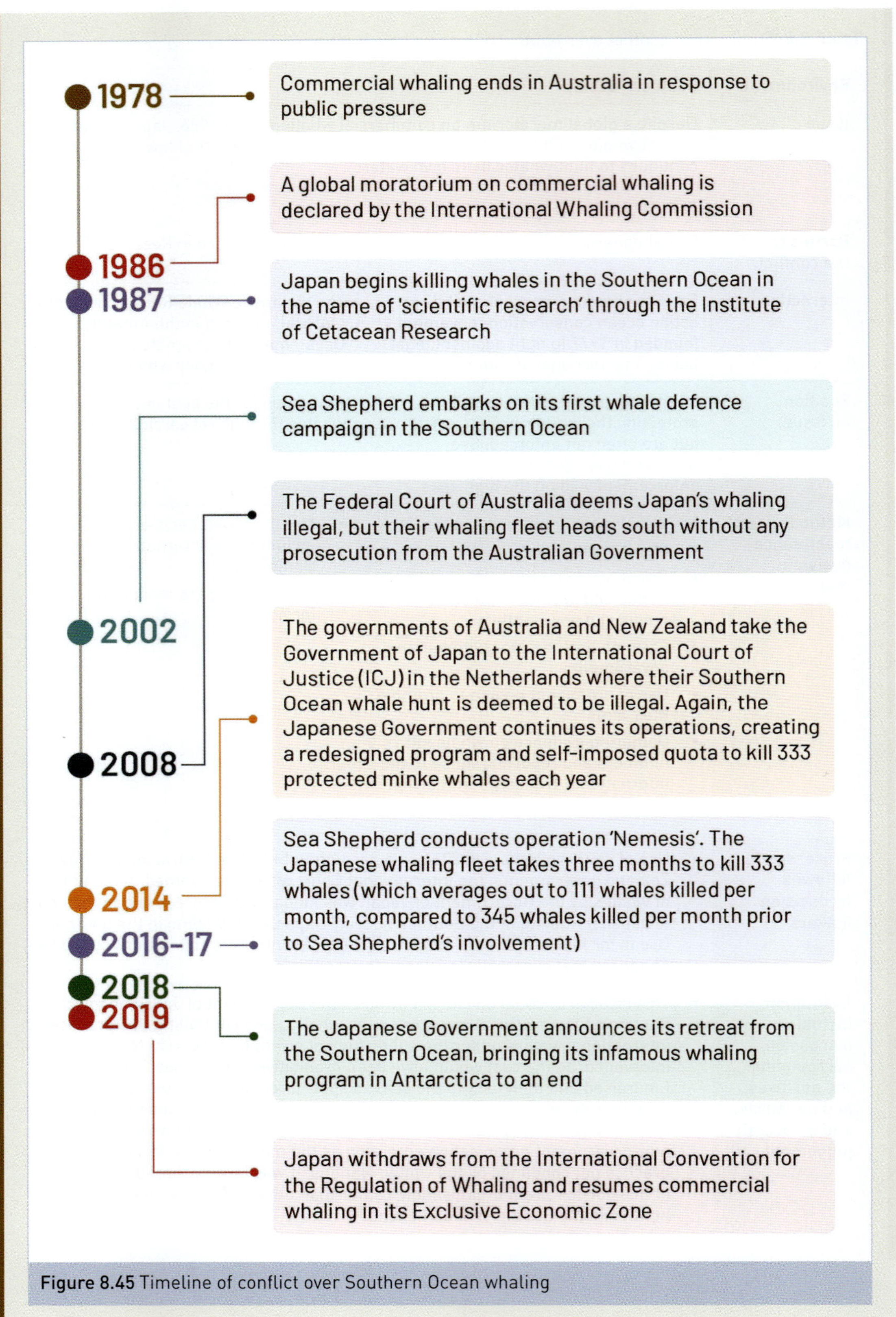

Figure 8.45 Timeline of conflict over Southern Ocean whaling

Fairfax Photo/JASON SOUTH

Figure 8.46 Sea Shepherd's Operation Nemesis, 2016–17

CASE STUDY

MURRAY–DARLING BASIN WATER ALLOCATIONS

The Murray–Darling Basin is a large area of eastern Australia, where water flows through a system of interconnected rivers and lakes. The two main rivers are the Murray River and the Darling River. The Darling begins in southern Queensland, where the Culgoa and Barwon rivers meet, while the Murray rises in the foothills of Mt Kosciusko. The two river systems meet at Wentworth in south-western New South Wales and meets the ocean at Goolwa in South Australia. All along this catchment the water is controlled by systems of weirs and dams that are designed to control the flow of water as is required for agriculture and, to a lesser extent, as flood mitigation.

The Basin includes most of New South Wales, some of southern Queensland, eastern South Australia, northern Victoria and all of the Australian Capital Territory.

The Murray–Darling Basin Authority (MDBA) is the body in charge of managing the waters of this basin and it is under growing pressure to manage increasing demands on the water. Higher and higher volumes of water are being required across the Basin, leading to significant competition from communities, farms, First Nations groups and the environment. Managing these competing interests doesn't come without challenges. Issues created can be put into three categories, which are all caused by changes in water flow and water needs:

- affecting humans – social and economic issues, and evolving water markets
- affecting the environment – ensuring there is enough water for nature to avoid fish deaths, salinity, algal blooms, blackwater events and acid soils
- affecting both – a challenging variable climate, including droughts, floods and bushfires.

The conflict we describe here relates to water allocations by the MDBA, with competing parties fighting it out for who gets what allowance and for how much. The Basin supports 7300 irrigated agriculture businesses, which make up 40% of Australian farms. All of these need water to produce the crops that feed Australians and help our economy.

Resource
Case Study: Murray–Darling Basin water allocations

Weblink
Murray–Darling Basin Authority

> Water in the Murray–Darling Basin can be bought and sold, either permanently or temporarily. This water is traded on markets – within catchments, between catchments (where possible) or along river systems. This form of trading allows water users to buy and sell water in response to their individual needs. Water trading has become a vital business tool for many irrigators and has evolved over time.

In 2018, the MDBA was accused of mismanaging the water of the basin by agricultural and environmental groups because of:

- mass fish death episodes of 2018, 2019 and 2023
- losing nearly two Sydney Harbours worth of water in deliberate overbank flooding at the 'Barmah Choke' (discussed later in this chapter) and as a result depriving the wheat and dairy farmers of the upper Murray of their entire water entitlement
- causing an ecological disaster in the flooding of the Barmah Forest for 141 days.

Groups opposing the MDBA say the motivation for sending water down the Murray River was to service the water allocations of the permanent plantations of almonds and fruit around the New South Wales–South Australia border, many of which are owned by foreign corporations.

Aggravating this situation was that water was travelling past farms that needed it and were told their annual water licences had no allocation, and forests were being inundated and drowned in unseasonal record flooding, yet the water kept travelling downstream to the farms that had purchased it through a system on the temporary water market, where the highest bidder wins.

Table 8.7 Summary of conflict over Murray–Darling Basin water allocations

Environment	The Murray–Darling Basin is a large area of south-eastern Australia, where water flows through a system of interconnected rivers and lakes, and where the two main rivers are the Murray River and the Darling River.	
Issue	Water allocations and inequality of these across the breadth of the Basin, and the side effects of environmental problems like fish deaths. In spring 2019, water was sent down the Murray to growers who had purchased allocations, flowing straight past farms that were not permitted to access it.	
Parties to the conflict	• Murray–Darling Basin Authority	• Agricultural groups • Ecologists
Interaction	Charged with managing the Murray Darlin Basin as an independent expertise-based statutory agency; it controls water flows and water allocations while caring for communities, businesses and the environment connected to the river basin.	Agricultural groups rely on water being delivered to them that they can access as per their allocation to grow their crops or animals. Ecologists measure the health of the environs along the river and pay attention to natural water flows and the health of the forests as a result of these occurring or not occurring.
Position on issue	A cornerstone of the strategy for managing water resources in the Basin is adaptive management – 'learning as you go' by trialling techniques, monitoring and making changes as needed.	The MDBA needs urgent review and the Basin Plan needs revision to ensure water continues to get where it is needed, and not to the highest bidder.
Methods used to influence decision makers	• Government media releases • Gathering evidence/presenting reports – in this case reports by an independent panel into fish deaths in the Lower Darling • Publications	• Media releases including television programs • Social media • Petition – to request a Royal Commission into the mismanagement of the Murray–Darling Basin • Environment Victoria campaigning for a review

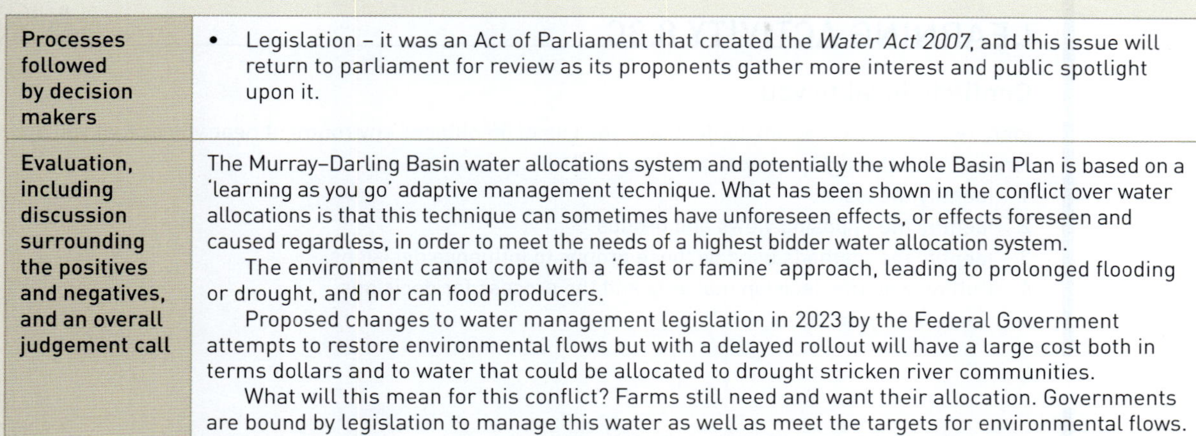

Processes followed by decision makers	• Legislation – it was an Act of Parliament that created the *Water Act 2007*, and this issue will return to parliament for review as its proponents gather more interest and public spotlight upon it.
Evaluation, including discussion surrounding the positives and negatives, and an overall judgement call	The Murray–Darling Basin water allocations system and potentially the whole Basin Plan is based on a 'learning as you go' adaptive management technique. What has been shown in the conflict over water allocations is that this technique can sometimes have unforeseen effects, or effects foreseen and caused regardless, in order to meet the needs of a highest bidder water allocation system. The environment cannot cope with a 'feast or famine' approach, leading to prolonged flooding or drought, and nor can food producers. Proposed changes to water management legislation in 2023 by the Federal Government attempts to restore environmental flows but with a delayed rollout will have a large cost both in terms dollars and to water that could be allocated to drought stricken river communities. What will this mean for this conflict? Farms still need and want their allocation. Governments are bound by legislation to manage this water as well as meet the targets for environmental flows.

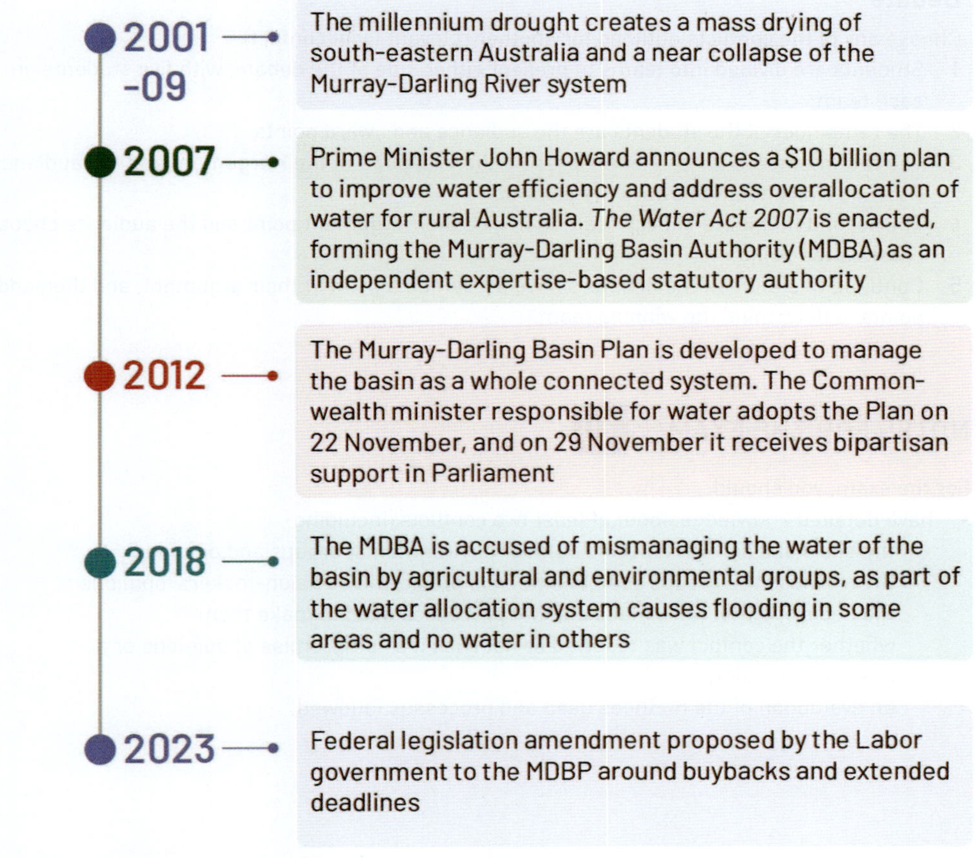

Figure 8.47 Timeline of conflict over Murray–Darling Basin water allocations

Resource
Additional Case Study: Menindee mass fish-kill incident

Worksheet
8.3c Conflicts local to you

LEARNING ACTIVITY 8.3C

Conflicts local to you

Identify a current or recent conflict over the use of an outdoor environment near you or that you are familiar with through personal or school visits:

1 Identify the issue.
2 Identify the opposing views and groups.
3 Identify the methods used by those groups to influence decisions.
4 Outline who the decision maker is and the process for decisions.
5 Share with your class!

Worksheet
8.3d Debate

LEARNING ACTIVITY 8.3D

Debate

Choose any of the conflicts outlined (or another relevant local conflict).

1 Students are divided into teams to present either side of the debate, with five students on each team.
2 The remainder of the students are the audience and award points.
3 The first members of both teams present the first point of their argument and the audience chooses which team to award the point to.
4 The second members of each team then give their argument point and the audience chooses who receives the point.
5 Continue this until all the debaters have presented a point of their argument, and then add the points to determine the winning team.

NOTES FOR THE EXAM

For the exam, you should:

- have detailed knowledge about at least two conflicts, including:
  - the basis of a conflict (specific outdoor environments, groups and opinions)
  - the methods the groups used to influence public and decision-makers' opinions
  - the decisions that were made and the processes used to make them
  - whether the conflict was resolved or not (was it a compromise of opinions or a one-sided result?)
  - an evaluation of the methods used and processes followed.

8.3 KEY CONCEPTS

- A conflict occurs where two different user groups of an outdoor environment have opposing views on how the location can be used.
- Individuals and groups use multiple methods to influence decision makers. The most common methods are:
  - formal petitions
  - public campaigns
  - street protests
  - development of collective action networks (e.g. environmental groups)
  - media-based activism
  - lawsuits
  - alternative knowledge creation (e.g. scientific research)
  - blockades
  - Environmental Impact Assessment objections
  - alternative proposals.
- Decision makers use processes to inform and make their decisions. Examples include:
  - community consultation
  - use of court systems
  - legislation (creating laws)
  - management plans
  - use of the political system
  - referenda.
- The Great Forest National Park conflict resulted in the park not being formed, as community consultation led both the government and opposition to choose not to support it.
- The removal of feral horses from the Alpine environment conflict was resolved through judicial (court) involvement and resulted in a compromised removal that allowed some horses to remain in the east, with shooting to be carefully managed but not disallowed.
- The creation of the Warburton Mountain Bike Destination faced one of the most thorough approval processes in Australia. It produced an environmental effects statement that, when coupled with extensive community consultation and submissions, resulted in the government approving the park with amendments of no trails in the Yarra Ranges National Park to protect the Leadbeater's possum and Mount Donna Buang wingless stonefly.
- Southern Ocean whale hunting has ceased by Japan's research vessels due to the Australian and New Zealand taking them to the International Court in 2014 and having the hunting ruled illegal. Although Japanese whalers continued to hunt, tireless direct action by the Sea Shepherd organisation meant that the whale harvest was seriously decreased and eventually Japan chose to cease its operations in 2018.
- Murray–Darling Basin water allocations are managed by the Murray–Darling Basin Authority (MDBA), formed under the *Water Act 2007* following the Millennium drought. In 2018, the MDBA was criticised for failing to meet the needs of water allocations in a catastrophe that involved water allocations being granted to the end of the river at the South Australia border, resulting in the flooding of forests at the Barmah Choke en route, with water also being unavailable to farms along the way. This conflict is unresolved but is being influenced by both agricultural and environmental groups, with the federal government to review the Basin plan, which is ongoing and begins in 2023 with Federal legislation amendments around buybacks.

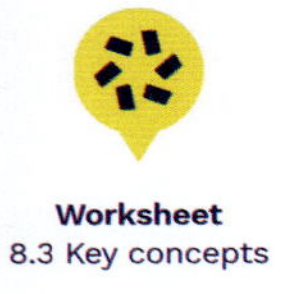

Worksheet
8.3 Key concepts

8.3 CONCEPT QUESTIONS

REMEMBERING

1 Name three methods used to influence decision makers.
2 What methods have the Australian Brumby Alliance used to influence decisions?

UNDERSTANDING

3 Using an example, describe how one conflict group uses one method to their advantage in influencing a decision maker.
4 What is a common method used across most conflicts to influence decision makers?
5 In your opinion, how successful is the method you named in Question 4 at achieving this influence?

APPLYING

6 Describe a conflict not listed in the textbook.
7 Using an example not presented in the text, explain a method used by a different group to influence decision makers in a conflict of use.

EXTENSION CHALLENGE

8 If you could create a process or group of people to make decisions around managing conflicts such as the ones studied in this chapter, what would the group look like? Who would its members be? How would it make decisions?

8.4 ENVIRONMENTAL POLITICS IN AUSTRALIA

KEY KNOWLEDGE

- an environmental issue in Australia and related policy from two federal political parties or representatives, including at least one of:
  - Labor party
  - Liberal-National Coalition
  - The Greens

KEY SKILLS

- analyse differing environmental politics in Australia

8.4.1 ENVIRONMENTAL POLITICS IN AUSTRALIA

As we studied in the last chapter, environmentalism and conservation actions really came alive between the flooding of Lake Pedder in 1972 and the stopping of the Franklin River dam in 1983 by the newly elected federal Labor government. The nation had been engaged by environmental campaigns encouraging care for outdoor environments, and were empowered to vote for conservation and see the environment afforded protection.

While environmental politics has not been as flashy since those events, environmental issues are still present in every Australian election. Australians care about the uniqueness and beauty of our outdoor environments, and political parties have listened, with each taking their own approach to implementing a strategy to mitigate environmental issues. Although sometimes utilised for political purposes, the environment will often benefit as these and other environmental issues are investigated, debated and, ideally, acted on. In the recent political landscape, the environment and climate change have become prominent in all political parties' policy framework. Social awareness of environmental issues has indeed necessitated political parties committing to the conservation of the environment. Environmental policy has even become part of the flagship statements that describe party direction.

Alamy/Michael Thomas

Figure 8.48 A voting booth during the Australian 2022 Federal elections in Deepdene, Melbourne, Victoria.

Left–right political spectrum

When learning about political parties and their views and philosophies that drive their policy directions, we use the left–right political spectrum as a handy guide. The left–right spectrum are labelled left-wing or right-wing due to their politics or opinions reflecting the established opinions of that place on the spectrum. The **political spectrum** dates back to 18th century France and the French Revolution. The Assembly president would be sat in the middle and those that supported the king sat to his right and supporters of the revolution to his left.

political spectrum a system of classifying different political positions upon axes that symbolise their position between socialist and capitalist political dimensions

Table 8.8 Political policy positions

Left	Right
Supports higher taxes on the rich	Supports lower taxes on businesses to help them grow
Believes in an equal society	Argues that social inequality is unavoidable
Believes in big government	Believes governments should play a limited role in people's lives

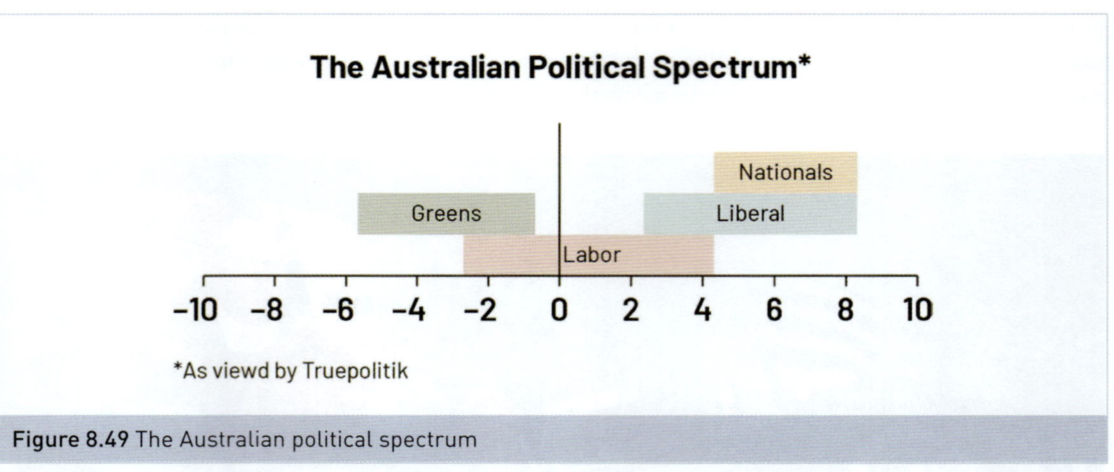

Figure 8.49 The Australian political spectrum

Australian Labor Party

The Australian Labor Party exists on the centre-left of the political spectrum, which fits with its beginnings as a party supported by trade unions and the everyday labouring worker. This **socialist** attitude and philosophy remains with the party today.

socialism
a political and economic theory of social organisation that advocates that the means of production, distribution and exchange should be owned or regulated by the community as a whole

The Australian Labor Party was one of the first major Australian political parties. The party entered national politics in the first federal elections of 1901 and gained many seats and considerable power. It governed intermittently during the first three decades of the 20th century, and again during World War II and the years immediately following, but it was then in opposition for more than 20 years. In 1972, the party won the federal election under Gough Whitlam and went on to enact sweeping reforms to protect Australia's environment, protecting the Great Barrier Reef, ratifying the RAMSAR and World Heritage Conventions and introducing Environmental Impact Assessments. In 1983, Bob Hawke's Labor Government won office, with his actions to prevent the damming of the Franklin River celebrated in this course. His government was also pivotal in banning mining in Antarctica and safeguarding the Daintree rainforests from logging.

Shutterstock/Naga11

Figure 8.50 Labor Prime minister Anthony Albanese

In 2007, the party came to power under Kevin Rudd and climate change became an issue that all parties needed to address. His successor, Julia Gillard, introduced a price on carbon as a mechanism to mitigate climate change, though this was repealed by the subsequent Coalition government led by Tony Abbott. While climate change policy has continued to focus on creating a way to trade away the pollution causing climate change, many leadership turnovers have occurred. This issue has been challenging for each political party we are studying.

THE LABOR WAY

Labor means reform, recovery and renewal. It means undoing the complacency and decline presided over by our opponents. It means recognising that the world has changed, and knowing that Australia must also change to guarantee prosperity and a good life for all our citizens. In the current world of uncertainty, Australia is looking to Labor to lead.

'ALP National Platform', 2021. www.alp.org.au

2022 KEY ENVIRONMENTAL POLICY PLEDGES, AUSTRALIA LABOR PARTY

Powering Australia

Labor's Powering Australia plan aims to create jobs, reduce power bills and reduce emissions through the use of renewable energy. Under this plan, renewable energy will be more affordable for Australian homes and businesses.

Protecting our unique environment

Environmental protection is a proud tradition of Labor. In government, Labor protected the Daintree, Kakadu, Great Barrier Reef, Franklin [River] and Antarctica. Landcare and the largest marine park network in the world were created by us.

Labor's plan to future-proof Australia's water resources

As part of Labor's Water for Australia plan, a National Water Commission will be established and the National Water Grid Investment Framework will be widened to finance a wider range of water supply projects.

https://alp.org.au/policies

Liberal Party of Australia

The Liberal Party of Australia is positioned as right-leaning on the political spectrum, in favour of a more **capitalist** society and a supporter of business. Liberal Party policy focuses strongly on wealth creation and people as individuals that deserve the right and support to determine their own future.

capitalism
an economic and political system in which a country's trade and industry are controlled by for-profit private owners, rather than by the state

THE LIBERAL WAY

Liberalism is a set of democratic values based upon a central belief in the rights, freedoms and responsibilities of all people as individuals and upon a conviction that those individual rights, freedoms and responsibilities are the surest foundation of strong community life.

'The Liberal way', FederalPlatform.pdf (liberal.org.au)

Malcolm Turnbull (2015 to 2018) was prime minister when Australia signed the Paris Agreement on climate change that would have steered our country in a positive direction of action on this issue. Unfortunately, the momentum that Turnbull had gathered was also his undoing, and his party sacked him from the top job before he could direct policy to counter climate change. His successor, Scott Morrison, went on to win the next election in 2019, but led the party to an election loss in 2022, in part because of his government's inaction on climate change.

The Nationals

The Nationals are a smaller political party that was formed to support the interests of regional and rural voters, and sits further to the right on the political spectrum than the Liberal Party. It has its own party leader and policies, but traditionally joins with the Liberal Party to form a coalition to govern together. Even though they are two parties with separate philosophies (Liberals with a more urban focus, the Nationals more rural) as a **coalition** they generally vote together and support policy on which they agree.

coalition
a group formed when two or more people or groups temporarily work together to achieve a common goal

COALITION (LIBERAL/NATIONALS) 2022 ENVIRONMENTAL ELECTION POLICIES

Forestry

A re-elected Coalition Government will:

- guarantee not to support any shutdowns of native forestry, and work to create permanent timber production areas
- invest $106.6 million to establish a National Institute for Forest Products Innovation, with a central hub in Launceston and up to five centres of excellence around Australia
- deliver $112.9 million in grants to accelerate adoption of new wood processing technologies
- invest $86.2 million in a new Plantation Establishment Program
- invest $4.4 million to strengthen Australia's fight against illegal logging and stop illegal timber imports from undercutting Australian producers.

Great Barrier Reef

The Coalition Government's $1 billion reef package includes:

- $579.9 million for water quality – working with land managers to remediate erosion, improve land condition and reduce nutrient and pesticide runoff
- $252.9 million for reef management and conservation – additional support for the Great Barrier Reef Marine Park Authority (GBRMPA) to reduce threats from crown-of-thorns starfish, implement advanced health monitoring systems and prevent illegal fishing
- $95.6 million to research and deploy world-leading reef resilience science and adaptation strategies
- $74.4 million for Traditional Owner and community-led projects including species protection, habitat restoration, citizen science programs and marine debris reduction.

A cleaner environment

A re-elected Coalition Government will:

- implement our Long-Term Emissions Reduction Plan to reach net zero emissions by 2050 through technologies, not taxes
- implement our new 10-year Threatened Species Strategy to 2031 by tackling feral pests and weeds, increasing the number of safe havens, helping more captive breeding programs and partnering with local communities in practical, on-ground action.

Adapted from www.liberal.org.au

The Australian Greens Party

Positioned to the left of the Australian Labor Party on the political spectrum, the Australian Greens Party has become a genuine third-party option in Australian politics. Although environment-based political groups had existed in various states throughout Australia in the 1970s and 1980s, in 1992 these groups merged their vision and resources to form the new Australian Greens political party (often called 'the Greens'). While the party's stance was developed around ecological sustainability, it holds this as the first of its four pillars, followed by grassroots participatory democracy, social justice, and peace and non-violence.

Although not classified as a major party, compared with the Labor and Liberal parties, the Greens are growing in size and today are too big to be called a minor party. Since its inception in Tasmania during the Franklin River dam issue, the party has grown and has consistently won seats in the federal parliament, particularly in the Senate. With some parliaments ruled by only slight majorities, often the votes of the Greens can be the difference between legislation passing or not.

Figure 8.51 Image from the Australian Greens Party website

2022 KEY ENVIRONMENTAL POLICY PLEDGES, AUSTRALIA GREENS PARTY

Phasing out coal, oil and gas

- Immediately ban the construction of new coal, oil and gas infrastructure, ensuring we can transition our economy to zero carbon energy while maintaining a safe climate
- Help out mining workers and communities by creating long-term, sustainable industries to assist in the move beyond fossil fuels and to ensure people do not lose work
- Phase out the mining, burning and export of thermal coal by 2030 to ensure we do our bit, so the world does not go over the 1.5 degree climate cliff.

100% clean green electricity

- Large-scale public investment in renewable energy and storage, to replace every coal-fired power plant in the country by 2030, ensuring we deal with the climate emergency in time
- Upgrading the electricity transmission and distribution grid, integrating more wind and solar energy while ensuring we keep the lights on
- The creation of a publicly owned non-profit power retailer, to push power prices down and end price gouging by the big energy companies.

Land restoration and carbon drawdown

- Ending land clearing and native forest logging so we can ensure we protect the carbon already in our native forests and grasslands
- Creating local forestry jobs through planting trees and restoring damaged forests and habitats
- Working with farmers to increase carbon sequestered on land, drawing down pollution while providing new revenue streams for farms
- Increasing the funding to Indigenous ranger programs.

A green Australia

- Stop the extinction crisis through setting a goal of zero extinction by 2030 and investing in a mass greening and restoration program, to ensure we are protecting habitats for our native animals and plants
- Green Australia by investing $24.4 billion over the next decade, including by restoring wildlife habitat, planting two billion trees by 2030 and re-establishing green space in our cities, regions and suburbs
- End native forest logging to help reduce fire risk, and transition workers into sustainable plantation forestry
- Save the Great Barrier Reef with grants to improve farming practices to stop runoff, stopping damaging maritime projects and banning offshore dumping to protect our oceans
- Save the Murray-Darling from corporate greed through water recovery targets and water buybacks.

The Australian Greens

Independents

While independent Members of Parliament (MPs) have always been part of Australia's political system, their popularity has never been greater than it is today. An independent MP is exactly that: independent. They are not beholden to follow the 'party line' or 'party thinking', or to vote 'the party way'. Rather, they aim to support their electorate and the communities that voted them in, having a strong focus on issues directly relevant to their electorate. In the 2022 federal election, 10 independents were voted into parliament.

HELEN HAINES, INDEPENDENT MEMBER FOR INDI

In 2022, Helen Haines won the 2022 seat of Indi as an independent for the second time. Before her, the independent Cathy McGowan held this seat for two terms of office. These independent candidates, along with 11 other independents voted into the 2022 Federal Parliament, are becoming more popular alternatives to the established political parties. Haines was listed by **Climate 200** as one of the 20 candidates it supported in the 2022 election, a group colloquially labelled as 'teal independents' as a way of referring to the group as a whole without drilling into idiosyncratic differences. They were largely disaffected Liberals with a focus on environmental issues – with blue for Liberal and green for the environment combining to make teal. Haines received 40.7% of the vote in the seat of Indi, 10.2% more than the votes for the Liberal party candidate.

Climate 200
a community crowd-funded initiative that supports political candidates committed to using a science-based response to the climate crisis, restoring integrity to politics and advancing gender equity

2022 KEY ENVIRONMENTAL POLICY PLEDGES, HELEN HAINES

Locally owned renewable energy

- A $300 million Local Power Fund to help every regional town build its own solar and community battery projects
- A new mechanism to attract private investment in locally-owned renewable energy projects
- A top-to-bottom review of the Government's existing energy agencies to give regional Australians a seat at the table.

Cheaper home batteries

- Lower the cost of batteries by up to $3000 with new incentives through the existing Small-Scale Renewable Energy scheme
- Offer no-interest loans of $5000 to households to purchase batteries, solar and electric-appliances

Helen Haines MP, https://www.helenhaines.org/issues/energy/

As Haines is an independent, she does not have published policies on all aspects of government. An alternative way to investigate her stance on issues is to look at her voting record during parliament. By doing this, we can see the following.

Helen Haines voted 'yes':

- that the federal government should amend its laws and policies to meet the objectives of the Paris Climate Agreement in September 2022.

Helen Haines voted a 'mixture of for and against':

- that the federal government should commit to achieving net zero emissions by 2050 as part of Australia's efforts to address climate change

Newspix/Aaron Francis

Figure 8.52 Helen Haines, independent member for Indi

Helen Haines voted 'generally against':

- that in light of the threat of climate change, the federal government should make as rapid a transition to renewable energy as possible
- increasing investment in renewable energy

Helen Haines voted 'consistently against':

- increasing state and territory environmental approval powers

LEARNING ACTIVITY 8.4A

How does your local federal member vote in Parliament?

Go to the website They Vote For You and enter your postcode into the opening window. This will then take you to a page that documents how the elected member for your area votes during siting time in Federal Parliament.

Questions

1 If your member is part of a major political party, note whether they always vote in accordance with their party. If they are an independent, note their voting preferences.

2 How does their voting align with your stances on issues?

Worksheet 8.4a How does your local federal member vote in Parliament

Weblink They Vote for You

Table 8.9 Environmental issues for government policy response

Issue	Description	Policy responses
Climate change	An increase in greenhouse emissions by developed nations since the Industrial Revolution (and, more recently, developing nations) has resulted in changes to the Earth's climate. They have been observed in every region and across the global climate system. In Australia, the effects have been observed in increased severe weather events and winter snow lines that are observably creeping higher in altitude by the decade. The Australian Government has ratified the Paris Agreement and reaffirmed its commitment to targets in subsequent COP meetings.	• Supports global emissions targets • Supports renewable energy targets
Habitat and species conservation	Native forest logging continues to be practised in Australia and has detrimental effects on species endemic to these environments. The pressures of urbanisation has also reduced habitat for native flora and fauna, and introduced pests and weeds.	• Supports policies and projects for habitat and species conservation. • Supports the Nature Repair Market Bill 2023 which in turn, protects native species
Water Management	In one of the driest continents on Earth, our nation is constantly faced with challenges of drought, water supply and salination. We need a sophisticated system of water management, the largest being the Murray–Darling Basin Plan. We've also seen pipelines to transport water proposed and become political issues.	• Supports amendment of Murray Darling Basin plan to improve accountability and support for impacted communities.
Great Barrier Reef	The reef is an outdoor environment unique to Australia and one of the best showcase reefs on the planet. It faces pressures of coral bleaching, pollution from agricultural runoff and damage by tourism. Its rescue is a galvanising issue we will see discussed more and more.	• Supports policies and projects for Great Barrier Reef protection

Table 8.10 Comparisons of examples of policy positions on environmental policies

Policy	ALP	Liberal National Coalition	Greens	Helen Haines (independent)
Greenhouse gas emission targets	• Net zero by 2050 • reduce Australia's emissions by 43% by 2030 – which will become Australia's target under the Paris Agreement	• Net zero by 2050, but through technologies not taxes • reduce Australia's emissions 26 to 28 per cent by 2030	• Net zero or net negative Australian greenhouse gas emissions by 2035 or sooner	• Net zero by 2050 • Haines voted 'yes' that the federal government should amend its laws and policies to meet the objectives of the Paris Climate Agreement
Renewable energy targets	• Renewable energy penetration to grow to 82% by 2030	• Across governments and the private sector over $40 billion has been invested in renewable energy in Australia since 2017	• A rapid transition to 100% • Large-scale public investment in renewable energy and storage, to replace every coal-fired power plant in the country by 2030 to deal with the climate emergency	• Supports locally owned renewable energy • Voted 'generally against': • that the federal government should make as rapid a transition to renewable energy as possible • increasing investment in renewable energy
Habitat and species conservation	• To improve the trajectories of priority threatened species by 2031 (policy, introduced by Liberals in 2021 and continued by Labor	• Threatened Species Strategy – further investment of $6 billion pledged for threatened species, habitat restoration, marine conservation and environmental projects	• Stop the extinction crisis through setting a goal of zero extinction by 2030 and investing in a mass greening and restoration program to ensure we are protecting habitats for native animals and plants	• Unknown
Water	• National Water Commission will be established, and the National Water Grid Investment Framework will be widened to finance a wider range of water supply projects	• $200 million pledge to the Environment Restoration Fund to help improve the water quality of the Yarra, Swan, Canning, Torrens, Brisbane, Georges, Hawkesbury and Nepean Rivers	• Save the Murray-Darling through water recovery targets and water buybacks	• Unknown
Great Barrier Reef	• Labor will protect the Great Barrier Reef through multiple measures including addressing climate change and improving water quality	• $1 billion Reef package to further support the Reef 2050 Long-Term Sustainability Plan	• Save the Great Barrier Reef with grants to improve farming practices to stop runoff, stopping damaging maritime projects and banning offshore dumping to protect our oceans	• Unknown

LEARNING ACTIVITY 8.4B

Environmental policies of Australian federal political parties

Research the environmental policies of two of the following Australian federal political parties or an independent at the most recent federal election:

- Liberal-National Coalition
- Australian Labor Party
- The Greens
- Independent MP

Any aspect of environmental policy can be investigated. Specific areas of the environment could include:

- greenhouse gas emissions
- renewable energy
- threatened species
- environmental law.

1 Compare each of the policies based on their level of commitment to each area of the environment. For example, how much are they prepared to do for the environment?
2 Prepare a scorecard of how each party/independent rates in relation to their policies.
3 Why do you think there are differences in the environmental policies of the different parties?

Worksheet 8.4b Environmental policies of Australian federal political parties

International agreements and their impact on Australian politics

In the late 1970s and into the 1980s, the world woke up to the concept that we are all connected on one planet, with one atmosphere. There was a massive issue with ozone depletion resulting in increased ultraviolet (UV) rays causing damage to the skin of humans and animals. Ozone acts as a protective layer of gas around the Earth, bouncing off UV rays, and the thinner layer made this very dangerous – for humans in particular. Scientists determined that the cause was ozone-depleting chemicals and these were banned by governments in 1987 as part of a worldwide agreement. Titled the Montreal Protocol, it was the first universally ratified treaty in United Nations history, involving 198 signatory countries. Ozone depletion was halted and the ozone layer continues to increase its volume in affected regions to this day as a result of this international government action. Once ratified, these types of agreements filter down to inform the practice of signatory nations and also place pressure on non-signatory nations. In more recent times, we have seen numerous agreements ratified by the United Nations that have become legally binding international treaties on climate change (see Figure 8.53).

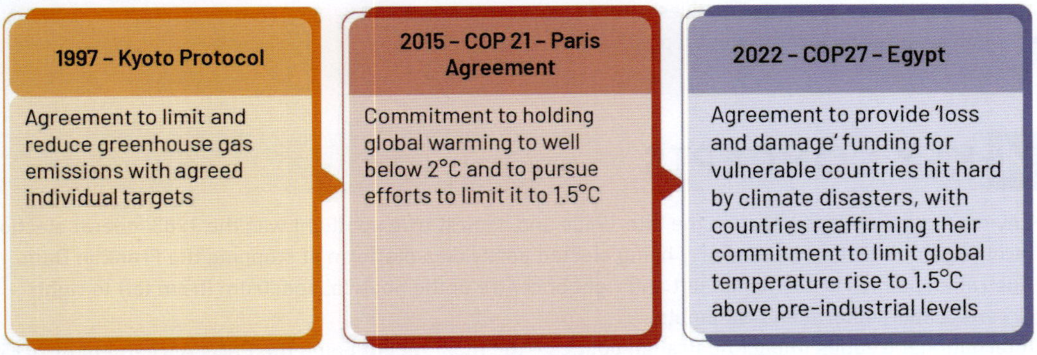

1997 – Kyoto Protocol
Agreement to limit and reduce greenhouse gas emissions with agreed individual targets

2015 – COP 21 – Paris Agreement
Commitment to holding global warming to well below 2°C and to pursue efforts to limit it to 1.5°C

2022 – COP27 – Egypt
Agreement to provide 'loss and damage' funding for vulnerable countries hit hard by climate disasters, with countries reaffirming their commitment to limit global temperature rise to 1.5°C above pre-industrial levels

Figure 8.53 International agreements on climate change

COP
Conference of the Parties to the United Nations Framework Convention on Climate Change (UNFCCC), the peak decision-making body for the world's climate change commitments

In particular, the most recent COP meeting in Egypt requested signatories to revisit and strengthen their 2030 targets in their national climate plans by the end of 2023, as well as accelerate efforts to phase down unabated coal power and phase out inefficient fossil fuel subsidies. For Australia, this international legally binding treaty informs our policy, with each party making their own judgements and plans about how to achieve the goal.

Alamy/urbanbuzz

Figure 8.54 Poster from COP27, November 2022

Alamy/Samantha Ohlsen

Figure 8.55 Logging and deforestation in Victoria, Australia

NOTES FOR THE EXAM 

For the exam, you should:

- be able to provide an overview of an environmental issue and related policy from two federal political parties or representatives.
- analyse differing environmental politics in Australia.

8.4 KEY CONCEPTS

- Environmental politics began in the late 1960s and its first true spotlight was the 1983 federal election involving Bob Hawke's Labor Party being elected partly on a platform of stopping the Franklin Dam.
- Australian politics is positioned left and right of centre of the political spectrum (from left to right: Greens, Labor, Liberal and Nationals).
- International treaties impact Australian politics – the most recent being the Paris Agreement of 2015 and its pressure on governments to act on greenhouse emissions.

- Climate change is caused by an increase in greenhouse gas emissions since the Industrial Revolution by developed nations, and more recently also developing nations.
- Native forest logging continues to be practised in Australia and has detrimental effects on species that are endemic to them. Furthermore, the pressures of urbanisation reduce habitat for native flora and fauna and introduce pests and weeds.
- As one of the driest continents on Earth, our nation is constantly faced with challenges of drought and water supply. We therefore require sophisticated systems of water management, the largest being the Murray–Darling Basin Plan. We've also seen desalination and pipelines to transport water proposed and become political issues.
- The Great Barrier Reef is an outdoor environment unique to Australia and one of the best showcase reefs on the planet. It faces pressures of coral bleaching, pollution from agricultural runoff and damage by tourism. Its rescue is a galvanising issue.
- The Australian Labor Party is positioned on the centre left of the political spectrum and is more in favour of a more socialist society. It has policy for Australia to be net zero by 2050, and to reduce Australia's emissions by 43% by 2030.
- The Coalition of the Liberal Party of Australia and the Nationals is positioned as right-leaning on the political spectrum, and is in favour of a more capitalist society. They pledged net-zero emissions by 2050 (through technologies, not taxes) and to reduce Australia's emissions 26–28% by 2030.
- The Australian Greens Party is positioned to the left of the Australian Labor Party on the political spectrum and is an environmental-based party with four pillars: ecological sustainability, grassroots participatory democracy, social justice, and peace and nonviolence. They pledged net zero emissions or net negative Australian greenhouse gas emissions by 2035 or sooner.
- Helen Haines is an independent 'teal' Member Of Parliament and supports the policy of net zero by 2050, as outlines in the Paris Agreement of 2015.

8.4 CONCEPT QUESTIONS

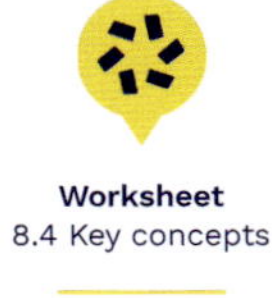

Worksheet
8.4 Key concepts

REMEMBERING

1 What are the three major parties and where are they positioned on the political spectrum?
2 What does it mean to be a socialist political party?
3 What does it mean to be a capitalist political party?

UNDERSTANDING

4 Discuss the possible reasons why each party has a different target for reducing greenhouse emissions on the way to net zero by 2050.
5 Explain an environmental issue and how each major party has directed policy to combat it.

APPLYING

6 Investigate and explain how a teal MP other than Helen Haines positions themselves on climate change policy.
7 Describe the three goals of Australia's Strategy for Nature 2019–2030.

EXTENSION CHALLENGE

8 The Australian Greens Party won 12.2% of the total vote at the 2022 federal election but won only four seats in the House of Representatives, whereas the Coalition won 35.7% of the vote and secured 58 seats. Why, with only triple the vote, could the Coalition win 14 times more seats in the lower house of parliament?

8.5 SOCIAL DEBATES ABOUT ENVIRONMENTAL ISSUES

KEY KNOWLEDGE

- the influence of social debates on relationships with outdoor environments, including one of:
  - climate change
  - renewable energy
  - water management

KEY SKILLS

- analyse the influence of social debates about environment issues on relationships with outdoor environments

8.5.1 SOCIAL DEBATES ABOUT ENVIRONMENTAL ISSUES

Australia has been facing multiple environmental issues in the past decade, many of which are complex and multifaceted. Whenever a complex issue arises with potential solutions, Australians enjoy nothing more than to settle in and debate the merits of different ways to approach the solution. Often we are heavily biased by our own point of view, which can be based on how a solution will affect our finances, how we like to use an outdoor environment or whether we to live in luxury while disregarding its effects on the planet.

Often before a debate reaches politics, ideas and issues have already begun to be discussed in conversations, tweets, blogs and social media posts, as well as at the family dinner table or on radio talkback shows. It is important to the fabric of a community to debate issues and share our opinions.

In this section, we will analyse how these social debates on climate change, renewable energy and water management influence our relationships with outdoor environments. We will build our understanding of how outdoor environments are threatened by these debates, assess possible causes of the issues and how people, groups or governments are debating the issues. We will then consider how humans are managing (or not managing) the issues and potential solutions, and how these debates can influence our relationships with outdoor environments.

Climate change

The level of carbon dioxide (CO_2) in the Earth's atmosphere has been rising steadily since the Industrial Revolution. Burning fossil fuels produces CO_2, which is released into the atmosphere. We are only beginning to find out how these changes in climate will affect humans and the planet. Our understanding of the climate is only basic, due to the many variables involved. We know that increased CO_2 holds in more of the heat radiated from the Earth's surface (called the greenhouse effect; see Figure 8.56 on the next page) and that sea levels are rising due to the melting of polar ice.

Our Australian climate is changing. Temperatures have risen over the past 100 years and have now reached unprecedented levels, which is consistent with global findings that recent years have been warmer than any multi-year period for at least 2000 years. Temperatures averaged over Australia have risen by 1.4°C since 1910. Human activity, particularly global emissions of greenhouse gases, has been the main driver of global temperature increases.

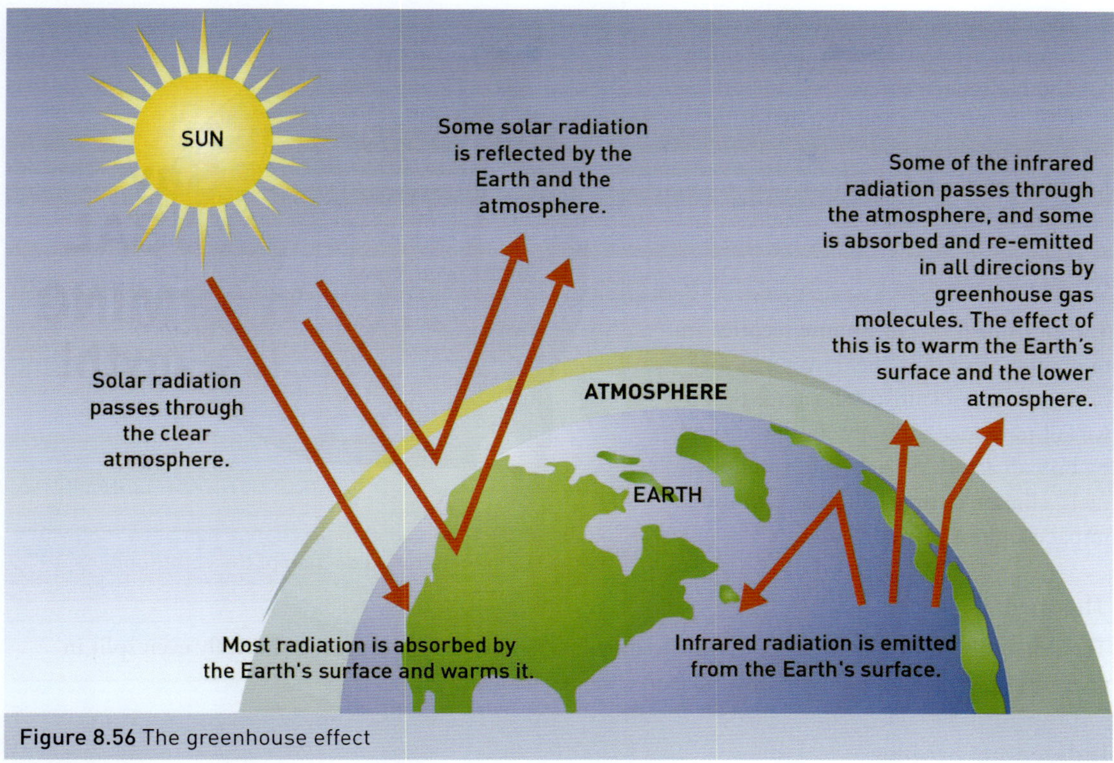

Figure 8.56 The greenhouse effect

However, Australian greenhouse gas emissions have fallen substantially since 2007, although the rate of change has decreased since 2013 (except for large falls in 2020 that are partly due to reductions resulting from the COVID-19 pandemic, which are likely to be temporary).

Changes in our land use, particularly in the forestry sector, from being a net source to a **net sink**, has made the biggest difference. The increasing proportion of Australian electricity generated from renewable sources has also contributed to the decrease.

net sink
where a process such as forestry absorbs more carbon than it emits

We have already observed major changes to climate in Australia, including an increased frequency and intensity in heat extremes (both on land and in the ocean), rising sea levels and an increased occurrence of dangerous fire weather in southern and eastern Australia that's well outside the range of historical records.

THE CLIMATE CHANGE DEBATE

Despite the changes observed around the planet, there is still debate over the validity of the theories that support global warming and associated climate changes. Debate often centres on whether humans have a major influence on the climate. It also questions the importance of the changes we are seeing and whether we need to do anything about them.

Opinions surrounding the climate change debate include the following.

Climate sceptics

Climate sceptics believe that:

- CO_2 is an inert gas and not a pollutant
- the Earth's atmosphere has warmed and cooled naturally over thousands of years and what is occurring now is another natural change
- warming is not caused by increased CO_2 levels through human activities like burning fossil fuels (coal, oil and gas) for energy production

climate sceptic
someone who believes that claims made by climate scientists and environmentalists that the climate is changing due to human activities are false or exaggerated

- the climate is too complex to predict
- humans are powerless to do anything about climate change
- CO_2 is actually good as it assists in greening the planet
- it is a waste of money to act on climate change.

greening
the process of transforming a space into a more environmentally friendly version with increased plant growth

mitigate
to attempt to slow, reduce or reverse the severity of something

Figure 8.57 Action on climate change has been met with fierce resistance from a section of the community

Climate scientists

The views held by greenhouse sceptics are in conflict with those held by a majority of the climate scientists who agree that if we fail to reduce carbon pollution caused by human activities such as burning fossil fuels, climate change will have profound impacts on our planet.

Weblinks
OzCoasts

Great Southern Star, coastal erosion'

Pacific Islands Climate Change

The media

The debate about this issue has been distorted by media reports that suggest a relatively even split in scientific opinion about human influence on climate change.

Indigenous peoples

Australia's climatic changes influence the way Indigenous peoples use traditional knowledge to read the natural variations of the outdoor environments in their Country. With rising temperatures and sea levels, ocean warming and shifting rainfall patterns, traditional seasons are changing or are being delayed, and affecting practices such as the frequency of cultural burning.

WHAT TO DO?

How we should respond to climate change is also a subject of debate. Humans have been able to adapt to environmental pressures throughout history and we need to be prepared for climatic changes in the future. A range of responses are possible:

- adapting
- mitigating the extent of climate change
- introducing economic incentives to modify human practices that contribute to climate change.

Some greenhouse sceptics are reluctant to respond at all to the threat of climate change. They argue that we do not need to take any action, as balance will be naturally restored over time. Most people believe that some combination of each type of response will need to occur. There is debate about what we do and how much to do.

Adapting to climate change involves changing the way we live so we can acclimatise to the effects on our environments. For example, sea level rises that are associated with rising atmospheric temperatures are forecast to threaten existing established coastlines through storm surges and coastal erosion. An adaptation strategy could involve building sea walls to protect existing property.

Changes do not occur suddenly, but there are small, gradual impacts. These small impacts do not require immediate action, so it can appear there is no urgency to respond.

Worksheets
Additional Learning Activity: 8.5a OzCoasts sea level rise mapping

Additional Learning Activity: 8.5b Coastal erosion

LEARNING ACTIVITY 8.5A

Solomon Islands living with sea levels that have risen

Watch the 'Pacific Islands Climate Change' episode of *Behind the News* on ABC.

1 What has been the result of sea level rise on the Solomon Islands?
2 Do you think something needs to be done to adapt to this? What could be done?

Worksheet 8.5c Solomon Islands living with sea levels

MITIGATION

Creating cleaner fossil fuel use or renewable energy alternatives to fossil fuels all together are the most significant opportunities to mitigate climate change. Methods of achieving this include renewable energy production (solar, wind and **biofuels**), nuclear energy, **clean coal** technologies, **carbon capture and storage** and greater use of less carbon-intensive fossil fuels (e.g. natural gas). Energy-saving techniques include improved building techniques (insulation and passive solar design), consuming less (buying fewer goods that require energy and natural resources to be made), and being more efficient in our behaviour (turning off and unplugging appliances). And, of course, we can try to protect our forests that remain massive storage houses of carbon.

New industries will appear and develop in response to these techniques, and some existing industries may decline (e.g. brown coal power generation). This is the source of debate as to whether to pursue such mitigation methods as the economic and social impact of required change may be seen by some as outweighing the threats posed by climate change.

biofuel
fuels made from renewable living raw materials (e.g. ethanol produced from common crops such as sugar cane and potato and added to petroleum)

clean coal
technology intended to enable continued use of coal as an energy source with reduced impact on the environment

carbon capture and storage
process of trapping CO_2 so that its effect on the climate is minimised

LEARNING ACTIVITY 8.5B

Alternative energy

1 Research two forms of alternative energy production available to Victorians.
2 Outline how energy is produced from these methods.
3 How could these methods of energy production contribute to mitigation of climate change?
4 List the advantages and disadvantages of these methods of energy production.

Worksheet 8.5d Alternative energy

ECONOMIC RESPONSES TO CLIMATE CHANGE

Economic responses to climate change have also caused debate. Examples of economic responses include a **carbon tax** and emissions trading schemes. A carbon tax is simply a tax that is charged to large CO_2 polluters, which is intended to promote less-polluting industrial processes. **Emissions trading** involves buying or allocating permits for set CO_2 emission levels produced. The more a company produces, the more permits that company requires. These permits can also be traded among companies. So, companies can make money by adopting processes that reduce CO_2 emissions, and then sell permits that are no longer required.

These responses have sparked debate that largely focuses on environment versus economy. It has been posited that any system that puts a price on greenhouse gas emissions would have a negative social effect because it could lead to an increase in the price of essential goods (e.g. electricity). This would affect households and place pressure on businesses, potentially leading to job losses. This fuels social debate as everyone is affected by climate change – not only by the extreme weather events or soaring temperatures, but also by the costs of mitigating it.

It's also worth noting here the strong and direct political consequences of climate change policy. Prime ministers Rudd, Gillard and Turnbull all campaigned to introduce a price on emissions, but failed because of mounting social debate and political pressure, ultimately resulting in that prime minister losing their position.

carbon tax
a tax charged to industries based on their level of greenhouse gas (primarily CO_2) production

emissions trading
a market-based approach used to control pollution by providing market incentives and requiring permits (which can be traded among companies) to use processes that produce CO_2

Figure 8.58 The government of Prime Minister Julia Gillard was the first to introduce a carbon emissions scheme in 2011. She lost the leadership of the party to Kevin Rudd in 2013, and Labor lost the 2013 election to the Coalition, which was led by Tony Abbot, who had campaigned strongly against the scheme, successfully labelling it a 'tax'.

Influence on relationships

How do debates about climate change influence our relationship with outdoor environments? That is, the influence on:

- the way outdoor environments are perceived
- how we interact with outdoor environments
- the impact that we have on outdoor environments.

Table 8.11 summarises the debate around one of these issues: whether climate change is human-made or naturally occurring.

Table 8.11 Climate change: human-made or naturally occurring?

Debate	Perception towards the environment	Interactions with the environment	Impacts of social debates on relationships
Climate change is being caused by human activity: • CO_2 causes temperature increase • Increased temperature creates impacts (held by environmental groups, 98% of climate scientists and political parties).	• Environment is not a limitless resource • Environment needs to be looked after • Environment needs to be preserved for future resource use (i.e. sustainability)	• Reduce fossil fuel use • Increase renewable energy use (e.g. solar energy and wind farms) • Improve building techniques (e.g. green building designs) • Improve town planning (e.g. sea-level overlays and energy ratings) • Protect forests to absorb CO_2 • Use carbon tax and emissions trading to encourage use of alternative energy source • Reduce visitation to outdoor environments in summer due to heightened risk of bushfires	• Increased natural habitat and biodiversity through protection of forests • Less CO_2 pollution due to alternative energy use • Decreased coal-powered energy production through more efficient housing • Increased challenges to many natural systems, with species and ecosystems forced to move, adapt or die out
Climate change is natural: • Human activities are not responsible for climate change • CO_2 temperature changes are natural • Science is flawed • Nothing can be done (held by 2% of climate scientists, energy-producing companies and some politicians).	• Environment is a resource to be used for production of energy for human activities	• Fossil fuels continue to be used for energy • There is no need to alter current human interactions	• Loss of biodiversity • Increased sea levels • Increased coastal erosion (e.g. dune destruction during storm surges) • Increased drought • Increased severe weather events • Decreased snow cover

The impact of social debates about climate change on our relationships with outdoor environments can be seen as a **positive feedback loop**, as illustrated in Figure 8.59.

positive feedback loop process in which the end products of an action cause more of that action to occur in a feedback loop

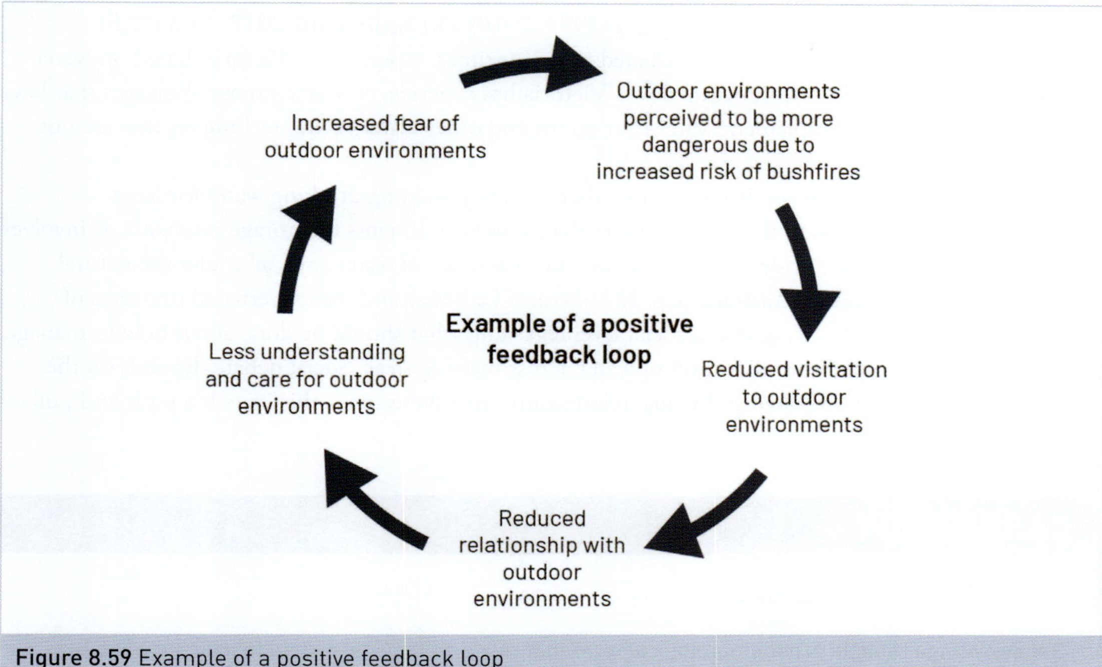

Figure 8.59 Example of a positive feedback loop

LEARNING ACTIVITY 8.5C

Philosophical chairs – climate change

Worksheet 8.5e Philosophical chairs

Divide the classroom into three sections, with students allocated equally among the sections. Each section is designated a different opinion regarding the cause of climate change:

- To avoid catastrophic climate change, Australia needs strong policy on reduction of greenhouse gases and carbon pricing schemes.
- Climate change has occurred throughout the planet's history and it will naturally and safely resolve without us changing our behaviours.
- We will innovate ourselves out of climate change through technology, and avoid dangerous climate change while continuing with our lifestyle behaviours.

The idea of this activity is for students to debate this issue using their own opinion or other evidence and try to convince classmates to move to their section of the room.

Rules

1. Only one student can talk at any time during a section's turn. No calling out!
2. Students raise their hands to be chosen to talk.
3. Sections take it in turns to talk.
4. Students can move to another section (after they have had their turn to speak on their allocated section). If they move, they need to state what has changed their mind.
5. Hint: You don't necessarily have to agree with the point of view of the section you are allocated. You might want to see if you can convince others to change their own viewpoint!

DEBATE ABOUT WATER MANAGEMENT

As we have investigated in earlier parts of Outdoor and Environmental Studies, one of the conditions that Australia's climatic variation results in is periods of prolonged less-than-average rainfall – or drought – as illustrated during the Millennium drought (1997–2010) and again in the 2017–19 drought.

Water-management issues have been debated in recent times, especially in Victoria, based on water shortages, allocations and environmental flows. Victoria has experienced water storage shortages that have forced the government to implement water restrictions and other water-saving techniques that are not very publicly popular.

Water management, however, is much more than simply providing drinking water for large populations, and requires us to think about more than water catchments for storage reservoirs. It involves management strategies for a wide range of uses in a diverse range of waterways, all under the natural climatic variation pressures of Australia (e.g. El Niño and La Niña) and the exacerbated pressures of climate change. Debates have raged at a social level regarding what should be done about how to manage the water crisis, particularly the allocation of water across massive areas. Social debate depends on the viewpoint of the debater and their individual relationship with the water – this issue is a push and pull of a finite resource.

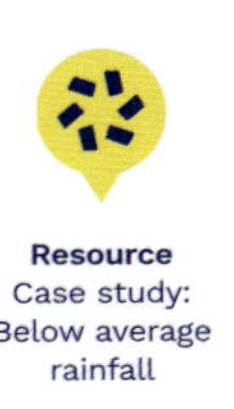

Resource
Case study: Below average rainfall

CASE STUDY

BELOW AVERAGE RAINFALL

'Below average rainfall' is a meteorological term that relies on long-term data to indicate what the average rainfall is for a location. In Australia, our long-term data indicates recurrent periods of severe drought. Australia has had such a strong history with droughts that they are named and engraved on our memories. Each has its causes, but all involve below average rainfall. Droughts over the past century or so include the following:

- 1895–1902 (Federation drought)
- 1914–15 drought (short, but the national wheat crop failed)
- 1937–45 (World War II drought, which included the 1939 Black Friday bushfires)
- 1965–68 drought
- 1982–83 drought (which included the 1983 Ash Wednesday bushfires)
- 1997–2009 (Millennium drought, which included the 2003, 2007 and 2009 Alpine bushfires, and the 2009 Black Saturday bushfires)
- 2017–19 drought (which included the 2019 Mallacoota and Corryong bushfires).

Figure 8.60 shows the national rainfall distribution during the 2017–19 drought.

QUESTIONS

1 With what frequency have droughts occurred in Australia over the past 120 years?
2 How often have they resulted in severe bushfires?
3 Do you think drought could be considered a normal part of Australia's climatic conditions? Why or why not?
4 How does below average rainfall impact agriculture, water management and recreation that relies on rainfall?
5 How can we plan for drought?

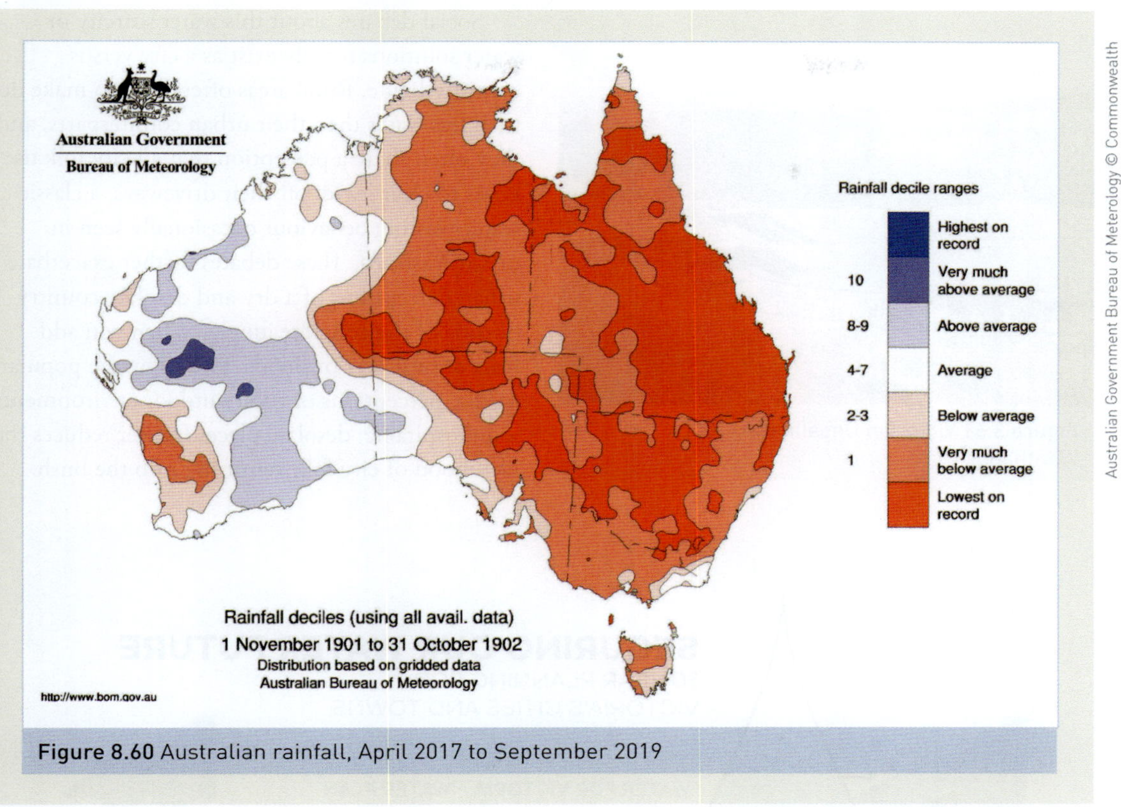

Figure 8.60 Australian rainfall, April 2017 to September 2019

Debate 1: Water for urban areas

Cities are thirsty places, demanding water for households, businesses and industry. During the Millennium drought, Melbourne needed more water than the storages could provide, so Victoria made large investments in the state's water security:

- The Sugarloaf Pipeline (2008) redirected water to link the Goulburn River from near Yea to the Sugarloaf Reservoir, north-east of Melbourne. In 2022, this project has also had a business case prepared to pump water in the opposite direction than intended, from south to north to support industry and farmers, particularly during dry conditions.
- The Victorian Desalination Plant, completed in 2012, can deliver up to 150 billion litres (150 gigalitres) of fresh drinking water per year to Melbourne.

Both of these projects caused massive debate. Indeed, social debate and political manoeuvring was tense – but both projects proceeded as water security was paramount. While the pipeline has not needed to flow from north to south, the desalination plant has been used to feed water into Melbourne's storage. While dam storage was at 98% in January 2023, following a very wet La Niña–influenced spring and summer, prior to 2022, storages were decreasing year on year – even with assistance from the desalination plant.

Shutterstock.com/Dorothy Chiron

Figure 8.61 Victorian Desalination Plant, Wonthaggi

Social debates about this water scarcity or water solutions mainly exist as a city versus country divide. Rural areas often have to make do with less water than their urban counterparts, and they often share a perception that all city folk use hoses to clear leaves off their driveways, a classic water-wasting behaviour occasionally seen in urban Australia. These debates further exacerbate urban perceptions of a dry and desolate country outside of the concrete jungle. When you add fears of bushfires or floods, fuelled by the popular media, perceptions of rural outdoor environments as inhospitable, desolate places further reduces the likelihood of city folk venturing into the bush.

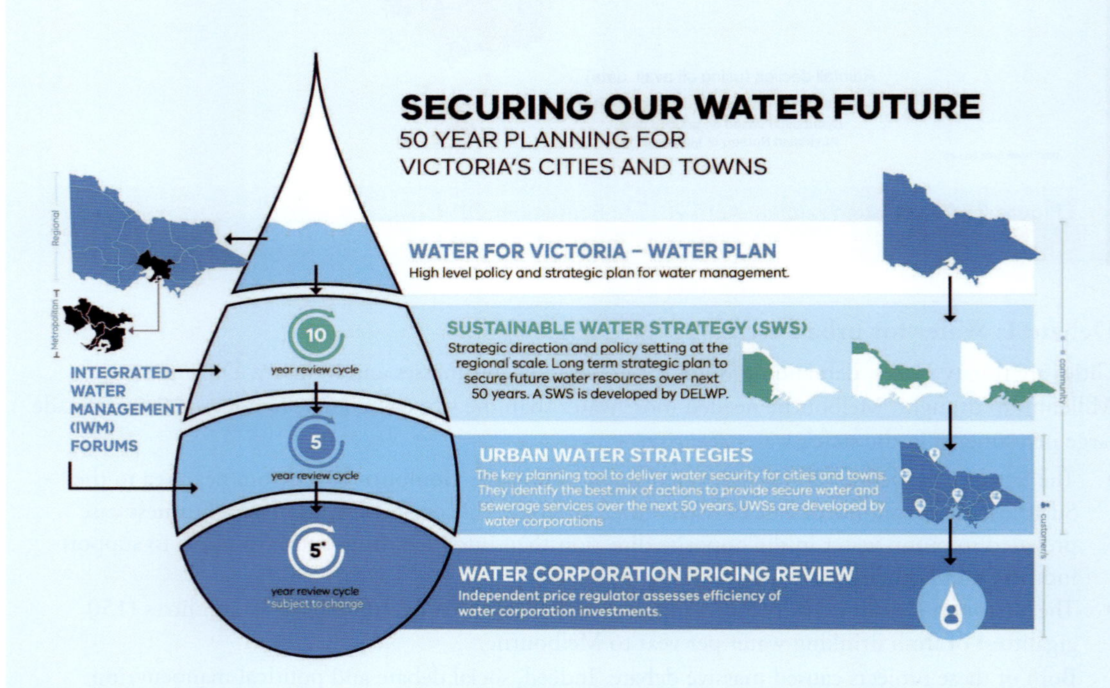

Figure 8.62 The state government's strategy for managing Victoria's water supply and use

Debate 2: Water for rural areas and agriculture

Water in rural Australia is always a popular discussion item, whether it is about local water management or quality, a dam being built that affects flow of water over one property to the next or social debate about government-managed water.

Here, we return to the Murray–Darling Basin conflict discussed earlier in the chapter. The Murray–Darling Basin provides water to 40% of Australia's farms to grow food for us to eat and export. These range from family-run farms growing wheat to multinational-run farms growing oranges or cotton. The debate, and the rub, of this situation is that everyone wants their share and it should be equitable, but many believe this is not the case.

Farmers pay for their allocation of water on an annual basis all along the Murray–Darling Basin system. Water in the Murray–Darling Basin can be bought and sold, either permanently or temporarily, on markets along the length of the system. In the event that there is water available in excess of environmental allocations, it is sold on the open market at market price (as all other allocations are). The water goes to the highest bidder, not necessarily to those farmers who may be suffering most from drought, for example.

Fairfax Photo/JOHN WOUDSTRA

Figure 8.63 International investment in irrigation properties in the lower Murray increases pressure on a bidding system, a system that could be seen as failing farmers.

Drought is tough on the whole river system: on farmers, on communities and on the environment. Achieving the balance of water needs and allocations has meant acquiring a portion of water from farmers, through purchase or through public investment into farm infrastructure, and returning it to the environmental side of the ledger. In 2019, there was debate to reverse the process of allocating water to the environment by instead returning it to irrigators. However, the Murray–Darlin Basin Authority (MDBA) made it clear this could only be done under certain circumstances:

> Regardless of the climate, the Water Act does not allow water earmarked for the environment to be returned to farmers either on loan or as a gift. There is provision, however, for environmental water to be sold on the open market, for anyone to purchase, strictly on condition that the environment doesn't suffer as a result.
>
> Source: Australian Government (Murray–Darling Basin Authority)

On one side of the social debate, you have farmers railing against the MDBA. They describe the system as broken, as it allows for top-dollar buyers to end up with the most water and the environment to receive its allocation, even if irrigators are suffering drought and not getting the water they need. They argue that water is lost to the system when the **Barmah Choke** is inundated; however, the MDBA counter that this water is not 'lost', but travels across the floodplain back into the river.

Barmah Choke the narrowest section of the River Murray where human-made weir gates have been added to keep the body of the river inside its river banks and travelling onward to other user areas. In the case of the Overbank flooding these artificial walls are too low to hold back the water and so the large flows being released to meet the needs of irrigators downstream means that a large proportion floods into the Barmah-Mellewa red gum forest, literally drowning it out of flood season in summer

The use of the Barmah Choke is controversial. It's the narrowest section of the River Murray, where human-made weirs keep the body of the river inside its banks before it's released to travel onward to other user areas. In the case of overbank flooding, these artificial walls are too low to hold back the water and so large flows being released upstream to meet the needs of irrigators downstream flood into the Barmah-Millewa red gum forest, literally drowning it. Meanwhile, farmers who could make use of the water are not permitted access to it.

There are also arguments by environmentalists campaigning for natural flows to be maintained year-round to avoid the catastrophes like the 2018, 2019 and 2023 **mass fish deaths** and unseasonal flooding of the Barmah forest.

Fairfax Photo/NICK MOIR

Figure 8.64 Between 2018 and the present, mass fish deaths have occurred in the lower Darling and upper Murray rivers due to a combination of drought, algal blooms and a sudden temperature drop, with social debates laying the blame squarely on the MDBA.

Resource
Additional Case Study: Drought reveals fault line in Murray-Darling Basin strategy

mass fish death when a large number of wild or farmed fish die suddenly and unexpectedly; this is more likely to happen in times of flood or drought

Debate 3: Water for Indigenous peoples' cultural flows

Indigenous peoples share a Kinship connection with all water systems of Australia. It is part of them and their beliefs, and has sustained them for tens of thousands of years. When Australia was colonised, control of the land and water began and Indigenous peoples have largely been excluded from the management of the water resources, and have had their connections to and sacred places within inland waterways disregarded.

Aboriginal Water Program
a program that works to include Aboriginal peoples in the way water is managed in Victoria and to reconnect communities to water for cultural, economic, customary and spiritual purposes

To counter this, the Victorian State Government in 2022 released the *Water is Life: Traditional Owner Access to Water Roadmap*. Part of the **Aboriginal Water Program**, the roadmap stirred up much debate surrounding Indigenous peoples' access to water, as agricultural users of water assert their rights to control the water, amid concerns there is already insufficient water to go round.

WATER DEBATE – HOW IT CAN INFLUENCE RELATIONSHIPS

As you have read, there are many examples of debate regarding the management of water in Victoria and these influence relationships with specific outdoor environments. Table 8.12 outlines the influence of debate on perceptions of, interactions with and impacts on the management of water.

Table 8.12 Debate over the supply of more water for agriculture versus the environment

Debate	Perceptions towards the environment	Interactions with the environment	Impacts of social debates on relationships
More water for agriculture	• Management is not natural and the river is dying, influenced by social debate and media coverage of fish kills and flooding forests, due to a water buying market dictating flows • The river more than ever remains a resource for us to use and prosper from	• Take as much water as possible. Increase water allocations when you can, so you're not left high and dry • There is reduced visitation of tourists or recreationalists due to fish death, poor river health, media coverage and social debate	**Positive feedback loop:** fewer people visiting the river outdoor environment may lead to fewer people caring for it and less protection
More water for the environment	• Environment not a limitless resource • Environment should be protected from human activities if possible	• Campaign for environmental flows • Lobby for government subsidies for water conservation techniques (e.g. tank rebate) • Divert floods to catchment areas	• Improve water storage levels • Maintain and improve environmental flows and management of water through Barmah Choke and reduce mass fish kills

Debates about renewable energy

fossil fuel
a deposit, such as petroleum, coal or natural gas, derived from the accumulated remains of ancient plants and animals and used as fuel

The vast majority of the energy that is consumed in Victoria is derived from **fossil fuels**, or decomposed organic matter that is mined and then burned to release heat energy. The big three are oil (and its derivatives of petrol and diesel), coal and natural gas. We have come to depend on fossil fuels to make electricity, heat homes and other buildings, drive cars and so on. However, the problem is in our dependence on, and overuse of, these energy sources, and the issues this has created. For example:

- We are consuming fossil fuels 100000 times faster than they are being produced. They cannot last forever (that is, they are non-renewable) and this is beginning to show in the decline of new deposits being discovered.
- Burning fossil fuels contributes more carbon dioxide emissions than any other human activity. We have already examined the potential impact of global warming and climate change on the outdoor environment and society earlier in this text. The vast majority of Victoria's electricity is currently produced by brown coal-powered generators. This technology is considered 'dirty' as it is a big polluter. About 40% of Victoria's carbon emissions come from coal-powered electricity generators alone.

Renewable energy is energy produced from sources that can be replenished or replaced. It includes energy that is generated from natural resources such as wind (wind farms), sunlight (solar panels), the water cycle (hydro-electricity), tidal and wave movement, geothermal energy and biofuels.

renewable energy energy that can be obtained from natural resources that can be constantly replenished

The Clean Energy Council argues:

> Right now, across Australia, renewable energy is powering millions of homes and businesses. It's creating thousands of jobs and reducing bills. It's not a passing phase, renewable energy is here now and it's growing faster than ever, and as with all growth and especially rapid-growth, debate as to its efficacy, its economics and its safety are also present.
>
> Clean Energy Council

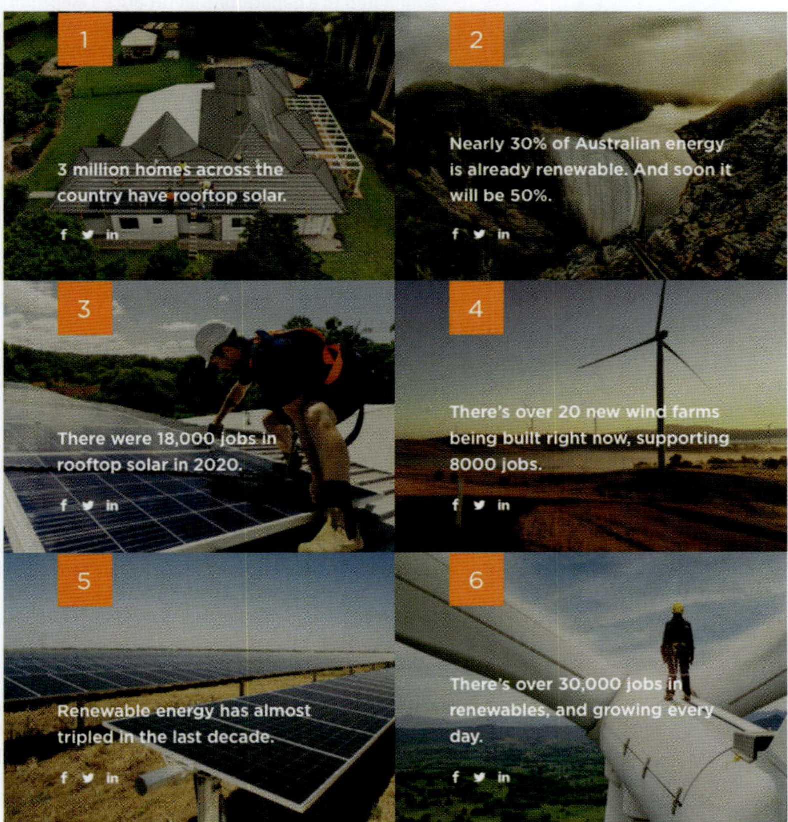

Figure 8.65 Poster from the Clean Energy Council's campaign, 'Ables not fables'

LEARNING ACTIVITY 8.5D

Clean Energy Council

Visit the Clean Energy Council website to examine in detail the current range of renewable energy resources available.

Worksheet 8.5f Clean energy council

Weblink Clean Energy Council

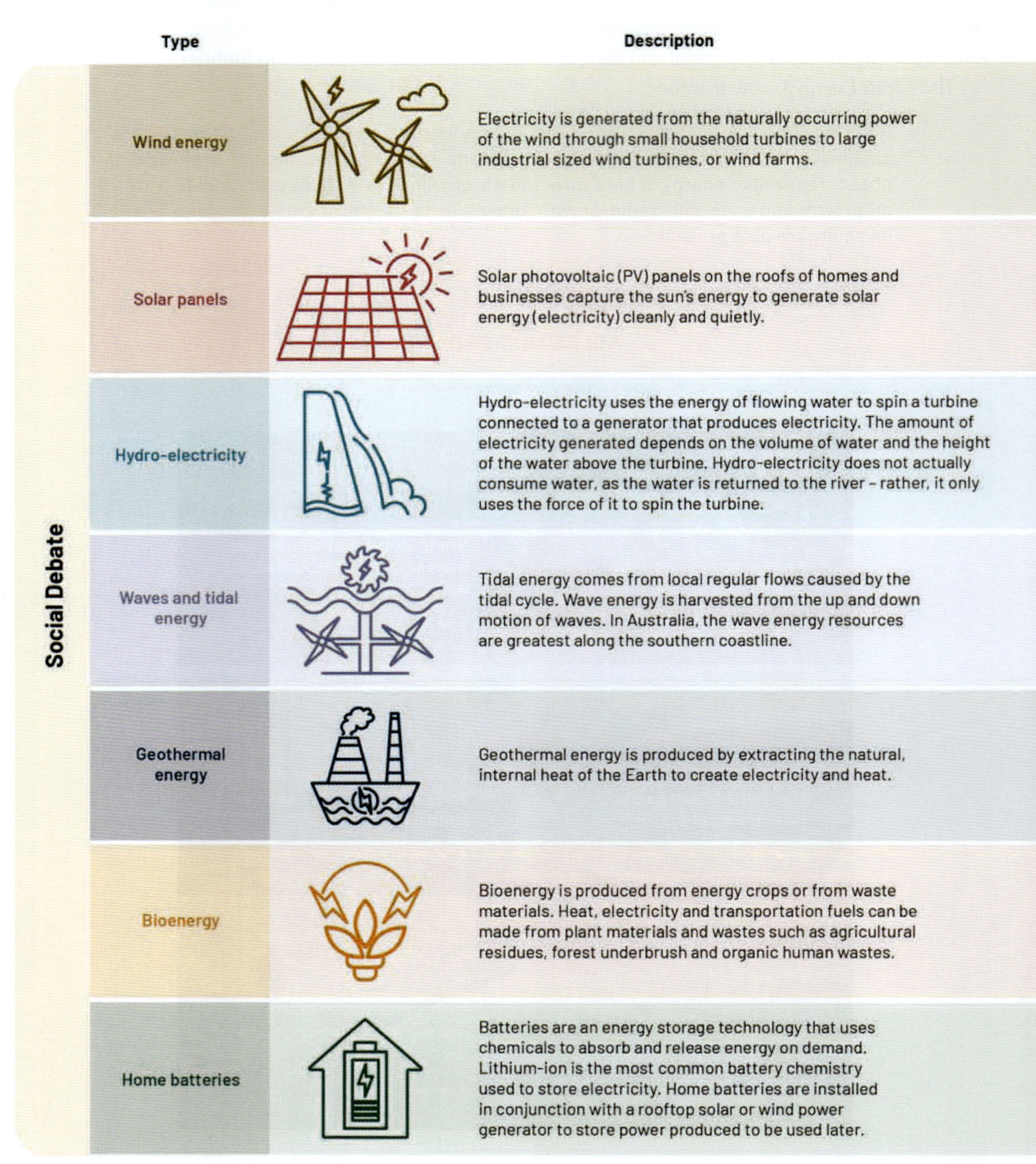

Social Debate

Type	Description
Wind energy	Electricity is generated from the naturally occurring power of the wind through small household turbines to large industrial sized wind turbines, or wind farms.
Solar panels	Solar photovoltaic (PV) panels on the roofs of homes and businesses capture the sun's energy to generate solar energy (electricity) cleanly and quietly.
Hydro-electricity	Hydro-electricity uses the energy of flowing water to spin a turbine connected to a generator that produces electricity. The amount of electricity generated depends on the volume of water and the height of the water above the turbine. Hydro-electricity does not actually consume water, as the water is returned to the river – rather, it only uses the force of it to spin the turbine.
Waves and tidal energy	Tidal energy comes from local regular flows caused by the tidal cycle. Wave energy is harvested from the up and down motion of waves. In Australia, the wave energy resources are greatest along the southern coastline.
Geothermal energy	Geothermal energy is produced by extracting the natural, internal heat of the Earth to create electricity and heat.
Bioenergy	Bioenergy is produced from energy crops or from waste materials. Heat, electricity and transportation fuels can be made from plant materials and wastes such as agricultural residues, forest underbrush and organic human wastes.
Home batteries	Batteries are an energy storage technology that uses chemicals to absorb and release energy on demand. Lithium-ion is the most common battery chemistry used to store electricity. Home batteries are installed in conjunction with a rooftop solar or wind power generator to store power produced to be used later.

Figure 8.66 A summary of the different types of renewable energy and the main points of debate for each

Debate

The environmental impact of wind energy is negligible, but debates have been heard over the noise produced by the rotor blades, visual impacts and deaths of birds that fly into the rotors (bird mortality).

The main debates with solar energy include

- you can't make solar power at night
- land disturbance/land use
- visual changes to the landscape
- impacts associated with hazardous materials
- energy required to manufacture and install solar components may also generate emissions.

Hydro-electricity does not produce significant greenhouse gas emissions but does have other major environmental impacts. The reservoirs often destroy vast areas of highly productive forest and wildlife habitat. The dams also damage freshwater ecosystems by blocking the movement of fish and other organisms.

Debates over tidal and wave energy focus on the aesthetic disturbance to coastal areas and the potential impact on marine ecosystems.

The primary impacts of geothermal plant construction and energy production are gaseous emissions, land use, noise and potential ground disturbances(also known as subsidence).

Bioenergy can have a negative environmental impact if there is too much CO_2 produced, contributing to climate change.

Debates surrounding the cost in dollars and for the environment of the production of the battery, its life span, and that only small recycling programs are currently in use.

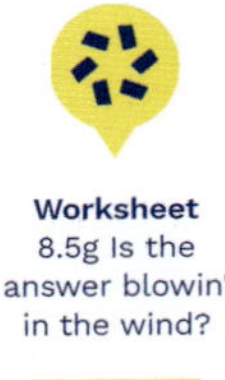

Worksheet
8.5g Is the answer blowin' in the wind?

LEARNING ACTIVITY 8.5E

Is the answer blowin' in the wind?

Not everyone is pleased with the increasing number of wind farms in Victoria. Investigate an interest group that is against this form of renewable energy.

1 What is the basis for their point of view?
2 What are some of the environmental impact issues associated with wind farms?
3 Provide an argument for or against the use of wind energy using information to justify your position.

SUMMARY OF THE RENEWABLE ENERGY DEBATE

Although the case for development and investment in renewable energy technology often seems universally accepted, there is a range of opinion regarding the various elements of renewables. Public support for renewable energy is increasing as awareness of environmental issues increases, and this is adding pressure on governments to act. International treaties such as the Paris Agreement of 2015 also stipulate targets for renewable energy uptake. In 2022, the Australian Labor Government committed to a renewable energy target (the proportion of energy produced by renewable sources) of 82% by 2030.

Alamy Stock Photo/Alex Gitlits

Figure 8.67 A solar farm outside of Melbourne,Victoria.

Alamy Stock Photo/talldwarf

Figure 8.68 A sign protesting against wind farms

INFLUENCE ON RELATIONSHIPS

Our relationship with outdoor environments can be influenced in many ways by the debate over renewable energy. This might be our relationship with a specific local environment or the global environment in general. Table 8.13 outlines the influence of debate on perceptions of, interactions with, and impacts on the use of renewable energy.

Table 8.13 Summary of the debate for and against renewable energy

Debate	Perceptions towards the environment	Interactions with the environment	Impacts of social debates on relationships
For renewable energy	• Environment not a limitless resource and must be protected • Climate change is an urgent threat requiring alternative methods of producing energy • Fossil fuels are a finite resource	• Reduction of mining of coal for energy production and its greenhouse gas emissions	• Net reduction of CO_2 emissions • Lessened impacts of climate change
Against renewable energy	• Environment is a resource for the production of energy for human use • The environment's visual integrity should be preserved • Biodiversity should not be threatened	• Continue mining coal for energy production • Support some renewable energies but not as replacement for fossil fuels	• Agricultural environments remain intact and free from wind or solar farm disturbances • Climate change impacts continue • There are negative visual impacts • Tourism will decrease • There will be a possible decrease in local land values • Noise pollution and possible health impacts may increase • Preserve bird species against possible bird collisions with turbine blades

Alamy Stock Photo/Alex Gitlits

Figure 8.69 The Portland Wind Farm. A similar, and controversial, 52-turbine installation was constructed at Bald Hills despite local opposition. An attempt was also made to stop the project in parliament, claiming it broke environmental law by threatening internationally and nationally listed migratory birds, including the endangered orange-bellied parrot.

NOTES FOR THE EXAM 

For the exam, you should:

- understand the basis of one specific environmental issue (including climate change, water management or renewable energy)
- analyse the influence of social debates about an environment issue on relationships with outdoor environments.

8.5 KEY CONCEPTS

- A social debate may include points of view, conversations or arguments that are expressed through different media – such as conversations, tweets, blogs and social media pages – by an individual or group on a particular topic by an expert of just someone with an interest.
- Climate change debate:
  - Temperatures have increased in Australia by 1.4°C since 1910, causing increased frequency and intensity of heat extremes, rising sea levels, dangerous fire weather and bushfires, and severe floods.
- Climate sceptics versus climate activists debate
  - We either stop climate change or we don't, but in the process we are affecting human relationships with outdoor environments by spotlighting their dangers due to climate change. This creates a positive feedback loop where fewer people visiting outdoor environments means fewer people to care for outdoor environments.
- Water management debate:
  - Water management is a response to a dry continent with persistent droughts and our insistence that we continue to grow food the way we do. Water for urban areas has been addressed in the past by the North-South pipeline and Victorian Desalination Plant.
  - Social debates involve a city versus country divide. The city has ways of getting water that the country doesn't.
  - The management of the Murray-Darling Basin has led to water allocations going to the highest bidder, pricing many irrigators out and causing mass movements of water transfer across the system inundating the Barmah. At other times, lower flows are contributing to mass fish deaths.
  - Water for Indigenous peoples' cultural flows is a new area of debate where concerns are that this represents another user of already scare water resources, which might otherwise be allocated to farmers.
- Renewable energy debate:
  - This involves whether to adopt clean alternatives for electricity production and phase out old fossil-fuel-fired power stations. Australia has emissions targets that need to be met by renewable energy targets.
  - The potential net reduction in CO_2 emissions would help to avoid catastrophic climate change.

There is an aesthetic effect when installing renewable-energy-producing technology, which can also impact habitats and migratory birds.

8.5 CONCEPT QUESTIONS

Worksheets
8.5 Key concepts

Additional Learning Activity: 8.5h Campaign to 'Save the Murray'

REMEMBERING

1 What is a social debate?
2 Provide an outline for one environmental issue in Victoria.
3 Describe the debate involving that issue, including the participants in the debate and their arguments.

UNDERSTANDING

4 Describe the influence of the debate on relationships with outdoor environments.
5 Explain the relationship between social debates and how we perceive outdoor environments.

APPLYING

6 Using an example not presented in the text, describe an environmental issue that you've studied in Victoria and what the debate is.

EXTENSION CHALLENGE

7 Why have the prime ministers of Australia been cursed when it comes to leading their political parties and the nation when backing solid climate-change-related policies?

Resource
Glossary – Chapter 8

Assessments
End of chapter exam
Glossary test

EXAM-STYLE QUESTIONS

1 a Define what it means for a human to be the custodian of a piece of land. (1 mark)

b Outline the requirements for a Traditional Owner group to be considered for formal recognition by the Victorian government. (5 marks; 1 mark per requirement)

c After a Traditional Owner group becomes a Registered Aboriginal Party, what is a further option for formal state or federal recognition? (1 mark)

2 Describe a recreational activity you have participated in this year and the environment which you participated in. (2 marks)

3 Analyse the impact of an economic relationship with an outdoor environment you have studied. (5 marks)

a Name a conflict you studied. (1 mark)

b Describe and evaluate the methods each opposing group used in your chosen conflict to influence decision makers. (8 marks)

c Describe the processes used to resolve or potentially resolve your chosen conflict. (2 marks)

4 Analyse the policy of two different federal political groups around an environmental issue. (4 marks)

5 Analyse the climate change debate. (7 marks)

CHECK YOUR KNOWLEDGE AND SKILLS

Resources
Key knowledge and skills checklist

Checklist for students to mark and date, making judgements of their knowledge and understanding.
(D = developing C = consolidating ER = exam ready)

D	C	ER	Check your knowledge and skills
☐	☐	☐	Describe what custodianship is
☐	☐	☐	Outline the process of formal recognition of Indigenous peoples' custodianship of land or land and water
☐	☐	☐	Compare examples of formalised Indigenous peoples' groups caring for outdoor environments
☐	☐	☐	Compare different human relationships with outdoor environments
☐	☐	☐	Outline the three types of interactions with outdoor environments: conservation, recreation and economic
☐	☐	☐	Outline the relationship of each type with outdoor environments – perceptions, interactions and impacts
☐	☐	☐	Describe two specific conflicts over uses of outdoor environments in detail
☐	☐	☐	Outline the methods used to influence the decision makers of conflicts specific to two conflicts
☐	☐	☐	Outline the processes used to resolve conflicts specific to two conflicts
☐	☐	☐	Evaluate the methods used by conflicting parties to influence decisions in their favour and the processes followed to resolve or potentially resolve said conflicts
☐	☐	☐	Analyse differing environmental politics in Australia
☐	☐	☐	Outline Australian federal politics including the history of environmental issues and the political spectrum of parties and representatives
☐	☐	☐	Understand international treaties on environmental issues and their influence on policy
☐	☐	☐	Outline a specific current environmental issue in Australia and the resulting political policies in place or proposed
☐	☐	☐	Analyse the influence of a social debate about one environmental issue on relationships with outdoor environments
☐	☐	☐	Describe what a social debate is
☐	☐	☐	Outline a specific environmental issue and the social debate involving it: climate change, renewable energy or water management

Unit 4
Sustainable Outdoor Environments

Unit 4 – Introduction

Unit 4 of the VCE Outdoor and Environmental Studies study design, 'Sustainable outdoor environments' explores the significance of sustainability in human relationships with outdoor environments, and the crucial need to balance human needs with the needs of these environments. This unit examines current policies and management strategies for achieving and preserving healthy and sustainable outdoor environments in contemporary Australian society.

Chapter 9 – The importance of outdoor environments explores the current condition of outdoor environments in Australia, and an understanding why these environments are important for both individuals and society. This chapter will examine sustainability and also investigate any current or potential threats to environments and suggest ways to improve environmental sustainability.

Chapter 10 – The future of outdoor environments examines a range of land management practices and what individuals can do to sustain these environments now and in the future. This chapter will also consider how to balance human needs with conservation efforts and what skills are necessary to be environmentally responsible citizens. Finally, we will examine current laws and conventions and suggest ways to improve them for better conservation outcomes.

Chapter 11 – Investigating outdoor environments focuses on planning and participating in outdoor experiences. Examine how to undertake an independent investigation including the creation of a logbook. This chapter will prepare students to complete their written report including conventions of report communication.

Through their studies of Unit 4 students will continue to develop practical skills and knowledge to participate safely and sustainably in outdoor experiences, and use their experiences and observations as the basis for reflection and analysis of the key skills and knowledge.

CHAPTER

9

UNIT 4 – AREA OF STUDY 1

The importance of healthy outdoor environments

KEY KNOWLEDGE

- the pillars of sustainability, the interdependence between these pillars and related critiques of sustainability
- observable characteristics to assess the health of outdoor environments, including:
  - quality of water, air and soil
  - species and ecosystem biodiversity
- the impact of threats on society and outdoor environments, including two of the following:
  - land degradation
  - introduced species
  - urbanisation
  - climate change
  - flood
  - fire
- the importance of healthy outdoor environments for individual physical and emotional wellbeing, and for society now and into the future
- local, national and international solutions and mitigation strategies to combat climate change across a range of environments.

KEY SKILLS

- analyse understandings of the interdependence of the pillars of sustainability and related critiques of sustainability
- evaluate the health of outdoor environments and create possible solutions to improve environmental health
- analyse threats to society and outdoor environments
- justify the importance of healthy outdoor environments for individuals and society
- analyse possible solutions and mitigation strategies for combating climate change.

VCE Outdoor and Environmental Studies Study Design 2024–2028, pp. 26-27.

Sustainability

Sustainability challenges society to use the Earth's limited resources efficiently and equitably to ensure the survival of current and future generations of all species. Through small everyday actions you can make a big difference, contributing to a more a sustainable world. The three pillars of sustainability – environmental, social and economic sustainability – work together to ensure the longevity and balance of our planet and its resources.

Observable characteristics

The environment gives us the essentials we need to survive, like clean air, water and food, and also offers us places to enjoy ourselves and feel inspired. The quality of air and water is crucial for environmental health, and if it's not maintained, both the environment and our recreational and economic activities are impacted. Soil formation takes a long time, but it can be quickly destroyed by harmful practices like deforestation. Biodiversity refers to all living organisms on Earth including different plants, animals, microorganisms, and ecosystems, and the genes they possess.

The impact of threats on society and outdoor environments

Understanding two significant threats, their causes, and some of their impacts on society and outdoor environments. Land degradation is caused by harmful human activities like unsustainable agriculture practices, while urbanisation resulting from increased population increases pressure on outdoor environments by creating a demand for land. Climate change is another significant threat that will have severe impacts on society and outdoor environments, affecting agriculture, human health, water and food security, and infrastructure.

The importance of healthy outdoor environments

Healthy outdoor environments are crucial as they provide clean air and water, recreation, and educational opportunities, enriching our lives. The aesthetic value of outdoor environments is based on their appearance and the emotional responses they evoke, such as a beautiful beach or mountain range. Outdoor environments also offer opportunities for positive risk-taking behaviours allowing individuals to explore their abilities and extend themselves. Access to nature plays a crucial role in promoting individuals and society's physical and emotional wellbeing.

KEY TERMS

aesthetic value
carbon sink
clear felling
climate feedback loop
critique
economic sustainability
ecosystem diversity
ecotourism
emissions
endemic
environmental buffer
environmental sustainability
genetic diversity
greenwashing
grey water systems
hydro-electricity
interdependence
intrinsic value
mariculture
multilateral
nature-deficit disorder
net zero
renewable
rotation planting
salinity
sequestration
smog
Snowy Mountains Scheme
social sustainability
species diversity
sustainable

Worksheets

Weblinks

Resources and templates

Assessments

To access resources above, visit **cengage.com.au/nelsonmindtap**

9.1 SUSTAINABILITY AND RELATED CRITIQUES

KEY KNOWLEDGE

- the pillars of sustainability, the interdependence between these pillars and related critiques of sustainability

KEY SKILLS

- analyse understandings of the interdependence of the pillars of sustainability and related critiques of sustainability

9.1.1 WHAT IS SUSTAINABILITY?

The concept of sustainability has been linked to many aspects of contemporary society, from its use to describe the packaging of our latest smartphone to its use in advertising to influence our choice of car or energy provider. The word 'sustain' is derived from the Latin sustinere, which means 'to hold up', and 'sustainable' can be defined as 'capable of being maintained in existence without interruption or diminution'. Other words synonymous with the term include 'prolong', 'support' and 'endure'. Over the years, the term '**sustainable**' has been associated with being environmentally friendly, **renewable** and green. These are all ambiguous terms that are used to refer to products and policies that claim to reduce harm to the environment. It is no wonder that confusion arises regarding its meaning.

Alamy Stock Photo/PURPLE MARBLES

Figure 9.1 A coffee cup made from plants

sustainable capable of being maintained in existence without interruption or diminution

renewable something that can be naturally replenished or replaced relatively quickly, typically within a human lifetime, without depleting its source

In essence, sustainability challenges society to use the Earth's limited resources efficiently and equitably to ensure the current and future survival of all species, while preserving healthy outdoor environments. The concept of sustainability has a long history, dating back to ancient civilisations, including the 60000-year-old practices of the Indigenous cultures of Australia. These cultures demonstrated a deep understanding of sustainability through their careful and controlled harvesting of resources, such as plants and trees, to ensure their availability for future generations. For example, the bark of specific trees was carefully harvested for the construction of canoes, shields, tools and other items, with great care taken not to damage the plant's system during the process.

The modern world has been heavily reliant on non-renewable energy sources, such as coal and gas, for an extended period of time. With the increasing global population and its impact on the Earth's resources, it has become crucial to sustainably manage these resources to ensure their availability for future generations, as well as mitigate the effects of climate change. The growth of population and technology has allowed us to better understand the impact of human activities on the environment and the challenges that come with economic and population growth.

Over the years, international collaborations aimed at raising environmental awareness and promoting sustainable practices have been established, including the Brundtland Commission in 1987, the Earth Summit in 1992, the World Summit in 2005 and the Earth Summit in 2012. These efforts aim to educate and engage the global community in finding solutions to protect the environment and promote sustainable development.

9.1.2 WHY IS SUSTAINABILITY IMPORTANT AND HOW CAN YOU BE MORE SUSTAINABLE?

We cannot maintain the Earth's ecosystems, or continue to function as we do, if more sustainable choices are not made. If harmful processes are maintained with no change, it is likely that we will run out of fossil fuels, huge numbers of animal species will become extinct and the atmosphere will be irreparably damaged.

Through small everyday actions, you can make a big difference and contribute to a more a sustainable world. Changing small behaviours such as putting rubbish into the correct bins for recycling, taking a reusable coffee cup when you leave home and only buying things that you need can have a lasting impact on the environment.

By taking the simple step of checking what food you already have in your home before shopping, you can help reduce waste and save resources. Every year, Victorians waste a staggering 250000 tonnes of edible food. This not only results in the waste of food itself but also the resources used to grow, transport and store the food. By being mindful of what you purchase, you can make a positive impact on sustainability.

Figure 9.2 A canoe scar-tree in Yugambeh Country in south-east Queensland. The tree trunk reveals the scar left by the careful removal of the outer bark to ensure the tree's survival.

Ditching single-use plastic products in favour of reusables is integral to sustainability. Using steel straws, cloth shopping bags, bamboo toothbrushes, compostable bin liners and glass coffee cups all mitigate the need for single-use items to be manufactured and then disposed of.

Using water-efficient shower heads and keeping showers under four minutes in length has a significant impact on the amount of water used in a household. Using the washing machine and dishwasher only when full and using cold water cycles also reduces water consumption and energy production, all of which help to preserve the Earth's natural resources.

Consider starting a compost heap or worm garden to reduce the amount of food waste and other compostable materials you throw away. Instead of putting fruit and vegetable scraps in the bin, which could end up in a landfill and contribute to the production of methane, composting recycles kitchen waste and provides nutrient-rich soil for your garden or pot plants. Composting can reduce the volume of kitchen waste by up to 95 per cent. Alternatively, many council areas have green bins for collecting kitchen and garden waste.

Resource
Case Study: More households are using organic waste collection

CASE STUDY

MORE HOUSEHOLDS ARE USING ORGANIC WASTE COLLECTION SERVICES

More Victorian councils introduced kerbside organic waste collection services. In 2019–20, 26.5% of Victorian households used a kerbside food and garden organics collection service, almost double the proportion from 2018–19. These collections have helped Victorian's reduce waste to landfill and the associated greenhouse gas emissions.

How much kerbside waste was recovered?
Diversion rate of
45.0%
after contamination

What's in our mixed recyclables?
Paper: 47.1%
Plastics: 9.5%
Glass: 27.4%
Metals: 2.7%
Contaminants: 13.3%

Figure 9.3 Infographic from Sustainability Victoria highlighting key information about Victoria's kerbside recycling program

Pillars of sustainability

Three pillars are commonly used to describe the concept of sustainability: social, economic and environmental. These pillars do not work in isolation but rather are connected to each other, and are all important in achieving overall sustainability. Another way to describe the three aspects of sustainability are 'people, profit and planet', reflecting the social, economic and environmental pillars.

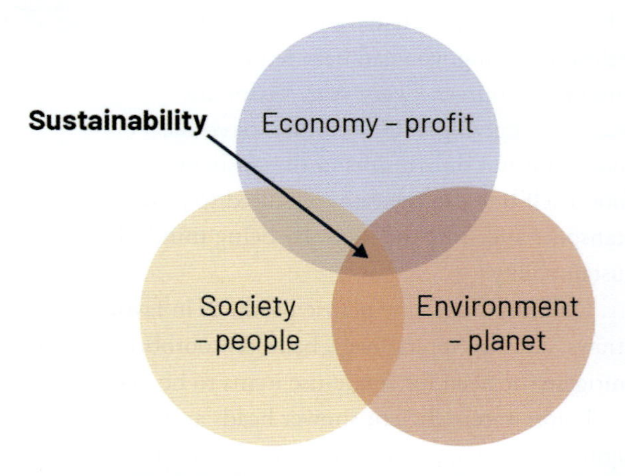

Sustainability
Economy – profit
Society – people
Environment – planet

Figure 9.4 The three pillars of sustainability

SOCIAL SUSTAINABILITY

social sustainability the concept of an inclusive and just society where every individual's needs are met, and where everyone has equal opportunities to contribute and participate in decision-making processes

The pillar of **social sustainability** includes environmental justice, human health, resource security and education, among other important social elements of society. Social sustainability encompasses a holistic approach to designing not just the physical environment, but also the social fabric of our communities. It involves creating infrastructure that supports various aspects of social and cultural life, such as community spaces, recreational facilities, social amenities and systems for citizen engagement. It also includes ensuring that there is adequate space for people to interact, and for places to evolve and grow over time. Social sustainability prioritises the wellbeing of individuals and communities, recognising that sustainable communities are built according to social and economic considerations, as well as environmental concerns. By prioritising social sustainability, we can create vibrant, liveable communities that promote wellbeing, happiness and a sense of belonging.

Social sustainability can be promoted in the following ways:

- Improving quality of life – acting to ensure people are living well; for example, that they have access to affordable housing, physical and mental medical support, education training opportunities, employment opportunities and, of course, safety and security.
- Improving equality – acting to reduce disadvantages to certain groups or helping certain groups to remove barriers to have more control of their lives.
- Improving social cohesion – acting to increase participation by individuals, as well as help groups to access public and civic institutions.

ECONOMIC SUSTAINABILITY

The pillar of **economic sustainability** includes job creation, profitability and proper accounting of ecosystem services for optimal cost–benefit analyses. To ensure economic sustainability, it's crucial to adopt practices that conserve and effectively manage these limited resources, such as reducing waste, promoting recycling and investing in renewable energy sources. Additionally, sustainable economic growth should prioritise equity and social welfare, rather than just maximising profits for a select few. This approach considers the long-term impacts on communities and the environment, leading to a more balanced and resilient economy. Moreover, by transitioning to sustainable practices, we can create new job opportunities, stimulate innovation and drive economic growth that benefits all members of society. In short, economic sustainability seeks to balance economic prosperity with environmental and social responsibility, ensuring that future generations have access to the resources they need to thrive.

economic sustainability the ability of economic systems to sustain a decent standard of living and operate in a way that promotes long-term environmental health and preserves natural resources for future generations

Economic sustainability can be promoted in the following ways:

- Devising less wasteful systems to reduce land use or making supply chains more efficient cuts down on the resources needed to produce a good or bring it to market.
- Prioritising low-impact economic development, through investing time and money in sustainable businesses, will help to create a waste-free world.
- Switching to renewable energy sources by converting operations to run on energy produced by solar or wind power rather than fossil fuels is one way an organisation can prioritise the future.

ENVIRONMENTAL SUSTAINABILITY

The pillar of **environmental sustainability** focuses on the wellbeing of the environment. This pillar includes water quality, air quality and the reduction of environmental stressors, such as greenhouse gas emissions. To achieve environmental sustainability, it is crucial to reduce our carbon footprint, conserve resources and minimise waste. This requires a shift in the way we produce and consume goods, as well as a heightened awareness about the impacts of our actions on the environment. Implementing renewable energy sources, improving waste management systems and promoting sustainable agriculture practices are all crucial steps in achieving environmental sustainability. Additionally, individuals can play a vital role by reducing their energy usage, conserving water and reducing their waste by practising recycling, composting and reuse. By working together to create a more sustainable future, we can ensure that future generations have access to a healthy and thriving planet.

environmental sustainability a concept that prioritises the wellbeing of the environment, encompassing aspects like water and air quality, as well as the reduction of environmental stressors such as greenhouse gas emissions

Renewable resources and energies can be used to promote environmental sustainability; renewable energy is sometimes referred to as sustainable energy. Literally 'to make new again', a renewable resource is one that is naturally replenished with time, with the growth of new organisms, natural recycling of materials or energy sources such as wind, geothermal and solar. A renewable resource is a natural resource that will replenish to replace the portion depleted by usage and consumption, either through natural reproduction or other recurring processes in a finite amount of time on a human time scale.

Promoting environmental sustainability

Environmental sustainability can be promoted in the following ways:

- Biomass – this is a renewable energy source that comes from organic matter such as crops, wood and waste from homes and industry. Biomass is a source of energy because it contains stored energy from the sun, which plants absorb during photosynthesis.
- Solar power – solar energy means converting the light heat from the sun into an energy source we can use, mostly electricity. Solar energy is rapidly expanding in popularity. Using photovoltaic panels, we can convert the sun's energy into usable electricity and power our homes or businesses.
- Wind power – wind energy generates electricity, which is free. Wind is a native source of energy that does not need to be mined or transported, taking two expensive costs out of long-term energy expenses.
- Hydropower – this is energy generated by the movement of water and has been used as a source of renewable energy for centuries. It involves capturing the energy of falling or flowing water, and using it to turn a turbine and generate electricity.
- Geothermal energy – this is produced from the heat of the Earth's core, and is a clean, renewable and abundant source of energy. This energy can be harnessed to generate electricity, heat homes and power various industrial processes.

By utilising these renewable resources and energies, we can help protect our environment and reduce our reliance on finite resources like fossil fuels, which release harmful emissions and contribute to climate change.

TOORA WIND FARM

The Toora wind farm is the second large-scale wind farm to be built in Victoria and the second major wind farm for Stanwell Corporation. It consists of 12 turbines with a combined capacity of 21 megawatts (MW). The turbines are made by Danish company Vestas and each has an electrical capacity of 1.75 MW. The turbines are mounted on tubular towers 67 metres high. The rotors have a total diameter of 66 metres. The blades pitch to optimise the power produced and to control the rotation speed. The turbines rotate at 19 to 21 revolutions per minute. Power is produced at wind speeds of between 3.5 m/s and 25 m/s. The Toora wind farm produces enough energy to supply more than 6600 homes and will abate the equivalent of up to 48000 tonnes of carbon dioxide per year.

Forest Lodge Farm, 2020

Trevor Pinder / Newspix

Figure 9.5 Toora Wind Farm on the South Gippsland Highway in the town of Toora

Examples to reduce land pollution to promote environmental sustainability include the following:

- Proper waste management – ensuring that waste is disposed of properly and safely is one of the most effective ways to prevent land pollution. This means properly disposing of hazardous and toxic waste, as well as general waste, in designated areas and in a manner that does not harm the environment.
- Reduce, reuse, recycle – implementing the 'reduce, reuse, recycle' model can help to reduce the amount of waste that ends up in landfills. This can be done by reducing the number of items we use and purchase, reusing items that still have a useful life and recycling those items that can no longer be used.
- Sustainable agriculture – agricultural practices can also contribute to land pollution. This can be mitigated by using sustainable agricultural methods that promote healthy soil, conserve resources and prevent soil erosion.
- Developing green spaces – creating green spaces, such as parks and gardens, can help combat the effects of land pollution and provide a healthy environment for people to enjoy.

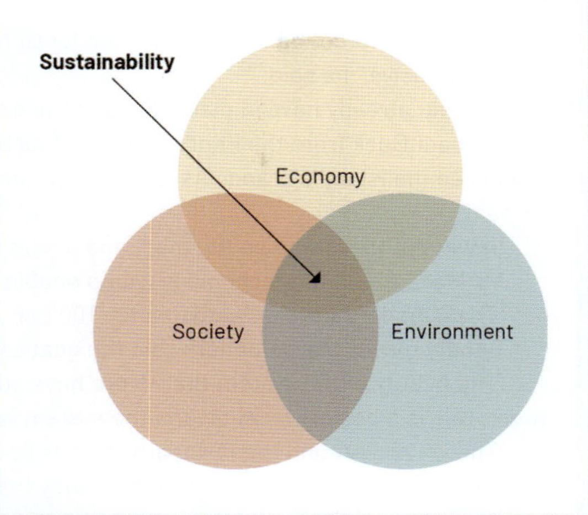

Figure 9.6 Interrelationship between the three pillars of sustainability

It is crucial to consider all three dimensions of sustainability, as they are interrelated and mutually reinforcing. One dimension cannot be fully achieved without considering the others. For instance, promoting economic sustainability through business growth and job creation must also take into account environmental sustainability to prevent further depletion of natural resources, and also ensure that these businesses and jobs promote social sustainability, such as fair wages, safe working conditions and respect for human rights. A balanced approach to sustainability that incorporates all three dimensions is necessary to ensure a long-term and resilient future for the planet and its inhabitants.

CASE STUDY

FLINDERS ISLAND GOING FOR GREEN WITH RENEWABLE ENERGY HUB, FAREWELLS DIRTY DIESEL

Flinders Island off Tasmania's north-east coast has traditionally been powered solely by diesel fuel. In December, Hydro Tasmania will flick the switch over to its Hybrid Energy Hub, which will enable the island to be powered by renewable energy.

When conditions are right, diesel generators will be switched off and the island will run 100 per cent on renewables. The system will use a combination of solar, wind, battery storage and enabling technologies to reduce the island's reliance on expensive shipped-in diesel and provide residents with a more reliable energy source.

Hydro Tasmania hybrid energy solutions manager Ray Massie said the project would reduce emissions by 60 per cent and put downward pressure on future power prices, improving economic conditions for business on the remote island.

'This will reduce emissions down to about 40 per cent of what has traditionally been generated from the power station. The use of diesel is a large expense,' he said.

Resource
Case Study: Flinders Island going for green with renewable energy hub, farewells dirty diesel

'It is future-proofing us against when fossil fuels eventually start to run out and therefore become more expensive,' he said.

'We will already have in place the technology to enable to us to continue on.'

Michael Buck from the Flinders Island Tourism and Business Association said the system enhanced the island's brand. 'As far as our branding is concerned, the clean green image fits,' he said.

'It will put the island on the map, and it sets an example of innovation.'

Mr Massie said the technology would enable the island to be solely powered by renewable energy.

'The islanders have an ambition for 100 per cent renewable energy,' he said.

Deputy mayor Marc Cobham said the goal could be achieved through the adoption of tidal energy.

'The beauty of the system that Hydro have put in is that when a wave tidal generation is developed more, this is a perfect location and the system will allow that to just plug in,' he said.

'The waters around the Furneaux Islands have been identified internationally as some of the best resources as far as tidal energy. So hopefully that will happen in the next few years.'

Mr Buck said the business community would benefit from an end to diesel fuel use.

'We would like 100 per cent renewable energy on the island, particularly given the cost of fuel and the cost of doing business here is higher than on the mainland,' he said.

'Anything that reduces the overall cost from a long-term point of view is a positive as far as the island business community is concerned.'

Adapted from 'Flinders Island going for green with renewable energy hub, farewells dirty diesel', Rhiannon Shine, 2017, ABC News

QUESTIONS

1 How is shifting power sources on Flinders Island a form of sustainability?
2 Outline two critiques of sustainable power generation.
3 Explain the three pillars of sustainability.
4 Explain, using examples from the article, how environmental and economic sustainability are interdependent.
5 How will the use of totally renewable energy sources on Flinders Island lead to increased social sustainability in the future?

NOTES FOR THE EXAM 

Each dimension of sustainability has the potential to improve outdoor environments, and when these pillars work in unison greater benefits are achieved. Ensure you understand each of the pillars of sustainability and can analyse the importance of each working together.

Ecological footprint

The concept of the ecological footprint is a useful tool for individuals to understand their personal impact on the environment and the planet's resources. It highlights the importance of reducing our consumption and waste to ensure the sustainability of the Earth's resources for future generations. By considering their personal ecological footprint, individuals can make informed decisions about their daily habits and work towards reducing their impact on the environment. Additionally, governments and corporations can also use this measure to assess the sustainability of their practices and policies, and make changes to promote sustainability. The ecological footprint highlights the **interdependence** of the three dimensions of sustainability and the need for a holistic approach to ensure a sustainable future.

interdependence of two or more things that are interconnected, relying on and influencing each other

An ecological footprint is calculated by completing a short questionnaire on a person's lifestyle, including their transport habits, diet and household energy use. This is quantified as how many global hectares would be required to support their consumption.

LEARNING ACTIVITY 9.1A

Your ecological footprint

Using the calculator on the WWF Australia website, calculate your ecological footprint.

1 Complete the survey.
2 What is your personal Earth overshoot day?
3 What does this mean?
4 How many Earths would be required if everyone lived like you?
5 How do your results compare with five other countries?
6 How can you lower your footprint score?
7 Using the fact and figures tab from this website, compare your data.
8 Using the solutions tab, analyse strategies to improve your results.

Worksheet
9.1a Your ecological footprint

Weblink
Calculate your ecological footprint

Critiques of sustainability

VAGUENESS OF THE TERM

In the past 'sustainability' has been **critiqued** as being one of the most 'jargonish jargon' words, with some describing sustainability as 'a good concept gone bad by misuse and overuse. It's come to be a squishy, feel-good catchall for doing the right thing'. In recent decades, sustainability has been used indiscriminately, from promoting government policies to selling cars. Robert Engelman, former president of the Worldwatch Institute, has been critical of the overuse of the term 'sustainability', and has coined the term 'sustainababble' illustrating how the term has lost both meaning and impact. Due to its vagueness in meaning, it can be applied to a multitude of facets in society with very little accountability or understanding of the consequence for the Earth's resources; therefore, it can be deemed meaningless – some say we have been **greenwashed**.

critique
the constructive and critical analysis of a concept

greenwashing
promoting the misleading or false perception that an organisation's products are 'green' or environmentally friendly

MEASURING SUSTAINABILITY

As the concept of sustainability is complex, it has complicated attempts to measure the effectiveness of sustainability practices. There are no set criteria or universal measurement indicators. The calculation of ecological footprints does provide a basic understanding of our reliance on the Earth's resources; however, this highlights the existence of problems rather than providing direction towards solutions. With no universal system to measure whether so-called 'sustainable' practices are effective in ensuring ongoing supply for future generations, it may be difficult to justify their continuation. With a spiralling world population, the demands on the world's resources are ever-increasing; therefore, it is difficult to know the scale of resources future generations will require, nor to measure if our current level of sustainability is sufficient.

DEVELOPING NATIONS

With the inequalities of economic growth around the world, it is difficult to expect all countries to adopt expensive renewable energy sources and impose restrictions on carbon **emissions** caused by the unsustainable use of fossil fuels by developed nations. Developing countries have larger emissions, causing

emissions
gases that result from energy production and other industrial processes

a greater environmental impact; however, this is not by choice. The notion of sustainability can be thought of as preserving the present, which is of no benefit for those who go hungry every night. It would be difficult to expect a society that is unable to meet its own basic needs to reduce consumption and be mindful of the needs of future generations. A key consideration of sustainable development is to ensure a fair and equitable access to resources for current generations to meet their most basic requirements; this can then be used as a platform to adopt more environmentally friendly practices in the future.

COST

grey water systems recycling systems that reuse wastewater generated from wash basins, showers and baths for uses such as toilet flushing and garden watering

Adopting sustainable practices is expensive. Installing rooftop solar panels and backyard **grey-water systems** may seem a straightforward ethical decision; however, the associated costs are beyond many household budgets, even with government subsidies. Yet, many sustainable practices are cost-effective over an extended period of time. A household solar panel system takes several years to pay for itself (depending on the size and quality of the system installed) and so will offer savings and a low environmental impact beyond these initial years.

Achieving true sustainability

Sustainability is a crucial issue that affects all aspects of society and our way of life. It is not just about preserving the environment, but also ensuring the wellbeing of current and future generations. The three pillars of sustainability – environmental, social and economic – work together to ensure the longevity and balance of our planet and its resources. The concept of ecological footprint gives us a way to measure our impact on the Earth's resources and the extent to which we are using them faster than they can be replenished. To achieve true sustainability requires a shift in the way we live and consume resources, a better understanding of the relationships between society, economy and the environment, and a commitment to creating a better future for all.

Shutterstock/Alex Farias

Figure 9.7 In all new housing estates in Victoria, a purple tap denotes a grey water system for water repurposing, and a gold tap is for clean water.

Worksheet 9.1b sustainability critiques

LEARNING ACTIVITY 9.1B

Sustainability critiques

1 Create a table and summarise each critique of sustainability.
2 Select an environment you have visited or studied and analyse the implications of sustainability critiques at this location.
3 Research the current government rebates to solar panel installation in Australia.
4 Using your local government website, research if there are rebates available to residents for water tanks, water saving devices or similar.
5 What are the biggest challenges the world faces in achieving true sustainability?
6 What changes need to be made in your own life and globally to promote the concept of sustainability?

9.1 KEY CONCEPTS

- The term 'sustainable' has been associated with being environmentally friendly, renewable and green. 'Sustainable' can be defined as 'capable of being maintained in existence without interruption or diminution'.
- The challenge for society is to enable the Earth's resources to be shared for the survival of current and future generations of all species, while ensuring ongoing healthy outdoor environments.
- First Nations' Australians have been using sustainable practices for millennia to ensure their ongoing survival.
- Common explanations of the term 'sustainability' include 'meeting the needs of the present without compromising the ability of future generations to meet their own needs'.
- Small daily actions can have significant sustainability outcomes; for example, short showers, recycling, composting, and reusable bags and coffee cups.
- Social, economic and environmental sustainability are commonly referred to as the three pillars of sustainability and are interrelated, meaning each one affects the others.
- Social sustainability includes environmental justice, human health, resource security and education. Social sustainability can be promoted through improvements to quality of life, equality and social cohesion.
- Economic sustainability includes job creation, profitability and proper accounting of the use of ecosystems. Economic sustainability can be promoted through streamlining supply chains to use fewer natural resources and creating a waste-free world.
- Environmental sustainability focuses on the wellbeing of the environment. This pillar includes water quality, air quality and reduction of environmental stressors. Environmental sustainability can be promoted through shifting to renewable energy production, avoiding land pollution and investing in long-term environmental planning.
- Critiques of sustainability include greenwashing, vagueness, difficulty in measuring impacts and cost.

9.1 CONCEPT QUESTIONS

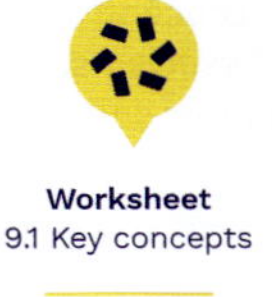

Worksheet
9.1 Key concepts

REMEMBERING

1 Describe sustainability.
2 Outline three examples of sustainability in an environment you have visited or studied.
3 Explain the three pillars of sustainability.

UNDERSTANDING

4 Explain why it is imperative we sustain outdoor environments now and into the future.
5 Analyse how the three pillars of sustainability are interconnected.

APPLYING

6 Provide three examples of critiques of sustainability in an environment you have visited or studied.
7 Analyse the critiques of sustainability you have provided.

EXTENSION CHALLENGE

8 Propose possible solutions to overcome the sustainability challenge in your household.
9 How can the solutions you have referenced above be more widely adopted in society to achieve a more sustainable society?

9.2 OBSERVABLE CHARACTERISTICS TO ASSESS THE HEALTH OF OUTDOOR ENVIRONMENTS

KEY KNOWLEDGE

- observable characteristics to assess the health of outdoor environments, including:
  - quality of water, air and soil
  - species and ecosystem biodiversity

KEY SKILLS

- evaluate the health of outdoor environments and create possible solutions to improve environmental health

9.2.1 EVALUATING THE HEALTH OF OUTDOOR ENVIRONMENTS

A thriving and sustainable outdoor environment is crucial for a good standard of living. The outdoors offers us the essential resources of life such as clean air, water and food, as well as opportunities for recreation and inspiration. Monitoring the wellbeing of outdoor environments is essential in order to identify any harmful practices and take action when needed. Given the intricate interactions within and between environments, assessing them can be a complex and labour-intensive process. To assist with decision-making, various environmental attributes have been established to evaluate the general state of an environment. These attributes can be utilised by government bodies, organisations, local communities and the public.

Environmental characteristics can be used as an early warning sign of environmental deterioration, such as increases in carbon dioxide (CO_2) or **salinity** levels, and to monitor positive effects of human activities, such as increasing biodiversity levels due to revegetation works.

salinity
the concentration of dissolved salts in water or soil

iStockphoto/Tenedos

Figure 9.8 Monitoring the health of a waterway

Environmental health is closely linked to the surrounding land use, and can be influenced by:

- urbanisation
- land clearing
- overgrazing
- pollution and erosion from livestock and mining practices
- chemicals used in industrial and farming practices
- modification of natural stream flows by dams and weirs
- climate (such as long periods of drought)
- invasion of waterways by exotic weeds.

It is possible to make judgements about the health of a specific outdoor environment by observation. The observable characteristics of healthy outdoor environments include:

- quality of water, air and soil
- species and ecosystems biodiversity.

Water quality

Water is a crucial component of a healthy and sustainable outdoor environment. Any alteration in water quality can have a negative impact on the environment and all its elements. Humans require water to meet our basic necessities, such as drinking and food production. If water quality is not preserved, not only will the environment suffer but the economic and recreational value of our water resources will also decline.

MEASURING WATER QUALITY

Testing water quality can indicate how healthy the water is in our rivers, oceans, creeks and wetlands. The Australian Government has set water quality standards for healthy aquatic ecosystems, known as the Australian and New Zealand Environment and Conservation Council (ANZECC) guidelines. The guidelines provide trigger values that are considered 'safe'. Below the level of the trigger value, there is a risk that life (plant or animal) will be harmed. The characteristics of water that can be tested include taste, odour and appearance (turbidity); temperature; and pH.

TASTE, ODOUR AND APPEARANCE (TURBIDITY)

The appearance, taste or odour of water from a water source offer some information about the quality and health of the environment. The advantage is that no additional equipment is required to take these readings. As our sense of taste is very sensitive, having a small taste of water will provide some sense of its quality.

Taste

Undesirable tastes may be caused by any of numerous organic compounds. These may be present naturally in the water or due to sewage or surface contamination sources. The odour of the water will provide some indication of the level of hydrogen sulphide (rotten egg smell), which, when present, will impart an undesirable odour and taste.

Turbidity

The appearance of the water will provide information on its turbidity or the measure of water clarity (how far light can travel through water). Eroding banks and soil runoff are major causes of elevated turbidity. The 'cloudiness' could be particles of clay, sand, silt, algae or plankton. Turbidity can affect water quality and water life because:

- not as much light gets into the water and so plants may not receive the light they need for growth
- the water warms due to the particles absorbing heat
- particles can get into gills of fish and other aquatic animals, making it harder for them to get oxygen
- habitat is lost as holes and crevices are filled with silt.

Temperature

Water temperature is a measure of how warm or cold the water is. Temperature affects the rate of photosynthesis by algae and larger aquatic plants, the metabolic rates of aquatic organisms and the sensitivity of organisms to toxic wastes, parasites and diseases. Aquatic species have evolved to live in specific water temperatures, and changes can affect how they function and how they could be affected by parasites and diseases. With extreme temperature change, many organisms will die. Changes in long-term temperature averages may cause differences in species that are present in the ecosystem. Warmer water holds less oxygen, which is necessary for the plants and animals living in the water. Temperature is affected by:

- removing trees and other plants, as there is less shade and more direct sunlight on the water
- particles in the water (turbidity), which can absorb heat and increase temperature
- runoff from roads that have been warmed by the sun or the release of warm water from industry and power plants, or cold water from dams.

Shutterstock.com/Ecopix

Figure 9.9 Turbidity is a measure of water clarity – how much the material suspended in water decreases the passage of light through the water. High turbidity reduces photosynthesis, clogs fish gills and affects fish breeding patterns.

pH of water

Power of hydrogen (pH) is a measure of the level of acidity or alkalinity in the water. It is measured on a scale of 1 (strong acid) to 14 (strong alkaline). Pure water has a pH of 7 (neutral).

Water with a pH of less than 4.8 or greater than 9.2 can be harmful to aquatic life. Most freshwater fish prefer water with a pH range between 6.5 and 8.4. pH is affected by:

- soils and rocks in the catchment
- chemicals released from industry
- carbon dioxide dissolved in water.

Plants use carbon dioxide for photosynthesis. During the day, when plants are photosynthesising, pH can go up. At night, when plants release carbon dioxide, pH goes down.

Air quality

Air is a vital component of our environment, consisting of a mixture of gases including nitrogen, oxygen and carbon dioxide. It plays a crucial role in sustaining life on Earth, protecting it from harmful ultraviolet radiation, moderating temperature extremes and supporting photosynthesis in plants. However, human activities such as industrial emissions, transportation and domestic wood burning have led to an increase in air pollution, which not only impacts the health and wellbeing of the community but also poses a threat to the health of outdoor environments. Some effects of air pollution, such as ozone depletion and the enhanced greenhouse effect, have become global issues that require collective action. Air pollution has been linked to respiratory problems, asthma and bronchitis. It is essential to monitor and mitigate its impact to maintain a healthy and sustainable environment.

MEASURING AIR QUALITY

There are scientific methods of studying the quality of air to determine how polluted it is. When we measure air quality, we measure the levels of the most widespread pollutants, including carbon monoxide, ozone, nitrogen dioxide and sulphur dioxide. The amount of pollution in the air is usually measured by the proportion of the total volume that it accounts for, and this is expressed as a percentage. These measures require specialised equipment; however, it is possible to make a judgement on the quality of air while in the outdoor environment by using your sense of sight and smell.

In Australia, air quality assessment incorporates various factors, including the amount of recent rainfall, the presence of smog, smoke or dust, and location and proximity to major infrastructure. Each of these factors contributes to air quality assessment in the Australian context.

Rainfall

Rainfall can have a positive impact on air quality by washing away pollutants from the atmosphere. It helps to settle particles and cleanse the air. Monitoring the amount of recent rainfall through sites such as Bureau of Meteorology can provide an indication of whether the recent rainfall may have led to a reduction in the amount of particles in the atmosphere. However, it's important to note that rainfall alone does not provide a complete assessment of air quality since other factors, such as emissions and whether conditions, also influence air pollution levels.

Smog

The presence of **smog**, smoke or dust are visible indicators of poor air quality. Smog refers to a combination of pollutants and fine particles, which create a hazy or foggy appearance in the atmosphere. It is one of the most widespread forms of air pollution in built-up environments. Smog occurs when emissions from industry, motor vehicles, incinerators, open burning and other sources accumulate under certain climatic conditions. These emissions are harmful to humans and to the health of outdoor environments. Smoke, often associated with bushfires or industrial emissions, can significantly degrade air quality. Dust particles from sources such as mining operations or arid (dry) regions can also impact air quality. Visual observation of smog, smoke and dust can raise awareness of reduced air quality.

smog
a type of air pollution characterised by a mixture of smoke and fog that is typically a result of human activities, particularly the burning of fossil fuels

Location

Location and proximity to major infrastructure – such as industrial areas, highways, rail networks or airports – can be an important determinant of air quality. These sources emit pollutants that can contribute to higher pollution levels in nearby areas. Monitoring stations are strategically placed across

Australia to measure pollutant concentrations, especially in urban areas with significant infrastructure. This helps to assess the impact of nearby sources on air quality and to identify pollution hot spots.

In addition to these factors, it's worth noting that Australian and state governments, through organisations like the federal Department of Climate Change, Energy, Environment and Water (DCCEEW) and state environmental agencies, actively monitor and assess air quality using sophisticated instrumentation and methods. They rely on air quality monitoring networks that provide data on pollutant concentrations, meteorological conditions and other relevant parameters.

Shutterstock.com/timallenphoto

Figure 9.10 Haze from bushfires across Victoria in 2019 caused a layer of smog to envelop Melbourne. The smog was made up of a mixture of smoke, fog and chemical fumes.

Smell

A judgement of air quality can be made on the smell of the air. Sulphur dioxide is an invisible gas, yet it has a nasty, sharp smell. The main sources of sulphur dioxide in the air are industrial activities that burn fossil fuels and motor vehicle emissions resulting from fuel combustion.

Worksheet 9.2a Measuring air quality

Weblink The EPA

LEARNING ACTIVITY 9.2A

Measuring air quality

The Victorian Environmental Protection Authority (EPA) measures a range of pollutants at each of its monitoring stations around Melbourne and throughout Victoria. Explore the EPA's most recent air quality recordings on the EPA website.

1 What is EPA AirWatch?
2 What is the current air quality in your area?
3 Summarise the air quality categories according to the EPA.
4 What air pollutants are most common in Victoria?
5 How can we reduce air pollution in Victoria?

Soil quality

Soil is often overlooked as a vital component of our environment, yet it plays a crucial role in supporting life on Earth. The rich and diverse soil ecosystem is the foundation for our food and fibre production, as well as for the wellbeing of plants, animals and other living organisms. Soil provides the essential nutrients, water and oxygen that sustain plant growth, protects against erosion, serves as a sponge for excess water, retains nutrients and is an **environmental buffer** in the landscape. It also acts as a filter, removing pollutants from water and air, and plays a critical role in the cycle of life by breaking down organic matter and returning its essential elements to the ecosystem. Soil formation occurs through natural processes such as weathering, erosion and decomposition, which gradually transform rock into the rich and fertile layer of earth that supports life.
Soil takes many years to form, but it can be destroyed very quickly by destructive practices such as deforestation. Soil holds 0.01% of the Earth's water.

Figure 9.11 Soil is composed of distinct layers ranging from organic upper layers (humus and topsoil) to lower layers (subsoil and bedrock).

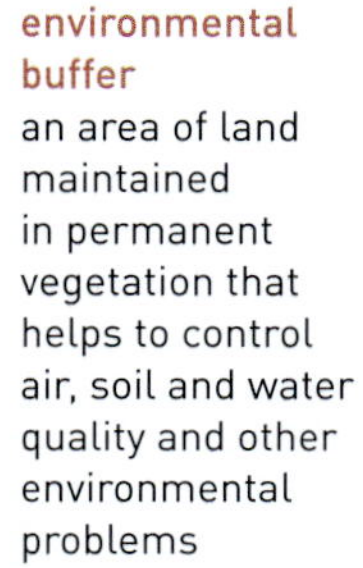

environmental buffer
an area of land maintained in permanent vegetation that helps to control air, soil and water quality and other environmental problems

In our efforts to meet the demands of an ever-increasing global population, we are having a devastating effect on soil quality (including widespread soil erosion). Some of the main contributors to this is deforestation and the overgrazing and over-cultivation of land, which is leading to salinated water supplies and desertification.

MEASURING SOIL QUALITY

The quality of soil is vital and is an indicator of healthy outdoor environments. The characteristics of soil that can be tested include:

1 colour
2 salinity
3 moisture
4 texture
5 structure
6 organic content
7 temperature
8 pH.

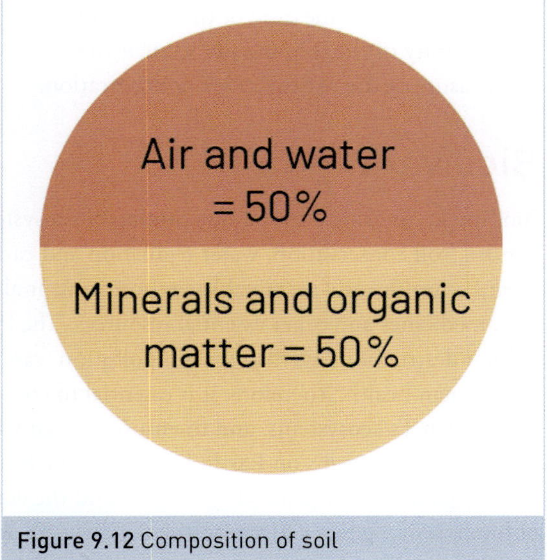

Figure 9.12 Composition of soil

Basic soil testing

Basic soil testing does not require any special equipment, but rather can be done by using your senses of touch and sight (observation). The basic tests listed on the next page will help you to understand the quality of the soil, and therefore help you to understand the health of the outdoor environment.

1 **Colour:** observing soil colour can provide information about its moisture and organic content. Place a handful of soil on a white background, such as a piece of paper. Dark-coloured soil with humus content is generally more fertile, while lighter-coloured soil is less fertile.
2 **Salinity:** excessive salt in soil can have negative effects on water sources such as groundwater and streams. High soil salinity can reduce crop yields, damage infrastructure and decrease biodiversity. Observe the vegetation in the area to gauge soil salinity.
3 **Moisture:** moisture in the soil plays a crucial role in transporting minerals and nutrients to plants. The type and amount of vegetation in an area can indicate the level of moisture in the soil. Thick and lush growth suggests high moisture, while sparse growth indicates lower moisture.
4 **Texture:** the texture of soil can impact the availability of air, nutrients and water. By determining the proportion of sand, silt and clay particles in soil, other soil properties can be estimated. To test soil texture, moisten some soil and squeeze it in your hand. If it holds together and forms a cast, it has a higher clay content.
5 **Structure:** soil structure refers to the arrangement of particles and pores in soil. Poor soil structure can reduce fertility, while well-structured soil can store and move water effectively. To gauge soil structure, insert a tent peg into the soil. The ease or difficulty of breaking the soil indicates its susceptibility to erosion and compaction.
6 **Organic content:** this is the amount of soil that is composed of living and dead plant matter in various states of decomposition.
7 **Temperature:** soil temperature affects various factors, including plant growth, germination, life cycle of small creatures, nutrient availability and decomposition rate. It is directly related to the temperature of the atmosphere. Check if the soil in the area is cool, icy or very hot.
8 **pH:** This is the measure of the acidity of alkalinity of the soil; the pH value is the measure of the hydrogen ion concentration.

Shutterstock.com/V.P.I

Figure 9.13 Observation plays a key role in assessing soil health

FAST FACT

It takes more than 500 years to form two centimetres of topsoil.

Biodiversity

Biodiversity is critical to the functioning of ecosystems, providing crucial services such as pollination, pest control, soil conservation, water regulation and carbon **sequestration**. A healthy and diverse ecosystem means a healthy and sustainable environment. It also contributes to human wellbeing, providing food, fuel, medicines and other essential resources. The loss of biodiversity, due to human activities such as habitat destruction, overfishing and pollution, can have serious consequences for both the environment and human health. Therefore, it is essential to conserve and protect biodiversity to maintain the balance of our planet's ecosystems, and to ensure a sustainable future for all living beings. Biodiversity is the total of all living organisms on Earth. It includes the full variety of all life forms – different plants, animals, microorganisms, the genes they contain and the ecosystems to which they belong. There are three levels of biodiversity: genetic diversity, species diversity and ecosystem diversity.

- **Genetic diversity** is the total genetic information contained in the genes of all species. It also refers to the variation in genetic information between species as well as the variations between individuals of the same species. The different colouring found within one species of kangaroos is an example of genetic diversity.
- **Species diversity** is the variety of species. It refers to both the number of species and the number of individuals within each species; for example, the different types of trees, shrubs, birds and insects you can find in your backyard, and the number of individuals in each species.

sequestration the process of capturing and storing carbon in sinks such as soils, forests and oceans

genetic diversity the total genetic information contained in the genes of all species

species diversity the number of species and the number of individuals within each species

- **Ecosystem diversity** is the variety of habitats, natural communities and ecological processes in the biosphere. Examples in Victoria include coastal, alpine, agricultural, urban and woodland ecosystems. One square metre of soil can hold a billion living things. These include insects, spiders, worms, centipedes, mites, fungi and bacteria.

ecosystem diversity the variety of habitats, natural communities and ecological processes in the biosphere

BIODIVERSITY IN AUSTRALIA

Because of the variety of ecosystems and the diverse and unique species within these ecosystems, Australia is considered to have high levels of biodiversity. Australia has about 450000 species, representing approximately 10% of the species estimated to inhabit the Earth. Australia has over twice the number of species of Europe, Canada and the United States of America combined. Approximately 80% of Australia's species are **endemic** to Australia – that is, they only occur in this country.

endemic a feature or species that is unique to a defined geographic location

Biodiversity loss in Australia

Biodiversity loss in Australia has had severe impacts on the country's ecosystems. The extinction of some species and decline of others not only affects the balance of nature but also disrupts the vital services the ecosystems provide, such as pollination, pest control and soil health. Australia has also lost more than of 10 species of birds, three species of frogs and at least 97 plant species. It is extremely likely that these figures are an underestimate as some species could have disappeared before they were collected, described and named.

Auscape International Pty Ltd / Alamy Stock Photo

Figure 9.14 Leadbeater's possum is listed under the federal *Environment Protection and Biodiversity Conservation Act 1999* as critically endangered.

The loss of biodiversity in Australia is attributed to a number of factors, including habitat destruction caused by urbanisation, deforestation and agriculture; the introduction of non-native species; pollution of soil, water and air; and the effects of climate change. These threats to the natural environment continue to pose a significant challenge to the preservation of biodiversity and the health of ecosystems.

Efforts to address these threats and conserve biodiversity must be taken seriously in order to ensure that future generations can enjoy and benefit from the diversity of life on Earth. This can be achieved through the implementation of conservation programs, protection of natural habitats and reduction of harmful practices that contribute to biodiversity loss.

The Australian Government's *Threatened Species Action Plan 2022–2032* maps a pathway to manage and restore Australia's threatened species and improve biodiversity across the nation. The action plan was developed with input from experts, the community, natural resource managers, scientists, conservation groups and First Nations peoples. Key objectives include:

- preventing new extinctions
- a commitment to protect and conserve more than 30% of Australia's land mass
- increased participation of First Nations peoples in the management and recovery of threatened ecological communities.

Measuring biodiversity

Making a judgement on the health of a specific outdoor environment using the observable characteristics of biodiversity requires an investigation of the amount and variety of different types of flora and fauna. Tools to measure biodiversity include:

- biodiversity audit tools, such as the Biodiversity Snapshots project developed by Museum Victoria
- the *Field Guide to Victorian Fauna* app, which is a great way to explore Victoria's unique and diverse wildlife
- the Department of Energy, Environment and Climate Action's biodiversity information products, which are used to measure biodiversity values at a single site and across Victoria's landscape.

Worksheet
9.2b observable characteristics investigation

Weblink
Prezi

LEARNING ACTIVITY 9.2B

Observable characteristics investigation

Shutterstock.com/Jordy scott

Figure 9.15 Use the observable characteristics to assess the health of outdoor environments.

During a practical lesson, investigate the observable characteristics (e.g. soil, air, water and biodiversity) at a school campus, local park or similar.

1. Choose three observable characteristics and describe how these may be assessed at this location.
2. Using your observations, analyse the health of each characteristic selected.
3. Make an overall evaluation on the health of this environment.
4. Propose possible solutions to improve the health of this environment.
5. Use images to support your investigation, presented using a template from the Prezi website or similar.

LEARNING ACTIVITY 9.2C

Identifying the characteristics of healthy outdoor environments

Use the information in the text and your own research to summarise different observable characteristics of healthy outdoor environments. Record your summaries in the table below. The table could be used to complete a field assessment of the health of a particular outdoor environment you have visited or investigated.

Characteristics	Description of characteristic	Explain two ways the characteristic can assess the health of outdoor environments
Water quality and adequacy: • taste • odour • appearance • temperature • pH • location		
Air quality and adequacy: • sight • smell • recent rainfall • location		
Soil quality and adequacy: • colour • moisture • texture • structure		
Amount of biodiversity: • range • number • location • ecosystems		

Worksheet 9.2c Identifying the characteristics of healthy outdoor environments

9.2 KEY CONCEPTS

- There are five observable characteristics to understand and apply to environments you have visited or studied: quality of water, quality of air, quality of soil, species biodiversity and ecosystem biodiversity.
- Environments provide us with the necessities of life, such as clean air, water and food, and offer places for recreation and inspiration.
- Water quality is essential for every aspect of environmental health. If it is not maintained, both the environment and our economic and recreational pursuits suffer.
- Water quality can be assessed through taste, odour, appearance (turbidity) and pH testing.
- Air is the colourless, tasteless, gaseous mixture (also known as the atmosphere) and is needed by almost all the living things on Earth.

- Air quality can be assessed through the amount of recent rainfall, visually through the presence of smog, smoke or dust, location and proximity to major infrastructure.
- Soil quality is vital for environmental health as it supplies nutrients, water and oxygen to plants.
- Soil takes many years to form, but it can be destroyed very quickly by destructive practices such as deforestation.
- Soil quality can be assessed through its colour, salinity, moisture, texture, structure, organic content, temperature and pH.
- Biodiversity is the total of all living organisms on Earth. It includes the full variety of all life forms – different plants, animals, microorganisms, the genes they contain and the ecosystems to which they belong.
- Species diversity is the variety of species and refers to both the number of species and the number of individuals within each species. Species diversity can be assessed by measuring the variety of species within a specific area.
- Ecosystem diversity is the variety of habitats, natural communities and ecological processes in the biosphere. Ecosystem diversity can be measured by assessing the differing ecosystems within Victoria.

Worksheet
9.2 Key concepts

9.2 CONCEPT QUESTIONS

REMEMBERING

1 Describe four ways to observe and assess outdoor environments.
2 Define 'biodiversity'.
3 Name, describe and provide an example of each of the three levels of biodiversity.

UNDERSTANDING

4 Why is it important to assess the health of outdoor environments?
5 Who sets water quality standards in Australia?
6 Outline the factors that could influence the health of an outdoor environment.

APPLYING

7 Why is maintaining a high quality of water, air and soil important?
8 Explain how location can influence all five observable characteristics.

EXTENSION CHALLENGE

9 For each of the four observable characteristics, investigate and suggest possible strategies to improve the health of an outdoor environment you have visited or studied.

9.3 THE IMPACT OF THREATS ON SOCIETY AND OUTDOOR ENVIRONMENTS

KEY KNOWLEDGE

- the impact of threats on society and outdoor environments, including two of the following:
  - land degradation
  - introduced species
  - urbanisation
  - climate change
  - flood
  - fire

KEY SKILLS

- analyse threats to society and outdoor environments

9.3.1 ENVIRONMENTAL THREATS IN AUSTRALIA

The isolation of Australia for over 40 million years has resulted in a unique and diverse outdoor environment that houses about 10% of the world's biodiversity. This rich variety of species, adapted to harsh climatic conditions, has created a fragile environment that is susceptible to change from external forces. Changes in climate, the introduction of new species and other environmental threats (e.g. outdated farming practices, industrial pollution and urbanisation) pose serious risks to the native flora and fauna and to human society, which depends on the productivity of the land.

Fairfax Photos/Mal Fairclough

Figure 9.16 Aerial photograph of the Victorian town of Charlton during major flooding in 2011. The town has been hit by further flooding events in the years since, including in late 2022.

The impacts of these environmental threats have already been devastating for Australia's outdoor environment. The introduction of new species has caused a decline in the population of native species, while industrial pollution and urbanisation have contaminated the soil, water and air. These threats, along with the impacts of climate change, are having a significant impact on the health of the outdoor environment and must be addressed to ensure the preservation of Australia's rich biodiversity for future generations.

Land degradation

Land degradation refers to detrimental changes in the condition of the land due to human interactions. These changes are linked to a reduction in the productive capacity of land and its economic value. Such interactions include older farming practices, industrial pollution and urbanisation. It is important to note that naturally caused undesirable changes are excluded as a cause of land degradation; however, human activities can indirectly affect phenomena such as floods and bushfires.

Over the past 200 years, land degradation has been occurring at an alarming rate across Australia. Land degradation is a serious threat to outdoor environments as reductions in the quality and quantity of productive land threatens our capacity to feed a growing national and world population. It also reduces native habitats and water resources, leading to reductions in biodiversity. The major types of land degradation are dryland soil salinity, soil contamination and erosion.

DRYLAND SOIL SALINITY

Salinity describes the salt content of soil or water – excessive concentrations are toxic to most life forms. Salinity is one of the major threats to the health and productivity of many catchments, as well as to the rural and urban communities that live in them. It affects rural landholders, urban developments, infrastructure (roads and bridges), water users and the environment.

Low concentrations of salt occur naturally in soils, rivers and streams, and in high concentrations in bays and oceans. Plants such as mangroves have a high salt tolerance and

Fairfax Photos/Pat Scala

Figure 9.17 The effects of erosion caused by salinity

are able to grow in salty environments around the coast. Salinity is caused by land clearing, cropping and irrigation practices (among other factors). Although the causes of rising soil salinity began soon after European settlement, the symptoms of this problem were not widely obvious until into the 20th century. When areas of land are cleared for farming, trees with deep root systems are replaced by crops. The water table begins to rise as the shallow root systems of crops cannot reach into the groundwater supply. Gradually, the rising groundwater brings salt to the surface. Most plants are salt sensitive and will die if the salt level escalates in their environment. Salty soil can only support salt-tolerant plants or, in severe cases, no vegetation at all. When vegetation dies, the topsoil becomes vulnerable to wind and rain erosion, potentially increasing the rate of valuable nutrients blowing away.

In irrigated farming areas, the salinity problem is exacerbated by excess surface water seeping into the groundwater table. The water table rises and salt moves up to the surface. Salt can also seep laterally into waterways and be transported to areas downstream, spreading degradation.

Salinity threatens wetlands, wildlife, farmland, drinking water, town services and infrastructure. Improvements in land management are helping to maintain sustainable yield and protect land from salinity. Accurate assessment of conditions for cropping, **rotation planting** of salt-tolerant perennial plants and land restoration are costly but necessary processes for the future.

rotation planting the successive planting of different crops on the same land to improve soil fertility and to help control insects and diseases

SOIL CONTAMINATION

Soil contamination is caused by over-fertilising crops, continuous legume cropping and wetland drainage for agriculture. All of these practices alter the acidity of the soil and can render the land useless for cropping or natural vegetation cover. The damage can be rectified by the costly application of lime to the soil, but it is preferable to address this problem by adopting prevention strategies. Rotation cropping, regular soil testing and moderating the application of fertilisers can help farmers achieve sustainable land use.

iStockphoto/Gudella

Figure 9.18 The effects of contaminated soil can be seen in this paddock

EROSION

Wind and water erosion are natural phenomena; however, land clearing, overgrazing of farm and pest animals, and salinity all contribute to the acceleration of erosion across the landscape. On coastal dunes, the impact of beach access by recreational users is damaging valuable sand dune vegetation. Plants and their root systems help the sands maintain their shape and form. Once the vegetation is removed or damaged, the sands literally blow away.

iStockphoto/Scorching Hot News

Figure 9.19 Eroded sand dunes

Steps taken to repair or halt the damage include revegetation programs, rest and rotation grazing periods for farmlands, control of pest species, increased farming of native animals and reduced farming of cows and sheep. It has been estimated that the impact of land degradation across Australia in terms of lost production value is in excess of $1.5 billion per year.

Introduced species

Invasive species can have a significant impact on native species and ecosystems by outcompeting and replacing them, altering food webs and affecting soil structure and water balance. They can also spread diseases and parasites to native species, and affect ecosystem processes like nutrient cycling, seed dispersal and pollination. Invasive species can also pose a threat to human health and the economy, affecting forestry, agriculture and tourism industries. In order to mitigate the impacts of invasive species, it is necessary to have a good understanding of their biology, ecology and range. This requires a combination of strategies including prevention, early detection and rapid response, control and management of invasive species.

Alamy Stock Photo/ Alexander Cimbal

Figure 9.20 A bushwalking boot washing station in the Brisbane Ranges National Park in Victoria, installed to reduce the risk of spreading cinnamon fungus

Invasive species also include disease-causing organisms such as fungi and viruses. The soil fungus *Phytophthora cinnamomi* (cinnamon fungus), responsible for a type of dieback in many native plants, was probably introduced to Australia through European settlement. More than 2800 weeds, between 100 and 400 marine species, 34 fish species, 25 mammal species, 26 bird species, six reptile species, one amphibian species and an unknown number of invertebrate species have been introduced to Australia since 1770. Species that have been introduced to the country have had a major impact, and some have resulted in dire consequences for Australia's land and waterways.

The introduced species that have the most devastating impact on Australian native species include blackberries, rabbits, foxes and cane toads. Because of the impact they have on native flora and fauna, these species are considered pests. They cause damage to land and water resources, can carry disease and prey on the native wildlife, and they may also compete with the native plants and animals for food and shelter.

CASE STUDY

MACQUARIE ISLAND IS ALIVE WITH WILDLIFE – AN EXEMPLAR FOR AUSTRALIA

Resource
Case Study: Macquarie Island is alive with wildlife - an exemplar for Australia

In 2014 Macquarie Island, a UNESCO World Heritage Site that lies about halfway between Tasmania and Antarctica, was declared pest free after a seven-year feral animal eradication project.

The removal of feral cats, rats, mice and rabbits from the subantarctic island is a great success story and shows how an island environment can be restored to its natural condition, giving native animals and plants the chance to not just survive, but flourish. There has been a resurgence of burrow-nesting birds, such as Antarctic prions and grey and blue petrels. They have all increased in numbers and are even starting to recolonise areas that suffered some of the most intense rabbit grazing damage on the island. ...

The eradication program was a multi-pronged attack, utilising shooting, aerial baiting, manual baiting, hunting dogs and the introduction of calicivirus to protect Macquarie Island's World Heritage values.

Macquarie Island was discovered in 1810 and over the next 60 years rats, mice, cats and rabbits were introduced to the island. At their height, it is estimated feral cats were killing 60000 seabirds a year. The cats were finally vanquished from the island in 2000, but not before they had

Figure 9.21 Before and after images of a hillside on Macquarie Islands Council website, posted February 10, 2021. Source:

helped drive two native bird sub-species into extinction – the Macquarie Island parakeet and the Macquarie Island rail.

In 2009 it was estimated that Macquarie Island played host to 130000 rabbits, 36000 rats and 103000 feral mice. Rabbits consumed the endemic mega-herbs and grasses, which in turn led to the loss of breeding habitat for nesting seabirds. Rats preyed on seabird chicks and eggs, killing petrel adults and chicks in their nest burrows and predating on blue petrels, making it impossible for this bird species to nest on the island.

...

A decade later, thanks to the incredible efforts of the eradication program, a major threat to these birds has been permanently removed and this magnificent island is once again a paradise for seabirds. The Macquarie Island experience shows that with vision and an effective suite of invasive species eradication measures, island environments can bounce back to their natural state and support the lives of unique native animals.

'Macquarie Island is alive with wildlife – an exemplar for Australia', Feral Herald blog on the Invasive Species Council website, posted February 10, 2021

BLACKBERRIES

The blackberry was introduced into Australia in the mid-1800s as a horticultural plant. Blackberry grows vigorously and can infest large areas quickly. It reduces native wildlife habitat, limits access to water sources, displaces native plants and provides habitat and food for introduced pest species such as rabbits and foxes. Blackberry impacts on society by reducing available grazing land and productivity due to competition for soil moisture and nutrients, and it reduces the aesthetic and conservation value of public lands, parks and reserves.

RABBITS

Rabbits were first introduced to Australia in the 18th century with the First Fleet as a food source, and became widespread after an outbreak caused by a release near Geelong in 1859. Rabbits contribute to soil erosion by burrowing, removing vegetation and disturbing soil. Economic damage by feral rabbits in Australia has been estimated at around $600 million annually in lost production.

DEER

Deer were introduced into Australia from Europe in the 19th century as game animals for hunting. Deer are a major emerging pest problem, causing damage both to the natural environment and agricultural businesses. Populations are expanding and deer are invading new areas. Feral deer can have a major impact on outdoor environments, trampling plants, fouling water holes and wallowing in sphagnum bogs and alpine peatlands that are nationally threatened ecological communities under the *Environment Protection and Biodiversity Conservation Act 1999*.

Figure 9.22 Sambar deer

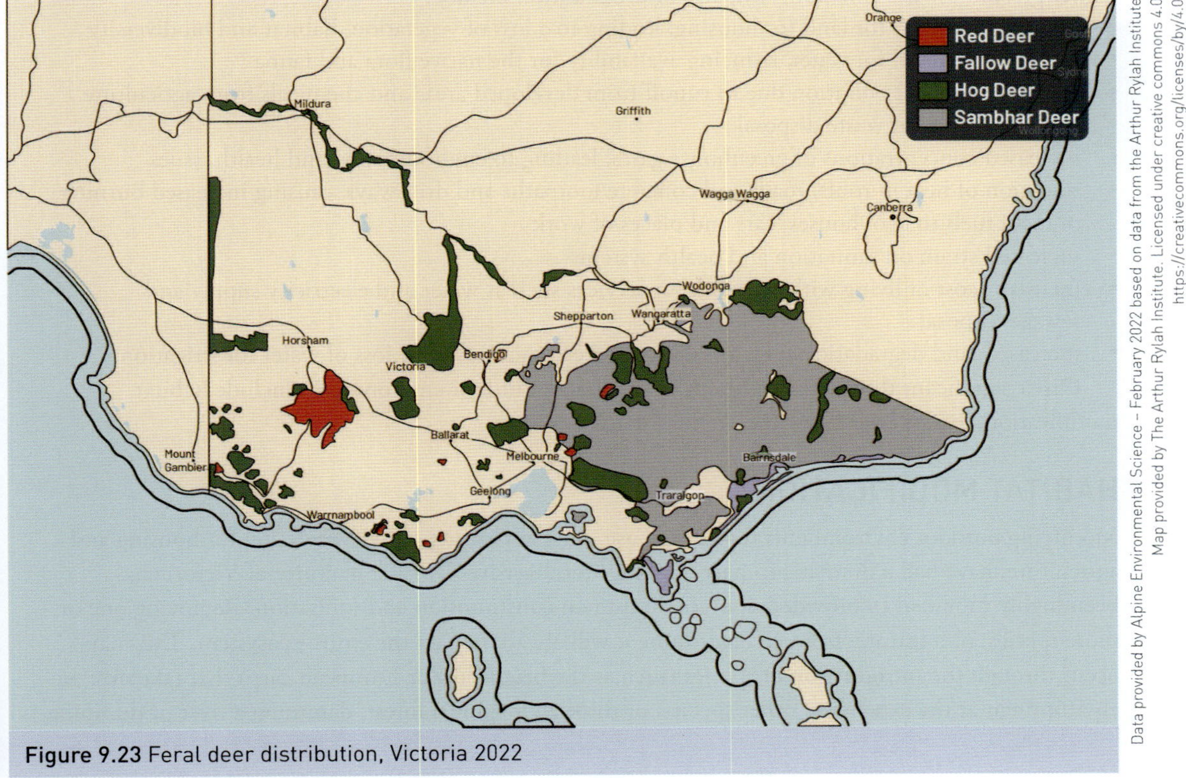

Figure 9.23 Feral deer distribution, Victoria 2022

MARINE PESTS

Victoria's marine environment is under serious threat from a range of introduced species. Species such as *Sabella worm*, Northern Pacific seastar and Undaria (a seaweed from Japanese waters) have reached Australia via the ballast water in ships. The Northern Pacific seastar (*Asterias amurensis*) is a major threat to the scallop industry and has wreaked havoc in Tasmanian fisheries for the past 30 years. It breeds prolifically, and some estimates indicated the population had reached 12 million two years after they were first detected in the bay. The seastar eats mussels, scallops and clams, threatening the **mariculture** industry.

mariculture
the cultivation of marine organisms for food and other products in the open ocean, an enclosed section of the ocean, or in tanks, ponds or raceways that are filled with seawater

FeralSCAN Pest Mapping app

Visit the FeralSCAN website and read the information about feral species in Australia. You can also download the free FeralSCAN Pest Mapping app, which can be used on outdoor experiences where you can record a sighting and/or damage to the environment.

Worksheet
9.3a FeralSCAN Pest Mapping app

Weblink
FeralSCAN

Urbanisation

Urbanisation refers to the growth of towns and cities and the increasing number of people who live in urban areas. It is driven by factors such as education and employment opportunities, access to healthcare and other urban amenities. In Australia, around 90% of the population lives in urban areas, concentrated mainly along our coastline. This growth results in increased demands for goods and services from outdoor environments and leads to urban sprawl, as suburban areas encroach on natural habitats. This, in turn, creates challenges for infrastructure development, such as public transportation, roads, schools, shopping centres, parks, sports fields and hospitals. Additionally, larger cities consume more energy, resulting in higher greenhouse gas emissions. Impacts of urbanisation include:

- increased demand for land for housing, leading to a loss of vegetation, habitat and biodiversity
- increased demand for water, reducing availability for the native flora and fauna
- air, water and soil pollution due to runoff from developed areas and chemicals from agriculture
- increased need for waste disposal
- increased concentrations of greenhouse gases, leading to environmental and health issues
- retention of heat in roofs, roads and concrete footpaths and driveways, causing increased burning of fossil fuels to cool houses, cars and places of work
- increased strain on sanitation and health systems
- increased cost of living with payment for services such as water and electricity supply and sewage disposal
- increased land degradation such as salinity and erosion caused by loss of water for irrigation.

Other significant threats to the health of Australian outdoor environments include habitat modification, loss of river flows and vegetation loss.

HABITAT MODIFICATION

Modifying outdoor environments for human use, such as clearing native vegetation for housing and agriculture, have had a dramatic impact on native biodiversity. The vast majority of Victoria has been highly modified to provide resources for human consumption and utilisation. Removing one or more species, or a feature, from an environment will always impact the entire ecosystem. This may occur through the introduction of species (such as the blackberry or European carp) that take over an environment at the expense of other species, or through logging a forest, damming a river or dumping chemicals into a creek. On a small or large scale, the impacts on entire ecosystems are extensive. Habitat modification accounts for much of the extinction of native Australian species. Over 310 species of native animals and over 1180 species of native plants are at risk of extinction.

LOSS OF RIVER FLOWS

In addition to drought (which is naturally occurring), **hydro-electricity** schemes, irrigation and town water supply systems have all drastically altered river flows within Victoria. Rivers and streams have been diverted and dammed (or, in some cases, both), causing a range of problems – from waterways drying up to complete inundation. Ecosystems have been severely affected or even destroyed. Since 1967, the **Snowy Mountains Scheme** has diverted 99% of the upper flows of the Snowy River for hydro-electricity, and into irrigation systems on the Murray and Murrumbidgee rivers. The impact on the river environment and the communities along the Snowy River and its tributaries has been dramatic. Since 2002, a policy change has led to an increase to 15% of original flow levels to increase the environmental health of the river and surrounding environment.

hydro-electricity the generation of electricity using water power

Snowy Mountains Scheme one of the most complex integrated water and hydro-electric power schemes in the world

VEGETATION LOSS

Victoria is losing around 2500 hectares of native vegetation every year. Clearing land for human settlements, grazing, mining, logging and growing crops have altered environments and, in some cases, totally removed all-natural vegetation and habitats for animals, birds and insects. Around 1% of multiple-use forests are harvested each year. Rotational regeneration efforts attempt to ensure that the industry is sustainable. It is projected that plantation timber will soon meet all demands for timber and forest products.

Clear felling is one of the most damaging practices that can occur in the outdoor environment – it wipes out trees of all ages and slows regeneration as plants struggle to survive without the sheltering protection of older, larger trees. Animal and bird habitats and nesting places are also lost, which can threaten species survival.

Bjorn Svensson / Alamy Stock Photo

Figure 9.24 Clearfell logging involves the removal of almost all vegetation from an area of forest, called a logging 'coupe'

clear felling the practice of cutting down all the trees on a site

Climate change

Climate change refers to the long-term alteration of global weather patterns due to a range of factors, including natural phenomena and human activities. While some changes in the climate have occurred naturally over time, the primary cause of the current climate change crisis is human-generated, resulting from the burning of fossil fuels and increased emissions of greenhouse gases such as carbon dioxide, methane, nitrous oxide and chlorofluorocarbons.

Over the past century, the increase in these gases has led to a rise in global temperatures, with the average air temperature increasing by 1.1°C since 1880. This shift in the climate has far-reaching impacts on society and the environment, including agriculture, human health, food and water security, and infrastructure. It is imperative that immediate action is taken to mitigate the effects of climate change, as failure to do so could result in devastating consequences.

POTENTIAL IMPACTS OF CLIMATE CHANGE ON SOCIETY

- Decreased availability of water for nature, human consumption, agriculture and manufacturing
- A severe restriction of irrigated cropping due to limited water in some years
- Melting of ice in the Arctic and Antarctic, leading to rising sea levels and widespread flooding, affecting millions of people and trillions of dollars' worth of buildings and roads
- Increased heat-related deaths and health problems, costing millions of dollars due to an increase in extreme heat days
- The permanent displacement of communities due to repeated natural disasters such as fire, flood and increasingly inhospitable conditions caused by extreme heat; this is already happening in Australia and around the world, leading to further global tensions around food supply and water
- Decreased snowfall, resulting in greater reliance on artificial snowmaking for recreational activities and a consequential loss of tourism revenue and jobs

POTENTIAL IMPACTS OF CLIMATE CHANGE ON OUTDOOR ENVIRONMENTS

- An increase in intensity of droughts and, in some areas, more floods
- A reduction in fish stocks due to reduction in stream flow and salt-water incursion into estuaries
- An increase in evaporation rates, leading to greater humidity and supercharged extreme rain events, monsoons, cyclone and floods
- An exacerbation of the melting of permafrosts and glaciers (permafrosts contain high levels of methane, a gas far worse than CO_2, which further increases climate change)
- An increased risk of bushfires
- Loss of coral reefs as seas become too warm and acidic
- Species with low adaptability or mobility facing extinction
- Ecosystems changing as new species of animals and plants appear or disappear

Society will need to learn to adapt to increased temperatures and the associated changes.

However, we must act now to reduce greenhouse gas emissions and adopt more environmentally friendly practices, such as renewable energies, to limit the severity of climate change impacts.

Floods

Floods are a natural part of the Australian ecosystem and play a crucial role in certain environments and dam inflows. However, they can also cause substantial harm to properties, infrastructure and human health, and result in loss of life. Climate change projections indicate that flooding is expected to become more severe. In Australia, floods are the most expensive natural disaster, with an average yearly cost of $8.8 billion for insured and intangible costs. The extent of flood damage depends on various factors, such as rainfall characteristics (intensity, frequency and duration), catchment conditions and the vulnerability of ecosystems and communities.

The extreme rainfall experienced in south-east Australia in 2022 was associated with a slow-moving weather system that formed after consecutive La Niña events. La Niña typically brings wetter conditions, while El Niño events typically bring drier conditions (see Chapter 4). These two contrasting weather patterns are due to natural variations in sea surface temperatures across the Pacific Ocean.

An increase in heavy rainfall will increase flood risk in cities, built-up urban areas and small catchments, where extreme rainfall over hours to a day can quickly become flash floods. It's more complex in rural areas and for larger river basins, where floods are driven by multi-day rainfall events and by the preceding soil moisture conditions.

The immediate effects of flooding can lead to tragic consequences, including loss of life and damage to homes, crops and livestock. Flooding can also result in health problems from waterborne diseases.

The long-term impacts of flooding can be equally devastating, including disruptions to essential services such as water, sanitation, electricity, transportation, communication, education and health care. Additionally, people can lose their sources of income and experience decreased buying power and a decrease in land value.

POTENTIAL IMPACTS OF FLOOD ON SOCIETY

Damage to public infrastructure affects a far greater proportion of the population than those whose homes or businesses are directly impacted by the flood. Flood damage to roads, rail networks and key transport hubs, such as shipping ports, can have significant impacts on regional and national economies.

Short-term downturns in regional tourism are often experienced after a flooding event. Images of flood affected areas often lead to cancellations in bookings and a significant reduction in tourist numbers.

At the same time, the environmental benefits of flooding can help the economy, leading to an increase in fish production, recharge of groundwater resources and maintenance of recreational environments.

POTENTIAL IMPACTS OF FLOOD ON OUTDOOR ENVIRONMENTS

In many natural systems, floods are crucial for maintaining ecological functions and promoting biodiversity. They establish connections between rivers and their surroundings, recharge groundwater resources, replenish wetlands, enhance the interconnectivity of aquatic habitats, and transport sediment and nutrients across the landscape. For several species, floods trigger essential life-cycle events, such as breeding, migration and dispersal. However, regions that have been significantly altered by human intervention are more vulnerable to the negative impacts of flooding.

Shutterstock/Powerhouse Productions

Figure 9.25 Flooding in Victoria, 2022

The adverse effects of floods include loss of habitat, spread of invasive plant species, release of pollutants into the environment, reduced fish populations, impaired wetland functions and the destruction of recreational areas. Coastal ecosystems, including marine production, heavily rely on the nutrients that are carried from the land during floods. However, floods also have negative effects on coastal marine environments by introducing excess sediment and nutrients, as well as pollutants such as chemicals, heavy metals and waste.

Fire

Bushfires are a type of wildfire that burns through vegetation and grasslands, causing widespread destruction and damage. They can be ignited by natural causes, such as lightning strikes, or human activities, such as campfires left unattended or deliberate arson. Climate change exacerbates the frequency and intensity of bushfires by contributing to prolonged drought conditions that dry out vegetation, making it more susceptible to ignition.

Australia has a long history of bushfires, with much of its landscape being naturally fire-prone. However, the recent increase in the severity of bushfires has had devastating impacts on both people and the environment. The 2019–20 Victorian bushfires alone burned an estimated 1.5 million hectares, impacting threatened species and their habitats, and leading to the population loss within more than 4400 species, with 244 having over 50% of their estimated state-wide habitat burnt, including rare and threatened species.

POTENTIAL IMPACTS OF FIRE ON SOCIETY

The impact of bushfires on society are immense and include loss of life, property and infrastructure. In terms of public health, bushfires can cause poor air quality, which can affect human and animal health, and can have long-lasting impacts to soil and water quality. Bushfires produce harmful smoke, which can cause fatalities, and fine particle air pollution, which directly threatens human health, even during relatively short exposures.

Fires do not only cause physical harm; many people experience mental trauma from the experience of emergency evacuation and losing homes, pets, belongings, livestock or sources of their livelihoods. Trauma is also associated with being trapped in high-risk areas, due to roads being blocked and interruptions to electricity, water and sewerage systems.

POTENTIAL IMPACTS OF FIRE ON OUTDOOR ENVIRONMENTS

Bushfires can have devastating impacts on plants, animals and ecosystems. After the initial devastation of the fires, impacts remain ongoing. When burned soils flow into streams and rivers, they fertilise water plants and algae. The extra nutrients can have benefits at moderate levels, but too much can over-fertilise and cause excessive algal growth. The same can be true in ocean environments, where smoke has shown to have a negative impact on marine ecosystems.

Bushfires have not only been made more likely and intense by climate change, but also add to it. Global warming is making bushfires burn more intensely and frequently, increasing Australia's annual greenhouse gas emissions and further contributing to global warming. This heightens the likelihood of recurring megafires that will release yet more emissions. This is commonly referred to as the **climate feedback loop**.

climate feedback loop a cycle that accelerates or decelerates a warming trend, such as bushfires creating conditions that encourage more intense bushfires

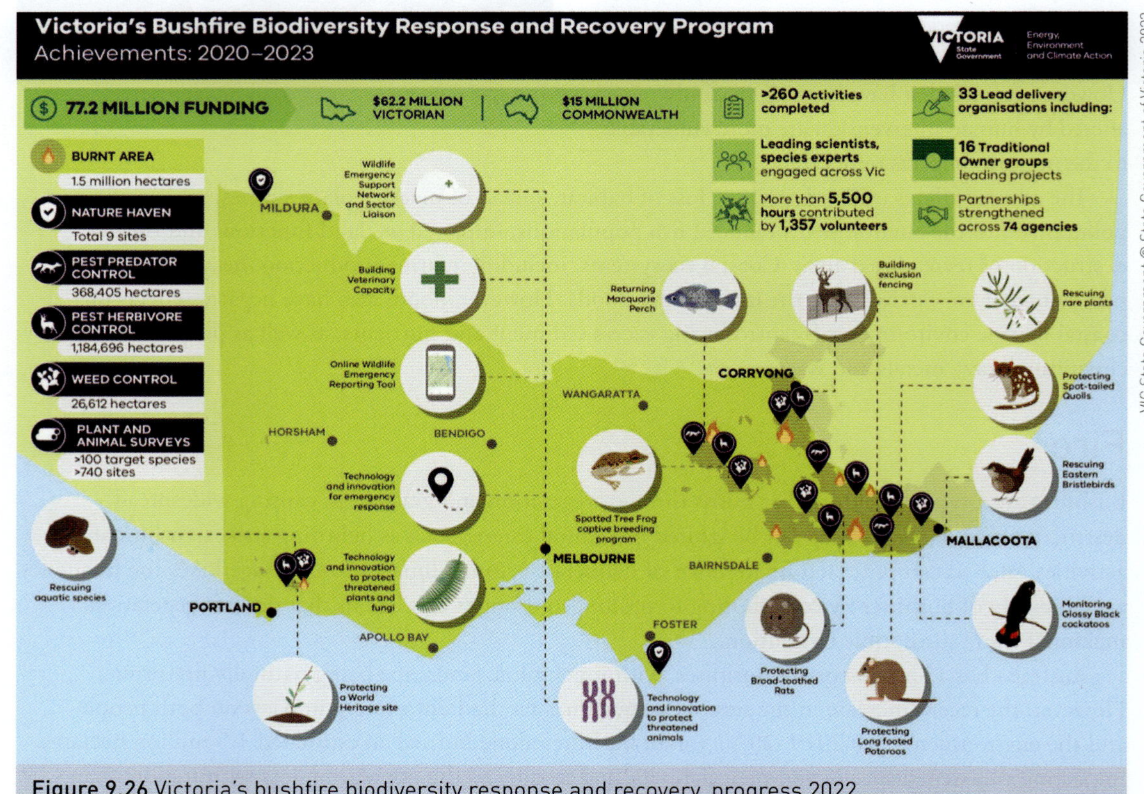

Figure 9.26 Victoria's bushfire biodiversity response and recovery, progress 2022

LEARNING ACTIVITY 9.3B

Threats to society and outdoor environments

Choose two threats explored in this module and complete a Venn diagram (see Figure 9.27) for both threats, detailing:

- description of the threat
- impact on the environment of selected threats
- impact on society (the population) of selected threats.

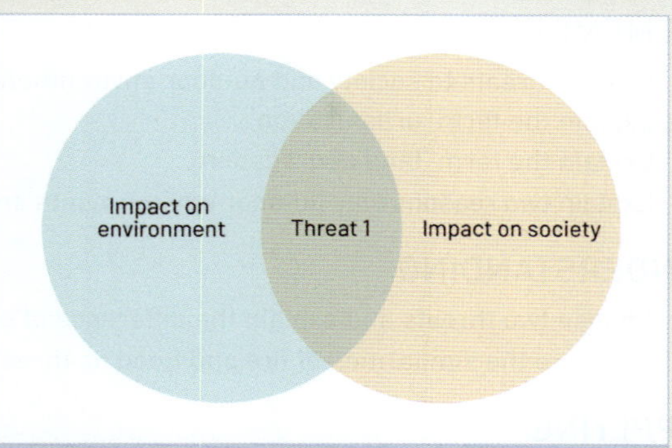

Figure 9.27 Threats to society and outdoor environments

Worksheet 9.3b Threats to society and outdoor environments

9.3 KEY CONCEPTS

- Land degradation refers to detrimental changes in the condition of the land due to human interactions.
- Land degradation is a serious threat to outdoor environments, as reductions in the quality and quantity of productive land threatens our capacity to feed a growing world population, as well as reducing native habitats and water resources, leading to reduction in biodiversity.
- Causes of land degradation include salinity, soil contamination, erosion and loss of native vegetation.
- Introduced species are animals, plants, parasites or disease-causing organisms that establish outside their natural range and become pests.
- Common introduced species to Australia include blackberry, rabbits, deer, foxes and cane toads.
- Urbanisation refers to the development of towns and cities, and the increasing number of people that live in urban areas.
- Growing population levels and development in urban areas increase pressure on outdoor environments.
- Impacts of urbanisation on the environment include demand for land, leading to a loss of vegetation, habitat and biodiversity; and demand for water, reducing availability for the native flora and fauna.
- Impacts of urbanisation on society include increasing concentrations of greenhouse gases, leading to environmental and health issues.
- Climate change is the significant and lasting change in weather patterns over an extended period of time.
- Evidence indicates that climate change will continue to cause significant impacts on society and outdoor environments, agricultural production, human health, water and food security, and infrastructure.
- Floods are part of Australia's natural ecology. They are important for some ecosystems and for boosting dam inflows
- La Niña events typically bring wetter conditions to northern and eastern Australia.
- Impacts of flooding to society include loss of tourism and damage to public infrastructure such as roads and rail networks.
- Impacts of flooding to the environment include a loss of habitat loss, dispersal of weeds and the spread of chemicals and heavy metals.
- Bushfires have always been a natural part of the Australian environment, and much of the Australian landscape is prone to bushfires.
- Climate change doesn't start bushfires, but it does cause them to become larger and more ferocious.
- Bushfires can cause poor air quality, which can affect human and animal health, and can have long-lasting impacts on soil and water quality.
- Fire impacts on society include loss of life, property and infrastructure, as well as reduced air quality and mental health pressures from emergency evacuations.
- Fire can have devastating impacts on plants, animals and ecosystems. After the initial devastation of the fires, impacts are ongoing.

Worksheet
9.3 Key concepts

9.3 CONCEPT QUESTIONS

REMEMBERING

1 List six threats to society and outdoor environments.
2 Explain the term 'urbanisation'.
3 Explain the term 'land degradation'.
4 Explain two reasons why outdoor environments are important to individuals and society.

UNDERSTANDING

5 Choose two threats and explain the difference of each threat on outdoor environments and society.
6 Compare the similarities of fire and flood as threats to society.

APPLYING

7 Referring to an environment you have visited or studied, explain the impact of introduced species within this specific environment.
8 Describe three strategies that you could implement to reduce the impact on outdoor environments of urbanisation.

EXTENSION CHALLENGE

9 Choose an outdoor environment you have visited or studied, investigate how one threat has impacted both society and this outdoor environment, and propose solutions to mitigate the effects of this threat.

9.4 THE IMPORTANCE OF HEALTHY OUTDOOR ENVIRONMENTS FOR INDIVIDUAL PHYSICAL AND EMOTIONAL WELLBEING, AND FOR SOCIETY NOW AND INTO THE FUTURE

KEY KNOWLEDGE

- the importance of healthy outdoor environments for individual physical and emotional wellbeing, and for society now and into the future

KEY SKILLS

- justify the importance of healthy outdoor environments for individuals and society

9.4.1 WHY ARE OUTDOOR ENVIRONMENTS SO IMPORTANT?

The importance of the outdoor environment cannot be overstated as it is intricately connected to our daily lives. Although 89% of Australians live in urban areas, the vast majority of products we consume are derived from or influenced by the outdoor environment. From clean air and water to recreational and educational opportunities, a thriving outdoor environment is crucial for our wellbeing and enriches our lives.

The value of outdoor environments is multifaceted and includes both tangible and intangible benefits. For instance, it provides us with passive or semi-passive enjoyment, such as recreation, aesthetic beauty and wilderness experiences. It also provides us with valuable consumer goods, including timber-based products, food and honey, charcoal, grazing land, mining resources and firewood. Additionally, outdoor environments play a critical role in ecological functions, including water catchment, soil protection and maintaining genetic diversity.

Over the past few generations, childhood has moved indoors. On average, today's kids spend up to 44 hours per week in front of a screen, and less than 10 minutes a day playing outdoors. And for too many kids, regular and safe access to nature is determined by race, income, identity, ability and the suburb or town that they live in.

WHAT IS NATURE-DEFICIT DISORDER?

Since 2005, the number of studies of the impact of nature experience on human development has grown from a handful to nearly one thousand. This expanding body of scientific evidence suggests that **nature-deficit disorder** contributes to a diminished use of the senses, attention difficulties, conditions of obesity, and higher rates of emotional and physical illnesses. Research also suggests that the nature-deficit weakens ecological literacy and stewardship of the natural world. These problems are linked more broadly to what health care experts call the 'epidemic of inactivity,' and to a devaluing of independent play. Nonetheless, we believe that society's nature-deficit disorder can be reversed.

'What is Nature-Deficit Disorder?' by Richard Louv, 15 October 2019 ©Richard Louv

nature-deficit disorder a metaphor to describe the human costs of alienation from nature

In this module, we will explore the importance of outdoor environments for individuals, and for society, by exploring the following:

- aesthetic value
- recreation and adventure
- physical and emotional wellbeing
- intrinsic value
- biodiversity
- education
- economic value
- future food and medicine sources
- scientific research.

Shutterstock/caseyjadew

Figure 9.28 Regular time outdoors helps children to thrive

Aesthetic value

The **aesthetic value** of outdoor environments is based on their visual appearance and the emotional responses they evoke. The diverse landscapes found in outdoor environments, such as mountain ranges, rivers and ancient rainforests, can elicit a wide range of emotional responses, including fear, appreciation, awe and contemplation. As discussed in Unit 1, personal responses to outdoor environments can vary greatly, and are influenced by the visual appeal and unique qualities of each individual environment.

aesthetic value a judgement of value based on the appearance of an object and the emotional responses it causes

If you have ever been in the bush and taken a photograph or conveyed an animated description of a spectacular mountain range, a waterfall, a snow gum or another aspect of your outdoor experience, you have been affected by the aesthetic value of the outdoor environment. Aesthetic value can refer to the ability of the outdoor environment to inspire creativity.

Recreation and adventure

Outdoor environments offer a vital escape from the daily pressures of life, allowing individuals to experience a simpler and more relaxed existence. They provide opportunities for positive risk-taking behaviours, enabling individuals to learn more about themselves and to push their limits. Activities such as camping, bushwalking, picnicking, boating, rock climbing, horse riding, skiing and fishing are enjoyed in outdoor environments. These activities can also provide employment opportunities in areas like **ecotourism**, environmental management and interpretation, as well as boosting sectors within the manufacturing and service industries that produce the necessary equipment and resources for these activities.

ecotourism responsible travel to natural areas that conserves the environment and improves the wellbeing of local people

Belinda Dalziel

Figure 9.29 A cross-country skier on the summit of Victoria's Mt Stirling

Physical and emotional wellbeing

Participating in passive or active recreational activities in the outdoor environment can influence a person's physical and emotional wellbeing. The physical benefits of exercising are well known; however, studies have shown that exercising in a natural setting is more effective than in a gym.

Parks Victoria, through its 'Healthy Parks Healthy People' resource, commissioned Deakin University to research the human health benefits of contact with nature. The findings showed that access to nature plays a vital role in human health, wellbeing and development. The research indicated that humans are, among other things, dependent on nature for psychological, emotional and spiritual needs that are difficult to satisfy by any other means.

Resource Case Study: Balance of nature: How spending time outdoors can change your life for the better

CASE STUDY

BALANCE OF NATURE: HOW SPENDING TIME OUTDOORS CAN CHANGE YOUR LIFE FOR THE BETTER

As significant greenspaces for relaxation and recreation, the health of our national parks is tied to our collective wellbeing.

'Self-care' is a buzzword that's worked its way into our vocabulary but with good reason: lately, the world has become a confusing, challenging place for many of us. According to a 2022 report released by the World Health Organization (WHO), global rates of depression and anxiety rose by more than 25% in the first year of the pandemic. So, it makes sense that people are searching for ways to improve their mental wellbeing. Apart from working less, eating nutritious food and prioritising sleep, what else can we do? With many workers at risk of burnout and a rising number of people looking to escape urban areas, research suggests that spending time in nature could be a vital (yet often overlooked) form of self-care

....

A 2019 study conducted with people impacted by bushfires sheds some light on the link between the natural environment and mental wellbeing. Dr Karen Block, Associate Director of the Child and

Community Wellbeing Program, Centre for Health Equity, in the Melbourne School of Population and Global Health, explains that researchers asked participants about how connected they felt to the natural environment. Their answers revealed the significant impact that nature can have on our mental wellbeing.

'People who felt more connected to the natural environment had lower rates of anxiety, depression and Post-Traumatic Stress Disorder. They also had higher rates of resilience and other positive wellbeing measures as well,' says Karen.

...

Importantly, you don't need to be a super fit rock climber or veteran thru-hiker to benefit from being in nature. From going on a family camping adventure in a national park, to bird-spotting in Melbourne's coastal wetlands, or going on a day hike to view spring wildflowers, there are many ways to enjoy the nature that's right on your doorstep. While most doctors recognise the therapeutic value of green spaces, most aren't quite at the stage of prescribing bushwalks. Thankfully there's nothing stopping you from scheduling a day in a national park yourself.

...

As the health of the natural environment is linked to our own wellbeing, psychologists around the world have been witnessing a sharp rise in people (especially young people) experiencing distress due to the impacts of climate change and biodiversity loss. The term 'eco-anxiety' has been coined to describe this growing concern for the health of our planet. And Victorians are no exception.

...

With a Volunteering Australia report indicating that people who volunteer are 'happier, healthier and sleep better than those who don't volunteer', could rolling up our sleeves to help to restore our national parks be a part of the solution when managing eco-anxiety? Dr Karen Block's research project indicates that any activities that boost your feelings of connection to nature can be advantageous. Add in the benefits that come with volunteering, and you could see a shift in your feelings of distress. The best part is that nature is highly accessible, even if you live in the heart of the city.

'Balance of nature: How spending time outdoors can change your life for the better', 11 October 2022 © Parks Victoria

QUESTIONS

1 Outline two benefits the participants felt when asked about the impact that nature can have on wellbeing.
2 Outline two ways individuals and families could connect to the natural environment to improve their wellbeing.
3 Explain two ways you have connected to outdoor environments and how they have influenced your wellbeing.

LEARNING ACTIVITY 9.4A

Logbook entry

Outline when you have enjoyed being in the outdoors as part of a practical experience. Include reflections on where you were, what activities you were doing, the weather and sounds you experienced. Analyse how this experience may have influenced your physical and emotional wellbeing.

Worksheet
9.4a Logbook entry

Intrinsic value

Humans tend to view outdoor environments in terms of their objective value. We often measure the usefulness of an area by considering what resources it can provide us and whether it meets our needs. However, it is important to realise and acknowledge that outdoor environments exist of their own accord and have a value in their own right, which is not dependent on human use or our relationship with that outdoor environment. The **intrinsic value** of an outdoor environment means that it is valued for what it is, rather than for what it can provide.

intrinsic value something that is prized for what it is, rather than for what it can provide

Biodiversity

Biodiversity is a key factor in the maintenance of environmental stability. It is the basis for the continuation of life and is important for the stability of specific species, ecosystems and regions. Biodiversity is also an essential element of the sustainability of human societies and outdoor environments. There are four major reasons for maintaining biodiversity:

1 All species have a right to exist. Compassion and moral responsibility require that humans respect non-human species.
2 Species reduction reduces the richness of a human experience. The aesthetic value of biodiversity lies in the beauty, symbolism and intrinsic worth of species.
3 Many species in the future may be found to have new properties in terms of food, medicine, and renewable and biological resources. These are just some of the utilitarian and economic values of biodiversity crucial to human survival.
4 Other species provide the life-support systems of our planet, maintain the atmosphere, influence the climate, generate and recycle essential nutrients within the topsoil, dispose of wastes, control pests and diseases, pollinate crops and provide a genetic store from which we may benefit in the future. These are the ecological values of biodiversity.

Reduced biodiversity results in ecosystems that are susceptible to environmental change. These ecosystems are impaired by stress and disruption.

The consequences of human activities such as farming and industrial development have been severe for many native species. Floral and faunal communities have been modified, populations have been displaced and species have become extinct.

The 15th Conference of the Parties (COP15) to the UN Convention on Biological Diversity, held in December 2022, marked a critical moment in the global effort to address the loss of biodiversity. The conference resulted in the adoption of a landmark agreement to protect and conserve at least 30% of the world's lands and oceans by 2030, referred to as the '30×30' target. This historic commitment represents the largest conservation effort in history and has the potential to make a significant positive impact on wildlife, address climate change and secure essential services provided by nature, such as clean water and crop pollination. By taking action to conserve biodiversity, COP15 has set the stage for a more sustainable future for both the planet and its inhabitants.

Education

Usually when we think of an educational setting, we immediately think of classrooms and lecture theatres. These provide controlled and efficient locations to learn, yet they are disconnected from the presented content. Learning while located physically within an environment allows visible examples to be experienced, enabling effective consolidation of the curriculum. Outdoor environments provide an endless array of educational opportunities. Education in outdoor environments may take the form of organised school excursions (such as your Outdoor and Environmental Studies trips), field trips undertaken by naturalist or bushwalking clubs or a personal trip to a favourite venue. Education within and about outdoor environments is important for the future of human society because people value places they have personally experienced. Consequently, people are more likely to want these places protected and conserved for future generations.

Belinda Dalziel

Figure 9.30 VCE Outdoor and Environmental Studies students learning about the endangered hooded plover at Point Leo on the Mornington Peninsula

Economic value

The Australian economy is heavily dependent on the country's thriving outdoor environment. Over the years, the exploitation of its abundant natural resources has contributed significantly to the high standard of living enjoyed by many Australians. From the gold rush of the 1850s to the current mining of minerals and energy resources such as iron ore, liquefied natural gas and coal, the environment has been a major source of wealth for the country. The legacy of being known as a major exporter of wool has transformed into a modern-day recognition of our expertise in extracting and exporting valuable resources from the land and sea.

Our environment also provides essential services, including clean water and water-cycle regulation, fertile soil, pollination, carbon storage and protection against natural hazards. These services, in conjunction with industries like tourism, agriculture and fishing, which rely on our natural capital, play a crucial role in maintaining our prosperity and overall wellbeing.

Australian Agriculture 2020–21

Total value of agriculture increased to $71 billion

New South Wales

NSW produced 13 million tonnes of wheat in 2020-21.

The Moree Plains shire produced NSW's largest wheat crop with 1.4 million tonnes.

South Australia

SA produced 52% of the nation's wine grapes (872,000 tonnes). District Council of Loxton Waikerie produced the state's largest crop of wine grapes with 193,500 tonnes.

Western Australia

WA had the nation's largest barley harvest with 4.6 million tonnes and continues to be the country's largest canola producer with 1.7 million tonnes.

The Esperance shire recorded the state's largest barley crop with 524,000 tonnes as well as the state's largest canola crop with 306,300 tonnes.

Victoria

Victoria has the country's largest dairy herd with 1.4 million head.

The Corangamite Shire has Victoria's largest dairy cattle herd with 224,300 head.

Tasmania

Tasmania continues to produce the nation's largest potato crop at 401,000 tonnes.

The Dorset Council produced the state's largest potato crop with 73,100 tonnes.

Queensland

QLD's beef herd is the largest in the country, accounting for almost half of Australia's total herd at 10.6 million head.

The Central Highlands Regional Council had Queensland's largest beef herd with 896,000 head.

Northern Territory

The NT had 53 million hectares of farmland and 1.7 million beef cattle. The gross value of mangoes increased 4% to $70 million.

Australian Capital Territory

The total value of nurseries in the ACT was $2.4 million and accounted for over a quarter (26%) of the ACT's total value of agriculture.

Figure 9.31 Australian agriculture by state/territory

Future food and medicinal sources

Securing reliable sources of nutritious food and effective medicines for an escalating global population will be a difficult challenge for future generations. It is estimated that the food supply will need to increase by 70% over the next 30 years to meet growing demand. Australians play an important global role in the production of food, with Australian farmers exporting around 60% of what they grow. Around 60 years ago, the answer to increasing food production was the introduction of chemical fertilisers – this worked, but at a great ecological cost. Scientific research, technology and, most importantly, the outdoor environment will play a crucial role in feeding future populations. Current research is investigating future food sources such as algae, insects, artificial meat grown in giant vats and 'greening' the world's arid areas for crop production.

Indigenous Australians have a rich history of utilising the therapeutic benefits of plants. In modern times, many medicines also come from the natural world, including over two-thirds of cancer treatments. Rainforest plants are a source of treatments for a variety of conditions, such as heart disease, malaria, arthritis and diabetes, as well as playing a role in producing antibiotics and antiseptics. Despite the proven success of these treatments, only a small fraction of rainforest plants have been studied for their medicinal properties, highlighting the vast potential for further discovery in the outdoor environment. One example of a plant product with both medical and economic benefits is tea tree oil, extracted from the *Melaleuca alternifolia* plant. This oil is a powerful antiseptic, germicide, antibacterial and fungicide, and is widely exported and utilised in products around the world. Supporting Australian-made tea tree oil products can help to boost the country's export economy.

LEARNING ACTIVITY 9.4B

Australian bush medicine

1 Visit the Australian Geographic website and search for 'Aboriginal bush medicines'.
2 Explore the 10 most common Aboriginal bush medicines.
3 Summarise three bush medicines, including:
   - name and image of species
   - location found in Australia
   - what the bush medicines aimed to treat
   - why each species is important to society.
4 Present your research to the class.

Worksheet
9.4b Australian bush medicine

Weblink
Australian Geographic

Scientific research

Through scientific research, natural sciences such as biology, chemistry, physics and astronomy have provided society with an ever-developing understanding of the complexities of outdoor environments. We have discovered how to utilise outdoor environments for many of our daily requirements, including food, medicine and building materials. Scientific research has provided us with technologies that enable the monitoring of natural processes. This has allowed us to understand natural and human-induced changes, and our impacts on the outdoor environment. For example, salinity has had a significant impact on productive land in Victoria. Scientific research that compares outdoor environments with salinity-affected environments has provided evidence that the loss of trees and other vegetation is a major contributing factor to salinity. Revegetation programs in salinity-affected areas are an effort to overcome this problem by attempting to return some naturalness to these modified environments.

NOTES FOR THE EXAM

When differentiating between the reasons why outdoor environments are important to the wellbeing of individuals and our broader society, ensure you have clear delineation between the two concepts. For example, 'education' empowers individuals such as VCE students to learn about the unique flora and fauna, increasing their understanding on how complex ecosystems co-exist. By contrast, outdoor environments are important for society as they provide the greater population with employment opportunities through primary industries such as horticulture and animal husbandry.

9.4 KEY CONCEPTS

- From clean air and water to recreation and educational opportunities, healthy outdoor environments are essential and enrich our lives.
- Aesthetic value is a judgement of value based on the appearance of an object and the emotional responses it causes, such as a beach or mountain range.
- Environments provide opportunities for individuals to undertake positive risk-taking behaviours that allow them to find out more about themselves and extend themselves; for example, rock climbing.
- Access to nature plays a vital role in human health, emotional wellbeing and development.
- Outdoor environments can provide an intrinsic value; that is, being prized for what it is, rather than the value it can potentially provide humans.
- Biodiversity is the basis for the continuation of life and is important for the stability of specific species, ecosystems and regions.
- Reduced biodiversity results in ecosystems that are susceptible to detrimental environmental change.
- Learning while being physically within the environment allows visible examples to be experienced, enabling effective consolidation of the curriculum. Consequently, people are more likely to want these places protected and conserved for future generations.
- Australia's economy and communities are dependent on our natural capital. Healthy ecosystems provide critical services such as fresh water, regulation of regional water cycles, soil fertility, crop pollination and carbon storage.
- Indigenous Australians have used the healing properties of plants for thousands of years. The majority of medicines in contemporary society are derived from the natural environment.
- Scientific research has provided us with technologies that enable the monitoring of natural processes.

Worksheet
9.4 Key concepts

9.4 CONCEPT QUESTIONS

REMEMBERING

1 Copy and complete the table below to summarise the reasons why healthy outdoor environments are important for individuals and for the future of society.

Reason	How this is important to individuals or society	Description	Example
Aesthetic value			
Recreation and adventure			
Physical and emotional wellbeing			
Intrinsic value			
Maintenance of environmental stability			
Education			
Economic value			
Future food and medicinal sources			
Scientific research			

UNDERSTANDING

2 Explain 'biodiversity'.

3 Describe three reasons for maintaining high levels of biodiversity.

4 Explain how to differentiate between individuals and society when discussing the importance of outdoor environments.

APPLYING

5 Choose an outdoor environment you have visited or studied and explain why education in this environment is important to improve your own physical and emotional wellbeing.
6 Analyse why outdoor environments are vital for scientific research for society now and into the future.
7 Give an example of how you have aesthetically experienced outdoor environments.

EXTENSION CHALLENGE

8 Investigate a specific environment you have visited or studied and explain three reasons why this outdoor environment is important for the future.

9.5 LOCAL, NATIONAL AND INTERNATIONAL SOLUTIONS TO COMBAT CLIMATE CHANGE ACROSS A RANGE OF ENVIRONMENTS

KEY KNOWLEDGE

- local, national and international solutions and mitigation strategies to combat climate change across a range of environments

KEY SKILLS

- analyse possible solutions and mitigation strategies for combating climate change

9.5.1 COMBATING CLIMATE CHANGE

Climate change refers to any long-term shifts in average weather patterns, including alterations in temperature, precipitation and wind patterns. The latest scientific evidence indicates that our planet is undergoing a rapid warming trend. The average global temperature has already increased by 1.1°C since the pre-industrial era.

It is widely accepted by scientists that human activities, particularly through the emission of greenhouse gases, are driving this warming trend. The rate of warming is also accelerating at an unprecedented pace compared to previous climate changes in history.

Mitigating the effects of climate change involves reducing the human impact on the Earth's climate system. This means reducing emissions of greenhouse gases, such as by reducing the burning of fossil fuels for energy, transportation and heat.

People have always adapted to climate change, and throughout history, communities have developed strategies to cope with weather extremes like fires, floods and droughts. The current global challenges posed by climate change include rising sea levels, shorter growing seasons and food insecurity. Addressing these challenges requires innovative solutions that build resilience and ensure long-term sustainability.

net zero the state where the amount of greenhouse gases emitted into the atmosphere is equal to the amount of greenhouse gases removed from the atmosphere

Net zero is a term that describes the state where the amount of greenhouse gases emitted into the atmosphere is equal to the amount of greenhouse gases removed from the atmosphere. It is like balancing a scale, where emissions represent one side of the scale and removals represent the other.

The goal of achieving net zero emissions is to stop the accumulation of greenhouse gases in the atmosphere, which contribute to climate change. However, because we have already emitted a significant amount of greenhouse gases, achieving net zero does not mean that we can stop all emissions immediately. Instead, we need to continue reducing emissions as much as possible and simultaneously removing the excess carbon dioxide from the atmosphere through various methods, such as reforestation, direct air capture, or carbon capture and storage. However, by reaching net zero emissions, we can at least limit the rise of global temperatures and mitigate the worst impacts of climate change.

NOTES FOR THE EXAM 

For this key knowledge, you are asked to analyse possible solutions to mitigate climate change, whether they be existing or into the future.

Alamy/Design Pics Inc

Figure 9.32 An aerial view of Tongass National Forest in Alaska in the United States. With their abundance of plants, forests often absorb more carbon than they release. These **carbon sinks** continually take carbon out of the atmosphere through the process of photosynthesis.

carbon sink
a place, such as a forest or an ocean, that continually takes carbon out of the atmosphere

Action to combat climate change on a local level

In Victoria, the government is taking action to reduce the state's emissions to net zero by 2050 and to ensure communities, businesses and institutions are better prepared to deal with the impacts of climate change.

Victoria's *Climate Change Act 2017* establishes a long-term target of net-zero greenhouse gas emissions by 2050, with five-yearly interim emissions reduction targets. In 2023, the Victorian Government had the following targets to cut the state's emissions:

- 2025: 28–33% below 2005 levels
- 2030: 50% below 2005 levels.

CASE STUDY

'ON YOUR MARKS, GIPPSLAND, GO!': AUSTRALIA'S FIRST OFFSHORE WIND INDUSTRY ONE STEP CLOSER

An Australian offshore wind industry is one step closer, with the Australian Government declaring an area in the Bass Strait off Gippsland as the first area suitable for developing offshore wind energy projects.

The declared area in Gippsland, Victoria, covers approximately 15000 square kilometres, and runs from offshore of Lakes Entrance in the east to south of Wilsons Promontory in the west.

The declaration comes after consultation with the Bass Coast, South Gippsland and Wellington local government communities, including First Nations peoples and industry.

This declaration confirms the area where developers will soon be able to apply for the first licences for offshore wind in Australia. Further consultations will take place on any proposed developments, once feasibility licences are issued.

There is potential for offshore wind projects in the area to support 3100 jobs over the next 15 years during their development and construction phases, and an additional 3000 ongoing operational jobs.

The declaration brings Australia one step closer to an offshore wind industry and a step closer to net zero by 2050.

'On your marks, Gippsland, go!' 19 December 2022 © Department of Climate Change, Energy, the Environment and Water

Resource
Case Study: 'On your marks, Gippsland, go!': Australia's first offshore wind industry one step closer

LEARNING ACTIVITY 9.5A

Local government climate change strategies

Using the 'Know your local council' website, find a link to your local council and research your local council's Climate Change Strategy. Include the name of your local council and a map of the area, and answer the following questions:

1 What are the strategies and aims of the council's climate change policy?
2 What are some of the key actions of the council's climate change policy?
3 How is the climate change policy being measured?
4 How can the community be involved to contribute to reducing the effects of climate change in this council area?
5 What do you think could be further done in your council area to mitigate the effects of climate change?

If your local council doesn't have a climate change mitigation strategy, or it is hard to access, base your research on the City of Melbourne.

Worksheet
9.5a Local government climate change strategies

Weblink
Know your local council
City of Melbourne

Action to combat climate change on a national level

Australia is already experiencing the impacts of climate change, which vary across the country. Under international climate agreements, at the time of printing, Australia has two targets to reduce greenhouse gas emissions:

- 5% below 2000 levels by 2020 (under the Kyoto Protocol)
- 26–28% below 2005 levels by 2030 (under the Paris Agreement).

Moreover, after a change of government in 2022, Australia turned a corner in national climate and energy policy. Australia has raised its expected emissions cuts from 30% by 2030 to 40% by 2030, although this still falls short of the legislated 43% by 2030 target.

Australia is committed to global cooperation to address climate change and is now undertaking stronger action on climate change. Australia has committed to update our Nationally Determined Contribution under the Paris Agreement and has a new ambitious 2030 target to reduce greenhouse gas emissions by 43 per cent below 2005 levels, putting Australia on track to achieve net zero emissions by 2050. This has now been enshrined in domestic law through the federal government's *Climate Change Act 2022*.

The Australian Government is working to mitigate the effects of climate change by reducing carbon emissions by:

- upgrading the electricity grid to support more renewable power
- reducing the price of electric vehicles
- supporting businesses and industries to innovate and adopt smarter practices and technologies
- encouraging businesses and consumers to reduce emissions
- regulating and reporting on greenhouse gas emissions
- helping the land and agriculture sector reduce greenhouse gas emissions
- partnering with our Indo-Pacific neighbours to reduce emissions
- helping negotiate and meet Australia's obligations under the Paris Agreement.

Resource
Case Study: Community battery grant guidelines now available

CASE STUDY

COMMUNITY BATTERY GRANT GUIDELINES NOW AVAILABLE

The $200 million Community Batteries for Household Solar program will enable Australians to store affordable solar energy for use during peak times, and to share excess power with other households in their area.

The Community Batteries for Household Solar program will help households and businesses to:

- lower electricity bills
- reduce pressure on the electricity grid for the broader community
- help us reach net zero by 2050.

As we ramp up renewable generation, we will need more energy storage to deliver secure and affordable energy for Australian businesses and households.

'Community battery grant guidelines now available'; 19 December 2022 © Department of Climate Change, Energy, the Environment and Water

QUESTIONS

1 Outline the national climate change mitigation strategy described above.
2 How will the Community Batteries for Household Solar program help to mitigate climate change?
3 What is Australia's national commitment to reduce greenhouse gas emissions?

LEARNING ACTIVITY 9.5B

Climate Active

Climate Active is a partnership between the Australian Government and Australian businesses to encourage voluntary climate action. Leading organisations are choosing to reduce their climate impact to zero by becoming carbon neutral. Visit the Climate Active website and search for information on how business can be 'climate active' in Australia.

1 What does it mean to be carbon neutral?
2 How can businesses across Australia achieve a climate active certification?
3 What are carbon offsets, and how do they mitigate the effects of climate change?
4 Outline ways to reduce and remove emissions.
5 How can carbon offsetting improve social, cultural and economic aspects of life?
6 Read one of the Climate Active feature stories and summarise how the business has helped to mitigate the effects of climate change.

Worksheet 9.5b Climate active

Weblink Climate Active

CASE STUDY

Torres strait islanders plead for climate action as government builds seawall

Global warming and rising seas are such a threat to Torres Strait communities that Australian governments have been forced to build a $15 million seawall to protect the remote, low-lying Boigu Island from ocean inundation and severe weather.

Indigenous Australians Minister Ken Wyatt on Friday announced the completion of the 1-kilometre wave return wall, along with the raising and 450-metre extension of an existing bund wall and upgrades to stormwater drainage on Australia's most northerly inhabited island.

'The new Boigu seawall is a practical response delivered by all levels of government working together to address the effects of climate change and the threat of inundation for our Torres Strait communities,' Mr Wyatt said.

The federal and Queensland governments, along with the Torres Strait Island Regional Council, are rolling out a $40 million coastal protection infrastructure program across five islands under Australia's National Climate Resilience and Adaptation Strategy.

Torres Strait Islanders want more than infrastructure. They are pleading with the federal government and international community to drastically increase efforts to reduce greenhouse emissions to slow rising sea levels, which they say will force them from their island homes. The sea has risen six centimetres in the Torres Strait in the past decade.

'Torres Strait Islanders plead for climate action as government builds seawall' Mike Foley; 1 April 2022, *The Sydney Morning Herald*

Resource Case Study: Torres Strait Islanders plead for climate action as government builds seawall

QUESTIONS

1 Explain the government bodies involved in addressing this issue.
2 Explain the National Climate Resilience and Adaptation Strategy.
3 Analyse the effectiveness of this strategy at mitigating the effects of climate change.

Action to combat climate change on an international level

The impacts of climate change are widespread and undeniable. According to the latest report from the Intergovernmental Panel on Climate Change (IPCC), human activities have led to a significant increase in the levels of CO_2 emissions, surpassing levels seen in the past two million years. These emissions have contributed to a global warming of approximately 1.1°C between 1850 and 2000. Without immediate and significant reductions in greenhouse gas emissions, the goals of limiting warming to 1.5°C or 2°C will become increasingly unattainable.

WHAT ARE UNITED NATIONS CLIMATE CHANGE CONFERENCES?

multilateral agreed upon or participated in by three or more parties, especially the governments of different countries

The United Nations Climate Change Conferences have evolved into prominent international gatherings over the past two decades. These **multilateral** conferences serve as the central platform for international discussion on climate change and bring together representatives from various governments worldwide.

These meetings are held under the jurisdiction of the Conference of the Parties (COP), which serves as the meeting of the parties to the Kyoto Protocol and the Paris Agreement. The conferences rotate among the five United Nations regional groups and are essential in addressing the global challenge of climate change.

Weblink
The Paris Agreement

Kyoto Protocol

The Kyoto Protocol, adopted on 11 December 1997 and ratified on 6 February 2005, is an international agreement that commits industrialised countries to limit and reduce their greenhouse gas emissions with specific, individually agreed targets. With 147 signatories as of 2020, the Protocol is a key component of the United Nations Framework Convention on Climate Change. The Protocol requires strict monitoring and reporting of actual emissions, with precise record keeping, and provides support to countries in adapting to the impacts of climate change.

Paris Agreement

The Paris Agreement is a globally binding treaty on climate change adopted by 196 signatories at the 21st Conference of the Parties (COP 21) in Paris on 12 December 2015. With its entry into force on 4 November 2016, the Agreement aims to limit the global average temperature rise to well below 2°C, and to pursue efforts to limit it to 1.5°C, compared to pre-industrial levels. This marks a significant milestone in the global response to climate change, as all nations have come together to make ambitious efforts to address the challenge of climate change and its impacts.

The Paris Agreement operates on a five-year cycle of increased action, with countries reporting transparently on their progress in mitigating climate change, implementing adaptation measures, and providing or receiving support. From 2024, countries will provide regular updates on the measures they have taken to address the challenge of climate change.

United Nations Climate Change Conference landmark agreement

The United Nations Climate Change Conference landmark agreement reached at COP27 in December 2022 underlines the importance of global cooperation in tackling the urgent threat of climate change. The decision to provide funding for 'loss and damage' acknowledges the devastating impacts that climate change is having on some of the world's most vulnerable communities and marks a crucial step forward in addressing the urgent need for climate action.

The reaffirmation of the commitment to limit global temperature rise to 1.5°C, in line with the Paris Agreement, demonstrates the resolve of countries to take action to address the global climate emergency. The call to revisit and strengthen 2030 targets and accelerate the phase-out of unabated coal power and inefficient fossil fuel subsidies shows a growing recognition of the need for ambitious and immediate action to reduce greenhouse gas emissions and transition to a more sustainable and resilient future.

UNITED NATIONS SUSTAINABLE DEVELOPMENT GOALS

The Sustainable Development Goals (SDGs) represent a collective global effort to create a better and more sustainable future for all. Adopted by the United Nations in 2015, the 2030 Agenda for Sustainable Development outlines a comprehensive plan for promoting prosperity and safeguarding the planet. The 17 SDGs serve as a rallying call for action by all nations, regardless of their level of development, to work together towards a common goal. The SDGs emphasise the interdependence of ending poverty, improving health and education, reducing inequality, promoting economic growth, combating climate change, protecting our oceans and forests, and other important issues. In short, they represent a commitment to creating a sustainable future for all.

Sustainable Development 13: Take urgent action to combat climate change and its impacts

Sustainable Development Goal 13 predicts that without urgent action on climate change:

- about one-third of global land areas will suffer at least moderate drought by 2100
- the sea level could rise 30 to 60 centimetres by 2100
- about 70–90 per cent of warm-water coral reefs will disappear if the 1.5°C threshold is reached; they will die off completely at the 2°C level
- between 3 billion and 3.6 billion people will live in contexts that are highly vulnerable to climate change
- by 2030, an estimated 700 million people will be at risk of displacement by drought alone.

Goal 13 has five targets to mitigate the effects of climate change globally:

1 Strengthen resilience and adaptive capacity to climate-related hazards and natural disasters in all countries.
2 Integrate climate change measures into national policies, strategies and planning.
3 Improve education, awareness-raising and human and institutional capacity on climate change mitigation, adaptation, impact reduction and early warning.
4 Implement the commitment undertaken by developed-country parties to the United Nations Framework Convention on Climate Change to a goal of mobilising jointly $100 billion annually by 2020 from all sources to address the needs of developing countries in the context of meaningful mitigation actions and transparency on implementation and fully operationalise the Green Climate Fund through its capitalisation as soon as possible.
5 Promote mechanisms for raising capacity for effective climate change-related planning and management in least developed countries and small island developing states, including focusing on women, youth and local and marginalised communities.

WHAT HAVE WE ACHIEVED SO FAR?

The Paris Agreement has inspired significant action towards mitigating climate change and has created opportunities for new low-carbon solutions . The adoption of carbon neutrality targets by countries, regions, cities and companies has grown, and zero-carbon solutions are increasingly competitive across 25 per cent of the emission-generating economic sectors. Despite the progress, however, much more needs to be done to reach the goals set forth in the Paris Agreement.

9.5 KEY CONCEPTS

- Climate change is any change in the climate, lasting for several decades or longer, including changes in temperature, rainfall or wind patterns.
- Mitigation is reducing the risk or severity of something occurring. The goal of climate change mitigation is to reduce the human impact within the Earth's climate.
- The Victorian government is taking action to reduce Victoria's emissions to net zero by 2050.
- Australia has raised its expected emissions cuts to 40 per cent by 2030.
- The Australian Government is working to mitigate the effects of climate change by upgrading the electricity grid to support more renewable power and reduce the price of electric vehicles.
- Changes in the Earth's climate have been observed in every region and across the whole global climate system.
- The Kyoto Protocol commits industrialised countries to limit and reduce greenhouse gas emissions with agreed individual targets.
- The Paris Agreement is a legally binding international treaty on climate change with the goal of limiting global warming to well below 2°C, and preferably to 1.5°C, compared to pre-industrial levels.
- Sustainable Development Goal 13: Take urgent action to combat climate change and its impacts has five key target areas for climate change mitigation.

9.5 CONCEPT QUESTIONS

Worksheet
9.5 Key concepts

REMEMBERING

1 Using an example of each, distinguish between local, national and international jurisdiction efforts to combat climate change.
2 Describe the terms 'climate change' and 'mitigation'.
3 What does 'COP' represent and what organisation facilitates its activities?

UNDERSTANDING

4 Explain the relationship between the Kyoto protocol and the Paris agreement.
5 How do greenhouse gas emissions contribute to climate change?
6 Describe the difference between climate change mitigation and climate change adaptation.
7 Visit the UN website on the Paris Agreement and research two countries that have signed on.

APPLYING

8 Complete the table below.

	Jurisdiction and name of climate change policy	Summary of climate change policy	Examples of climate change mitigation strategies from this policy
Local			
National			
International			

9 Referring to the examples you have completed in the table above, suggest possible improvements to each of these strategies in mitigating climate change.

EXTENSION CHALLENGE

Living climate change on the coastline of Chile

'We are seeing changes that we have never seen before – excessive rain, inexplicable things,' says Alejandro Naiman, a hake fisher and community leader from El Manzano in southern Chile. Small-scale fishing is the lifeline for most people living in this coastal community, which is now threatened by the changing climate. Dramatic environmental changes are affecting the availability and abundance of species, forcing artisanal fishers and small-scale fish farmers to modify their livelihoods.

To mitigate the impact on their livelihoods, people from El Manzano and other coastal communities have come together to learn new ways to add value to their products and diversify their production. 'People talk about climate change; we are living it every day,' says Juan Torrejón, head of the local aquaculture association in Tongoy. 'We have a proposal for a processing plant, so our associates and cooperative are not just individual producers, but we can also offer products.'

They are also considering developing sustainable tourism as an alternative source of income. Matías Torres, a fisher from Coliumo, explains, 'Coliumo offers more than fishing. There is diving, nature trails, trekking and gastronomic products with a local identity. All these activities together allow the people of Coliumo to have alternatives to better adapt to the effects of climate change.'

'Living Climate Change on the coastline of Chile', Bringing Data to Life: SDG human impact stories from across the globe, United Nations.

QUESTIONS

1. Outline the UN position in global climate change mitigation.
2. Analyse the climate change mitigation strategy being used by the fishers of southern Chile.
3. How are these people making adaptions to their livelihoods due to the effects of climate change?
4. Suggest how other nations may adopt climate change mitigation strategies in their countries.

UNIT 4 AREA OF STUDY 3 LOGBOOK ENTRY REMINDER

Have you had an outdoor experience yet? If so remember to complete your logbook entry. Use chapter 11, learning activity 11.1B and 11.3A to guide your data collection and report writing.

Resource
Glossary

Assessments
End of chapter exam

Glossary test

EXAM QUESTIONS

1 Explain environmental sustainability and provide two examples you have observed in outdoor environments you have visited or studied. (5 marks)

2 A range of threats impact both society and the environment. Choose two of the following and analyse how each threat impacts both society and the environment. (6 marks)

- land degradation
- introduced species
- urbanisation
- climate change
- flood
- fire

3 Describe a climate change mitigation policy and provide an example of this strategy in action. (3 marks)

4 Name an outdoor environment you have visited or studied:

a Provide and justify two reasons why this outdoor environment is important for individuals' physical and emotional wellbeing. (5 marks)

b Justify two further reasons why this outdoor environment is important for society. (5 marks)

5 Analyse the role of sustainability in maintaining outdoor environments for the future. (3 marks)

6 Identify an outdoor environment that you have visited or studied this year. A range of environmental observable characteristics can be used to assess the health of this outdoor environment. Select two observable characteristics that could be used to assess the level of health of this outdoor environment (5 marks):

a quality of water

b quality of air

c quality of soil

d species biodiversity

e ecosystem biodiversity

9780170477123

CHECK YOUR KNOWLEDGE AND SKILLS

Resources
Key knowledge and skills checklist

Checklist for students to mark and date, making judgements of their knowledge and understanding.
(D = developing C = consolidating ER = exam ready)

D	C	ER	Check your knowledge and skills
☐	☐	☐	Explain the term 'sustainability' and provide examples of sustainability in a range of environments
☐	☐	☐	Analyse critiques of sustainability
☐	☐	☐	Describe three pillars of sustainability – economic, environmental and social
☐	☐	☐	Analyse how the pillars of sustainability are interconnected
☐	☐	☐	Describe five observable characteristics to assess the health of outdoor environments
☐	☐	☐	Use observable characteristics to assess and evaluate the health of an environment
☐	☐	☐	Propose possible solutions to improve the health of outdoor environments based on your assessment, using the observable characteristics
☐	☐	☐	Understand two threats to outdoor environments and society
☐	☐	☐	Analyse the impact of each threat on society and an environment you have visited or studied
☐	☐	☐	Describe several reasons why outdoor environments are important for individuals and society, now and into the future
☐	☐	☐	Using environments you have visited or studied, justify why these outdoor environments are important for individuals and society
☐	☐	☐	Explain the term 'climate change and mitigation'
☐	☐	☐	Discuss a local climate change mitigation policy, including an example of a strategy that has been implemented to reduce greenhouse gas emissions
☐	☐	☐	Discuss a national climate change mitigation policy, including an example of a strategy that has been implemented to reduce greenhouse gas emissions
☐	☐	☐	Describe global conventions on climate change
☐	☐	☐	Discuss an international climate change mitigation policy, including an example of a strategy that has been implemented to reduce greenhouse gas emissions
☐	☐	☐	Analyse local, national and global climate change mitigation strategies

CHAPTER 10

UNIT 4 – AREA OF STUDY 2

The future of outdoor environments

KEY KNOWLEDGE

- Indigenous and non-Indigenous peoples' land management strategies for achieving and maintaining healthy and sustainable outdoor environments
- Acts or conventions related to the management and sustainability of a specific outdoor environment, species or ecological community, including two of the following:
  - *Flora and Fauna Guarantee Amendment Act 2019* (Vic)
  - Ramsar Convention (international treaty, 1971)
  - *Environment Protection and Biodiversity Conservation Act 1999* (Cth)
  - *Victorian Environmental Assessment Council Act 2001* (Vic)
  - *Planning Environment Act 1987* (Vic)
- community actions undertaken to sustain healthy outdoor environments, including two of the following:
  - regenerative farming
  - Trust for Nature
  - Landcare
  - community groups such as 'Friends of …'
- individual actions undertaken to promote and sustain healthy outdoor environments, including two of the following:
  - environmental activism
  - environmental advocacy
  - ethical and sustainable consumerism
  - green home design.

KEY SKILLS

- analyse management strategies for maintaining outdoor environments
- evaluate the effectiveness of specific Acts and conventions related to managing and sustaining outdoor environments
- propose changes to current Acts and conventions to further improve the health of a specific environment, species or ecological community
- evaluate the effectiveness of community actions undertaken to sustain healthy outdoor environments
- compare a range of individual actions to sustain healthy outdoor environments.

Indigenous and non-Indigenous peoples land management strategies for achieving and maintaining healthy and sustainable outdoor environments.

Indigenous land and sea management, also known as 'caring for Country,' aims to protect, maintain, heal and enhance healthy and diverse ecosystems, landscapes and cultural values. Public land is managed by governments and their agencies, such as Parks Victoria or the Department of Energy, Environment and Climate Action (DEECA). Victoria's parks protect some of the state's most precious and unique natural landscapes and provide critical habitat for many threatened species.

Acts or conventions related to the management and sustainability of a specific outdoor environment species or ecological community

Environmental Acts and conventions are regulations established by governments or their agencies to minimise human impact on outdoor environments. These Acts usually include action statements that provide detailed guidelines for conservation management practices for specific endangered species. Proposed changes to enhance conservation efforts may involve better surveillance and monitoring of environments to ensure that management strategies are being followed.

Community actions undertaken to sustain healthy outdoor environments

Community groups are formed by people with similar interests to sustain the environment in Victoria and Australia. They educate the public, advocate for conservation policies and carry out projects to improve the environment. These groups often focus on specific issues like wildlife conservation, habitat protection or restoring degraded areas.

Individual actions undertaken to promote and sustain healthy outdoor environments

Throughout history, combinations of environmental activism and advocacy have played crucial roles in saving a range of outdoor environments. Environmental activism is defined as the actions of individuals or groups that aim to protect or support the environment, whereas environmental advocacy is activities that aim to protect the environment by influencing business and government decisions.

KEY TERMS

action statement
advocacy
caring for Country
conservation covenant
conventions
Crown land
cultural burning
energy efficiency
environmental activism
Environmental Acts
fair trade
greenwashing
green wedge
land management
management plan
monoculture
passive solar
public land
regenerative farming
taxon
transboundary wetlands
upcycling

Worksheets

Weblinks

Resources and Templates

Assessments

Nelson MindTap

To access resources above, visit **cengage.com.au/nelsonmindtap**

10.1 INDIGENOUS AND NON-INDIGENOUS PEOPLES' LAND MANAGEMENT STRATEGIES

KEY KNOWLEDGE

- Indigenous and non-Indigenous peoples' land management strategies for achieving and maintaining healthy and sustainable outdoor environments

KEY SKILLS

- analyse management strategies for maintaining outdoor environments

Introduction

In Outdoor and Environmental Studies, we explore the diverse relationships between human society and outdoor environments. There are varying perspectives on the appropriate use of natural resources.

To ensure the preservation of these environments for future generations, we employ sustainable practices and implement policies. These strategies involve the efficient utilisation of physical and human resources for **land management**. Policies serve as guidelines for those involved in environmental management and may take the form of formal plans or political positions.

land management the responsibility of managing the use and development of land resources

Our goal is to either restore damaged outdoor environments or to maintain their health through effective strategies and policies. In this chapter, we explore a range of contemporary land management practices in different environments and investigate local and individual actions to sustain outdoor environments, now and into the future.

10.1.1 INDIGENOUS LAND MANAGEMENT STRATEGIES

In their book, *Caring for Country*, Emilie Ens and Tein McDonald assert that Indigenous peoples in Australia 'hold a unique understanding of the relationship between humans and the natural world, viewing it as a reciprocal and interconnected partnership, in contrast to the Western paradigm of a linear and exploitative relationship between humans and the environment'.

This indigenous concept of '**caring for Country**' is currently being practised throughout Australia by Indigenous communities, groups and organisations. Indigenous land and sea management serves multiple purposes, including the preservation of customary practices, promotion of community wellbeing, conservation of biodiversity and the development of sustainable economic activities.

caring for Country the management of land and waterways by Indigenous peoples, and the sustainable land management practices and initiatives led by them

The overarching aim of Indigenous land management is to maintain, protect and enhance the ecological health and diversity of ecosystems, promote the productivity of landscapes and conserve cultural values and traditions. This approach not only benefits the environment and its inhabitants, but also preserves the rich cultural heritage of Indigenous peoples and contributes to their social and economic prosperity.

Table 10.1 Examples of contemporary Indigenous land management initiatives

	Examples of land management
Commercial and economic	• Bush harvesting of plant foods, medicines and seed for sale • Harvesting for commercial wildlife industries (e.g. crocodile egg harvesting) • Pastoral and related activities (e.g. mustering and sale of feral animals) • Plantations (e.g. firewood and sandalwood) • Art and craft production • Cultural ecotourism • Rehabilitating and revegetating mine sites or other disturbed areas • Land restoration and other natural resource management services carried out under contract arrangements • Employment in Indigenous and co-managed parks and protected areas
Spiritual and cultural	• Protecting and maintaining cultural sites, stories and songlines • Recognising important cultural areas • Performing cultural or customary activities
Natural resource management	• Weed control and monitoring • Feral animal control and monitoring • Fire management • Monitoring and management of threatened species and ecological communities • Conservation of natural water bodies • Soil erosion control and soil rehabilitation • Native nursery, seed collection and planting • Visitor and tourist management (e.g. track maintenance and signage)
Conservation	• Creating seasonal harvest calendars • Surveying catchments • Hunting for feral animals such as foxes, camels, buffalo, pigs or cats that threaten the delicate ecosystem of the bush • Tracking endangered species • Recording (new) plants • Protecting biodiversity • Removing seeds and weeds, including invasive pests like African Buffel grass • Removing rubbish left by tourists at camping spots, and ghost nets, plastic and other marine debris from seas and beaches • Returning threatened species to their native habitat • Managing controlled burns and setting fire breaks to prevent devastating bush fires and to protect outstations and sacred sites • Conducting fisheries surveillance and compliance patrols • Helping with sustainable water management, including animal rescue
Education	• Teaching government departments and tourists about their connection with the land, the seasons and bush foods • Assisting with providing cross-cultural education and capacity building within their communities • Taking Aboriginal children out on Country so they can learn from their elders

Conservation management strategy in focus

Traditional landowners often collaborate with government agencies and non-Indigenous organisations to conserve and care for the land. One example is the collaboration between the CSIRO and Indigenous groups to create Indigenous seasons calendars.

Indigenous individuals and groups are actively engaging with the CSIRO to share and learn about their knowledge and management practices. As Australia's First Scientists, Aboriginal and Torres Strait Islander peoples possess a comprehensive understanding of the changing seasons and the signs of their progression.

This seasonal knowledge plays a crucial role in various activities on their traditional lands and recording this information provides a means for Indigenous knowledge holders to showcase and communicate their connection to, use of and management of their traditional lands.

Over the past 15 years, the CSIRO has been working with Indigenous partners to co-design, refine and implement a season calendar methodology for documenting and presenting this seasonal understanding. The resulting co-created season calendars have proven to be valuable tools for representing Indigenous connections to their traditional lands.

Worksheet
10.1a Research Indigenous Australians' seasonal calendars

Weblinks
CSIRO
Bureau of Meterology's Indigenous Weather Knowledge

LEARNING ACTIVITY 10.1A

Research Indigenous Australians' seasonal calendars

Visit the calendars page of the Indigenous knowledge section of the CSIRO website to find out more about their work with Indigenous groups to create indigenous seasonal calendars.

1 Explain why Indigenous people are called Australia's First Scientists.
2 What are Indigenous seasonal calendars and what do they represent?
3 Discuss how season calendars assist Indigenous land management.
4 Explore the Bureau of Meteorology's Indigenous Weather Knowledge page on its website to research further examples of Indigenous seasons.

Natural resource land management strategy in focus

Another important collaboration between a government body and local Indigenous groups is Indigenous ranger projects. These initiatives are designed to facilitate the sharing of traditional knowledge by Indigenous peoples, alongside conservation training, to empower them to take an active role in the management and preservation of their land, sea and cultural heritage. As part of these programs, Indigenous rangers undertake projects targeted at:

- bushfire mitigation
- endangered species protection
- biosecurity compliance
- establishing partnerships with educational, research and charitable groups and commercial businesses
- engagement with schools.

As is noted by the National Indigenous Australians Association, the Indigenous rangers' 'work is highly regarded by Indigenous communities across Australia for its positive impacts on the environment, employment, and social, cultural, and economic benefits'.

CASE STUDY

MURRAY-DARLING BASIN INDIGENOUS RIVER RANGERS PROGRAM

To mark World Ranger Day, in July 2021, the Australian Government announced the establishment of five new Indigenous river ranger groups, following applications to the Murray-Darling Basin Indigenous River Rangers grant round.

The $3.1 million program empowers Indigenous organisations to improve waterway health, manage country and sustain the Basin's valuable environmental assets. The rangers, ranger coordinators and support staff will use their knowledge and connection to country to support environmental and cultural outcomes across the breadth of the Basin. The new rangers will begin working in mid-2021 for a period of 12 months.

First people of the millewa-mallee aboriginal corporation

The Lindsay, Mulcra and Wallpolla Islands River Ranger Floodplain Management Program will engage a team of Indigenous rangers in the Mildura region, in north-west Victoria. The rangers will work on floodplain protection, a native fish breeding program, weed management, replanting native wetland plants and cultural mapping in the Murray-Sunset National Park. The new ranger group will work in partnership with traditional owners, the Mallee Catchment Management Authority, local education and training providers and other key stakeholders. The project will build on environmental work conducted in the area by the local community, government and Indigenous cultural rangers.

The new rangers group will be led by a program coordinator and team leader, providing up to six new jobs for Indigenous people in the Mildura area.

Resource
Case Study: Murray-Darling Basin Indigenous river rangers program

Cultural burns management strategy in focus

Many Aboriginal peoples across Australia have long utilised fire as a means of managing landscapes. Cultural use of fire has been vital in maintaining a balanced ecological environment, and the cessation of cultural burning practices after European colonisation has resulted in negative impacts on biodiversity and increased the risk of bushfires.

In modern times, Western science and traditional management practices of Indigenous Australians are being combined to help manage the environment. Controlled cultural burns can have positive effects on the environment, including reducing greenhouse gas emissions.

Cultural burns are conducted during the appropriate season and under favourable conditions, and typically employ 'cool' fires, which are smaller and have low intensity, reducing the risk of high flames that can burn entire trees and

Joe Sambono

Figure 10.1a A low-intensity burn on Lardil Country (Mornington Island, North Queensland)

forest canopies. Protecting the canopy is a priority in cultural burning as it contains valuable resources such as insects, birds' nests, bats and shade. Removing the canopy also changes the surrounding ecosystem by allowing sunlight to penetrate and dry out the soil. Unlike hazard-reduction burns, which involve fire lines that can create walls of flames, cultural burns typically use spot ignitions to create a mosaic of fires, leaving space for wildlife to escape.

In 2021, **cultural burning** was integrated into Victoria's existing burning program and further plans are in place. These burns will be conducted in areas nominated by traditional owners, who will lead the process in partnership with Forest Fire Management Victoria.

Joe Sambono

Figure 10.1b Grass and trees after a low-intensity burn

cultural burning a traditional practice developed by Indigenous peoples to enhance the health of the land and its people; it includes burning (or prevention of burning) for the health of particular plants and animals

Worksheet 10.1b Joint land management arrangements

Weblink Forest and Reserves Victoria

LEARNING ACTIVITY 10.1B

Joint land management arrangements

Using the Forest and Reserves Victoria website, investigate a joint land management arrangement between Parks Victoria and Indigenous groups and answer the following questions.

1 Explain the joint management strategy across Victoria with Traditional Owners.
2 What are the objectives of Traditional Owner land management boards in Victoria?
3 Select one of the four Traditional Owner groups and explain the environment they manage and the strategies they use to sustain outdoor environments.

10.1.2 NON-INDIGENOUS PEOPLES' LAND MANAGEMENT STRATEGIES

Public land management

public land land managed by governments and their agencies to protect sites of environmental and cultural value, and to provide opportunities for community and recreational use

management plan a document that contains guidelines on how an area of public land is managed; it articulates the vision, goals, outcomes, measures and long-term strategies for parks within planning areas

When we participate in outdoor activities, we often travel to venues that are considered **public land**. As they are not privately owned or managed, these areas are usually under the control of the government and its agencies, such as Department of Energy, Environment and Climate Action (DEECA).

The Victorian Government is responsible for preserving and maintaining public lands in a sustainable manner to ensure that future generations can also enjoy their natural beauty and benefits. As such, DEECA, through its various arms, implements **management plans** and policies that regulate the use and access to these lands.

For example, Parks Victoria is responsible for managing over 4 million hectares of parks, gardens and reserves, which are protected for their scenic, cultural, scientific and recreational values. This includes iconic areas such as the Grampians National Park (Gariwerd), Wilsons Promontory National Park and the Port Phillip Bay Parklands.

Water authorities are responsible for managing Victoria's water resources and protecting water-related ecosystems, while committees of management play a crucial role in the maintenance and management of land held in public trust, such as community parks and public gardens.

In all cases, the aim of public land management is to balance the competing needs and interests of different user groups, while protecting and conserving the natural and cultural values of these lands. The Victorian Government works with local communities and stakeholders to manage public lands in such a way that everyone has the opportunity to enjoy and benefit from them.

Although there are many categories of public land that are managed for a range of purposes, we will concentrate mainly on the type of venues you are likely to visit and enjoy for recreation. The Victorian Government currently holds and manages 8.8 million hectares of public land across Victoria, representing 40% of the state. Of the total government land in Victoria, approximately:

- 7.7 million hectares is reserved Crown land
- 300000 hectares is unreserved Crown land
- 200000 hectares is freehold land
- 600000 hectares is roads (mix of Crown and freehold land).

Crown land is held by the State of Victoria on behalf of the British Crown. Crown land can be either reserved for a particular public use or unreserved, which means it has not been set aside for a particular public use. Most of the land held by the Victorian Government is public land reserved for national parks and state forests.

Crown land
land that is owned by a state/territory government or the Commonwealth

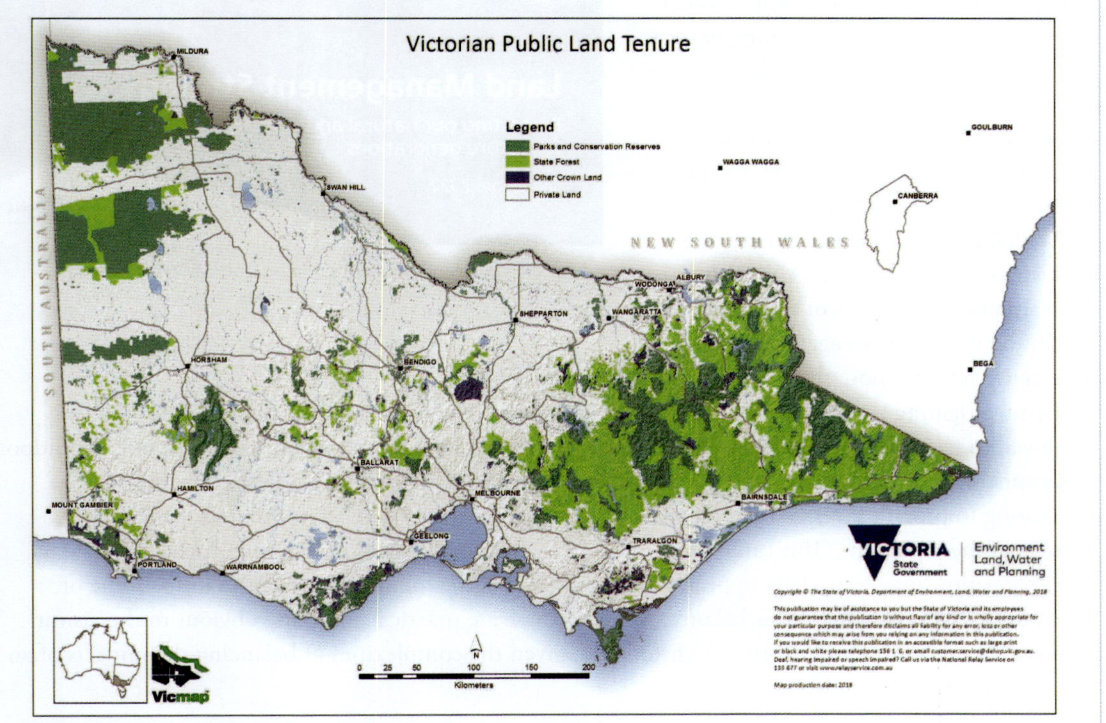

Figure 10.2 A map of Victoria showing the areas of different land ownership

Parks Victoria manages national parks, state parks, coastal parks, marine national parks, marine sanctuaries and wilderness parks reserved under the *National Parks Act 1975* (Cth), as well as metropolitan, reservoir and regional parks, heritage gardens, conservation reserves and recreation reserves managed under other legislation.

Victoria's parks are some of the most precious and valuable assets in the state, preserving its spectacular natural landscapes, rich biodiversity, and important cultural and historic sites. They serve as vital habitats for numerous threatened species and house a significant number of Aboriginal cultural heritage sites, providing a tangible connection to the state's history and traditions. The parks' stunning beauty and recreational opportunities draw almost 100 million visits annually, making them a cornerstone of regional tourism and an important contributor to the overall wellbeing of communities.

Figure 10.3 Parks Victoria's key Land Management Strategy

Which people manage this land and what strategies do they use? Although there are many categories of employees within the management agencies, we often collectively refer to those that enact these strategies as rangers.

By investigating the role of a ranger, we can identify an array of management strategies that are employed to try to balance the impacts of the various interactions that humans have with public outdoor environments. For example, a ranger in a popular national park such as Wilsons Promontory will be attempting to provide a venue for a range of recreational pursuits while attempting to manage the impacts of these pursuits. This can be considerably challenging in areas of high use, such as the Tidal River recreation zone, especially in peak visitation periods. The following sections examine some of the strategies used to achieve this balance. Many of these are practical and quite obvious to the visitor; however, a ranger's job is far from straightforward, given the complexities of balancing the impacts of an array of human interactions.

ROLE OF A PARK RANGER

A ranger's work is as varied as the environments they manage. Many of their responsibilities fall into two categories:

- managing natural value – making sure the natural environment and cultural sites are protected
- visitor services – helping visitors to enjoy and understand parks.

Table 10.2 A summary of management strategies for protecting natural values and enhancing visitor services

Management strategies for natural values	Management strategies for visitor services
• Protect, enhance and manage natural, recreational and cultural assets • Identify weeds and pest animals and eradicate or control those that hold the most threat to native plants and animals • Identify and protect populations of threatened or endangered animals • Assist with fire suppression (firefighting) • Promote and maintain historic assets • Develop cooperative relationships with local Indigenous groups • Work with volunteers on projects such as weed control, maintaining tracks and other infrastructure • Grant permits and oversee researchers studying within the park • Monitor and grant permits to businesses operating within parks, such as hydro-electricity, cafes and beekeeping	• Maintain and develop visitor facilities such as toilets, picnic and camping areas, and trails for various users • Maintain and improve regulatory, interpretive and educational signage • Respond to emergency situations such as 'search and rescue' • Deliver information, events and services to customers, schools and other stakeholders

Adapted from Parks Victoria (c) Victoria State Government

Newspix/Jeff Darmanin

Figure 10.4 A Parks Victoria ranger

> On a broader level, the role of a park ranger involves extensive planning, researching, strategic thinking and people management to effectively balance conservation and recreational values of each asset.
>
> 'Role of a Park Ranger', Parks Victoria

Private land management

The management of private land, which covers nearly 60% of the land in Victoria, is a complex and multifaceted issue that is influenced by several factors. Landowners, as the primary decision-makers, have the ultimate control over how their land is managed, but they are also impacted by organisations that seek to promote healthy outdoor environments. These organisations often provide education, resources and support to landowners, empowering them to make informed decisions about their land.

Many private landowners are passionate and knowledgeable about their properties, and are becoming increasingly active in the management of outdoor environments. They recognise the value of their land and the important role they play in conserving the environment and preserving its natural beauty. With a rise in awareness about environmental issues, many landowners are adopting sustainable and regenerative practices, such as reducing tillage, increasing crop residue retention, regularly soil testing and liming, and building soil carbon. These practices not only improve the health of the land but also benefit the surrounding ecosystems.

Despite limited public access, private lands play a crucial role in the conservation of Victoria's natural resources. They serve as critical habitats for wildlife and play a vital role in maintaining the region's biodiversity. By supporting the efforts of private landowners, organisations and government programs aim to ensure that the health of these lands is preserved for future generations. Land management practices are evolving, with a rise in sustainable and regenerative practices to improve the health of private land across Australia.

Table 10.3 Summary of private land management strategies common to Australian farms

Private land management concern	Private land management strategy
Managing groundcover and soil degradation	Ground cover levels of at least 50% to 70% (depending on location) are needed to protect the soil surface from wind and water erosion. Maintaining good ground cover also improves drought resilience by ensuring pastures can respond quickly to rain.
Managing soil acidity	Up to 50% of Australia's agricultural land has soil acidity, meaning the pH is at or below 5.5. Soil acidity is a natural process accelerated by agriculture. Reducing ammonium-based fertilisers and regular pH testing of soil in addition to liming – applying crushed limestone – neutralises acid in the soil.
Rotating crops and diversifying	Planting a variety of crops can have many benefits, including healthier soil and improved pest control. Crop diversity practices include intercropping (growing a mix of crops in the same area) and complex multiyear crop rotations.
Reducing and eliminating tillage	Traditional ploughing (tillage) prepares fields for planting and prevents weed problems but can cause soil loss. No-till or reduced-till methods, which involve inserting seeds directly into undisturbed soil, can reduce erosion and improve soil health.
Integrating livestock and crops	Combining smart integration of horticulture (fruit and vegetable crops) and agriculture (cow and sheep) can make farms far more efficient and profitable
Adopting agroforestry practices	Planting native trees and shrubs into farms can provide shade and shelter to protect plants, animals and water resources, as well as prevent soil erosion and salinity
Managing water runoff	Keeping year-round growth of native vegetation and plant density greatly improves water runoff
Managing salinity	This involves striking a balance between the volume of water entering and leaving the groundwater system. The water table can be lowered by planting, regenerating and maintaining native vegetation, and maintaining good ground cover

LEARNING ACTIVITY 10.1C 

Land use of environments in your local area

Use the Victorian Resources Online on the Agriculture Victoria website to investigate the land use of environments in your local area of Victoria.

1 What proportion of land use in your selected environment is used for the following purposes?
   - Conservation
   - Forestry
   - Agriculture
   - Horticulture

Worksheet
10.1c Land use of environments in your local area

Weblink
Victorian Resources Online

THE ROLE OF NON-GOVERNMENT ORGANISATIONS IN SUPPORTING PRIVATE LAND MANAGEMENT

Many non-government organisations work with private land holders to assist in long-term land management. For example, they may employ scientists and ecologists to deliver practical on-ground conservation programs based on the latest science. Their ongoing fund-raising projects bring significant investment to target feral animals, weeds and inappropriate fire regimes.

Many conservation organisations are also carrying out cutting-edge scientific research on the private land properties they own and manage.

Organisations that work to manage private land and assist private landowners include the following:

- Australian Wildlife Conservancy – acquires land and works with other landholders to establish sanctuaries for the conservation of threatened wildlife and ecosystems
- Bush Heritage Australia – owns and manages reserves throughout Australia
- Nature Conservancy – supports other conservation non-government organisations to buy and manage high priority land
- Tasmanian Land Conservancy – conserves, enhances and protects Tasmania's natural environment by purchasing and managing land
- Trust for Nature – enables people to bequeath land or money for conservation and for the purchase of Victoria's bush lands.

NOTES FOR THE EXAM 

For the exam, you should:

- understand Indigenous land management strategies to maintain, sustain and range outdoor environments
- understand non-Indigenous land management strategies to maintain, sustain and range outdoor environments
- analyse Indigenous and non-Indigenous land management strategies to maintain and sustain outdoor environments.

10.1 KEY CONCEPTS

- Indigenous land and sea management is also known as 'caring for Country' and aims to protect, maintain, heal and enhance healthy and ecological diverse ecosystems, and productive landscapes.
- Examples of Indigenous land management practices include commercial and economic activities, natural resource management, cultural and spiritual woks, environmental work and education.
- Public land is not privately owned and is managed under the control of governments and their agencies such as the Department of Energy, Environment and Climate Action (DEECA) though Parks Victoria.
- Parks Victoria manages national parks, state parks, coastal parks, marine national parks, marine sanctuaries and wilderness parks.
- Victoria's parks protect many of the state's most precious places – significant and spectacular natural landscapes that have unique and intrinsic value. Parks are the stronghold of the state's flora and fauna, providing critical habitat for many threatened species.
- Public land managers are usually referred to as rangers.
- A ranger's role is as varied as the environments they manage. Many of their responsibilities fall into two categories: managing natural value, to ensure the natural environment and cultural sites are protected; and operating visitor services, to help visitors enjoy and understand parks.
- Public land management strategies include: protect, enhance and manage natural, recreational and cultural assets; identify weeds and pest animals and eradicate or control those that hold the most threat to native plants and animals; identify and protect populations of threatened or endangered animals; and deliver information, events and services to customers, schools and other stakeholders.
- Sixty per cent of land in Victoria is privately owned and includes large areas of land with significant conservation value.
- Private landowners play an important role in land conservation across Victoria.
- Private land management strategies include managing soil acidity, rotating and diversifying crops, eliminating tillage, increasing ground cover, reducing salinity and managing water runoff.
- Organisations that work to manage private land and assist private landowners include Australian Wildlife Conservancy, Bush Heritage Australia, Nature Conservancy and Trust for Nature.

Worksheet
10.1 Key concepts

10.1 CONCEPT QUESTIONS

REMEMBERING

1. Explain the term 'caring for Country'.
2. Outline two Indigenous land management strategies.
3. Outline two non-Indigenous land management strategies.

UNDERSTANDING

4. Describe cultural burns.
5. Describe two non-Indigenous land management strategies observed in an outdoor environment you have visited or studied.
6. Explain the role of a ranger.

APPLYING

7. Analyse how two non-Indigenous land management strategies work to sustain healthy outdoor environments.
8. Using examples, compare and contrast the differences between public and private land management strategies.
9. Analyse how two Indigenous land management strategies work to sustain healthy outdoor environments.

EXTENSION CHALLENGE

10 Research an Indigenous land management strategy that is working with another land manager (public or private) to manage and sustain a specific environment. Cover the following:
- location
- detail of land management strategy
- environment or species aiming to conserve
- analysis of the effectiveness of this strategy.

10.2 ACTS OR CONVENTIONS RELATED TO THE MANAGEMENT AND SUSTAINABILITY OF ENVIRONMENTS

KEY KNOWLEDGE

- Acts or conventions related to the management and sustainability of a specific outdoor environment, species or ecological community, including two of the following:
  - *Flora and Fauna Guarantee Amendment Act 2019* (Vic)
  - Ramsar Convention (international treaty, 1971)
  - *Environment Protection and Biodiversity Conservation Act 1999* (Cth)
  - *Victorian Environmental Assessment Council Act 2001* (Vic)
  - *Planning Environment Act 1987* (Vic)

KEY SKILLS

- evaluate the effectiveness of specific Acts and conventions related to managing and sustaining outdoor environments
- propose changes to current Acts and conventions to further improve the health of a specific environment, species or ecological community

Environmental Acts and conventions

Environmental Acts or conventions are measures implemented by governments or their agencies and are designed to reduce human impact on outdoor environments. In this section, we examine two prescribed Acts or conventions and how each functions to manage and sustain a specific species and/or ecological community.

Environmental Acts measures implemented by governments or their agencies designed to reduce human impact on outdoor environments

In Australia, there are three levels of government: federal, state/territory and local. Each level has the authority to create and enforce laws in the areas under their jurisdiction. These laws may take the form of by-laws for local government authorities or Acts of Parliament for state and federal governments. Legislation is enacted when the government develops policies on specific issues and formalises them through the legislative process.

A treaty is a written agreement between nations that establishes a relationship governed by international law. Both parties agree to abide by the terms of the treaty, and if either fails to fulfil their obligations, they may be held accountable under international law. Treaties may also be referred to as **conventions**, although this term is sometimes used for agreements prior to the formal execution of a treaty.

conventions agreements used in multilateral government decisions

In this section, we consider some examples of Acts and conventions that relate directly to the natural environment.

10.2.1 FLORA AND FAUNA GUARANTEE AMENDMENT ACT 2019 (VIC) (FFGAA)

This Act amended the *Flora and Fauna Guarantee Act 1988* to provide a modern and strengthened framework for the protection of Victoria's biodiversity. It is the primary Victorian legislation for safeguarding endangered species and ecosystems while effectively addressing potential threats.

The FFGAA underscores the significance of proactive measures to prevent additional species from facing endangerment. The Act underscores the value of collaborative strategies for biodiversity preservation and acknowledges the vital role of government agencies and the community in conservation efforts.

OBJECTIVES OF THIS ACT

i To guarantee that all taxa of Victoria's flora and fauna, other than taxa specified in the Excluded list, can persist and improve in the wild and retain their capacity to adapt to environmental changes; and

ii to prevent taxa and communities of flora and fauna from becoming threatened and to recover threatened taxa and communities so their conservation status improves; and

iii to protect, conserve, restore and enhance biodiversity, including -

- A flora and fauna and their habitats; and
- B genetic diversity; and
- C ecological communities; and
- D ecological processes; and

iv to identify and mitigate the impacts of potentially threatening processes to address the important underlying causes of biodiversity decline; and

v to ensure the use of biodiversity as a natural resource is ecologically sustainable; and

vi to identify and conserve areas of Victoria in respect of which critical habitat determinations are made.

Authorised Version *Flora and Fauna Guarantee Amendment Act 2019* No. 28 of 2019, www.Legislation.vic.gov.au

CHANGES UNDER THE FFGAA

The Amendment Act:

- introduces principles to guide the implementation of the FFG Act, including consideration of the rights and interests of Traditional Owners and the impacts of climate change
- requires consideration of biodiversity across government to ensure decisions and policies are made with proper consideration of the potential impacts on biodiversity
- clarifies existing powers to determine critical habitat and improves their protection by encouraging cooperative management
- gives effect to a consistent national approach to assessing and listing threatened species using the Common Assessment Method (CAM), which will reduce duplication of effort between jurisdictions and facilitate the monitoring and reporting of species' conservation status
- modernises the FFG Act's enforcement framework, including stronger penalties.

...

Species will now be listed under the FFG Act in the following categories of threat:

- extinct
- extinct in the wild
- critically endangered
- endangered
- vulnerable
- conservation dependent.

Victoria's Framework for Conserving Threatened Species © State Government of Victoria 2023

ACTION STATEMENTS

Once a species, community or threatening process is listed, the *Flora and Fauna Guarantee Amendment Act 2019* requires that an **action statement** be prepared by the Department of Energy, Environment and Climate Action (DEECA).

AAP Photo/Melbourne Zoo

Figure 10.5 The Baw Baw frog (*Philoria frosti*), listed as critically endangered under the FFGAA

action statement
a detailed document that describes the recommended conservation management practices for specific threatened species

Action statements provide information on a species, reasons for their decline, and threats that affect it. The purpose is to report on past management actions and to establish a set of new management actions. It may also include additional information on what needs to occur in the future to conserve and manage a **taxon** or community, or a threatening process.

taxon
a group of any taxonomic rank (such as a species, family or class); taxa are ranked based on their shared traits – the lower down the hierarchy you go, the more specific the grouping characteristics become

Specifically, action statements contain:

- a summary of our knowledge of the listed species including description, distribution, habitat, life history and ecology and important populations
- conservation status both state-wide and federally
- key threats to populations; for example, loss of habitat, habitat degradation, inappropriate fire regimes, and pest plants and animals
- a summary of past management actions
- intended management actions to help secure populations and enable their long-term persistence
- other considerations including socioeconomic matters and climate change implications.

Management plans will be developed for taxa, communities or threatening processes that require a more in-depth assessment of conservation requirements.

Action statements resulting from the FFGAA describe the conservation strategies that should be applied to protect native fauna and flora such as the Baw Baw frog (see Figure 10.5).

LEARNING ACTIVITY 10.2A

Species investigation

Using the DEECA website, navigate to the Action statement for Victoria's faunal emblem, Leadbeater's possum.

1. What is the current Victorian and national conservation status of this species?
2. Briefly describe this species, including an image.
3. Briefly describe the distribution where this species is located, and the type of habitat it lives in.
4. Suggest reasons for its extinction or decline in numbers.
5. Outline three threats to these species.
6. Outline three management actions (including who carries them out) that can be adopted to assist these species.
7. Analyse the effectiveness of this action plan in sustaining outdoor environments.
8. Propose two changes to improve the management of the Leadbeater's possum.

Worksheet
10.2a Species investigation

Weblink
DEECA

Ramsar Convention (international treaty, 1971)

The Ramsar Convention on Wetlands of International Importance is the earliest of the contemporary global intergovernmental environmental agreements. During the 1960s, countries and non-governmental organisations were worried about the escalating loss and deterioration of wetland ecosystems, particularly for migratory waterbirds. Therefore, the treaty was negotiated and eventually adopted in Ramsar, Iran in 1971. It was put into effect in 1975.

The Ramsar Convention is an international (or intergovernmental) treaty that 'provides the framework for national action and international cooperation for the conservation and wise use of wetlands and their resources'. Wetlands are defined in the treaty as 'swamps, marshes, billabongs, lakes, salt marshes, mudflats, mangroves, coral reefs, fens, peat bogs or bodies of water – whether natural or artificial, permanent or temporary'. Wise use refers to 'the conservation and sustainable use of wetlands and their resources, for the benefit of humankind'.

Under the three pillars of the convention, the parties have committed themselves to:

- work towards the wise use of all their wetlands through national land-use planning, appropriate policies and legislation, management actions and public education
- designate suitable wetlands for the List of Wetlands of International Importance ('Ramsar List') and ensure their effective management
- cooperate internationally concerning **transboundary wetlands**, shared wetland systems, shared species, and development projects that may affect wetlands.

transboundary wetlands
wetlands that exist across one or more borders

The purpose of the treaty is to ensure migratory birds have habitats maintained at both ends of the migratory path, and that countries have designated wetlands of international significance, which they protect. For a wetland to be designated to this list, it must satisfy one or more of the nine criteria for identifying wetlands of international importance. These can be found on the Australian Government Ramsar website.

Weblink
Ramsar

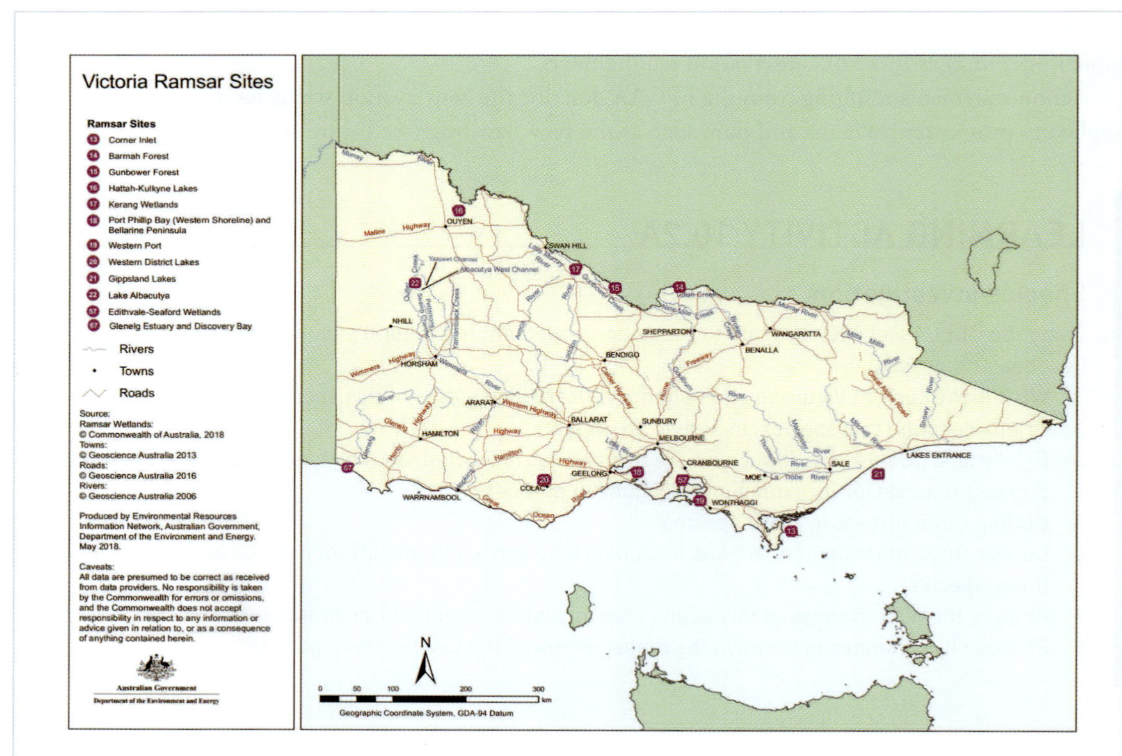

Figure 10.6 Victorian Ramsar sites

RAMSAR SITES

There are 12 Ramsar sites in Victoria (see Figure 10.6):

1 Barmah Forest
2 Corner Inlet
3 Edithvale-Seaford Wetlands
4 Gippsland Lakes
5 Glenelg Estuary and Discovery Bay
6 Gunbower Forest
7 Hattah-Kulkyne Lakes
8 Kerang Wetlands
9 Lake Albcutya
10 Port Phillip Bay (western shoreline) and Bellarine Peninsula
11 Western District Lake
12 Western Port.

Each Ramsar site in Victoria has:

- an ecological character description, which documents the ecological character of the Ramsar site at the time it was listed under the Ramsar Convention
- a Ramsar information sheet, which provides essential data on the Ramsar site and is required by the Ramsar Convention
- a Ramsar site management plan, which sets out the management objectives and strategies for each Ramsar site. There has been a recent process of renewing site management plans. For seven sites, the management plans were embedded in regional waterway strategies that were developed by relevant catchment management authorities over 2013–14. The other four Ramsar sites have stand-alone plans; three of these were renewed over 2015–16, and one is currently in the process of being renewed.

CASE STUDY

GLENELG ESTUARY AND DISCOVERY BAY

Situated approximately 340 kilometres west of Melbourne on the border with South Australia, the Glenelg Estuary and Discovery Bay Ramsar Site was listed in 2018.

Glenelg Estuary and Discovery Bay:

- has diverse aquatic habitats, including intertidal sandy beaches, estuarine habitat, freshwater swamps and permanent lakes
- supports nationally threatened coastal saltmarsh, and eight nationally or internationally listed species of conservation significance, such as the eastern curlew and Australasian bittern, which are both critically endangered
- provides feeding, spawning and nursery habitat for 28 fish species
- supports 24 bird species that migrate annually from the northern hemisphere
- has rare geological and geomorphic features.

This Ramsar site has a range of distinct wetland types that are both hydrologically and ecologically connected. The connection between the marine, estuarine and freshwater components is significant for fish migration and reproduction. This includes the mulloway (or butterfish), which feeds in the Glenelg Estuary and migrates up to 400 kilometres to the Murray Mouth to spawn. The site also supports a large population of the ancient greenling (*Hemiphlebia mirabilis*), the only member of this damselfly family globally.

Resource
Case Study: Glenelg Estuary and Discovery Bay

The Gunditjmara Aboriginal people have a living association with the Ramsar site, which has great cultural significance for them, as it is part of their Koonang (sea) and Bocara Woorrowarook (river forest) country. The site also supports recreational and tourism activities, including sightseeing, walking, camping and recreational fishing.

The site covers 22289 hectares and includes part of the Lower Glenelg National Park, Discovery Bay Coastal Park and all the Nelson Streamside Reserve, which are managed by Parks Victoria.

Management strategies aim to:

- manage visitor activities in the Lower Glenelg National Park and the Discovery Bay Coastal Park to minimise disturbance of shorebirds and beach nesting birds
- investigate options to mitigate the impacts of climate change (sea level rise) on coastal habitats and improve resilience
- protect high-priority locations from extensive shoreline erosion
- develop and implement measures to control carp within the Glenelg Estuary and prevent movement into the Long Swamp Complex.

Jean-Marc La Roque/AUSCAPE

Figure 10.7 Discovery Bay

Worksheet 10.2b Ramsar research

Weblink DEECA

LEARNING ACTIVITY 10.2B

Ramsar research

Choose a Ramsar site from the list of 12 above that is close to your school or to an environment that you have visited or studied.

Using the DEECA website, navigate to the Ramsar site you have selected and respond to the following:

1. Where is this Ramsar site located? Include a map.
2. Who manages this Ramsar site?
3. Outline the species conserved within this Ramsar site.
4. Outline three threats to these species.
5. Outline three management actions (including who carries them out) that can be adopted to assist these species.
6. Evaluate the effectiveness of the Ramsar convention in sustaining outdoor environments.

Environmental Protection and Biodiversity and Conservation Act 1999 (Cth)

The *Environment Protection and Biodiversity Conservation Act 1999* (Cth) (EPBC) is the key environmental legislation in the Australian government. According to the Department of Climate Change, Energy, the Environment and Water, EPBC provides a critical legal framework 'to protect and manage nationally and internationally important flora, fauna, ecological communities and heritage places – defined in the EPBC Act as matters of national environmental significance'.

1 The objects of this Act are
- i to provide for the protection of the environment, especially those aspects of the environment that are matters of national environmental significance; and
- ii to promote ecologically sustainable development through the conservation and ecologically sustainable use of natural resources; and
- iii to promote the conservation of biodiversity; and to provide for the protection and conservation of heritage; and
- iv to promote a co-operative approach to the protection and management of the environment involving governments, the community, land-holders and indigenous peoples; and
- v to assist in the co-operative implementation of Australia's international environmental responsibilities; and
- vi to recognise the role of Indigenous people in the conservation and ecologically sustainable use of Australia's biodiversity; and
- vii to promote the use of Indigenous peoples' knowledge of biodiversity with the involvement of, and in co-operation with, the owners of the knowledge.

Environment Protection and Biodiversity Conservation Act 1999
No. 91, 1999, Federal Register of Legislation, www.Legislation.vic.gov.au

WHAT IS PROTECTED UNDER THE EPBC

The EPBC provides legislative protection over the following:
- World Heritage areas
- national heritage places
- wetlands of international importance (listed under the Ramsar Convention)
- listed threatened species and ecological communities
- listed migratory species (protected under international agreements)
- Commonwealth marine areas
- Great Barrier Reef Marine Park
- nuclear actions (including uranium mines)
- water resources (that relate to coal seam gas development and large coal mining development).

NOMINATING A SPECIES, ECOLOGICAL COMMUNITY OR KEY THREATENING PROCESS UNDER THE EPBC

Anybody can nominate a native species, ecological community or threatening process for listing under any of the categories specified in the EPBC.

An agreement has been struck between the federal government and all states and territories in Australia to establish a common assessment method to determine whether a species is threatened.

To make this determination, the Threatened Species Scientific Committee undertakes a rigorous scientific assessment of the species or ecological community's threat status.

Resource
Case Study: Environment Protection and Biodiversity Conservation Act 1999

CASE STUDY

ENVIRONMENT PROTECTION AND BIODIVERSITY CONSERVATION ACT 1999

White Star-Bush

EPBC ACT LISTING STATUS – CRITICALLY ENDANGERED; DATE EFFECTIVE 5 OCTOBER 2022

Neil Maclachlan / Alamy Stock Photo

Figure 10.8 Emerald Star Bush

Description

The white star-bush, also known as the emerald star-bush, is a slender upright shrub growing to 1.5 metres tall, with dark green circular leaves, reaching 30 × 10 mm in size. White star-bush typically flowers from early October to late November. The taxon is a fire-sensitive obligate seed regenerator (OSR). Intense fire therefore results in mortality.

Distribution

The taxon is endemic to Victoria and is only known from the south-eastern slopes of the Dandenong Ranges from Belgrave, Monbulk, Emerald and Avonsleigh.

Threats

- Habitat loss and habitat degradation in response to agricultural and urban intensification
- Fire management activity including mineral earth breaks, repeat slashing and high-frequency fuel reduction burning on both private and public land

Conservation and management priorities

- Implement formal conservation arrangements, management agreements and/or covenants on private land. For crown and private land, investigate inclusion of habitat into reserves if possible.
- Investigate invasive species impacts (including from grazing, trampling and predation).
- Identify and remove weeds in the local area of known sites using appropriate methods and manage sites to prevent introduction of invasive weeds that threaten or could become a threat to the white star-bush.
- Fire impacts: develop and implement a suitable fire management strategy for the white star-bush.

Adapted from 'Conservation Advice Asterolasia asteriscophora subsp. Albiflora', 21 April 2022

Victorian Environmental Assessment Council Act 2001

The Victorian Environmental Assessment Council (VEAC) was established under the *Victorian Environmental Assessment Council Act 2001*. The Council comprises five members, who are required to have a diverse range of experience, skills and knowledge in areas related to the management of public land and natural resources.

The Council's role is to conduct investigations and assessments, and to provide expert advice to the Victorian Government on the protection and sustainable management of the environment and natural resources on public land. Any investigations and recommendations must consider social impacts, resource use and the needs of the environment. Consultation with the community and community responses to reports are key requirements of any investigation undertaken by VEAC. The VEAC serves as a source of guidance and recommendations, but does not have any decision-making authority.

VEAC replaced the Environment Conservation Council (ECC), which in turn succeeded the Land Conservation Council (LCC). The organisation's evolution reflects the shifting focus towards preserving the environment and natural resources for future generations.

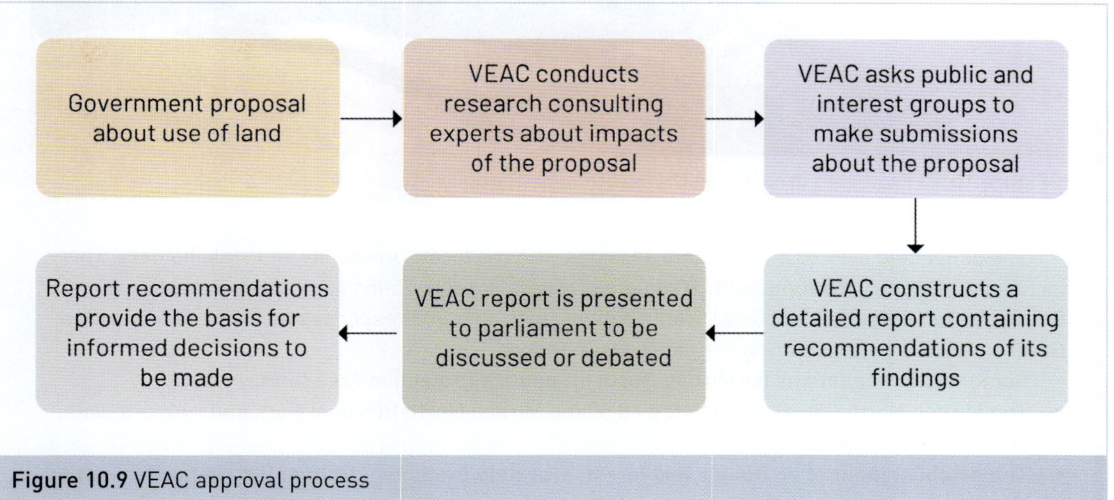

Figure 10.9 VEAC approval process

CASE STUDY

ASSESSMENT OF THE VALUES OF IMMEDIATE PROTECTION AREAS IN THE STRATHBOGIE RANGES AND MIRBOO NORTH, JULY 2022

In November 2021, the Victorian Environmental Assessment Council (VEAC) was requested by the Minister for Energy, Environment and Climate Change to carry out an assessment of the values of state forests in the areas identified as Immediate Protection Areas (IPAs) in the Strathbogie Ranges and Mirboo North.

There were five purposes of the assessment:

1 Identify the biodiversity, ecological, and geological and geomorphological values of the specified area.
2 Identify the cultural heritage, social and economic values of the specified area.
3 Identify the current and likely future threats to those values, including climate change.
4 Identify the typical land use categories commensurate with the identified values.
5 Assess the potential economic implications of proposed land use changes recommended by the EPCE and provided to the Council.

Resource
Case Study: Protection areas in the Strathbogie Ranges and Mirboo North

Alamy Stock Photo/Jason Edwards

Figure 10.10 Protected area within the Strathbogie Rangers

The panel held over 50 engagement sessions with the local community and stakeholders. This included on Country meetings with Traditional Owners, local drop-ins and targeted meetings. Engage Victoria also saw important contributions, with over 700 responses received. The Victorian Government is now considering the panel's final report.

Some recommendations for Mirboo North Immediate Protection Area (IPA):

- The Mirboo North IPA, along with the adjoining Mirboo North Regional Park and Lyrebird Walk, are significant forests.
- The establishment of an IPA for this forest, providing protection from native timber harvesting, represents significant efforts by local community members and groups to advocate for their local forest.
- An assessment of the status of the Strzelecki koala should take place to inform recognition of the Strzelecki koala as a distinct genetic unit within the Victorian conservation framework, including future management actions and conservation advice.

Planning Environment Act 1987 (Vic)

Legislated in the Victorian parliament in 1987, the *Planning Environment Act* established a central framework for planning the use, development and protection of land in Victoria.

The Act outlines the procedures for preparing and amending the Victoria Planning Provisions and planning schemes, including how to obtain permits under schemes, settle disputes, enforce compliance with planning schemes and permits, and other administrative procedures.

According to Planning Victoria, the main functions of the Act are to:

- set the broad objectives for planning in Victoria
- set the main rules and principles for how the Victorian planning system works
- set up the key planning procedures and legal instruments in the Victorian planning system
- define the roles and responsibilities of the minister, councils, government departments, the community and other stakeholders in the planning system.

OBJECTIVES OF PLANNING IN VICTORIA

The objectives of planning in Victoria are set out in the PE Act. They are:

i to provide for the fair, orderly, economic and sustainable use and development of land;
ii to provide for the protection of natural and man-made resources and the maintenance of ecological processes and genetic diversity;
iii to secure a pleasant, efficient and safe working, living and recreational environment for all Victorians and visitors to Victoria;
iv to conserve and enhance those buildings, areas or other places which are of scientific, aesthetic, architectural or historical interest, or otherwise of special cultural value;
v to protect public utilities and other assets and enable the orderly provision and coordination of public utilities and other facilities for the benefit of the community;
vi to facilitate development in accordance with the objectives set out in paragraphs a), b), c), d) and e);
vii to facilitate the provision of affordable housing in Victoria;
viii to balance the present and future interests of all Victorians.

Planning and Environment Act 1987 © Copyright State Government of Victoria

The Act has undergone several amendments since its inception. The most recent amendment was the *Planning and Environment Amendment (Distinctive Areas and Landscapes) Act 2018*. This amendment establishes the declaration of distinctive areas and landscapes and requires the preparation and implementation of a Statement of Planning Policy for each declared area to ensure consistent decision-making by public entities. The amendment serves to preserve and protect the unique character and beauty of these areas for future generations.

WHY CONSIDER ENVIRONMENT PROTECTION IN PLANNING?

Effective land use planning is crucial for protecting the environment and avoiding adverse impacts on communities. The surrounding environment can also have an impact on land uses, such as contamination from land and groundwater, or the migration of landfill gas. By considering these potential risks in the planning process, land use planning helps to identify and mitigate harmful outcomes, promoting sustainable and responsible land use practices.

WHAT ARE GREEN WEDGES?

Melbourne's planning policies have tried to protect non-urban areas for over 40 years. Early policy documentation stated:

> Land use, resources, terrain, vegetation and habitat vary extensively throughout the non-urban areas. It is intended that the basic attributes and resources contained within the areas shall be preserved to a maximum degree, and that environment management policies shall be specifically oriented towards this objective.
>
> Melbourne and Metropolitan Board of Works, page 54 (1971)

Since the early 1970s, planning policies have worked to protect the economic, social and environmental landscape values of Victoria's non-urban areas through what are known as 'green wedge' areas.

green wedge land defined in a metropolitan fringe planning scheme as being outside an Urban Growth Boundary

Green wedges are the non-urban or built-up areas of metropolitan Melbourne that lie outside the Urban Growth Boundary. These designated areas effectively form a ring of green spaces around Greater Melbourne. There are 12 green wedge areas across 17 municipalities, which cover a wide range of different uses and unique environments, from the coastal areas of the Mornington Peninsula to the green hills of the Yarra Valley.

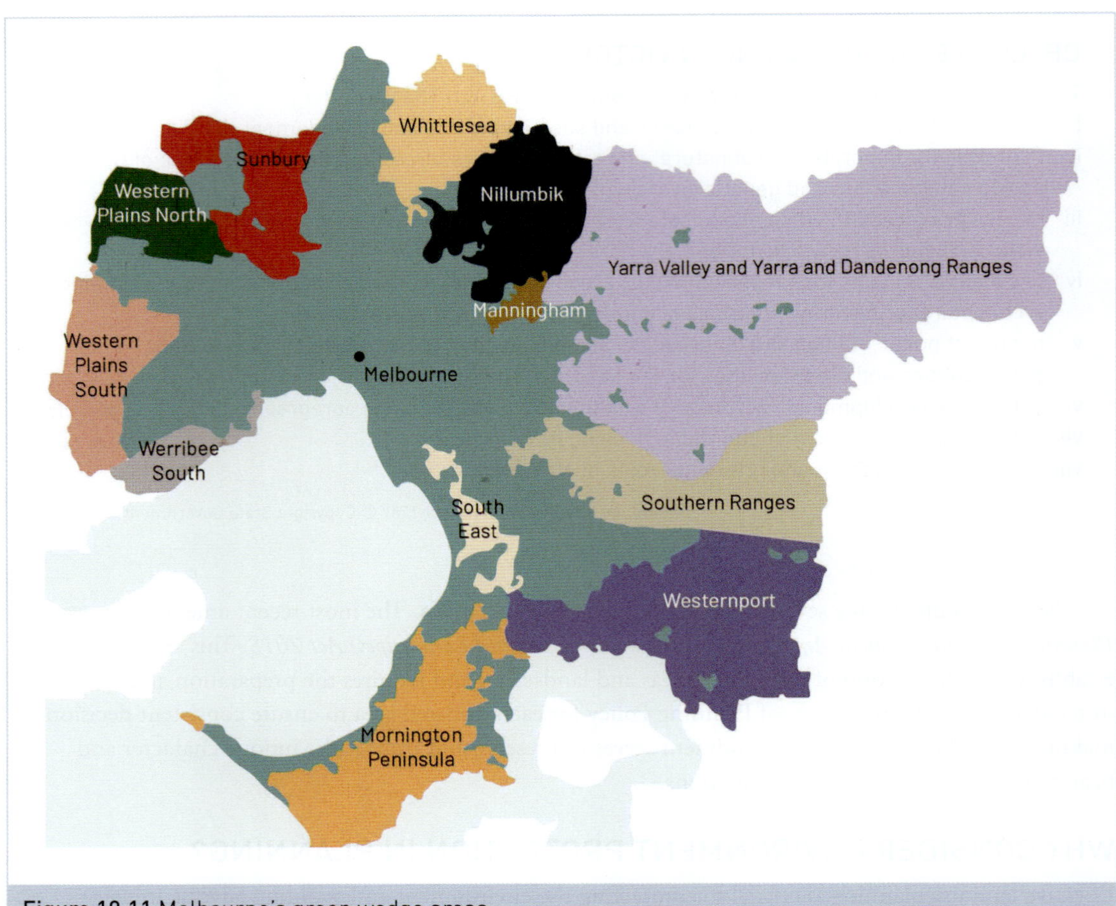

Figure 10.11 Melbourne's green wedge areas

Green wedge areas contain a mix of agriculture and low-density activities such as:

- major infrastructure that supports urban areas, including airports and water treatment facilities
- major quarries used in the building industry
- cultural heritage sites
- biodiversity conservation areas
- water catchments.

About one-third of the total green wedge area is public land, including national parks, other parks, reserves and closed protected water catchments.

Figure 10.12 Green wedges provide important green spaces across greater Melbourne

LEARNING ACTIVITY 10.2C 

Green wedge research task

Visit the Planning page on the Victorian Government webpage and search for 'green wedge'. Research a green wedge area near you, or an environment of interest to you, and write a report that includes:

1 location and map of green wedge area
2 land use history of this location
3 environmental pressures to this green wedge area
4 significant flora and fauna in this green wedge area
5 summary of the management plan for this green wedge area.

Your research could be displayed using a visual presentation tool such as Prezi.

Worksheet 10.2c Green wedge research task

Weblink Vic Gov Planning

Australia's Strategy for Nature 2019–2030

Although not a prescribed Act or convention, Australia's national biodiversity strategy and action plan, *Australia's Strategy for Nature 2019–2030*, serves as an important guide for nature conservation across the country. This overarching framework brings together national, state and local strategies, laws, policies and actions aimed at preserving nature. The strategy is based on scientific principles, but also incorporates an innovative approach that goes beyond simply protecting nature. Instead, it focuses on promoting adaptation, resilience and responsible natural resource management in urban, rural and natural environments, both on land and at sea.

NATIONAL GOALS

The strategy has three priority focus areas, or goals, underpinned by 12 objectives. The goals work together in continuous loops designed to reinforce each other. By connecting people with nature, we enhance their desire to care for nature, which in turn builds knowledge that can be shared to improve our care for nature and the benefits we receive from connecting with nature. Each objective has a number of progress measures, which will be used to track and report on the success of the Strategy.

Goal 1: Connect all Australians with nature

Most Australians care about nature. However, modern life means many of us have become disconnected from nature in our daily lives and sometimes we forget its importance. Connecting Australians with nature is essential to our long-term mental and physical health, economic prosperity and national identity. Indeed, maintaining the connection with nature is central to Aboriginal and Torres Strait Islander culture and identity.

Australia's Strategy for Nature 2019-2030 © Commonwealth of Australia 2019

Goal 2: Care for nature in all its diversity

Nature is an asset from which all Australians benefit. While building the resilience of our native species and the health of our landscapes, seascapes and aquatic environments is challenging, the benefits of success greatly outweigh the costs.

Australia's Strategy for Nature 2019-2030 © Commonwealth of Australia 2020

Weblink
Australia's Nature Hub

Goal 3: Share and build knowledge

Effective management and protection of nature in Australia is best supported by an evidence-based approach built on sound knowledge. Decisions should draw on scientific information and data, including from biodiversity, taxonomic, ecological and sociological sciences, nature management methods and approaches, and traditional ecological values and knowledge. There is significant effort around Australia to extend this knowledge base, and to monitor, evaluate and report on actions at regional and national levels. Australia's Nature Hub has a 'submit an action' function where the public can make recommendations or observations on Australia's nature. The website also has interactive links to policies, programs and legislation that all contribute to enhance Australia's strategy for nature.

Australia's Strategy for Nature 2019-2030 © Commonwealth of Australia 2021

PROPOSED CHANGES TO ACTS OR CONVENTIONS

The Acts or conventions selected for this study contribute to the management and sustainability of outdoor environments. Each Act or convention has its own specific goals and objectives, as well as established processes and action or management plans. These Acts and conventions each play a vital role in balancing the needs of humans and conservation efforts to sustain the environment. Further research into two of these Acts and conventions is required to propose changes and improvements to enhance conservation efforts.

Some examples of proposed changes to enhance conservation efforts may include:

- employing specialised rangers or expertise dedicated to ensure the survival of critically endangered species or ecological communities
- improving surveillance and monitoring of environments to ensure management strategies are being adhered to
- increasing funding to monitor species, ecological communities or environments listed within each Act or convention
- increasing habitat for species listed within each Act or convention
- revisiting the action or management plan for further strategies
- increasing captive breeding of listed species
- enhancing protection measures.

Auscape International Pty Ltd / Alamy Stock Photo

Figure 10.13 Once thought to be extinct, the critically endangered Leadbeater's possum was rediscovered in 1961.

LEARNING ACTIVITY 10.2D

Worksheet
10.2d Acts and conventions

Investigate Acts and conventions

Research two of the Acts or conventions discussed in this chapter that could be applied to environments you've visited or studied. For each Act or convention, answer the following questions:

1. What is the name of the Act or convention and what date did it come into force?
2. What are its key aims or objectives?
3. What locations does it apply to?
4. Who enforces it?
5. How does it sustain the outdoor environment? (Be specific – research a species or ecological community, including the name of species, its listing status, threats to the species or community, and its management plan.)
6. Analyse how effective it is in its objectives to sustain specific outdoor environments.
7. Propose changes to this act or convention to further sustain the species or ecological community you have researched.

NOTES FOR THE EXAM

For the exam, you should:

- understand two Acts or conventions from the prescribed list
- understand how each Act of convention manages and sustains a specific outdoor environment, species or ecological community
- propose changes of each Act or convention to further improve the health of a specific environment, species or ecological community.

10.2 KEY CONCEPTS

- Environmental Acts or conventions are measures implemented by governments or their agencies and are designed to reduce human impact on outdoor environments.
- The *Flora and Fauna Guarantee Amendment Act 2019* (FFGAA) is designed to provide a modern and strengthened framework for the protection of Victoria's biodiversity.
- The three objectives of the FFGAA are to:
  - guarantee that all taxa of Victoria's flora and fauna can persist and improve in the wild and retain their capacity to adapt to environmental changes
  - prevent taxa and communities of flora and fauna from becoming threatened and to recover threatened taxa and communities so their conservation status improves
  - protect, conserve, restore and enhance biodiversity.
- Action statements are detailed documents that describe the recommended conservation management practices for specific threatened species.
- The Ramsar Convention 1971 is an international (or intergovernmental) treaty that provides the framework for national action and international cooperation for the conservation and wise use of wetlands and their resources.
- Three pillars of the Ramsar convention are to:
  - work towards the wise use of all their wetlands through national land-use planning, appropriate policies and legislation, management actions and public education
  - designate suitable wetlands for the List of Wetlands of International Importance

  - cooperate internationally concerning transboundary wetlands, shared wetland systems, shared species and development projects that may affect wetlands.
  - The *Environment Protection and Biodiversity Conservation Act 1999* (EPBC) is the Australian Government's central piece of environmental legislation and provides a legal framework to protect and manage nationally and internationally important flora, fauna, ecological communities and heritage places.
- Three objectives of the EPBC are to:
  - provide for the protection of the environment, especially those aspects of the environment that are matters of national environmental significance
  - promote ecologically sustainable development through the conservation and ecologically sustainable use of natural resources
  - promote the conservation of biodiversity.
- Using the Common assessment method, species can be listed as: Vulnerable, Endangered, Critically Endangered, Extinct in the Wild, Extinct or Conservation Dependent.
- The Victorian Environmental Assessment Council (VEAC) was established under the *Victorian Environmental Assessment Council Act 2001*.
- Role of the VEAC is to conduct investigations and assessments, and provide advice to the Victorian Government relating to the protection and ecologically sustainable management of the environment and natural resources of public land.
- The *Planning Environment Act 1987* (Vic) establishes a framework for planning the use, development and protection of land in Victoria.
- Three main functions of the *Planning Environment Act* are to set the:
  - broad objectives for planning in Victoria
  - main rules and principles for how the Victorian planning system works
  - key planning procedures and legal instruments in the Victorian planning system.
- Changes to enhance conservation efforts may include:
  - specialised rangers or expertise dedicated to ensure the survival of critically endangered species or ecological communities
  - improved surveillance and monitoring of environments to ensure management strategies are being adhered to
  - increased funding to monitor species, ecological communities or environments listed within each act or convention.

Worksheet 10.2 Key concepts

10.2 CONCEPT QUESTIONS

REMEMBERING

1 Identify two Acts or conventions you have studied, including the full name and date of each.
2 For each of these Acts or conventions, explain three of their aims or objectives.
3 Outline the common classification of species, ecological communities or environments in Australia.
4 Outline three improvements that could enhance conservation efforts of species and ecological communities.

UNDERSTANDING

5 Choose one Act or convention related to an environment you have visited or studied.
6 Research two species or ecological communities this Act or convention manages, including:
  - name of species or community
  - habitat and/or geographical distributions
  - listing status
  - threats
  - management and action plans.

APPLYING

7 Referencing the management or action plan for your selected species or community from Question 4, evaluate the effectiveness of this Act or convention in managing and sustaining a specific outdoor environment you have visited or studied.

EXTENSION CHALLENGE

8 Referencing the management or action plan for your selected species or community from Question 4, propose changes to this Act or convention to better manage the species or ecological community within the specific outdoor environment you have visited or studied.

Questions you may consider:
- What practical measures could increase conservation efforts?
- How may these conservation efforts be monitored or measured?
- How can the proposed changes assist to remove a species from a listing status?
- Are there examples of measurable changes in other countries for improved conservation management that could be applied to the species or community you have investigated?

10.3 COMMUNITY ACTIONS UNDERTAKEN TO SUSTAIN HEALTHY OUTDOOR ENVIRONMENTS

KEY KNOWLEDGE

- community actions undertaken to sustain healthy outdoor environments, including two of the following:
  - regenerative farming
  - Trust for Nature
  - Landcare
  - community groups such as 'Friends of ...'

KEY SKILLS

- evaluate the effectiveness of community actions undertaken to sustain healthy outdoor environments

Community action

Community groups work to conserve and improve the health of the environment by educating and raising awareness among the public, advocating for conservation policies and undertaking practical projects that enhance the environment. Community groups often focus on a specific area or issue, such as wildlife conservation, habitat protection or the restoration of degraded areas. These groups play a vital role in protecting the environment by providing on-the-ground support, engaging with local communities and collaborating with other organisations and government agencies to achieve common goals. They can also bring local knowledge and perspectives to conservation initiatives, making them more effective and relevant to the community. The support and involvement of community groups is critical in the overall success of conservation efforts and the protection of Victoria and Australia's outdoor environments.

In this section we need to examine two of the four prescribed community actions and evaluate how each community action promotes and sustains outdoor environments.

Regenerative farming

Australia's agriculture history is marked by extensive clearing of forests, overgrazing and soil degradation. In times of drought, the land becomes extremely vulnerable to erosion and soil loss, transforming into a dust bowl. Dominant non-Indigenous land management practices have left the environment fragile, and traditional farming practices have been proven to be unsustainable, given Australia's nutrient-poor soils.

monoculture
the cultivation of a single crop in an area

regenerative farming
a system of farming principles and practices that seeks to rehabilitate and enhance the ecosystem of the farm by placing a premium on soil health

Much of Australia's farming processes are based on a **monoculture** approach. This involves compartmentalising landscapes and manipulating them to mass-produce a particular species for human consumption. Crops, horticulture, pasture for livestock (including the dairy and wool industries) and aquaculture are all usually based on this approach.

Regenerative farming is a system that prioritises soil health and encompasses a range of practices aimed at rehabilitating and enhancing the ecosystem of the farm. It focuses on aspects such as water management, fertiliser use, carbon sequestration and biodiversity. By doing so, it promotes increased yields, higher health and vitality for farming communities, and improved resilience to climate instability. In addition to capturing carbon and aboveground biomass, regenerative farming works to enhance the ecosystem services, enrich the soil and improve watersheds.

To make improvements in soil structure and quality, and reduce topsoil erosion, is to change the way we manage our landscape. Having an appropriate amount of ground cover in the form of grass and vegetation helps bind soils together and protects the topsoil from wind erosion. Once the ground cover is gone, the landscape becomes a dust bowl. Denuded landscapes become further susceptible to animal plagues such as locusts. Regenerative farming entails a change in perspective, moving away from the traditional approach of relying heavily on fertilisers and other inputs to boost crop yields. Instead, it prioritises the adoption of sustainable farming practices that align with the capacity of the farm and the type of crops being produced. Without regenerative farming methods, producing one calorie of food requires the use of 10 calories of fossil fuels. Regenerative farming places a strong emphasis on soil health, biodiversity, preserving native grass cover and implementing resilient systems within the farm. As a result, it promotes natural solutions to pest problems and reduces the impact of salinity.

Table 10.4 Regenerative farming practices and their benefits

Examples of regenerative farming practices	Benefits of regenerative farming practices
Applying organic composts, fertilisers	Reducing input costs
Implementing time-controlled planned grazing	Improved soil health – structural
Fencing off waterways and implementing water reticulation for stock	Regenerating, rather than degrading, the natural resource base
Investing in revegetation	Building a landscape that is more resilient, especially to climate extremes (such as flood, drought and fire), and is able to recover more quickly
Changing crop rotations	Increased productivity, leading to increased profits
Incorporating green manure or under-sowing of legumes	Facilitating healthy nutrient cycling
Reducing or ceasing synthetic chemical inputs	Producing healthier, more nutritious food and livestock, and therefore healthier people

Shutterstock/William Edge

Figure 10.14 Grass fed cattle grazing in south-west Victoria

Trust for Nature

Established in 1972, Trust for Nature is a not-for-profit organisation that works to protect native plants and wildlife in cooperation with private landowners. Trust for Nature's vision is for a future in which Victoria's nature is valued, protected and thriving. According to Trust for Nature, flora and fauna on much of Victoria's privately owned land may not be getting necessary protection.

Trust for Nature was established under the Victorian Conservation Trust Act 1972 (Vic) to enable people to contribute to nature conservation by donating land or money. Trust for Nature's purpose is to work with Victorians to protect nature on private land forever. Trust for Nature protects 437 threatened plants and animals, protects over 110000 hectares across Victoria and works with over 140 organisations, farmers, zoos and Aboriginal groups to achieve its aims.

Trust for Nature is now one of Victoria's primary land conservation organisations, with several tools to help people protect biodiversity on private land.

Since 1972, Trust for Nature has developed conservation covenants as a way to protect native plants and wildlife on private land. It has now protected more than 72905 hectares through more than 1567 perpetual conservation covenants. Trust for Nature's work wouldn't be possible without the dedication of volunteers. Many of Trust for Nature's conservation reserves are cared for by volunteer Committees of Management.

The main components of Trust for Nature's conservation program are:

- conservation covenants
- a stewardship program
- property purchase and ongoing management (Trust for Nature properties)
- a 'revolving fund'.

conservation covenant
a management agreement placed on a property's title to ensure native plants and wildlife on the property are protected forever

CONSERVATION COVENANTS – PROVIDING PERMANENT PROTECTION

The **conservation covenant** program was developed by Trust for Nature so that landowners would be able to permanently protect native wildlife and plants on their properties. This remains one of the most effective ways to protect wildlife and native plants on private land. The *Victorian Conservation Trust Act 1972* backs conservation covenants (there are currently more than 1567 conservation covenants). The agreement is voluntary and negotiated between Trust for Nature and each individual landowner. Trust for Nature then works with private landholders to provide ongoing support through its stewardship program. A specific management plan for each covenanted property is developed to maintain and further improve wildlife.

STEWARDSHIP – THE SUPPORT PROGRAM

Trust for Nature helps landowners to better manage their land through its stewardship program, once a property has had a covenant placed on it. Trust for Nature helps covenantors maintain and improve the health of native wildlife and plants on their property by offering conservation management advice and property maintenance tips to landowners. This advice may include weeding, fencing and controlling of invasive species such as rabbits and deer. Trust for Nature may also offer information about land-management incentives available, technical advice, practical assistance with habitat and species monitoring.

TRUST FOR NATURE PROPERTIES

Trust for Nature buys and maintains properties that have a high conservation value to protect native wildlife and plants, and to allow for regeneration of damaged habitats and protection for the future. On some of these properties, private land conservation practices are in place. The 'State-wide Conservation Plan, 2030' identifies 18 landscapes across Victoria that will make the greatest contribution to private land helping to save vulnerable plants and animals. Trust for Nature properties that are purchased work to mitigate habitat fragmentation, and improve wildlife corridors between private and public land across Australia.

REVOLVING FUND – A CYCLE OF SUCCESS

Trust for Nature's Revolving Fund purchases private properties across Victoria that are high in conservation values. This land is then restored, improving the biodiversity, and resold to new owners with ongoing protection through a conservation covenant. The revolving fund has purchased and protected nearly 7000 hectares across Victoria.

The Revolving Fund enables the protection of native vegetation in places where other methods have been found to be less effective. It also helps to introduce new participants to conservation via land protection. The funds revolve in perpetuity as all the money that is generated through the sales is returned to replenish its reserves and enable future purchases.

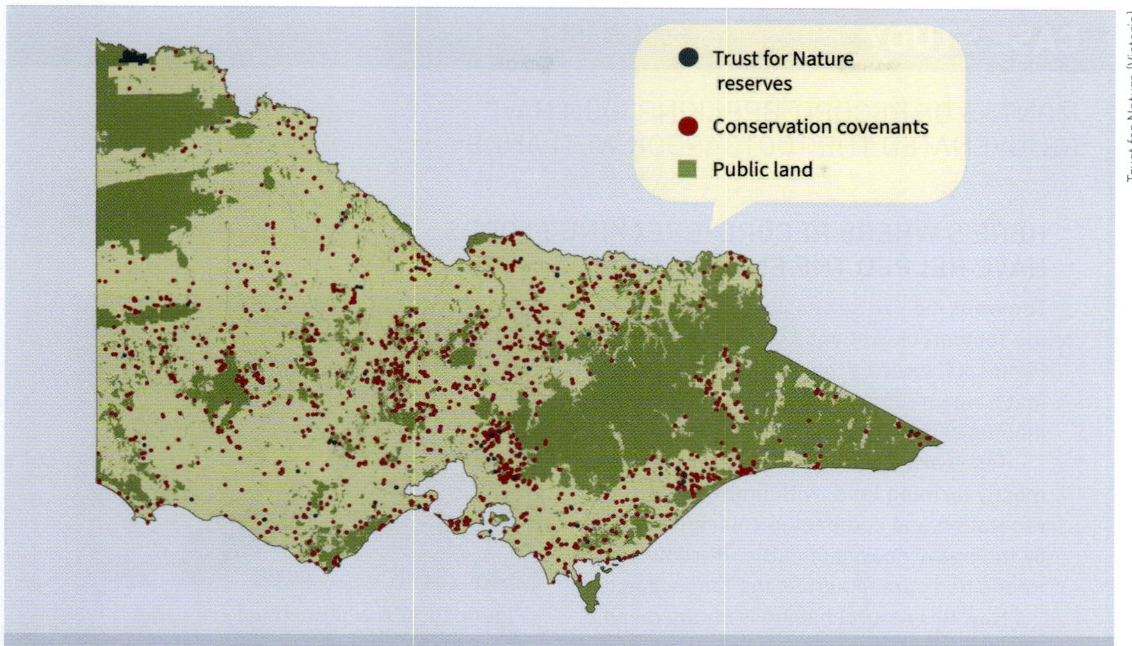

Trust for Nature (Victoria)

Figure 10.15 Victorian Trust for Nature reserves and land under conservation covenants from 2022

CASE STUDY

TRUST FOR NATURE – BUSH FOR BIRDS

Over 480 hectares of significant habitat important to the critically endangered regent honeyeater have been protected with Trust for Nature conservation covenants in north-east Victoria. The regent honeyeater was once widespread across south-east Australia, but its numbers have been in decline due to land clearing. An estimated 400 adult birds remain across four key breeding areas, with one at Chiltern in Victoria. Landholders in this area will work with Trust for Nature and other key conservation groups to protect and enhance the regent honeyeater's habitat. Activities will include fencing of remnant vegetation, weed control, planting native trees and establishing paddocks trees.

Shutterstock.com/Ken Griffiths

Figure 10.16 Regent honeyeaters

Resource
Case Study: Trust for nature - bush for birds

Resource
Case Study: Months of record-breaking rain have rejuvenated the Tootgarook Wetlands

CASE STUDY

MONTHS OF RECORD-BREAKING RAIN HAVE REJUVENATED THE TOOTGAROOK WETLANDS

HERE'S WHERE RECORD-BREAKING FLOODS HAVE HELPED THE ENVIRONMENT

The white-bellied sea eagle has returned to the Tootgarook Wetlands on the Mornington Peninsula, lured by the abundance of prey that have streamed into waterways following months of heavy rain.

Figure 10.17 A white-bellied sea eagle soars above the Tootgarook Wetlands.

iStockphoto/Liz Leyden

Ten years ago, this 600-hectare site was mostly farming land, with limited native vegetation and compacted soil caused by the heavy hooves of cattle. Introduced weeds had spread across the landscape.

Now this natural habitat is on the mend, and pest plant species are being drowned out by downpours that have rejuvenated the expansive swamp.

Much of the Tootgarook Wetlands is privately owned but protected by a Trust for Nature covenant, or legal contract, that prevents development and agriculture. The non-profit group has been managing the land with the aim of restoring the natural environment.

Trust for Nature south central area manager Ben Cullen said the white-bellied sea eagle hadn't been seen in the area for many years but was now back, searching for swamp skinks and other reptiles.

Cullen has seen other birds, including the endangered Australasian bittern, also return to the site, where native phragmites (reeds) are expanding in area across the former farmland.

'The water has spread out to other areas, which means there's more habitat,' he said. 'There's more chance of food supplies and hopefully even increasing genetic capabilities because we're expanding on the viable area that these species can breed in.'

Cullen said the wetland, which is bordered by roads and houses, was crucial in filtering water before it flowed into Port Phillip Bay.

'Here's where record-breaking floods have helped the environment' Benjamin Preiss and Miki Perkins, January 2, 2023, The Age

Worksheet
10.3a Trust for Nature

Weblink
Trust for Nature

LEARNING ACTIVITY 10.3A

Trust for Nature

Use the Trust for Nature website to research an example of the work currently being undertaken or previously undertaken by Trust for Nature.

1 Where is the program being implemented In Victoria?
2 What are the aims of this project?
3 What are the species or communities it is targeting for improvement?
4 What component of Trust for Nature is being undertaken to improve the health of this specific environment (e.g. conservation covenant)?
5 Analyse how this example of Trust for Nature is effective in sustaining outdoor environments.

Landcare

Landcare was established in 1986 in Victoria and became a national movement in 1989. The formation of Landcare brought farmers and conservationists together to resolve environmental concerns. Landcare has evolved to become one of the largest volunteer movements in Australia, where thousands of people and community groups work together towards sustainable land use practices. These include enhancing biodiversity; conserving land, waterways and coasts; and building resilience in Australia's food and farming systems.

In Victoria, Landcare began when farming neighbours recognised they could be more effective if they addressed natural resource management concerns together. Landcare Victoria's vision is for rich and diverse Victorian landscapes, supporting productive farming, healthy ecosystem's and thriving communities. Landcare Victoria supports 600 groups and thousands of individuals.

Landcare has achieved success in facilitating sustainable land management practices on both public and private land, from the bush and the coast to urban and peri-urban areas of Victoria. Members coordinate their management efforts with surrounding landowners. The government contributes knowledge, expertise and some funding to assist in the combined efforts of Landcare members.

Figure 10.18 Members of Little River Landcare Group's 'Positive Farming Footprints' training program visit the Evergraze research site.

HOW DOES LANDCARE SUSTAIN HEALTHY OUTDOOR ENVIRONMENTS?

According to Landcare, their groups undertake wide ranging activities and long-term projects, including the following:

- planting native trees, shrubs and grasses to create habitat for native animals to improve biodiversity
- assisting landholders and farmers to care for the soils on their property to help prevent salinity and erosion
- assisting landholders and farmers to use energy and water-usage efficiencies, such as farming effluent recycling, sediment control and solar panel installation
- getting the community involved in restoring and protecting their local environment, with planting days, weed and rubbish removal, and installing bird nesting boxes

- organising volunteers in the community to help to resolve local environmental issues, working in partnership with local councils, natural resource management agencies, farming groups, business and industry, and researchers
- mitigating climate change by protecting and stabilising beaches and sand dunes to protect fragile coastlines
- consulting Traditional Owners about local land management and decision-making
- making 'bug hotels' to encourage bees and insects to pollinate gardens
- helping to educate private landowners on ways to better look after their land and support species conservation, especially if the land is part of a habitat corridor
- building fences to protect vegetation and livestock, and keep out feral pests.

Resource Case Study: Yinnar-Yinnar South Landcare group - Billy's Creek revegetation project

CASE STUDY

YINNAR-YINNAR SOUTH LANDCARE GROUP – BILLY'S CREEK REVEGETATION PROJECT

Yinnar-Yinnar South Landcare group is planting a forest in the Billy's Creek Valley in Jeeralang, Victoria. This project aims to increase biodiversity and to be part of a potential continuous future corridor, or bio-link, between the Tarra Bulga and Morwell National Parks. This project aims to restore tree canopies and enhance wet forest conditions to improve the habitat for great glider, powerful owl and koala populations. This site is protected from logging and farming, and is set aside for its ecological value. Activities of this Landcare project include:

- blackberries sprayed in the riparian zone
- seed collected and regerminated at the group's greenhouse to replant at this site in the future
- Billy's creek riparian zone revegetated
- immediate removal of deer, feral sheep and goats in this environment advocated.

Worksheet 10.3b Landcare research

Weblink Landcare Victoria

LEARNING ACTIVITY 10.3B

Landcare research

Visit the Landcare Victoria website and access the project section. If possible, select a current project for an area you have studied or visited. Complete research on a Landcare group, considering the following:

1. Where is the outdoor environment?
2. Who are the members of the group?
3. What are the environmental issues involved?
4. Briefly describe their project.
5. How does this project sustain healthy outdoor environments?

Community groups such as 'Friends of …'

The Victorian Environment Friends Network represents the common interests of all Friends groups in Victoria. Its vision is to protect, restore and enhance the Victorian natural environment through community volunteer groups working for their special places and native species.

Groups operate in conjunction with a relevant management authority, which is usually Parks Victoria, a local council or other organisations such as Trust for Nature.

The Victorian Environment Friends Network aims to:

- provide support for the reserve or species
- assist with special projects selected by the Friends in consultation with the relevant management authority
- bring into contact people with a common interest in the reserve or species
- foster public awareness of the reserve or species
- support the effective management of native flora and fauna in Victoria.

LEARNING ACTIVITY 10.3C

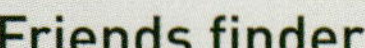

Friends finder

Visit the Victorian Environment Friends Network website, select a 'Friends of ...' group and complete the following:

1 Describe the environment the group is sustaining.
2 What other groups or organisations does this group partner with, if any?
3 Outline the conservation efforts being made by this 'Friends of ...' group.
4 Evaluate the effectiveness of this group in sustaining this outdoor environment.

Worksheet 10.3c Friends finder

Weblink Victoria Environment Friends Network

NOTES FOR THE EXAM

For the exam, you should:

- understand two community actions from the prescribed list to sustain healthy outdoor environments
- evaluate two community actions undertaken to sustain healthy outdoor environments.

10.3 KEY CONCEPTS

- Like-minded individuals form community groups to sustain a range of environments across Victoria and Australia.
- Regenerative farming is a system of farming principles and practices that seeks to rehabilitate and enhance the entire ecosystem of a farm by placing a premium on soil health.
- Regenerative farming practices include applying organic composts and fertilisers, fencing off waterways to stock and implementing time-controlled grazing.
- Trust for Nature is a not-for-profit organisation that works to protect native plants and wildlife in cooperation with private landowners.
- Components of Trust for Nature's conservation program include:
  - conservation covenants
  - a stewardship program
  - property purchase and ongoing management (Trust for Nature properties)
  - a 'revolving fund'.
- Conservation covenants are management agreements placed on a property's title to ensure native plants and wildlife on the property are protected forever.
- Landcare was established in 1986 in Victoria and became a national movement in 1989.
- The formation of Landcare brought farmers and conservationists together to resolve environmental issues.
- Landcare groups voluntarily coordinate their management efforts with surrounding land owners. The government contributes knowledge, expertise and some funding to assist in the combined efforts of Landcare members.

- Landcare activities to sustain healthy outdoor environments include:
  - planting native trees, shrubs and grasses to create habitat for native animals to improve biodiversity
  - working with landholders and farmers to care for soils to help prevent salinity and erosion
  - making 'bug hotels' to encourage bees and insects to pollinate gardens.
- The Victorian Environment Friends Network represents the common interests of all Friends groups in Victoria.
- Its vision is to protect, restore and enhance the Victorian natural environment through community volunteer groups working for their special places and native species.

Worksheet 10.3 Key concepts

10.3 CONCEPT QUESTIONS

REMEMBERING

1 Outline what is meant by 'community actions'.
2 List four community actions undertaken to sustain healthy outdoor environments.
3 Outline outdoor environments you have visited or studied where community actions take place.

UNDERSTANDING

4 Describe two community actions, including:
   a the name of the community action and when it was established
   b why the community action was established
   c the aims and or objectives of the community action
   d the geographical location of this community action.

APPLYING

5 For each of the community actions you selected in Question 4, explain three management strategies or techniques this action undertakes to sustain a specific outdoor environment you have visited or studied.
6 Choose one community action you have researched and evaluate the effectiveness of this action in sustaining a specific outdoor environment you have visited or studied.

EXTENSION CHALLENGE

7 Referencing two community actions, investigate how members within the community could be involved in this action to sustain outdoor environments:
   a How can they join this community group?
   b Are there membership obligations?
   c How does their involvement contribute to improving outdoor environments?

10.4 INDIVIDUAL ACTIONS TO PROMOTE AND SUSTAIN OUTDOOR ENVIRONMENTS

KEY KNOWLEDGE

- individual actions undertaken to promote and sustain healthy outdoor environments, including two of the following:
  - environmental activism
  - environmental advocacy
  - ethical and sustainable consumerism
  - green home design

KEY SKILLS

- compare a range of individual actions to sustain healthy outdoor environments

In this section, we delve into specific, tangible actions that can be taken to promote and sustain healthy outdoor environments. These actions refer to practical techniques that contribute to sustainability. The goal is to ensure an equitable, healthy future for all people and their environments by responsibly utilising natural resources while preserving the ecological balance.

It is not just the responsibility of governments and land managers to maintain healthy and sustainable environments – individuals play a crucial role in ensuring the preservation of our fragile and unique environments for present and future generations. There are numerous actions that individuals can take to contribute to sustainability. Although this section will explore only a small sample of these actions, they are an important aspect of promoting a sustainable future.

Throughout history, environmental activism and advocacy have played key roles in conserving important areas, including the Great Barrier Reef, south-west Tasmania's forests and rivers, K'Gari (Fraser Island) and many others. These campaigns have influenced public and political views on forests, wild rivers, mines and mining practices, and have led to significant changes in these areas.

NOTES FOR THE EXAM 

For the exam, you should:

- examine two prescribed individual actions and compare how these individual actions promote and sustain outdoor environments.

Environmental activism

Historically, environmental activism in Australia began with the recognition of the country's unique landscapes and biodiversity by early European naturalists. However, it was in the 1960s and 1970s that many individuals came together to form grassroots environmental groups in response to the catastrophic loss of biodiversity and government plans to further damage the natural environment for economic gain. One of the first such groups was the United Tasmania Group (UTG), formed in 1972 with the goal of stopping the damming of Lake Pedder in south-west Tasmania. The UTG was the world's first green political party, fielding candidates in the Tasmanian state election in 1972. Members of the UTG formed the Tasmanian Greens and were key in the foundation of the Australian Greens political party.

During this period, the Tasmanian Wilderness Society (later known as the Wilderness Society) also played a pivotal role in saving the Franklin River from hydro-electric power in 1983, becoming a catalyst for change in Australia's environmental policies.

Environmental activism is defined as the actions of individuals or groups that aim to protect or support the environment. This includes increasing individual or group awareness about the health of the environment through the use of direct action, such as demonstrations or lobbying to pressure governments or organisations to enact policy changes. Environmental activism can also involve vigorous campaigning to bring about political or social change.

environmental activism the actions of individuals or groups that aim to protect or support the environment

The influence of environmental activism in recent times has been far-reaching, and instrumental in bringing about change. A study conducted between 2010 and 2020 revealed that environmental groups in Australia advertised more than 24000 events through social media platforms alone that aimed to engage individuals and raise awareness about environmental issues. Climate change, anti-nuclear testing and biodiversity loss are among the key areas that have garnered widespread attention and activism in recent years. The numerous events organised by these groups – such as film screenings, seminars, protests and clean-up days – have played a crucial role in bringing public attention to the pressing environmental concerns that threaten our planet. As a result of such activism, governments and organisations have been forced to adopt and implement policies to address these environmental challenges. The impact of environmental activism has been substantial, and it continues to be a driving force for change, shaping the way society views and approaches environmental issues.

Some of the largest groups you can join to become an environmental activist in Australia include:

- Australia Conservation Foundation
- World Wide Fund for Nature
- Friends of the Earth
- The Wilderness Society
- Greenpeace
- School Strike 4 Climate
- Australian Youth Climate Coalition

Worksheet
10.4a Environmental activism in action

Weblink
Schools Strike 4 Climate

LEARNING ACTIVITY 10.4A

Environmental activism in action – School Strike 4 Climate

School Strike 4 Climate promotes non-violent, student-led environmental activism across Australia. Its aims include:

- net zero by 2030, meaning no new coal, oil or gas projects to be established, including the Adani coal mine
- 100% renewable energy generation by 2030.

Individuals can join and become environmental activists by attending an action, hosting an action, joining a team, donating, subscribing and following the group on social media platforms, and using the hashtag #DontNabOurFuture.

Visit the Schools Strike 4 Climate website and research:

1 an upcoming action by School strike for climate
2 how a local group organises a School Strike 4 Climate event.

Alamy/Steven Sklifas

Figure 10.19 School Strike 4 Climate protest

Resource
Case Study: Thousands protest across Australia against Adani Carmichael mine

CASE STUDY

THOUSANDS PROTEST ACROSS AUSTRALIA AGAINST THE ADANI CARMICHAEL MINE

Thousands of people have protested across Australia against the establishment of the controversial Adani coal mine in Queensland. Some activists travelled across Australia to call for a halt to the coal mine. Other activists walked the streets of most capital cities across Australia holding signs reading: 'There is no Planet B' and 'The climate is changing – why aren't we?' Activists are protesting against the environmental damage caused by mining operations, and the greenhouse gas emissions that are created by the use of coal from the mine to generate power. They also argue that Adani's mine, rail and port project is destroying the ancestral lands, waters and cultures of Indigenous peoples in the region. Adani does not have the consent of the local Wangan and Jagalingou peoples for the project.

The role of individuals in shaping the landscape of environmental activism is significant. Through their commentary and social media presence, they can draw attention to environmental issues and spur discussions within communities. In recent years, a new wave of environmental activism has emerged in Australia, known as 'Generation Greta', in reference to the influence of young activist Greta Thunberg. This generation of activists is characterised by their outspokenness and efforts to sway decision-makers on environmental issues, particularly those related to climate change.

Shutterstock.com/Spill Photography

Figure 10.20 Protests against the Adani Carmichael mine have been held across Australia in recent years

LEARNING ACTIVITY 10.4B 

Environmental activists

Research an influential young environmental activist and answer the following questions:

1 What is the key environmental issue this individual is campaigning to change?
2 Summarise the influence this individual has had in creating environmental change.

Worksheet 10.4b Environmental activists

Environmental advocacy

Environmental **advocacy** refers to efforts aimed at protecting the environment by influencing business and government decisions. The most significant tool in these campaigns is people power – the driving force created by many people speaking and acting together towards a common goal.

advocacy an action that speaks in favour, supports and pleads on behalf of others or issues

Environmental advocacy is a crucial aspect of environmental protection, as it aims to bring about change through the collective efforts of individuals and organisations. By assembling communities and raising awareness about environmental issues, environmental advocacy can create pressure on decision-makers to implement policies and practices that prioritise the health of the environment.

The main goal of environmental advocacy is to achieve a more sustainable future, one in which natural resources are managed in a responsible and equitable manner. This is accomplished through the use of various strategies and tactics, such as educational outreach, community organising and direct action.

Environmental advocacy can take many forms, including lobbying, public speaking, letter-writing campaigns and demonstrations. In order to be effective, environmental advocacy campaigns must engage a large and diverse audience, and effectively communicate the importance of environmental protection to a wide range of stakeholders.

Furthermore, environmental advocacy can serve as a catalyst for social and political change, inspiring individuals and organisations to work together towards a common goal, and fostering a sense of community and solidarity in the face of environmental challenges.

HOW DOES EDUCATION PROMOTE ENVIRONMENTAL ADVOCACY?

Education about environmental issues can encourage conservation and advocacy activities that lead to reductions in pollution, waste, deforestation and biodiversity loss. These activities can range from volunteering, political engagement and advocacy to donations and green purchasing. These impacts are positive. Data shows that, as educational attainment increases, people are up to two to four times more likely to participate in environmental conservation and advocacy activities. People with higher levels of education were also more likely to report that they could be encouraged to get more involved in environmental advocacy.

Resource
Case Study: Advocacy in action: Australian Conservation Foundation

Weblink
Australian Conservation Foundation

CASE STUDY

ADVOCACY IN ACTION: AUSTRALIAN CONSERVATION FOUNDATION

ABOUT THE AUSTRALIAN CONSERVATION FOUNDATION

ACF is Australia's national environment organisation.

We are more than half a million people who speak out for the air we breathe, the water we drink, and the places and wildlife we love. We are proudly independent, non-partisan and funded by donations from our community.

Imagine a world where water flows clean. Where everyone shares abundant energy from the sun and wind. A world where forests, rivers, people, oceans and wildlife thrive – a tomorrow even more beautiful than today.

This is the world that we can see. This is the world we're creating.

In order to get there, nature needs us, now.

Australian Conservation Foundation

QUESTION

Visit the Australia Conservation Foundation website. Explain three ways you could join in advocating for the environment in Australia.

Ethical and sustainable consumerism

Sustainable consumerism involves adopting lifestyles and behaviours that promote the efficient use of resources while ensuring access to basic needs and quality of life. This approach aims to balance economic prosperity with social equity and environmental responsibility. By making conscious choices, consumers can play a crucial role in promoting sustainable production and consumerism patterns, reducing their environmental impact and contributing to a more equitable and sustainable world.

In Australia, there is a growing awareness of the need for sustainable consumerism, and many consumers are making changes to their spending habits in line with this goal. The shift towards sustainable consumerism is driven by a desire to meet the global sustainability challenge and create a more resilient future for all. Whether it be through choosing environmentally friendly products, supporting local businesses or reducing waste, sustainable consumerism is a crucial step towards creating a more sustainable world.

Adamcalaitzis | Dreamstime.com

Figure 10.21 Sourcing local and seasonal produce from farmers markets is an example of sustainable consumerism.

Sustainable consumerism practices include the following:

- Buying local and seasonal produce – local produce such as fruit and vegetables are ripened on the farm and often contain more nutrients. Local produce also travels less distance to the consumer's plate, reducing transport emissions, and is often more seasonal (e.g. berries are seasonal crops of spring and summer in Victoria).
- Following zero-waste principles – this seeks to reduce consumerism and ensure products are made to be reused in order to maximise recycling. The five key concepts of this feature are to refuse, reduce, reuse, recycle and rot (composting).
- Changing energy consumption habits – this involves moving away from using electricity at peak times. Most energy retailers have peak times to use power (commonly 4pm to 10pm) and off-peak times (commonly 10pm to 3pm). Using power in off-peak times reduces power bills and helps to keep the power grid stable. Conscious consumers can also opt to use fans instead of air-conditioners, and reduce heat from direct sunlight by using shades on windows (especially north and west facing).

LEARNING ACTIVITY 10.4C

Eating seasonal produce

Visit the Australian Food Guide website and search for the Victorian Seasonal Food Guide. Use the guide to investigate changes you can make to ensure you are eating produce seasonally. Make a summary table of the fruits and vegetables you often consume, note which are seasonal to where you live and share this with your household.

Worksheet
10.4c Eating seasonal produce

Weblink
Australian Food Guide

greenwashing where a company falsely portrays itself as more environmentally friendly than it actually is, in an effort to bolster its reputation and increase sales

upcycling the process of transforming by-products, waste materials or unwanted products into new materials or products of better quality

fair trade a set of standards in the supply chain of a product, such as fairer conditions, improved working conditions and better prices for producers

Ethical consumerism is the practice of aligning one's spending with one's values. This includes considering factors such as fair working conditions and sustainable production, packaging and transport practices throughout a company's entire supply chain.

iStockphoto/lechatnoir

Figure 10.22 Shopping for second-hand clothing is an example of ethical consumerism.

Ethical and sustainable consumerism reduces impacts on people and the environment. People who shop this way are known as 'conscious consumers'. Ethical products are made without exploiting workers, don't have negative impacts on the environments in which products are made and avoid cruelty to animals. Sustainable products don't deplete non-renewable resources and are made in a socially responsible way.

However, some companies engage in '**greenwashing**', or falsely portraying themselves as more environmentally friendly than they actually are, in an effort to bolster their reputation and increase sales. Conscious consumers will develop the skills to critique the claims made by manufacturers and so make their own judgments about the environmental credentials of such organisations.

Conscious consumers may adopt practices including:

- buying second-hand
- avoiding bottled water
- avoiding cling film (plastic sandwich wrap)
- **upcycling**
- shopping seasonally and locally
- eating less meat
- using reusable shopping bags
- switching to **fair trade** products such as coffee and chocolate.

Worksheet 10.4d Incorporating ethical and sustainable consumerism in your life

Weblink Shop Ethical

LEARNING ACTIVITY 10.4D

Incorporating ethical and sustainable consumerism in your life

Using the Shop Ethical website and researching other websites, investigate five changes you could incorporate to be a more sustainable and ethical consumer.

Copy the table below and add your findings.

Product or service you currently use or consume	Ethical or sustainable change to this product or service	How this change in purchasing is an example of ethical or sustainable consumerism

CASE STUDY

TRIPLE J'S HOTTEST 100 PARTNERS WITH THE AUSTRALIAN CONSERVATION FOUNDATION

Triple J has teamed up with the Australian Conservation Foundation for the 2023 Hottest 100.

This is our most critical decade for climate action in Australia.

Catastrophic floods and bushfires, blistering heatwaves and severe drought: extreme weather is becoming more frequent and unpredictable due to climate change, and the damage is unfolding right in front of our eyes. But we know there is hope when we all come together and speak up for our climate and our environment.

Figure 10.23 Triple J's Hottest 100 T-shirt

Each T-shirt sold will fund ACF action to protect our environment and tackle climate change. Triple J's Hottest 100 T-shirt is made from 100% recycled materials and all profits will go to ACF.

Adapted from ABC/triplej (www.abc.net.au/triplej/hottest100/22/acf)

Resource
Case Study: Triple J's Hottest 100 Partners with the Australian Conservation Foundation

SPOTLIGHT

Patagonia – 'Worn wear' and '1% for the planet'

Patagonia has a unique program via its main website where consumers can shop sustainably and purchase second-hand Patagonia gear. This helps to reduce landfill waste from clothing and encourages people to buy less and trade in any outdoor gear when they no longer have a use for it. People can drop off their unwanted Patagonia gear at a retail shop or via the online platform in return for a Patagonia voucher.

Since 1985, Patagonia has also pledged 1% of sales to the preservation and restoration of the natural environment. More than $140 million in cash and in-kind donations have been made to domestic and international grassroots environmental groups.

Figure 10.24 Patagonia's '1% for the Planet' campaign

Green home design

Green home design is the practice of creating structures, such as homes, offices and factories, using environmentally responsible and resource-efficient processes and materials. This type of design not only commits to sustainable building techniques, but also to designs that promote resource-efficient living practices within the structures.

As of 2022, new homes and major renovations in Victoria must meet the minimum energy performance requirements set by building regulations and achieve a minimum 6-star performance standard. These regulations, codes and standards help to improve sustainability.

energy efficiency using less energy to perform the same task and eliminating energy waste

The Nationwide House Energy Rating (NatHERs) is a star rating system that ranges from 0 to 10 and measures the **energy efficiency** of a house based on its design, including the roof, walls, floor and windows. Houses with higher NatHERs are naturally more comfortable in both winter and summer, making them easier and more cost-effective to heat and cool.

Sustainability Victoria recommends the following considerations when building or renovating a home to meet green standards.

SPOTLIGHT

Summary of key principles for energy-efficient design

Sustainability Victoria offers the following advice for designing energy-efficient homes.

Block orientation

When choosing a block for your home, make sure it will allow you to place the living areas where you spend the most time on the north side of the house. This will take advantage of the winter sun, keeping these rooms warm, light and bright. Avoid or minimise windows facing west or east and make sure they are well shaded for summer.

Building materials

Does your builder:

- buy local materials to reduce carbon footprint
- select materials that don't have toxic elements and can be recycled and reused
- source timber from sustainable plantations and buy local materials to reduce carbon footprint?

Efficient lighting

Smarterlighting design and taking advantage of natural light will save you money on your ongoing energy bills.

Energy-efficient appliances

When choosing fridges, televisions, washing machines, dishwashers and so on, choose energy efficient appliances with the highest energy star ratings.

Energy-efficient hot water systems

A solar hot water system might be more expensive to buy and install, but the running costs will be significantly lower. The Victorian Government is providing a rebate on solar hot water systems for eligible households.

Insulation

Insulating your home's ceiling, walls and floors can save you up to 45% on the cost of running your home. Adding as much insulation as possible, while making sure it is installed correctly with no gaps and proper waterproofing, is a sound investment.

Solar power

Installing a solar PV system will allow you to generate renewable energy and reduce your electricity bills. The Victorian Government is providing a 50% rebate (up to $2225) for eligible households.

Ventilation

Both windows and exhaust fans can help control ventilation to maintain air quality while minimising air leakage.

Water

Save water by considering water recycling systems and choosing water saving fittings and appliances. Grey water systems are mandatory in most new housing systems. Energy saving showerheads can also receive a rebate in some council areas.

Window frames, glazing and shading

Smarter window design and external shading can make your home bright and comfortable all year round. Double glazing, which has become less expensive, will reduce heat losses from your home, and reduce heat gain through your windows in summer.

Zoning and efficient heating and cooling

Designing zones in your home will allow you to efficiently heat and efficiently cool rooms individually. Doors are a great way to zone areas of your home, for example, between corridors and bedrooms or living areas.

Adapted from 'Key Principles of Energy Efficient Design', 22nd September 2023, Sustainability Victoria

LEARNING ACTIVITY 10.4E

Passive design investigation

Visit the Passive House page on the Australian Government's Your Home website and answer the following questions:

1 Explain passive design concepts.
2 Explain where the passive design concept originated from.
3 Outline the three levels of passive house certification in Australia.
4 What does 'retrofitting' mean and what is an example of this?

Worksheet 10.4e Passive design investigation

Weblink Passive House

Resource
Case Study:
Green home

passive solar building design principles that promote the sun's energy to heat and cool living spaces without the use of active mechanical systems

CASE STUDY

GREEN HOME

Earthshack is lovingly handcrafted in the rolling green hills of Lardner in West Gippsland. A menagerie of passive solar design, permaculture and eco-friendly living. We aim to inspire and encourage your journey towards creating a sustainable home for you and your family.

Earthshack (www.earthshack.com.au)

Earthshack is an 8.4 star energy efficient home that has been well designed and owner built. Sustainable features including straw-bale, **passive solar** design, thermal mass, glazing, efficient room layout, insulation, convective cooling cupboard and natural finishes. With a year-round temperature range of 16.5°C to 22.5°C without cooling, this home is the ultimate off-grid comfy home.

Figure 10.25 Earthshack, Gippsland

Earthshack (www.earthshack.com.au)

Table 10.5 Earthshack's sustainable design features

Energy system	
2.97 KW solar system	• Faces north for maximum winter electricity • In summer we have too much electricity
Large battery bank to store energy	• When the sun goes down, our electricity comes from here
Passive solar design	
Long side faces north	• Takes advantage of climate to maintain year-round comfortable temperature in home
Large north-facing windows	• Winter sun moves low in the sky to the north • Winter sun comes into house
Small south-facing windows	• Glass is a weak point, so minimised where solar gain isn't possible
Shading (veranda or eave) over windows	• Blocks summer sun and lets in winter sun (free heat and light)
Living areas and kitchen to the north	• These rooms need maximum sunlight and warmth for sitting around
No hallway and diagonally opposite windows	• Fast natural ventilation with maximum air flow when windows are opened
Straw-bale walls	• Huge amount of insulation value – this stops heat transfer through the wall • Stops summer heat coming inside • Stops winter heat from oven moving out
Concrete floor to the north	• Thermal mass to store free winter heat

Water	
2 × 22000 litre water tanks	• One tank is downhill to catch the water • When we have too much electricity, water is pumped uphill to the 'header' (high) tank • When taps are turned on in house, the water flows downhill with gravity
Hand basin fills toilet cistern	• When hands are washed water flows into cistern ready for next flush. Toilet is then flushed with soapy hand water instead of clean drinking water
Electricity saving	
Cut-off switch at front door	• Turns off all electric items (not fridge and freezer). Also turns off standby power
Waste	
Huge underground worm farm	• Showers, toilet, sinks and washing machine all feed this system. Produces no smell and less greenhouse gases to traditional septic system
Cooling cupboard	• Cool air from south side of house is piped into a cupboard that stores fruit and veg. Zero electricity
Cooking	
Wood-fired oven	• Heats home as well. Only used in winter. Uses local timber mill off-cuts and windfall branches from paddocks
Natural/low toxic finishes	
Chalk-based wall paint on timber work	• Less VOCs (volatile organic compounds) and more sustainable product than traditional paint

Earthshack (www.earthshack.com.au)

QUESTIONS

1 What are the functions of a cut-off electrical switch and a cool cupboard?
2 Outline five examples of Earthshack's sustainable features.
3 For each example above, analyse how this feature promotes sustainability of the Earth's resources.
4 Make suggestions about how some of Earthshacks' sustainability features may be applied to your home and research the cost.
5 Visit the Your Home page on the Australian Government's website and navigate to the case studies to further develop your understanding of best practice sustainable home design features.

NOTES FOR THE EXAM

For the exam, you should:

- compare a range of individual actions to promote and sustain healthy outdoor environments, including at least two of the following:
  - environmental activism
  - environmental advocacy
  - ethical and sustainable consumerism
  - green home design.

If we are to maintain healthy outdoor environments, both in Australia and on a global scale, it is imperative that individuals and society adopt environmentally responsible behaviour. Through your studies in Outdoor and Environmental Studies and hands-on experiences, you will gain a deeper understanding of the current state of the environment and acquire the knowledge and skills necessary for interacting sustainably with outdoor environments.

Healthy outdoor environments and biodiversity are crucial for the physical and emotional wellbeing of individuals. You will learn to recognise the potential impacts on society from significant environmental threats and to identify strategies that can be implemented to address these issues. As you continue to build upon your knowledge and skills, it is your responsibility to put your understanding into action and become an environmentally responsible citizen in the future.

10.4 KEY CONCEPTS

- Throughout history, environmental activism and advocacy have played crucial roles in saving the Great Barrier Reef, significant areas of south-west Tasmania's forests and rivers, and K'Gari (Fraser Island).
- Environmental activism is defined as the actions of individuals or groups that aim to protect or support the environment.
- Examples of environmental activism include direct action such as demonstrations or lobbying to pressure governments or organisations to enact policy changes, and vigorous campaigning to bring about political or social change.
- Environmental advocacy is activities that aim to protect the environment by influencing business and government decisions.
- Environmental advocacy can take many forms, opening doors to careers focused on conservation, alternative energy sources, policy-making, consulting and research.
- Ethical consumerism is the practice of spending money in a way that aligns with your values, while sustainable consumerism is about doing more and better with less.
- Examples of ethical and sustainable consumerism include buying local and seasonal produce, changing energy consumption patterns, buying second-hand products and switching to fair trade products.
- Green home design is the practice of creating structures (such as homes, offices and factories) using processes and materials that are environmentally responsible and resource-efficient.
- Key principles of energy-efficient design include block orientation, building materials, efficient lights and appliances, insulation and ventilation.

10.4 CONCEPT QUESTIONS

Worksheet
10.4 Key concepts

REMEMBERING

1 Outline what is meant by 'individual actions to sustain outdoor environments'.
2 List four individual actions to sustain healthy outdoor environments.
3 Explain ethical and sustainable consumerism.
4 Explain the concept of green home design.

UNDERSTANDING

5 Compare and contrast the terms 'environmental advocacy' and 'environmental activism'.
6 Describe two individual actions, including a description of the action.
7 Explain each action in detail, including species or environments they may work to sustain.

APPLYING

8 Using the Venn diagram below, or similar, compare the two individual actions you have investigated in sustaining outdoor environments.

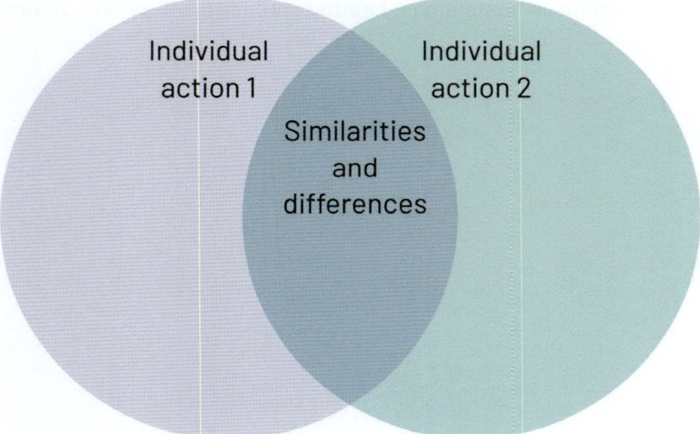

EXTENSION QUESTION

9 Referencing two individual actions, investigate how you could be involved in this action to sustain outdoor environments.
   a How could you join this community group?
   b Are there membership obligations?
   c How could your involvement contribute to improving outdoor environments?

UNIT 4 AREA OF STUDY 3 LOGBOOK ENTRY REMINDER

Have you had an outdoor experience yet? If so remember to complete your logbook entry. Use chapter 11, learning activity 11.1B and 11.3A to guide your data collection and report writing.

Resource
Glossary – Chapter 10

Assessments
End of chapter exam questions

Glossary test

EXAM-STYLE QUESTIONS

1 Analyse one Indigenous and one non-Indigenous management strategy to maintain the health of outdoor environments. (4 marks)

2 Describe a community action you have studied and how this group works to sustain healthy outdoor environments. (3 marks)

3 Evaluate the effectiveness of another community action undertaken to sustain healthy outdoor environments. (4 marks)

   a Identify an Act or a convention that could apply to an outdoor environment you have studied. (1 mark)

   b Evaluate the effectiveness of this Act or convention in sustaining the outdoor environment you have studied. (4 marks)

   c Propose changes to this Act or convention to improve the health of the outdoor environment you have studied. (3 marks)

4 Describe an individual action you have undertaken or studied that supports the health of outdoor environments. (2 marks)

5 Compare two individual actions in sustaining a healthy outdoor environment. (4 marks)

CHECK YOUR KNOWLEDGE AND SKILLS

Resources
Key knowledge and skills checklist

Checklist for students to mark and date, making judgements of their knowledge and understanding.
(D = developing C = consolidating ER = exam ready)

D	C	ER	Check your knowledge and skills
☐	☐	☐	Explain a range of Indigenous land management strategies
☐	☐	☐	Analyse how Indigenous land management strategies maintain specific outdoor environments
☐	☐	☐	Explain a range of non-Indigenous land management strategies
☐	☐	☐	Analyse how non-Indigenous land management strategies maintain specific outdoor environments
☐	☐	☐	Understand two Acts or conventions related to the management of outdoor environments
☐	☐	☐	Describe aims and objectives of each Act or convention
☐	☐	☐	Explain how each Act or convention manages a specific species or ecological community in an environment you have visited or studied
☐	☐	☐	Evaluate the effectiveness of each Act or convention in managing and sustaining a specific outdoor environment you have visited or studied
☐	☐	☐	Propose changes to each Act or convention to better manage and improve the health of a specific species or ecological community in an environment you have visited or studied
☐	☐	☐	Understand two community actions undertaken to sustain outdoor environments
☐	☐	☐	Explain the strategies of each community action in sustaining a specific environment you have visited or studied
☐	☐	☐	Evaluate the effectiveness of community action in sustaining a specific outdoor environment you have visited or studied
☐	☐	☐	Understand two individual actions undertaken to sustain outdoor environments
☐	☐	☐	Explain the strategies of each individual action in sustaining environments
☐	☐	☐	Compare effectiveness of individual actions in sustaining outdoor environments

CHAPTER 11

UNIT 4 – AREA OF STUDY 3

Investigating outdoor environments

KEY KNOWLEDGE

- outdoor and environmental concepts related to the human relationships with, and health and sustainable use of, the selected outdoor environments
- the nature of primary and secondary data relevant to the investigation
- conventions of report communication, including appropriate structure (Introduction, Body and Conclusion), terminology and representations of the data.

KEY SKILLS

- plan for and conduct an independent investigation into a range of outdoor environments
- collect relevant data in a range of outdoor environments, authenticated through use of a logbook
- evaluate and report data and information, including findings and implications.

KEY TERMS

independent investigation
investigation
logbook
ongoing
primary data
report conventions
secondary data
written report

Resources and Templates

- Planning and recording outdoor experiences: Year-long plan template p. 570
- Planning and recording outdoor experiences: Logbook entry template p. 574
- Planning and recording outdoor experiences: Tips for a successful logbook template p. 583

Nelson MindTap

To access resources above, visit **cengage.com.au/nelsonmindtap**

On completion of this chapter, students should be able to plan and conduct an independent investigation that evaluates selected outdoor environments.

The written report document's findings form an independent investigation of at least two visited outdoor environments. The written report is assessed in Unit 4 at a time determined by the teacher and is assessed at 40 marks.

WHAT IS AREA OF STUDY 3?

This area of study is an ongoing culmination of learning in the class, student-led investigation, outdoor experiences and visited outdoor environments from the beginning of Unit 3 and is assessed in Unit 4.

Your teacher will select four key knowledge points across Units 3 and 4 and establish a plan to visit two outdoor environments over the year.

The key knowledge points selected must be able to be reflected in the chosen outdoor environments. During visits to a minimum of two chosen outdoor environments, you will take examples of primary data to document within your logbooks.

Students will keep an ongoing logbook from the start of Unit 3 that documents both primary data and a summary of secondary data during a range of practical outdoor experiences. These experiences can be undertaken on campus, during school hours, as overnight experiences or as journey-based experiences.

Within this chapter we will unpack conventions of report writing, including primary and secondary data, and examples of logbooks.

11.1 OUTDOOR AND ENVIRONMENTAL CONCEPTS

KEY KNOWLEDGE

- outdoor and environmental concepts related to the human relationships with, and health and sustainable use of, the selected outdoor environments

KEY SKILLS

- plan for and conduct an independent investigation into a range of outdoor environments

Outdoor and environmental concepts encompass the four selected key knowledge points across Unit 3 and 4.

11.1.1 WHAT IS AN INDEPENDENT INVESTIGATION?

An **independent investigation** is your own gathering of evidence documented into your **logbook**. This investigation will be driven by the selected key knowledge points and environments visited, as well as the specified key knowledge and key skills within Area of Study 3.

You will use the evidence recorded in your logbook to produce a **written report** that demonstrates the application of the key skills and key knowledge to the selected outdoor environments over the year.

independent investigation
a student-driven research task based on the key knowledge points and skills within outdoor environments

11.1.2 HOW AND WHEN WILL AREA OF STUDY 3 BE ASSESSED?

At the beginning of Unit 3 your teacher will select and communicate to you four key knowledge points from Unit 3 and 4 that will be assessed within Unit 4 AOS 3. The selected outdoor environments should draw on the four selected key knowledge points across Units 3 and 4.

You must retain a logbook for **ongoing** assessment throughout the year, containing observations of outdoor environments and experiences you have investigated. Your logbook must have regular entries to validate your learning on at least two visited outdoor environments.

The following two examples are of planners to guide your ongoing **investigation**.

logbook
a document of evidence of practical outdoor experiences, outdoor environments visited, primary and secondary data, and other valuable information to be added to throughout the year

written report
a form of writing that is written concisely and clearly that examines experiences or events from a research investigation.

ongoing
continuing to develop and progress

investigation
observe, study or carry out an examination in order to establish facts and reach new conclusions

Planner example 1

Outdoor experience and environment 1 ____________________
Unit 3 AOS 1 Key knowledge • ____________________
Unit 3 AOS 2 Key knowledge • ____________________
Outdoor experience and environment 2 ____________________
Unit 4 AOS 1 Key knowledge • ____________________
Unit 4 AOS 2 Key knowledge • ____________________

Planner example 2

Outdoor experience and environment 1

____

Unit 3

Key knowledge

- ______________________________

Key knowledge

- ______________________________

Unit 4

Key knowledge

- ______________________________

Key knowledge

- ______________________________

Outdoor experience and environment 2

____

Unit 3

Key knowledge

- ______________________________

Key knowledge

- ______________________________

Unit 4

Key knowledge

- ______________________________

Key knowledge

- ______________________________

The timing of the assessment for this investigation will be selected by your teacher at some point in Unit 4 when you have enough data from your ongoing investigation throughout the year.

Resource
Planning and recording outdoor experiences: Year-long plan template

LEARNING ACTIVITY 11.1A

Your ongoing investigation

This learning task is to begin to frame your ongoing investigation that is Unit 4 AOS 3. This task begins at the start of Unit 3, where your teacher will have selected four key knowledge points across Unit 3 and 4 for you to initiate your investigation.

You may then like to frame your investigation with a leading or core question(s) to help guide your research; this will be dependent on the key knowledge selected and the outdoor environments you visit along with types of outdoor experiences.

As an example, we have selected four key knowledge points across Unit 3 and 4 and applied them to be intended to be investigated at both visited outdoor environments.

A key investigative question to frame this example investigation may be: 'To observe and understand how relationships have changed with these environments overtime and how we can assess the health and better manage these environments into the future'.

Outdoor experience and environment 1
Mornington Peninsula

Unit 3

Key knowledge

- *relationships with outdoor environments expressed by specific Indigenous peoples' communities before and after European colonisation*

Key knowledge

- *conservation, recreation and economic relationships with outdoor environments*

Unit 4

Key knowledge

- *observable characteristics to assess the health of outdoor environments, including:*
  - *quality of water, air and soil*
  - *species and ecosystem biodiversity*

Key knowledge

- *Acts or conventions related to the management and sustainability of a specific outdoor environment, species or ecological community, including two of the following:*
  - *Flora and Fauna Guarantee Amendment Act 2019* (Vic)

Outdoor experience and environment 2
Lysterfield Park

Unit 3

Key knowledge

- *relationships with outdoor environments expressed by specific Indigenous peoples' communities before and after European colonisation*

Key knowledge

- *conservation, recreation and economic relationships with outdoor environments*

Unit 4

Key knowledge

- *observable characteristics to assess the health of outdoor environments, including:*
  - *quality of water, air and soil*
  - *species and ecosystem biodiversity*

Key knowledge

- *Acts or conventions related to the management and sustainability of a specific outdoor environment, species or ecological community, including two of the following:*
  - *Flora and Fauna Guarantee Amendment Act 2019* (Vic)

Complete your own template and key investigative question to help guide your ongoing investigation .

Outdoor experience and environment 1 __________
Unit 3 Key knowledge • __________ Key knowledge • __________
Unit 4 Key knowledge • __________ Key knowledge • __________
Outdoor experience and environment 2 __________
Unit 3 Key knowledge • __________ Key knowledge • __________
Unit 4 Key knowledge • __________ Key knowledge • __________

Based on the key knowledge you are given by your teacher, brainstorm the aspects of what you want to know about these environments. Here are some guiding questions that may assist you:

1 **Outdoor Environment and Human Relationships**

How do people relate with these specific outdoor environments, and how does this connection impact the sustainability of these environments?

2 **Sustainable Use of Outdoor Environments**

How can outdoor relationships be balanced with the preservation of natural resources and ecosystems in the selected environments?

3 **Data Collection for Understanding**

- What methods are best suited for gathering data on how people relate with and utilise the chosen outdoor environments?
- How can historical data and existing records enhance our understanding of the relationships between human activity and environmental impact?

4 **Effective Communication of Investigation**

- How can research findings be effectively presented to highlight the connections between human relationships and sustainable practices in outdoor environments?
- Adapt and use these simplified questions as a starting point for your research investigation.

Belinda Dalziel

Figure 11.1 Students listen to a safety briefing before a sunrise surfing session at Shoreham – an example of primary data collection.

11.1.3 WHAT IS A LOGBOOK?

A logbook is a record of events, experiences and information presented in a systematic order. For the purpose of this study, the logbook is a tool used to collate and document evidence of your ongoing investigation of the four selected key knowledge points in addition to the specific key knowledge and skills in Unit 4 Area of Study 3.

As indicated, your teacher will advise you of the key knowledge points you will investigate for the year, as well as the prescribed key knowledge points for Area of Study 3. Two outdoor environments are to be visited to allow you to gain enough information to gather primary data for your logbook entries.

The environments you visit and the outdoor experiences you undertake throughout the year will provide you with sufficient data for your logbook entries; this will assist you in your final written report for Area of Study 3.

Key components of a logbook

Logbooks can be kept as a hard copy or on a device; your teacher must sight your logbook regularly for ongoing authentication purposes throughout the year. They are to be kept as a source of your primary data, supported by a summary of secondary data for your written report.

The mode of logbook entries may include handwritten notes, photos, drawings or sketches, transcribed audio taken from a guest speaker (permission sought prior) and transcribed audio taken from your own observations.

Resource
Planning and recording outdoor experiences: Logbook entry template

LEARNING ACTIVITY 11.1B

Create your own logbook template

Whether you source a physical book, type and print sheets, or keep your logbook in electronic form, here are suggested inclusions for each visit:

- date of visit
- location visited
- Country
- environment type at location visited (e.g. alpine, coastal)
- outdoor experiences that occurred or activities undertaken
- flora at location visited
- fauna at location visited
- key knowledge points from Unit 3 or 4 applied
- environmental pressure and threats observed (e.g. urbanisation, significant invasive species)
- sustainability measures observed (e.g. self-composting toilets)
- examples of primary data you collected
- images and sketches
- other information gathered during outdoor experience
- what secondary data you can add to this logbook entry (data to support the selected key knowledge points)
- references (where required)
- possible limitations you encounter (e.g. becoming unwell and not being able to go on an outdoor experience).

Belinda Dalziel

Figure 11.2 Students making notes in their logbooks during an outdoor experience

Example of a logbook entry

Date	
Location	Mornington Peninsula and Western Port Bay, inclusive of Balnarring, Point Leo and Shoreham
Environment type	Coastal, marine and grassy woodland
Country	Boon Wurrung Country
Outdoor experiences that occurred	• Sea kayaking • Surfing
Flora	• Green comb spider orchid • Black sheoak • Small flower flax-lily
Fauna	• Hooded plover (endangered) • Powerful owl • Southern brown bandicoot • Growling grass frog
Environmental pressure and threats observed	• Introduced species (foxes, cats, rabbits and blackberry) • Population-based pressures • Tourism • Farming • Climate change (rising sea levels and severe weather events)
Key knowledge points from Unit 3 or 4 applied	Unit 3 AOS 1 • Relationships of non-Indigenous peoples with specific outdoor environments as influenced by and observed in local or visited outdoor environments during historical time periods: • Early colonisation (1788–1859) • Pre-Federation (1860–1900) • Post-Federation (1901–1990) Unit 3 AOS 2 • Conservation, recreation and economic relationships with outdoor environments
Examples of primary data you collected	• Recorded images of the infrastructure built during these time periods • Observed environmental impacts of infrastructure • Recorded notes from information signs • Transcribed audio notes from guest speaker (e.g. a park ranger at the site) • Participated in recreation activities (sea kayaking and surfing). These sessions were facilitated by experts in recreation activities • Recorded notes during practical experience on safety briefings, equipment and environmental impacts
Images and sketches	Added within logbook
Other information gathered during outdoor experience	
What secondary data you can add to this logbook entry	• Summary of classroom learning, including your own notes taken during class time • Your own research on the selected key knowledge points • Credible investigative websites such as CSIRO, Bureau of Statistics, and Department of Energy, Environment and Climate Action

Note: this is a brief sample of possible logbook template. Students and teachers will need to build their own templates, depending on outdoor experiences and key knowledge points selected.

11.2 PRIMARY AND SECONDARY DATA

KEY KNOWLEDGE

- the nature of primary and secondary data relevant to the investigation

KEY SKILLS

- collect relevant data in a range of outdoor environments, authenticated through use of a logbook

When conducting your ongoing investigation, you will be directly involved in the data-collection process through the collation of primary data. In addition to this, you will also strengthen your investigation and logbook entries with secondary data. Both sets of data will directly reflect the four key knowledge points selected across Units 3 and 4 and you should directly enter both sets of data into your logbook.

11.2.1 PRIMARY DATA

primary data data collected in the field or outdoor environments that is reflective of the key knowledge points and skills on which you are reporting (note: multiple sources of data collection strengthen the credibility of your findings)

Primary data is information you collect in outdoor environments you visit. Multiple sources of primary data should be experienced, observed and collected during practical experiences throughout the year as part of your ongoing investigation.

Primary data sources are objects and documents created or written during the time being investigated; for example, during the outdoor experience or very soon afterwards. Examples of primary data for logbook entries may include:

- photographs
- hand-drawn sketches
- notes transcribed from guest speakers (or recorded where appropriate)
- notes from information signs in outdoor environments
- reflections from participation in outdoor experiences
- notes from observed environmental impacts.

Belinda Dalziel

Figure 11.3 Students observing the Gunnamatta outfall, taking notes from information signs and drawing sketches of the environmental impacts into their logbooks – examples of primary data

Table 11.1 shows an example of a year-long plan of outdoor environments and experiences, selected key knowledge points, and possible primary and secondary data sources to be gathered while on outdoor experiences.

Table 11.1 Example of year-long plan for outdoor experiences

Sample of possible key knowledge points selected	Possible class practical outdoor experience	Examples of primary data collected for logbook
Unit 3 AOS 1 Relationships of non-Indigenous peoples with specific outdoor environments as influenced by and observed in local or visited outdoor environments during historical time periods: • Early colonisation (1788–1859) • Pre-Federation (1860–1900) • Post-Federation (1901–1990)	Visited a post-Federation historical site such as a large infrastructure building or development	• Recorded images of the infrastructure built during these time periods • Observed environmental impacts of infrastructure • Recorded notes from information signs • Transcribed audio notes from guest speaker such as a park ranger at the site
Unit 3 AOS 2 Conservation, recreation and economic relationships with outdoor environments	Bushwalking and conservation activities on outdoor experience such as the Grampians or Wilsons Promontory	• Notes from your participation and reflection on bushwalking and tree planting at this environment • Photos of participation and environmental impacts • Guest speaker notes from park ranger or similar • Recorded notes from information signs
Unit 4 AOS 1 Observable characteristics to assess the health of outdoor environments, including: • quality of water, air and soil • species and ecosystem biodiversity	Local park investigation	• Recorded images and sketches of creek, land, and flora and fauna to reflect observable characteristics • Recorded notes from information signs • Recorded notes taken from observations
Unit 4 AOS 2 Community actions undertaken to sustain healthy outdoor environments, including two of the following: • regenerative farming • Trust for Nature • Landcare • community groups such as 'Friends of …'	Site visit to Trust for Nature land holding	• Photos and sketches of site • Recorded notes from guest speaker from Trust for Nature or similar • Recorded notes taken from observations

Figure 11.4 Primary data collected on a class visit to a shoreline

11.2.2 SECONDARY DATA

secondary data information that someone else has collected and made available (note: when using secondary data, it is imperative that you reference the source and determine if the source is credible)

Secondary data is information that someone else has previously collated and made available, such as information facilitated by your teacher or discovered via your own independent research. Secondary data is added to your logbook to enhance the primary data you have collected when on an outdoor experience. Secondary data is directed by the selected key knowledge points across Units 3 and 4, as well as the key knowledge points for Unit 4 AOS 3.

Collecting secondary data complements the primary data you have collected and may also confirm, modify or contradict your findings during primary data collation. You must ensure you source secondary data accurately, and critically evaluate this data where it is relevant to your investigation. Secondary data examples for logbook entries could include:

- a summary of classroom learning, including your own notes taken during class time
- your own research on the selected key knowledge points
- information from credible investigative websites, such as CSIRO, Bureau of Statistics and the Department of Energy, Environment and Climate Action.

It is important to note that secondary data must be accurately sourced within your logbook and must be able to be justified.

LEARNING ACTIVITY 11.2A

Sourcing and documenting primary and secondary data

Complete the following tasks to assist you in planning your data collection:

1. Detail how you are going to document primary data during your outdoor experiences.
2. Outline if you need any additional mechanisms to assist you in primary data collection.
3. Outline possible sources of secondary data you may use in your investigation.
4. What are the advantages and disadvantages of the sources of data you have referenced above?

11.3 CONVENTIONS OF REPORT COMMUNICATION

KEY KNOWLEDGE

- conventions of report communication, including appropriate structure (introduction, body and conclusion), terminology and representations of the data

KEY SKILLS

- evaluate and report data and information, including findings and implications

11.3.1 WRITING A REPORT

Research without report writing has little purpose. The written report is the means by which you will communicate the findings of your ongoing investigation to your teacher. Research investigations provide you with the opportunity to present your work to an audience in summary form, perhaps focusing upon your findings and conclusions, and the possible implications and applications of these. In this section, we provide an overview of basic report writing structure and **report conventions**.

report conventions the structure of written reports and the process by which the information is gathered and disseminated

Introduction

The introduction is where you introduce the reader to the broader context of your investigation, environments visited and practical outdoor experiences, and then narrow down to the four key knowledge points you intend to address. The introduction should:

- discuss the importance or significance of the investigation to be reported
- define the purpose of the report; for example, key knowledge points across Units 3 and 4 you addressed and outdoor environments you visited
- inform the reader of any limitations to the report, or any assumptions you made.

Body

The body or main section is where you describe what you did during the ongoing investigation. You need to describe the outdoor environments you visited and outdoor experiences you had that enabled you to complete your logbook entries and subsequently this written report.

This section is where you interpret and evaluate your investigation. To do this, you need to summarise your key results, justify any unexpected findings and explain how your results relate to your investigation as stated in the introduction of the report.

You must also include any unplanned changes to the original process that occurred during the investigation. These changes should have been documented in your logbook. The body contains the main substance of the report, organised into sections that may contain headings and subheadings rather than paragraphs. You may want to organise your report into sections in the body related to the outdoor environments and experience visited by correlating the key knowledge points that you were investigating for that experience. You could then add the primary data you collected during each of these outdoor experiences.

The body of a report may include the following:

- a description of the key knowledge points and environments that are being investigated
- several references to primary and secondary data relevant to the investigation
- a description of what you did and why, such as the use of surveys, photographs and summaries of information signs or interviews
- a discussion and evaluation of the data collected - this should comment on the findings of your report.

Conclusion

This summarises the key findings from the body section. Relate your conclusion to the context of your investigation and arrange your points logically so that major conclusions are presented first.

Reports also may require a discussion of implications and recommendations. Implications are subjective opinions about what action you think could be followed. These must be realistic, achievable and clearly relate to the conclusion of the report. A conclusion never introduces any new ideas or results. Rather, it provides a concise summary of those ideas that have already been presented in the report. When writing a conclusion, you should:

- briefly restate the purpose of your investigation
- identify the main findings (e.g. the reflection of each key knowledge point addressed)
- note the main limitations that were relevant to the interpretation of the results
- summarise what the investigation has contributed to your broader understanding of the key knowledge points selected.

Reference list

This must contain all the material cited in the report. It must be accurate and consistent with a standard referencing style determined by your teacher.

Written report checklist

Read the assessment criteria clearly and clarify what needs to be in the report and what type of report it is to be. Have you:

- followed the structure, using the correct headings?
- provided an introduction?
- discussed and evaluated the results and findings, including the use of primary and secondary data?
- come to a conclusion?
- made some recommendations?
- provided references?
- included any appendices?
- checked punctuation and spelling?

NOTES FOR THE EXAM 

For the exam, you should:

- plan and conduct an ongoing investigation, reflecting on four key knowledge points across Units 3 and 4
- visit two outdoor environments where you have documented primary data into your logbook
- supplement your logbook with secondary data
- evaluate your report data, discuss findings and state conclusions (and possibly implications or recommendations).

LEARNING ACTIVITY 11.3A

Written report practice

This learning task is designed to be completed multiple times throughout the year to prepare you to write your final written report for Unit 4 Area of Study 3.

Unit 4 AOS 3 requires you to investigate four key knowledge points from Units 3 and 4, while visiting two outdoor environments. For this practice task, you will only write your report on one or two key knowledge points and build it in scope as the year progresses. Prior to completing this task, you should complete at least one outdoor experience and have made at least one entry of primary and one entry of secondary data into your logbook. The information that you have entered will support the selected key knowledge point from Unit 3 or 4 your teacher has selected for you and your class.

This is only a template to guide your written report skills. Use the subheadings of introduction, body and conclusion to guide you, and try to build your skills in writing a report.

The prompts within each section are to guide you – you may add more! Good luck!

Ensure you have your logbook available when completing this practice written assessment.

Introduction

- Introduce the reader to your investigation.
- Outline the teacher-selected key knowledge point.
- Outline the outdoor environment visited.
- State any limitation when conducting the investigation (e.g. severe weather).

Body

- Describe why the outdoor environment was chosen and how it reflects the key knowledge points; for example:

  Lysterfield Park was a venue for this outdoor experience as it reflects recreation, economic and conservation relationships.
- Describe the outdoor environment visited, including geographical location, Country, biome, and flora and fauna.
- Describe outdoor experience(s) undertaken.
- Outline what you did and why; for example:

  For example, our class participated in removing rubbish from vegetation as a form of conservation relationship and listened to a ranger presentation about other conservation efforts within the park.
- Include multiple examples of primary data referenced.
- Include multiple examples of secondary data referenced.
- Discuss any unplanned changes.
- Discuss and evaluate the data collected – the findings of your report.

Conclusion

- Summarise key findings.
- Discuss any limitations.
- Provide recommendations.
- Include references section.

11.3.2 TIPS FOR A SUCCESSFUL LOGBOOK

Examples of primary data	• Recorded images of the infrastructure built during these time periods • Observed environmental impacts of infrastructure • Recorded notes from information signs • Transcribed audio notes from guest speaker such as a park ranger at the site • Participated in recreation activities (sea kayaking and surfing). These sessions were facilitated by experts in recreation activities. Notes were taken during practical experience on safety briefings, equipment and environmental impacts

Observation area (teacher-generated prompts)	Primary data (observations during outdoor experiences)	Secondary data (follow up on observations) and extra information from class that is relevant
What plants did you observe?	• Green comb spider orchid • Black sheoak • Small flower flax-lily	
What animals did you observe?	• Hooded Plover (endangered) • Powerful owl • Southern brown bandicoot • Growling grass frog	
Environmental pressure and threats observed	• Introduced species (foxes, cats, rabbits and blackberry) • Population-based pressures • Tourism • Farming • Climate changes (rising sea levels and severe weather events)	

Observation area (teacher-generated prompts)	Primary data (observations during outdoor experiences)	Secondary data (follow up on observations) and extra information from class that is relevant
Key knowledge points from Unit 3 or 4 applied	**Unit 3 AOS 1** relationships of non-Indigenous peoples with specific outdoor environments as influenced by and observed in local or visited outdoor environments during historical time periods: • Early colonisation (1788–1859) • Pre-Federation (1860–1900) • Post-Federation (1901–1990) **Unit 3 AOS 2** conservation, recreation and economic relationships with outdoor environments	
Other information gathered during outdoor experience		
What secondary data do you need to follow up on?		Summary of classroom learning, including your own notes taken during class time Your own research on the selected key knowledge points Credible investigative websites such as CSIRO, Bureau of Statics, Department of Energy, Environment and Climate Action

Resource
Planning and recording outdoor experiences: Tips for a successful logbook template

GLOSSARY

CHAPTER 1

aestivate
to spend a period in dormancy; similar to hibernation

ballast water
water taken on board to provide stability for ships

biomes
a large, naturally occurring community of flora and fauna occupying a major habitat

biodiversity
the number and variety of organisms found within a specified area

biosphere
the parts of Earth where life dwells

built environments
areas that have been created or modified by people, including buildings, parks and transport systems

catchment area
the area of land where water from precipitation drains into a body of water

cinnamon fungus
a soil-borne water mould that produces an infection that causes root rot or dieback

climate
the prevailing weather conditions of a region

colonise
to populate an area

Country
the term often used by Indigenous peoples to describe the lands, waterways and seas to which they are connected. The term contains complex ideas about law, place, custom, language, spiritual belief, cultural practice, material sustenance, family and identity (Source: AIATSIS)

custodianship
the responsibility for taking care of or protecting something

drought
a long period of abnormally low rainfall, especially one that adversely affects growing or living conditions

dredging
an excavation activity or operation usually carried out to gather up bottom sediments and dispose of them at a different location

endemic
a feature or species that is unique to a defined geographic location

ephemeral
temporary or intermittent

evaporation
the change of a liquid into a vapour

exotic species
a species living outside its native distributional range

experiential education
an engaged learning process whereby students 'learn by doing' and by reflecting on the experience

exploitation
making use of and benefitting from resources, often in an unsustainable way and accompanied by environmental degradation

extinct
no longer existing or living

extrinsic motivation
motivations to engage in an activity because we want to earn a reward or avoid punishment

fragmentation
the reduction or breaking up of one area of habitat into several smaller separate areas

geology
the scientific study of the origin, history and structure of the Earth

interrelationships
the way in which two or more things affect each other because they are related in some way

intertidal
an area that is above water at low tide and under water at high tide

intrinsic motivations
motivations to engage in an activity that we get from within ourselves

irrigation
the artificial application of water to arable land for agricultural use

Kinship
an Indigenous person's relationship and responsibilities to other people, to their Country and to natural resources

land reclamation
the process of creating new land from the ocean, riverbeds or lakes

managed parks
areas of public land that are controlled by and are the responsibility of governments

microclimate
the prevailing weather conditions of a small, specific place within a larger area

nature
the living things, the ecosystems and the processes that form them, and the places in which we find all of these

outdoor environments
areas of the natural world, as a whole or in a particular geographical area

outdoor experiences
activities completed outside, most commonly in natural settings

perennially
lasting or active through the year or through many years

private land
land that is not owned by a government

remnant vegetation
small patches of native plants that remain after conversion of landscapes to agricultural or other use

riparian vegetation
plant habitats and communities along a waterway's margins and banks

salinity
the concentration of dissolved salts in water or soil

sclerophyll
an Australian vegetation type having plants (typically eucalypts, wattles and banksias) with hard, short and often spiky leaves

sedge
a grass-like plant with triangular stems and inconspicuous flowers, typically growing in wet ground

songlines
songlines are ancient paths that crisscross the land, often spanning vast distances, and are believed to have been created by ancestral beings during the Dreamtime or creation period

sphagnum bogs
species of mosses; alpine sphagnum bogs are found in permanently wet sites in high rainfall alpine, sub-alpine and montane areas of NSW, ACT, Victoria and Tasmania

subtidal
an area that is permanently covered with water

sustainability
the ongoing capacity of Earth to maintain all life

swales
shallow troughs between sand dunes

topography
the landforms or surface features of a region

urban environments
areas of permanent infrastructure designed to support higher population densities, such as cities and towns

urbanisation
the physical growth of urban areas as a result of rural migration

weed
a type of plant that can invade and establish itself in an ecosystem where it was not originally present, regardless of whether it is a native or non-native species, and deprive other plants of space and food

wilderness
an environment that is big, remote and untouched (or relatively untouched) by humans

CHAPTER 2

absolute risk
the uppermost limit of risk in a particular situation or activity, assuming safety has not been considered

advertising
the practice and techniques employed to bring attention to a product or service

competence
the ability of someone to be able to deal with the situation they are in, which comes from the skills and experience that they have

contemporary
events or actions that have occurred within the past 15 years

cultural background
patterns of thinking, feeling and acting that stem from the social context of one's life experience, such as ethnicity, race, socioeconomic status, gender, language, religion, sexual orientation and geographical area

deep ecology
a conservationist philosophy that regards humans as one of many equal components of a global ecosystem

depiction
how someone or something is represented in words or images

gender
the state of being on the male–female continuum, but also often used with reference to social and cultural differences

mainstream media
traditional forms of mass media, as television, film, radio, magazines and newspapers

perceived risk
the subjective assessment that a person makes about the risk they are about to face in a particular situation

physical ability
the quality of being able to perform some type of physical action

portrayal
the way in which something is represented

real risk
the risk that actually exists for a particular situation or activity, given that safety has been considered and controls put in place

relationship
the way in which two or more people or things are connected, or the state of being connected

response
the feeling or emotion that an outdoor environment or outdoor experience creates in your mind

risk
the potential to lose something that you value measured against the possibility of gaining something you value

social media
forms of electronic communication through which users create and share information, or participate in social networking

socioeconomic status
an individual's or family's economic and social position in relation to others based upon income, education and occupation

sustainability
the ongoing capacity of Earth to maintain all life

technology
the application of scientific knowledge for practical purposes to extend our human abilities and to manipulate nature to satisfy our wants and needs

CHAPTER 3

code of conduct
a set of rules outlining the responsibilities of, or proper practices for, an individual, group or organisation undertaking a particular activity in the outdoors

global positioning system (GPS)
a satellite-based navigation system that can determine accurate and precise locations and give directions to other destinations

hazard
something with the potential to cause loss or injury

likelihood
the possibilities, high or low, that someone will come into contact with the hazard

minimal impact strategies
practices that aim to have as little environmental impact as possible

mud maps
rough sketches of a place or journey that show key features and likely routes, but are not drawn to scale

risk
the potential to lose something that you value measured against the possibility of gaining something you value

risk assessment
a systematic process of analysing the potential risks that may be involved in an activity

risk rating
the result of using a likelihood and consequence matrix to determine the severity or rating of a risk

stakeholder
a person, group or organisation that has interest or concern in an issue

topography
the landforms or surface features of a region

topographic maps
maps showing detailed graphical representations by contour and lines of features that appear on the Earth's surface

CHAPTER 4

abiotic
a non-living feature in an environment

aquaculture
the farming of aquatic organisms, including fish, molluscs, crustaceans and aquatic plants, with some sort of intervention in the rearing process to enhance production

atmosphere
the thin layer of gases that surrounds the Earth

biogeochemical
the cycle in which simple substances and chemical elements are transferred between living elements and the environment

biorhythms
cyclic pattern of changes in activity of living organisms

biotic
a living feature in an environment

bush food
the term for all foods native to Australia, such as seeds, starchy roots, fruits, vegetables, spices and nuts, as well as meat, insects, fish and seafood

carbon dioxide (CO_2)
a colourless, odourless gas that is the fourth-most abundant in the atmosphere (approximately 0.04%); it is produced from the burning of fossil fuels and acts as a greenhouse gas

carnivore
an animal or plant that feeds on animals

circadian rhythms
a 24-hour cycle in the physiological processes of living organisms

climax vegetation
vegetation that establishes itself in an area over a long time in the absence of any major disturbances

customary laws
a series of laws or rules informed by Kinship, enforced by sanctions and used as a way of resolving disputes

diurnal
animals and plants that are active during the day

El Niño
extensive warming of the central and eastern tropical Pacific, associated with an increased probability of drier conditions in Australia

fire intensity
the amount of heat energy released by a fire; higher intensity equals a hotter fire

fossil fuels
a deposit, such as petroleum, coal or natural gas, derived from the accumulated remains of ancient plants and animals and used as fuel

fuel loads
a measure of how much fuel is present and available to burn in the form of grass, undergrowth or forest litter

geochemical cycles
the circulation of biological, geological and chemical substances

greenhouse gas
a gas in an atmosphere that absorbs and emits radiation; examples include carbon dioxide, methane, nitrous oxide and ozone

herbivore
an animal that feeds on grass and other plants

immigration
the movement of a species into another country or region to which they are not native in order to settle there; species may be seeking better resources or mating partners, or escaping danger and environmental changes

impervious surfaces
areas that have been covered by any material that impedes the infiltration of water into the soil

intertidal zone
the area of foreshore and seabed exposed to air at low tide and submerged at high tide (i.e. the area between low-tide and high-tide marks)

kinetic
relating to or resulting from movement

La Niña
extensive cooling of the central and eastern tropical Pacific Ocean, associated with increased probability of wetter conditions in Australia

land management
the responsibility of managing the use and development of land resources

matter
in physics, that which has both a mass and volume, which occupies space and which possesses a rest mass (as distinct from energy)

metaphor
when two unlike things are compared with each other because of something they have in common

middens
mounds of discarded shells built up over time; some of the most visible archaeological evidence of Indigenous peoples' campsites and diets

migration
the movement of a species from one location to another in response to changes in habitat

net sink
where a process such as forestry absorbs more carbon than it emits

neutral phase
where warming winds towards the western Pacific keep the central Pacific Ocean relatively cool, associated with average rainfall and temperatures for eastern Australia

nocturnal
animals and plants that are active at night

other special rights
land or forest subject to Native Title determinations, registered Indigenous Land Use Agreements and legislated special cultural use provisions

outdoor environments
environments that have minimal influence from humans, but may also include those that have been subjected to human intervention

perspective
a particular attitude towards or way of regarding something; a point of view

photosynthesis
a process used by plants to convert light energy into chemical energy to grow

precipitation
when water is released from clouds in the form of rain, freezing rain, sleet, snow or hail

renewable energy
energy that can be obtained from natural resources that can be constantly replenished

respiration
the physiological process that enables animals to exchange carbon dioxide

sacred places
areas or natural features in the landscape that are significant under Aboriginal tradition, including rocks, waterholes, trees, plains and billabongs

sustainable tourism
a type of tourism that reduces the negative impacts of tourism on the environment, society and economy, while maximising positive ones; by conducting responsible travel, sustainable tourism aims to meet the needs of present tourists without compromising those of future generations to enjoy the same natural, cultural and social assets

totemism
an Indigenous term referring to spiritual and physical connection between humans, animals and the natural world that is governed by a complex belief and values system

transpiration
the evaporation of water into the atmosphere from the leaves and stems of plants

trophic level
a feeding level; organisms that form one link in a food chain – producers, consumers or decomposers

vocational
relating to an occupation or employment

CHAPTER 5

carbon tax
a tax charged to industries based on their level of greenhouse gas (primarily CO_2) production

community-based environmental group
a group of people who share a common interest in a local environment and work together to develop plans and goals to protect it

community engagement
the process of individuals and groups working together as a result of a special or shared interest

contrasting
of two things that appear as opposites, or strikingly different from each other

conservation
the preservation, protection, management or restoration of the natural environment, inclusive of ecosystems, vegetation, wildlife and natural resources, such as soil and water

conservation value
the significance of a natural environment, habitat or species, based on its contribution to biodiversity, ecosystem services, scientific research, cultural heritage or aesthetic value

direct impacts
the impacts of technology on an environment caused by the action itself

ecological community
naturally occurring and unique groups of plants and animals; their presence can be determined by factors such as soil type, position in the landscape, climate and water availability

ecosystem services
the benefits provided to humans through transforming environmental resources such as vegetation and waterways into essential goods such as food and clean water

emissions
gases that result from energy production and other industrial processes

fledgling
a young bird that has just taken flight

green infrastructure
natural or semi-natural systems (e.g. parks, wetlands, green roofs, trees or rain gardens) designed to provide ecosystem services and benefits to urban communities

habitat fragmentation
habitat that is divided or broken down into smaller habitats (e.g. when a road is constructed in a swamp and the swamp is separated into two)

habitat restoration
the rehabilitation of an outdoor environment to re-create the original ecological conditions and promote the recovery of native species

indirect impacts
the impacts of technology on an environment caused by the production or disposal of the technology itself; it can be immediate (e.g. pollution) or delayed (e.g. climate change from greenhouse gas emissions)

land-use planning
minimises negative effects on communities and the environment, preventing incompatible developments in close proximity. This includes reducing issues like noise, odours, dust, air pollution and stormwater contamination

non-renewable
a resource that does not renew itself at a sufficient rate for sustainable economic extraction in meaningful human time frames

population density
the number of people living in a given area

protected areas
involves identifying areas of high conservation value and reducing impacts on these areas through zoning

regenerative farming
a system of farming principles and practices that seeks to rehabilitate and enhance the entire ecosystem of the farm by placing a heavy premium on soil health, with attention also paid to water management, fertiliser use, carbon sequestration and more

remnant vegetation
small patches of native plants that remain after conversion of landscapes to agricultural or other use

responsible pet ownership
balancing of the needs of pets with the needs of the environment in which the pets exist (e.g. keeping cats from roaming, keeping dogs on a lead in nature reserves and bushland, and horse riding only in designated areas)

technology
the application of scientific knowledge for practical purposes to extend our human abilities and to manipulate nature to satisfy our wants and needs

threatened species management
actions such as breeding programs, invasive species control and habitat restoration used to protect and recover threatened or endangered species

urban sprawl
the rapid expansion of the geographic extent of cities and towns, often characterised by low-density residential housing and increased reliance on private vehicles for transportation

urbanisation
the development and physical growth of towns and cities, including residential areas, as people move to these locations

CHAPTER 6

Beaufort Wind Scale
a measure that relates wind speed to observed conditions at sea or on land

cold front
the boundary between warm air and relatively cooler air, which can produce gusty winds in summer (leading to increased fire danger) and damaging winds and heavy rain in winter

contingency plan
a plan designed to take into account a possible future event or circumstance

El Niño
extensive warming of the central and eastern tropical Pacific, associated with an increased probability of drier conditions in Australia

extreme weather
weather that is unexpected, unusual, severe or unseasonal based on conditions recorded in the past

field sketch
a simplistic drawing of a particular location

high-pressure system
an area of high pressure relative to its surroundings, which usually means dry, settled weather and light winds

isobars
the plain lines curving across a weather map, connecting points with the same air pressure; they can also be used to interpret air flow around weather systems

La Niña
extensive cooling of the central and eastern tropical Pacific Ocean, associated with increased probability of wetter conditions in Australia

low-pressure system
an area of low pressure relative to its surroundings, which usually means cold, wet and windy weather

peer-led activities
where students assume leadership roles and support each other in planning and facilitating an experience

remote camera
a rugged and weatherproof camera designed to take motion-activated images and videos without the need for a person to be present

route card
a document used as an aid in navigation by detailing information about a planned route

scat
wild animal poo

trough
an elongated area where atmospheric pressure is low relative to its immediate surroundings, which usually causes cloud or even showers and thunderstorms to develop

warm front
the boundary where warm air progressively displaces cool air, which usually brings warmer air and, at times, steady rainfall, grey skies and more humid conditions

weather forecast
a prediction of the conditions of the atmosphere for a given location and time

weather map
a visual representation of the locations and movement of weather patterns

weather pattern
a period of time when the weather is consistent

wind chill
the lowering of body temperature due to the passing flow of lower-temperature air

wind gust
a brief increase in the speed of the wind, usually for less than 20 seconds

CHAPTER 7

adversary
an opponent in a contest or fight

alienation
transfer of the ownership of property rights

anthropocentric
regarding humankind as the central or most important element of existence

contemporary
events or actions that have occurred within the past 15 years

El Niño
extensive warming of the central and eastern tropical Pacific, associated with an increased probability of drier conditions in Australia

endemic
a feature or species that is unique to a defined geographic location

environmental activism
the actions of individuals or groups that aim to protect or support the environment

fire-stick farming
the consistent and repeated use of fire to clear vegetation and create open forests to ensure food for both people and wildlife

hunting and gathering
of a community or society that hunts animals for meat and other useful materials, and gathers wild fruit, vegetables, roots, nuts, grasses and other edible plants

hydro-electricity
the generation of electricity using water power

impacts
what happens as a result of our relationships with outdoor environments

industrialisation
the development of industry on an extensive scale

interactions
what we do in, and with, the outdoor environments

Kinship
an Indigenous person's relationship and responsibilities to other people, to their Country and to natural resources

La Niña
extensive cooling of the central and eastern tropical Pacific Ocean, associated with an increased probability of wetter conditions in Australia

metaphor
a figure of speech where a word or phrase is used to describe an object in question and the word or phrase has no literal connection to the object but aids in understanding the object

middens
a mound or deposit containing shells, animal bones and other refuse that indicates the site of a human settlement

mobile
communities that move across large distances and to many different locations within their Country, mostly in arid environments

nation building
the process of constructing a national identity including the development of national myths as well as major infrastructure development

neutral phase
where warming winds towards the western Pacific keep the central Pacific Ocean relatively cool, associated with average rainfall and temperatures for eastern Australia

perceptions
what we think about outdoor environments

pyrolysis
the thermal decomposition of organic matter in a low-oxygen environment

reconciliation
restoration of friendly relations, especially between Indigenous and non-Indigenous individuals and communities in Australia

relationship
the way in which two or more people or things are connected, or the state of being connected

sedentary
of communities or community members that remain in a single permanent location (referred to as tribal base camp or village) with fresh water and plentiful year-round food resources

semi-sedentary
communities that move from one location to another and back again in regular cycles within their Country

soil salinity
the salt content in the soil; the process of increasing the salt content is known as salination

terra nullius
land that is legally deemed to be unoccupied or uninhabited

volatile
of a substance; easily evaporated to produce a flammable vapour

CHAPTER 8

Aboriginal Water Program
a program that works to include Aboriginal peoples in the way water is managed in Victoria and to reconnect communities to water for cultural, economic, customary and spiritual purposes

Barmah Choke
the narrowest section of the River Murray where human-made weir gates have been added to keep the body of the river inside its river banks and travelling onward to other user areas. In the case of the Overbank flooding these artificial walls are too low to hold back the water and so the large flows being released to meet the needs of Irrigators downstream means that a large proportion floods into the Barmah-Mellewa red gum forest, literally drowning it out of flood season in summer

biofuel
fuels made from renewable living raw materials (e.g. ethanol produced from common crops such as sugar cane and potato and added to petroleum)

capitalism
an economic and political system in which a country's trade and industry are controlled by for-profit private owners, rather than by the state

carbon capture and storage
process of trapping CO_2 so that its effect on the climate is minimised

carbon tax
a tax charged to industries based on their level of greenhouse gas (primarily CO_2) production

clean coal
technology intended to enable continued use of coal as an energy source with reduced impact on the environment

Climate 200
a community crowd-funded initiative that supports political candidates committed to using a science-based response to the climate crisis, restoring integrity to politics and advancing gender equity

climate change
a change in global or regional climate patterns, in particular a change apparent from the mid to late 20th century onwards and attributed largely to the increased levels of atmospheric carbon dioxide produced by the use of fossil fuels

climate sceptic
someone who believes that claims made by climate scientists and environmentalists that the climate is changing due to human activities are false or exaggerated

coalition
a group formed when two or more people or groups temporarily work together to achieve a common goal

commodity
something that can be used for commercial advantage (i.e. it can be bought and sold)

COP
Conference of the Parties to the United Nations Framework Convention on Climate Change (UNFCCC), the peak decision-making body for the world's climate change commitments

conflict
a serious disagreement or argument, typically a protracted one; a serious incompatibility between two or more opinions or interests

conservation
the preservation, protection, management or restoration of the natural environment, inclusive of ecosystems, vegetation, wildlife and natural resources, such as soil and water

Country
the term often used by Indigenous peoples to describe the lands, waterways and seas to which they are connected. The term contains complex ideas about law, place, custom, language, spiritual belief, cultural practice, material sustenance, family and identity (Source: AIATSIS)

cultural burning
a traditional practice developed by Indigenous peoples to enhance the health of the land and its people; it includes burning (or prevention of burning) for the health of particular plants and animals

custodianship
the responsibility for taking care of or protecting something

economic
relating to or based on the production, distribution and consumption of goods and services

emissions trading
a market-based approach used to control pollution by providing market incentives and requiring permits (which can be traded among companies) to use processes that produce CO_2

extinguished
in relation to native title, that native title holders are no longer able to fully exercise their traditional rights in an area as the result of governments granting freehold land or leases, or constructing public works

fossil fuel
A deposit, such as petroleum, coal or natural gas, derived from the accumulated remains of ancient plants and animals and used as fuel

greening
the process of transforming a space into a more environmentally friendly version with increased plant growth

human condition
all of the characteristics and key events of human life, including birth, learning, emotion, aspiration, morality, conflict and death

interest groups
groups of individuals with similar values who aim to promote their views about an issue (note that government or government agencies are not considered to be interest groups)

Kinship
an Indigenous person's relationship and responsibilities to other people, to their Country and to natural resources

management plan
a document that contains guidelines on how an area of public land is managed; it articulates the vision, goals, outcomes, measures and long-term strategies for parks within planning areas

mass fish death
when a large number of wild or farmed fish die suddenly and unexpectedly. They are more likely to happen in times of flood or drought

mitigate
to attempt to slow, reduce or reverse the severity of something

net sink
where a process such as forestry absorbs more carbon than it emits

political spectrum
a system of classifying different political positions upon axes that symbolise their position between socialist and capitalist political dimensions

positive feedback loop
process in which the end products of an action cause more of that action to occur in a feedback loop

recreation
pastimes that are a diversion from day-to-day routines, including active or passive activities that provide the participant with fun, relaxation, enjoyment or fitness

renewable energy
energy that can be obtained from natural resources that can be constantly replenished

socialism
a political and economic theory of social organisation that advocates that the means of production, distribution and exchange should be owned or regulated by the community as a whole

sovereignty
a state or a governing body [that] has the full right and power to govern itself without any interference from outside sources or bodies

statutory authorities
a government organisation established to exercise specific powers; for example, to manage outdoor environments such as a state parks or national parks

synergy
the interaction of two or more elements that produces an effect that's great than the sum of its parts

terra nullius
land that is legally deemed to be unoccupied or uninhabited

CHAPTER 9

aesthetic value
a judgement of value based on the appearance of an object and the emotional responses it causes

carbon sink
a place, such as a forest or an ocean, that continually takes carbon out of the atmosphere

clear felling
the practice of cutting down all the trees on a site

climate feedback loop
a cycle that accelerates or decelerates a warming trend, such as bushfires creating conditions that encourage more intense bushfires

critique
the constructive and critical analysis of a concept

economic sustainability
the ability of economic systems to sustain a decent standard of living and operate in a way that promotes long-term environmental health and preserves natural resources for future generations

ecosystem diversity
the variety of habitats, natural communities and ecological processes in the biosphere

ecotourism
responsible travel to natural areas that conserves the environment and improves the wellbeing of local people

emissions
gases that result from energy production and other industrial processes

endemic
a feature or species that is unique to a defined geographic location

environmental buffer
an area of land maintained in permanent vegetation that helps to control air, soil and water quality and other environmental problems

environmental sustainability
a concept that prioritises the wellbeing of the environment, encompassing aspects like water and air quality, as well as the reduction of environmental stressors such as greenhouse gas emissions

genetic diversity
the total genetic information contained in the genes of all species

greenwashing
promoting the misleading or false perception that an organisation's products are 'green' or environmentally friendly

grey water systems
recycling systems that reuse wastewater generated from wash basins, showers and baths for uses such as toilet flushing and garden watering

hydro-electricity
the generation of electricity using water power

interdependence
of two or more things that are interconnected, relying on and influencing each other

intrinsic value
something that is prized for what it is, rather than for what it can provide

mariculture
the cultivation of marine organisms for food and other products in the open ocean, an enclosed section of the ocean, or in tanks, ponds or raceways that are filled with seawater

multilateral
agreed upon or participated in by three or more parties, especially the governments of different countries

nature-deficit disorder
a metaphor to describe the human costs of alienation from nature

net zero
the state where the amount of greenhouse gases emitted into the atmosphere is equal to the amount of greenhouse gases removed from the atmosphere

renewable
something that can be naturally replenished or replaced relatively quickly, typically within a human lifetime, without depleting its source

rotation planting
the successive planting of different crops on the same land to improve soil fertility and to help control insects and diseases

salinity
the concentration of dissolved salts in water or soil

sequestration
the process of capturing and storing carbon in sinks such as soils, forests and oceans

smog
a type of air pollution characterised by a mixture of smoke and fog that is typically a result of human activities, particularly the burning of fossil fuels

Snowy Mountains Scheme
one of the most complex integrated water and hydro-electric power schemes in the world

9780170477123

social sustainability
the concept of an inclusive and just society where every individual's needs are met, and where everyone has equal opportunities to contribute and participate in decision-making processes

species diversity
the number of species and the number of individuals within each species

sustainable
capable of being maintained in existence without interruption or diminution

CHAPTER 10

action statement
a detailed document that describes the recommended conservation management practices for specific threatened species

advocacy
an action that speaks in favour, supports and pleads on behalf of others or issues

caring for Country
the management of land and waterways by Indigenous peoples, and the sustainable land management practices and initiatives led by them

conservation covenant
a management agreement placed on a property's title to ensure native plants and wildlife on the property are protected forever

conventions
agreements used in multilateral government decisions

Crown land
land that is owned by a state/territory government or the Commonwealth

cultural burning
a traditional practice developed by Indigenous peoples to enhance the health of the land and its people; it includes burning (or prevention of burning) for the health of particular plants and animals

energy efficiency
using less energy to perform the same task and eliminating energy waste

environmental activism
the actions of individuals or groups that aim to protect or support the environment

Environmental Acts
measures implemented by governments or their agencies designed to reduce human impact on outdoor environments

fair trade
a set of standards in the supply chain of a product, such as fairer conditions, improved working conditions and better prices for producers

greenwashing
where a company falsely portrays itself as more environmentally friendly than it actually is, in an effort to bolster its reputation and increase sales

green wedge
land defined in a metropolitan fringe planning scheme as being outside an Urban Growth Boundary

land management
the responsibility of managing the use and development of land resources

management plan
a document that contains guidelines on how an area of public land is managed; it articulates the vision, goals, outcomes, measures and long-term strategies for parks within planning areas

monoculture
the cultivation of a single crop in an area

passive solar
building design principles that promote the sun's energy to heat and cool living spaces without the use of active mechanical systems

public land
land managed by governments and their agencies to protect sites of environmental and cultural value, and to provide opportunities for community and recreational use

regenerative farming
a system of farming principles and practices that seeks to rehabilitate and enhance the ecosystem of the farm by placing a premium on soil health

taxon
a group of any taxonomic rank (such as a species, family or class); taxa are ranked based on their shared traits – the lower down the hierarchy you go, the more specific the grouping characteristics become

transboundary wetlands
wetlands that exist across one or more borders

upcycling
the process of transforming by-products, waste materials or unwanted products into new materials or products of better quality

CHAPTER 11

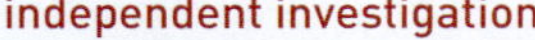

independent investigation
a student-driven research task based on the key knowledge points and skills within outdoor environments

investigation
observe, study or carry out an examination in order to establish facts and reach new conclusions

logbook
a document of evidence of practical outdoor experiences, outdoor environments visited, primary and secondary data, and other valuable information to be added to throughout the year

ongoing
continuing to develop and progress

primary data
data collected in the field or outdoor environments that is reflective of the key knowledge points and skills on which you are reporting (note: multiple sources of data collection strengthen the credibility of your findings)

report conventions
the structure of written reports and the process by which the information is gathered and disseminated

secondary data
information that someone else has collected and made available (note: when using secondary data, it is imperative that you reference the source and determine if the source is credible)

written report
a form of writing that is written concisely and clearly that examines experiences or events from a research investigation

INDEX

Note: Page numbers followed by *f* or *t* indicate material in figures or tables respectively

D

E

F

J

K

L

M

P

R

S